The Clinical
NANOMEDICINE
HANDBOOK

The Clinical
NANOMEDICINE
HANDBOOK

Sara Brenner, MD, MPH - Editor

CRC Press
Taylor & Francis Group
Boca Raton London New York

CRC Press is an imprint of the
Taylor & Francis Group, an **informa** business

CRC Press
Taylor & Francis Group
6000 Broken Sound Parkway NW, Suite 300
Boca Raton, FL 33487-2742

First issued in paperback 2017

ISBN-13: 978-1-4398-3478-7 (hbk)
ISBN-13: 978-1-138-07578-8 (pbk)

Library of Congress Cataloging-in-Publication Data

The clinical nanomedicine handbook / editor, Sara Brenner.
   p. ; cm.
   Includes bibliographical references and index.
   ISBN 978-1-4398-3478-7 (alk. paper)
   I. Brenner, Sara, 1980- editor of compilation.
   [DNLM: 1. Nanomedicine--methods--Handbooks. QT 29]

R857.N34
610.28--dc23   2013027125

Visit the Taylor & Francis Web site at
http://www.taylorandfrancis.com

and the CRC Press Web site at
http://www.crcpress.com

Contents

Preface

Nanomedicine is the medical application of nanotechnology in the prevention, diagnosis, and treatment of diseases. *The Clinical Nanomedicine Handbook* is intended to serve as an authoritative reference for clinicians, including physicians, nurses, healthcare providers, dentists, scientists, and researchers involved in clinical applications of nanotechnology. Although many texts and publications have been released on various topics at the intersection of nanotechnology, biology, and medicine, none have approached the convergence of these fields from a distinctly clinical vantage point.

The American Society for Nanomedicine and the National Institutes of Health (NIH) highlight the emergence of nanomedical innovations as a necessary enabler in the development and deployment of preventive, diagnostic, and therapeutic medical tools of the twenty-first century. These will be driven by the application of nanotechnology and engineering principles to clinical interventions, which manipulate single molecules or molecular assemblies in cells, drugs, or medical devices.

At the intersection of traditionally siloed disciplines, nanomedicine is blazing a path for truly innovative, cutting edge, preventive interventions, rapid diagnostics, and effective treatments. Current research and applications in specialties are covered in this handbook, which is sure to expand in the coming years. Potential applications, as cited by the NIH, include nanotechnologies that enable physicians to identify and destroy primary cancer cells before metastasis, molecular procedures that can remove a dysfunctional cellular component and replace it with an engineered biological machine, and nanoscale pumps that deliver targeted drug therapy. The NIH also predicts that nanomedicine will enable the development of novel tools that will allow scientists to build synthetic biological devices, such as high-throughput sensors to scan for the presence of infectious agents or metabolic imbalances, as well as nano-enabled therapeutics aimed at rapidly addressing the identified problem.

From an economic perspective, nanotechnology and its affiliated applications are projected to impact every known industry and create entirely new industrial clusters such as nanomedicine and nanobioscience. Nanotechnology accounted for nearly $300 billion in 2009 and is projected to hit $2.6 trillion by 2014. It is predicted that the economic impact of nanotechnology will approach the size of the information technology and telecom industries combined with the potential to be 10 times larger than the biotechnology economy in the next 5 years. In terms of revenue, 16% of goods in healthcare and life sciences will incorporate nanotechnology. In other words, nanobioscience products alone are expected to be a $416 billion industry by 2014.

Cover Art Credits

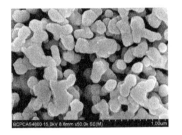

Silver np 4 (Blue image wedges): Cluster of green-route synthesized silver nanoparticles under a TEM. Center for Converging Technologies, University of Rajasthan, Jaipur, India. Photographed by Sahil Tahiliani & Abhijeet Mishra. 2011.

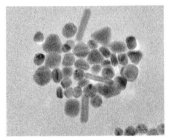

Silver np 1 (Green image wedges): Electron micrograph of chemically synthesized silver nanoparticles of various shapes i.e. rods, sphere, pyramidical etc. Center for Converging Technologies, University of Rajasthan, Jaipur, India. Photographed by Sahil Tahiliani & Ruchir Priyadarshi. 2012.

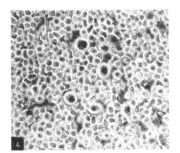

Nci-vol-2505-300 (Orange image wedges): Electron micrograph of macrophages in the brain before and after in vitro infection by HIV-I. Laboratory of Tumor Cell Biology. National Cancer Institute. Dr. Suzanne Gartner (photographer). 1990.

SEM blood cells (Red image wedges): Scanning electron microscope image of normal circulating human blood cells. National Cancer Institute. Bruce Wetzel and Harry Schaefer (photographers). 1982.

Editor

Dr. Sara Brenner is a preventive medicine and public health physician at the College of Nanoscale Science and Engineering (CNSE), State University of New York, serving as the assistant vice president for NanoHealth Initiatives and an assistant professor of nanobioscience. Her research and initiatives aim to develop novel nanotechnology applications in the life sciences, including medicine and public health.

She leads health and safety research related to nanoparticle and engineered nanomaterial exposures in the workplace, consumer marketplace, and environment. She is chair of the steering committee for the NanoHealth and Safety Center at CNSE, a public–private partnership that is addressing gaps in our understanding of the safety and risk associated with the unique characteristics of nanoscale materials. Dr. Brenner's research team incorporates theory from many disciplines such as physics, engineering, biology, genetics, toxicology, medicine, public health, epidemiology, industrial hygiene, and environmental science to advance risk assessment and reduction strategies for occupational exposures, monitoring of materials that may impact population health and public safety, and the development of industrial practice standards for product safety. She is also the CNSE program director of the MD/PhD program in medicine and nanoscale science or engineering, a program that she helped cofound with SUNY Downstate Medical Center. It is the first dual-degree clinical training program in nanomedicine that aims to produce a new, hybrid generation of physician researchers.

She is also the recipient of the Albany-Colonie Chamber of Commerce Women of Excellence Award 2012 Emerging Professional.

Contributors

Himanshu Aggarwal, MD
Division of Urology
Department of Surgery
Albany Medical Center
Albany, New York

Sara Brenner, MD, MPH
SUNY College of Nanoscale Science and
    Engineering
Albany, New York

W. John Byrne, MD
The Vascular Group, PLLC
The Institute for Vascular Health and
    Disease
Albany Medical College
Albany, New York

Jason Chouake, MD
Division of Dermatology
Department of Medicine
Albert Einstein College of Medicine
Bronx, New York

John Danias, MD, PhD
Department of Ophthalmology
SUNY Downstate Medical Center
Brooklyn, New York

Eman Elhawy, MD
Flushing Hospital Medical Center
Flushing, New York

Adam Friedman, MD
Division of Dermatology
Department of Medicine
Albert Einstein College of Medicine
Bronx, New York

Barry A. Kogan, MD
Division of Urology
Department of Surgery
Albany Medical Center
and
Urological Institute of Northeastern
    New York
Albany, New York

Heidi M. Mansour, PhD
Division of Dermatology
Department of Medicine
Albert Einstein College of Medicine
Bronx, New York

Manish Mehta, MD
The Vascular Group, PLLC
The Institute for Vascular Health and
    Disease
Albany Medical College
and
Vascular Research and Registry
Albany, New York

Adnan Nasir, MD
Department of Dermatology
The University of North Carolina at
    Chapel Hill
Chapel Hill, North Carolina

Kathleen J. Ozsvath, MD
The Vascular Group, PLLC
The Institute for Vascular Health and
    Disease
Albany Medical College
Albany, New York

Chun-Woong Park, PhD
Division of Dermatology
Department of Medicine
Albert Einstein College of Medicine
Bronx, New York

Philip S.K. Paty, MD
The Vascular Group, PLLC
The Institute for Vascular Health and
    Disease
Albany Medical College
Albany, New York

David Schairer, MD
Division of Dermatology
Department of Medicine
Albert Einstein College of Medicine
Bronx, New York

Yaron Sternbach, MD
The Vascular Group, PLLC
The Institute for Vascular Health and
    Disease
Albany Medical College
Albany, New York

John B. Taggert, MD
The Vascular Group, PLLC
The Institute for Vascular Health and
    Disease
Albany Medical College
Albany, New York

Ellis H. Tobin, MD
Department of Medicine
Albany Medical College
and
Upstate Infectious Diseases
    Associates
Albany, New York

Julielynn Wong, MD, MPH
Center for Innovative Technologies and
    Public Health
Toronto, Ontario, Canada

1

Nanotechnology Applications for Infectious Diseases

Ellis H. Tobin, MD

CONTENTS

1.1 Introduction

It has long been an axiom of mine that the little things are infinitely the most important.

Sir Arthur Conan Doyle (1900)

Infectious diseases (IDs) are the clinical manifestations that result from infections due to a myriad of pathogenic viruses, bacteria, fungi, and parasites, whereas the discipline of ID is the research and clinical practices having to do with host–pathogen interactions, diagnosis, treatment, and prevention of infection. The application of nanotechnology to ID heralds a much anticipated interface, one with potentially far-reaching implications. Indeed, some of the earliest applications of nanotechnology to medicine were in the field of ID: the use of gold nanoparticles (GNPs) for immunolabeling of salmonella surface antigens (Faulk and Taylor 1971), the encapsulation of amphotericin and doxorubicin within liposomal nanoparticles to treat fungal infections and AIDS-associated Kaposi's sarcoma, respectively (Lopez-Berestein et al. 1989; Bogner et al. 1994), and the use of pegylated interferon to treat hepatitis C (Zeuzem et al. 2000). It can even be argued that one of the earliest examples of the application of nanotechnology to ID, although neither were called that back then, dates to the early Renaissance, when Paracelsus used colloidal gold to treat tuberculosis (TB) and syphilis (DeWitt 1918; Dykman and Khlebtsov 2010).

This chapter explores the nanotechnology–ID interface, two disciplines accustomed to dealing with matter on a small scale. It endeavors to make the argument that the union of these disciplines will have a large impact on healthcare in technologically advanced as well as resource-limited parts of the world. The contextual framework of the chapter is one that introduces the nanoscientist and non-ID practitioner to the profession of ID and the ID clinician to fundamental concepts of nanoscience that have direct applications to their medical specialties. An emphasis is placed on nanotechnology applied to the detection and analysis of microbial pathogens and biomarkers of infection. Many of these applications, in this study, are in the proof-of-concept stage of development, while others are in various phases of clinical research. Several nanotechnology applications, liposomal amphotericin for example, are in current clinical practice.

Nanotechnology is a scientifically diverse discipline that initially encompassed engineering, materials science, physics, and chemistry. It has expanded to involve the biological sciences, where the fields of nanobiotechnology and nanomedicine are rapidly emerging. Nanotechnology exploits the complex and remarkably unique properties of matter at the nanoscale (Planinsic, Lindell, and Remskar 2009; Kim, Rutka, and Chan 2010). Although detailed descriptions of these fundamental physicochemical properties are beyond the scope of this chapter, simplified illustrations are provided so that a working knowledge of the nanobiotechnology–ID interface can be gained. There is little doubt that the practice of medicine will be profoundly influenced by nanotechnology, and the confluence of biology and nanotechnology will have a dramatic impact on ID.

1.1.1 Basic Nanoscience Concepts

To begin a discussion of nanotechnology with relevance to ID, it is useful to introduce a few basic concepts pertaining to the properties of matter at the nanoscale. Nanostructures possess a very large ratio of surface area to volume. This concept can be intuited at the macroscopic (bulk) scale by considering the ratio of surface area to volume of a single pad of sticky notes. Now consider the ratio when all the individual notes from the pad are stuck to the sides of a refrigerator. Amplify that relationship a million times as you contemplate the ratio of surface area to volume of nanosticky notes. A nanoparticle may consist of just a few atoms, or in the case of nano-thin films and filaments, it may be just a few atoms thick. Given the surface area-to-volume relationship discussed earlier, it is easy to appreciate that a large fraction of the atoms that make up the nanostructure reside on its surface (Eustis and EL-Sayed 2006). The behavior of these surface atoms confers many of the unique properties associated with matter at the nanoscale. To give nanoscale dimensions an ID perspective, it will be helpful to consider that human immunodeficiency virus (HIV) particles have a diameter of approximately 120 nm, and the diameter of nanocrystal quantum dots (QDs) and magnetic nanoparticles (MNPs) range between 1 and 40 nm.

Surface atoms are relatively reactive, having fewer neighbors to share chemical bonds (Roduner 2006). This facilitates the attachment of a variety of molecules (e.g., antibiotics, nucleotides, proteins, antibodies, and aptamers) to nanostructured surfaces by chemical and electrostatic means. The optical and electronic properties of the surface atoms behave differently than at the bulk scale. This is, in part, due to quantum confinement, a term used to describe the effects on electron energy levels as they are squeezed into the boundary limits of nanostructures (Roduner 2006). Quantum confinement influences how nanoparticles interact with electromagnetic energy, and imparts unique luminescent and electric conducting properties that can be exploited in biomedical science.

In addition to their unique optoelectronic properties, nanoparticles can be engineered to develop distinctive magnetic and superparamagnetic behaviors (Gossuin et al. 2009). They can be made small enough to contain single magnetic domains that impart large magnetic moments, a property that lends itself to many biomedical applications. Examples of nanoparticle properties and behaviors and how they interface with ID will be provided throughout this chapter. Before arriving at the interface, however, a brief description of ID and the scope of ID are in order.

1.2 Specialty of Infectious Diseases

For a period of time from approximately 1965 to 1985, there was considerable optimism that medical science had succeeded in controlling the emergence and spread of major IDs, and as a result, the need for ID specialists would rapidly decline (Ervin 1986; Fauci 2001; Petersdorf 1986; Spellberg et al. 2008). Great progress had been made in childhood immunizations and in the eradication or significant reduction of IDs such as smallpox, poliomyelitis, TB, and guinea worm (Aylward et al. 2000). Unfortunately, the optimistic expectations made at the time were somewhat naïve and definitely short lived. IDs continue to be the most prevalent health-care problem throughout the world (Fauci 2001). The ID optimists of the mid-twentieth century did not anticipate the following: the ongoing global wave of emerging and reemerging IDs as illustrated in Table 1.1 (Jones et al. 2008; Morens, Folkers, and Fauci 2008; Olano and Walker 2011); the increasing prevalence of an elderly population with waning immunity (U.S. Department of Health and Human Services 2011); the proliferation of immunosuppressed patients receiving cytoreductive chemotherapies, immune-modulating drugs, hematopoietic stem cell, and solid organ transplantations (Barshes, Goodpastor, and Goss 2004; Wahrenberger 1995); the expansive use of indwelling catheters, prostheses, and implantable medical devices that interferes with host defense barriers and invites the formation of biofilms (Donlan and Costerton 2002; Zimmerli, Trampuz, and Ochsner 2004; Baddour et al. 2010; Hooton et al. 2010; Hodgkiss-Harlow and Bandyk 2011; WHO 2011b); the initial expansion, and later contraction, of newly discovered antimicrobials and their attendant overuse (Boucher et al. 2009; Gyssens 2011; Jabes 2011); the mathematical complexities of antimicrobial pharmacokinetics (PKs) and pharmacodynamics (PDs) (Drusano 2007); the increase in type and variety of drug-resistant microorganisms, and the ease and speed with which they are transmitted within health-care settings and between distant geographic communities (Kunin 1993; Small 2009; Rice 2009; Parija and Praharaj 2011); the realization that ID agents are capable of causing certain types of cancer, gastric ulcers, and other similar chronic medical conditions (Suzuki, Saito, and Hibi 2009; Hannu 2011; de Martel et al. 2012; Zaghloul and Gouda 2012); the changing climate, geologic, and demographic patterns that have altered the interplay between humans, reservoirs, and vectors of IDs (WHO 2009; Shuman 2010; Ermert et al. 2012); the persistence of crushing poverty, wars, floods, and drought that promotes starvation, poor

TABLE 1.1

Emerging Infectious Agents Described Since 1967

Year	Infectious Agent
1967	Marburg virus
1969	Lassa virus
1971	JC virus
1972	Norovirus
1973	Rotavirus
1975	Parvovirus B_{19}
1976	*Vibro vulnificus, Cryptosporidium parvum*
1977	Ebola virus, *Clostridium difficile, Legionella pneumophila*, Hantaan virus, *Campylobacter spp*
1979	*Cyclospora cayetanensis*
1980	HTLV-1
1981	*Staphylococcus aureus* toxic shock syndrome toxin
1982	*Borrelia burgdoferi*, HTLV-II, prion diseases
1983	HIV-1, *Helicobacter pylori*, hepatitis E virus
1984	*Haemophilus influenzae aegyptus*
1985	*Enterocytozoon bieneusi*, Borna disease virus
1986	*Chlamydia pneumoniae*, HIV-2
1987	Dhori virus
1988	Human herpes virus 6, Barmah Forest virus
1989	*Rickettsia japonica*, hepatitis C virus
1990	Hepatitis E virus, *Balamuthia mandrillaris*
1991	Guanarito virus, *Encephalitozoon hellem, Ehrlichia chaffeensis*
1992	*Vibrio cholerae* 0139, *Bartonella henselae, Rickettsia honei, Tropheryma whippelii*
1993	Sin Nombre hantavirus
1994	*Anaplasma phagocytophilum*, Hendra virus, Sabia virus, human herpesvirus 7, human herpesvirus 8
1996	Andes virus, Creutzfeldt-Jakob disease (new variant)
1997	*Rickettsia slovaca*, Influenza A H5N1[a]
1998	Menangle virus, *Brachiola vesicularum*
1999	*Ehrlichia ewingii*, Nipah virus
2000	Whitewater Arroya virus
2001	Human metapneumovirus
2002	*Cryptosporidium hominis*
2003	SARS-coronavirus
2004	Monkeypox virus, human corona virus $NL6_3$
2005	Human bocavirus, human coronavirus HKU1[b], HTLV-3, and HTLV-4
2009	Influenza A H1N1 ("swine flu")

Source: Olano, J.P. and D.H. Walker, *Arch. Pathol. Lab. Med.*, 135, 83–91, 2011.

Abbreviations: HIV, human immunodeficiency virus; HTLV, human T-cell lymphotropic virus; SARS, severe acute respiratory syndrome.

[a] Bird flu.

[b] Identified in Hong Kong.

sanitation, and limited access to health care and increases the vulnerability to agents of ID (U.S. Department of State 2000; United Nations General Assembly 2001; Black, Morris, and Bryce 2003; Jones et al. 2003; Hunt 2006; Hotez et al. 2009; Grace et al. 2012); and the limited availability of complex, and expensive diagnostic and therapeutic platforms within health-care systems (Barker et al. 2006; Yager et al. 2006). All of the aforementioned issues contribute to IDs being the leading cause of morbidity and mortality on a worldwide scale.

The specialty of ID is unique in that it is neither a single organ nor a single organ system focused. Infections occur at all anatomical sites and cause localized as well as systemic illness. Unlike all other forms of disease, infections are due to transmissible agents, many of which are zoonotic in origin (Grace et al. 2012). Many IDs are unique to specific geographic regions and seasons of the year, while others are emerging and reemerging on a regular basis (Fauci 2001; Olano and Walker 2011). The immunologic interaction of a host with a particular ID agent may be so subtle as to go completely unnoticed in one individual, yet cause dramatic and life-threatening illness in another. The specialty of ID is a discipline that integrates clinically relevant aspects of virology, bacteriology, mycology, parasitology, venereology, immunology, antimicrobial pharmacology, clinical microbiology, epidemiology, and public health. Most importantly, as presented in Table 1.2, it requires a detailed knowledge of noninfectious illnesses that can masquerade and be mistaken for an ID.

ID is, in a sense, an egalitarian medical discipline; everyone develops infections of one sort or another at some point in their lifetime. However, while most infections are minor and inconsequential, the magnitude of potentially lethal, acute, and chronic IDs is enormous. This is particularly true for the severely malnourished, the immunocompromised, the pharmacologically immune suppressed, and the people residing in resource-limited

TABLE 1.2

Common Noninfectious Illnesses That Mimic Infectious Diseases

Illness	Symptom and Sign
Drug hypersensitivity	Fever, rash, meningitis, myocarditis, hepatitis, diarrhea, joint inflammation, leukocytosis, eosinophilia, cytopenia
Autoimmune diseases	Fever, rash, adenopathy, joint inflammation, opthalmic inflammation, cough, pleuritis, pneumonitis, myocarditis nephritis, meningoencephalitis
Neoplastic diseases	Fever, skin and soft tissue inflammation, bone destruction, adenopathy, pain, cough, pneumonitis, carditis, hepatitis, diarrhea, bowel obstruction, meningitis, leukocytosis, cytopenia
Sarcoidosis	Fever, rash, joint inflammation, adenopathy, pneumonitis, carditis, hepatitis, meningitis
Inflammatory bowel disease	Fever, sterile skin abscesses, diarrhea, hepatitis, cholangitis
Gout and pseudo-gout	Fever, rash, joint inflammation
Thyroiditis	Fever, rash
Pancreatitis	Fever, leukocytosis, sepsis-like syndrome
Veno-occlusive disease	Fever, skin and soft tissue inflammation, diarrhea, leukocytosis
Pulmonary embolism	Fever, shortness of breath, pulmonary infiltrates, sepsis-like syndrome
Subarachnoid hemorrhage	Fever, headache, meningitis
Kawasaki's disease	Fever, conjunctivitis, adenopathy, rash, myocarditis, meningitis, leukocytosis
Factitious fever	Fever, myriad self-inflicted signs and symptoms

parts of the world. ID physicians encounter patients with all manner of infections acquired either in the community or in the health-care setting. Since IDs cross all demographic boundaries, the ID practitioner is likely to interface with the entire spectrum of the health-care community, including nurses, pharmacists, clinical laboratory technologists, epidemiologists, radiologists, pathologists, dermatologists, and all medical, surgical, and dental subspecialists. For those ID clinicians involved in epidemiology and public health, interactions with construction engineers, health-care administrators, and government officials are commonplace. One should anticipate a time in the not-too-distant future when nanoscientist will be added to the list of individuals to whom ID practitioners relate.

The clinical practice of ID specialists is as varied and diverse as the geographic location in which they choose to engage their skills: tropical, temperate, regional, international, inner city, suburban, industrial, rural, clinic setting, or hospital based. Although ID specialists in the United States are comprised mostly of internal medicine and pediatrics physicians who obtain 2–3 years of additional training beyond their residency and are required to appear for examinations leading to board certification, any physician with an interest in microbiology, antimicrobial pharmacology, or public health can develop IDs expertise in his or her field (Kass 1987; Knobler et al. 2006).

Training and practice within the ID specialty can take several pathways, including, but not limited to, basic or epidemiological research, clinical care, and public health. In general, the basic research track in the subspecialty typically entails advanced training in experimental cell and molecular biology, microbiology, genetics, immunology, host–pathogen interactions, or antimicrobial pharmacology. The public health track focuses on the epidemiology of infections, the identification and investigation of outbreaks, methods to reduce transmission, and infection prevention. The clinical ID track expands the foundations of internal medicine and pediatrics, concentrates on the signs and symptoms of illness that may suggest the presence of a particular infection and its pathophysiology, investigates demographic and immunologic features of the host in an effort to ascertain clues that allow formulation of a differential diagnosis, pursues cost-effective diagnostic modalities, identifies strategies for optimizing therapy, and promotes measures for prevention of IDs. The emerging technological advances in nanoscience will, no doubt, impact all aspects of IDs research, clinical practice, and public health.

1.3 Scope of Infectious Diseases

In considering the role that nanobiotechnology is likely to play in the ID realm, a brief discussion of the scope and magnitude of IDs in human populations is in order. Figure 1.1 depicts major IDs categories, those impacting the largest number of people, and provides a perspective on their prevalence and distribution between resource-scarce and industrialized parts of the world. The greatest burden of IDs is borne by people residing in the poorest regions, where a combination of illiteracy, overcrowding, malnutrition, lack of basic sanitation, and dysfunctional government infrastructure promotes public health neglect and the transmission of infections (WHO 2008). Moreover, in these poverty-stricken regions, the morbidity and mortality associated with IDs contribute significantly to the loss of productive years of life, resulting in further economic stagnation and societal instability (U.S. Department of State 2000; United Nations General Assembly 2001). The potential role that nanobiotechnology may play in confronting some of these problems is being addressed

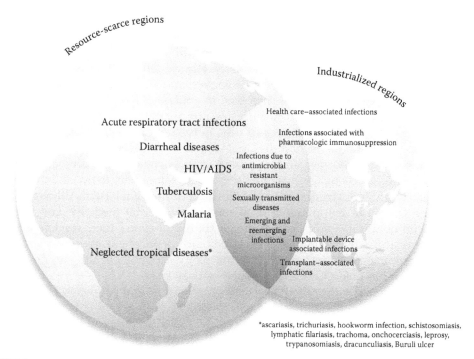

FIGURE 1.1
Major infectious diseases categories and burden relative to geographic prevalence.

by international agencies, including the United Nations Millennium Project (Juma and Yee-Cheong 2005) and the Foresight Project (Barker et al. 2006). In 2011, the First Workshop on Nanomedicine for Infectious Diseases of Poverty convened in South Africa, and the report of that conference offers an excellent description of an ID–nanotechnology interface of great potential (Collins 2011). The hope is that the rewards of nanobiotechnology will benefit both the industrialized and developing regions of the world.

While the burden of IDs has a global disproportionality, it is correct to say that infections significantly impact those residing in industrialized parts of the world as well. Health care–associated infections (HAIs), emerging and reemerging infections, antimicrobial resistance, pharmacological immunosuppression, implantable medical devices, and sexually transmitted diseases (STDs) figure prominently in the overall burden of diseases affecting people residing in technologically advanced regions (Fauci 2001; Fonkwo 2008; WHO 2011b). Indeed, as depicted in Figure 1.1, many ID problems are shared among developing and industrialized nations.

1.4 Magnitude of Infectious Diseases

As presented in Table 1.3, IDs are the leading cause of death throughout the world (WHO 2008). Approximately 15 million people die each year as a direct result of infection. This staggeringly large number does not take into account the contribution by IDs to the burden of noncommunicable diseases such as cardiovascular disease, diabetes, cancer, cirrhosis, nephritis,

TABLE 1.3

Leading Causes of Mortality and Burden of Disease Worldwide 2004

Mortality	%	DALYs[a]	%
1. Ischemic heart disease	12.2	1. Lower respiratory infections	6.2
2. Cerebrovascular disease	9.7	2. Diarrheal diseases	4.8
3. Lower respiratory infections	7.1	3. Depression	4.3
4. COPD	5.1	4. Ischemic heart disease	4.1
5. Diarrheal diseases	3.7	5. HIV/AIDS	3.8
6. HIV/AIDS	3.5	6. Cerebrovascular disease	3.1
7. Tuberculosis	2.5	7. Prematurity, low birth weight	2.9
8. Trachea, bronchus, lung cancers	2.3	8. Birth asphyxia, birth trauma	2.7
9. Road traffic accidents	2.2	9. Road traffic accidents	2.7
10. Prematurity, low birth weight	2.0	10. Neonatal infections	2.7

Source: World Health Organization, Health Statistics and Informatics, 2004.

Abbreviations: HIV, human immunodeficiency virus; AIDS, acquired immunodeficiency syndrome.

[a] DALYs (disability-adjusted life years) = the sum of the years of life lost due to premature mortality, and the years lost due to disability or poor health in a population.

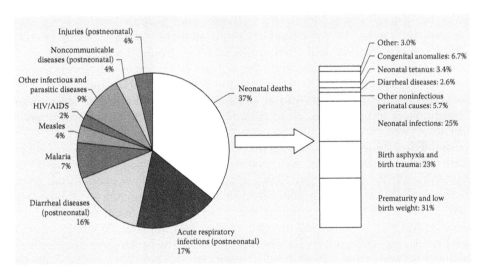

FIGURE 1.2
Distribution of causes of death among children younger than 5 years and within the neonatal period, 2004. (From World Health Organization, Health Statistics and Informatics, 2004. With permission.)

and injuries (Fauci 2001; WHO 2008). In addition to mortality, IDs significantly contribute to the loss of healthy and productive years of life, and to the development of certain cancers such as cervical carcinoma due to human papilloma virus (HPV), hepatocellular carcinoma due to hepatitis viruses B and C, Kaposi's sarcoma due to human herpes virus type 8, and various types of hematologic cancers due to human T-cell leukemia virus and Epstein-Barr virus (Fauci 2001; Suzuki, Saito, and Hibi 2009; de Martel et al. 2012; Zaghloul and Gouda 2012).

Figure 1.2 identifies IDs as being the leading cause of mortality in children under the age of 5 years, resulting in approximately 8 million deaths per year (Rudan et al. 2007; WHO 2008). Infections that lead to pneumonia and diarrhea are responsible for nearly half of these deaths, followed by neonatal infections, malaria, measles, and HIV/AIDS. An additional 5.5 million deaths per year due to infections occur in young adults (WHO 2008). It is important

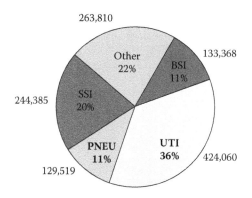

FIGURE 1.3

Health care–associated infection rates in U.S. hospitals among adults and children outside of intensive care units, 2002. *Abbreviations:* BSI, blood stream infections; UTI, urinary tract infections; PNEU, pneumonia; SSI, surgical site infections. (Adapted from Klevens, R.M. et al., *Public Health Rep.*, 122, 160–166, 2007. With permission.)

to emphasize that these figures are heavily weighted toward the poorest countries of the world, most notably sub-Saharan Africa and South-East Asia. In adults aged 15–59 years, IDs account for between 29 and 62% of deaths in these resource-limited regions. Since this age group encompasses most of the potential labor force in these countries, disability and death due to IDs represent a societal, and often overlooked, catastrophe (Fonkwo 2008).

The magnitude of these bewilderingly large numbers assumes heartbreaking proportions when one considers that 1500 people die due to IDs every hour, and over half of them are young children. Through innovation, miniaturization, and point-of-care capabilities, nanobiotechnology has the potential to make health care, in all of its forms, more accessible and affordable.

HAIs, defined as the presence of ID agents or toxins acquired in a health-care setting, are significant problems in both resource-limited and industrialized parts of the world (Klevens et al. 2007; WHO 2011b). In the United States, as shown in Figure 1.3, greater than 1.5 million HAIs occur annually, and their annual direct medical costs approach $45 billion (Klevens et al. 2007; Scott 2009). It is worth noting that most HAIs are associated with the use of intravascular and bladder catheters, breathing tubes, and implantable medical devices, all of which breech natural immunity barriers and promote the development of biofilms. This is particularly relevant, since nanotechnology may someday provide a means of developing biocompatible materials capable of preventing biofilm formation.

1.5 Nanoscience Principles, Methods, and Terminology Relevant to Infectious Diseases

> Nano-Haiku
>
> Nano prefixed words.
> Imagination salad.
> Fine nano-jargon.
>
> Anon

Credit should go to the nanoscientists who give playful descriptive names to a variety of nanoparticles and nanodevices that have found applications in biotechnology. They join the ranks of microbiologists and ID taxonomists who assign inventive nomenclature to many of the microorganisms and IDs encountered in clinical practice. Consider the following examples: aspergillus, Legionnaires' disease, dengue fever, chikungunya virus, *loa loa*, buckey balls, QDs, nanowires, nanotubes, nanorods, and magnetic nanocannonballs. For the purposes of this chapter, the terminology, methods, tools, and principles most relevant to the nanoscience–ID "nano-nexus" are presented.

Since nanobiotechnology is a new and rapidly emerging discipline, it is reasonable to assume that most ID specialists are unacquainted with many of its principles, methods, and technical terms. Furthermore, much of the current scientific literature pertaining to the nanotechnology–ID interface assumes some familiarity with complex physicochemical phenomena relevant to nanoscience. This chapter will help bridge the ID–nanoscience technology gap. It is intended to ease the transition into nanomedicine for ID practitioners, facilitate their understanding of clinically relevant literature, and enhance comprehension of an innovative science that offers great promise in their practice specialty. The terminology and concepts presented herein are by no means exhaustive, and references providing detailed descriptions of nanoscience fundamentals will be cited. For the nanoscientist, apologies are offered in advance for any oversimplifications of complex chemical and physical principles. Since nanobiosensing will be emphasized in this chapter, it will be helpful to briefly describe conventional approaches to the diagnosis of IDs as a comparative framework.

1.6 Current Approach to Diagnosis of Infections

The clinical manifestations of infection are often variable and nonspecific: malaise, fever, or pain, for example. Symptoms and signs of infection can be localized to a specific anatomical region, generalized, or a combination of both. Moreover, symptoms and signs of infection can be subtle and chronic, or acute and severe. When evaluating a patient experiencing an illness that suggests the presence of an underlying ID, there is a broad differential diagnosis to entertain. Once the noninfectious etiologies are weighed and deemed unlikely to be causing illness, a long inventory of potential viral, bacterial, fungal, and parasitic organisms must be considered. This inventory of potential microbial pathogens can usually be narrowed down on the basis of findings established by the clinical history and physical examination performed in the context of demographic, geographic, seasonal, and host immunologic associations. A further refinement of the differential diagnosis can often be made on the basis of routine hematologic and metabolic laboratory data and, when called for, radiologic findings.

This approach to the differential diagnosis constitutes the cognitive, stepwise progression taken by ID practitioners in their search for the cause of an illness, and it leads to the point where a choice—often empiric—of therapeutic options is made. It is frequently the case that the diagnosis is initially provisional, with confirmation ultimately being established or presumed by either growth and identification of the pathogen in culture or absent growth, with detection of surrogate biomarkers reflecting a particular microbiologic etiology. Quite often, a definitive diagnosis is neither immediately forthcoming nor firmly established, and in these cases, diagnostic assumptions have to be made on the basis of illness progression. Thus, in the clinical practice of ID—if resources allow—much time and effort is devoted to the detection and identification of microbial pathogens that may be responsible for illness in an individual patient or, in the case of an epidemic, a population of patients. Once a pathogen is identified, definitive diagnosis can be established, specific therapies administered, and appropriate infection control measures instituted.

With the exception of nucleic acid amplification methods, the traditional approaches to the laboratory detection of microorganisms have not changed significantly in more than 50 years (Gill, Fedorko, and Witebsky 2005). These detection methods, including microscopy, culture, and immunoassays, may fail to establish the ID diagnosis for a variety of reasons (Kaittanis, Santra, and Perez 2010; Theron, Eugene Cloete, and de Kwaadsteniet 2010; Shinde, Fernandes, and Patravale 2012). Microorganisms may be sequestered in cells or tissues not readily accessible for specimen procurement. At the time of patient evaluation, the pathogen may no longer be viable, having been eradicated by host immune mechanisms or antimicrobials. Even if viable at the time biological specimens are obtained, microorganisms may be fastidious and grow poorly in culture, grow very slowly, or not at all grow.

In the absence of culture detection of microbial pathogens, the identification of surrogate biomarkers (e.g., nucleotides, peptides, cellular antigens, antibodies, and toxins) may or may not identify the cause of infection. Immunoassays directed at antigen detection may lack sufficient sensitivity and specificity to establish a definitive diagnosis. Serologic methods measuring specific host antibodies that develop in response to infection may also lack sufficient sensitivity and specificity. Moreover, since serologic assays rely on the detection of a host antibody response, they tend to provide a retrospective diagnosis, not necessarily one that is contemporaneous with illness onset (Halfpenny and Wright 2010).

Developments in molecular testing methods have addressed some of the limitations of traditional microbiology techniques (Olano and Walker 2011; Wolk, Mitchell, and Patel 2001). These analytical tools greatly enhance the ability to detect many types of microorganisms in a timely fashion, particularly those that may not readily grow in culture. Limitations of these methods include laboratory contamination with amplified DNA that can cause false positive results, inability to distinguish living from dead microorganisms, and the need for reagents and equipment that are often unavailable in many diagnostic settings (Archibald and Reller 2001; Wolk, Mitchell, and Patel 2001; Mabey et al. 2004).

Many of the traditional microbiologic detection schemes are limited by relatively slow turnaround time, the need for costly reagents, expensive and complicated equipment that require the expertise of highly trained personnel. As a consequence, access to many of these methods of detection is limited to locations with abundant technological resources (Mabey et al. 2004; Chin, Linder, and Sia 2006; Petti et al. 2006; Kiechle and Holland 2009; Warsinke 2009; Lee et al. 2010). The development of nanobiotechnology and its applications to biosensing strategies offer great promise in addressing many of the limitations of traditional clinical and public health microbiology detection schemes (Jain 2005, 2007; Rosi and Mirkin 2005; Vo-Dinh 2007; Morrison et al. 2008; Hauck et al. 2010; Kaittanis, Santra, and Perez 2010; Lee et al. 2010; Tallury et al. 2010; Theron, Eugene Cloete, and de Kwaadsteniet 2010; Tibbals 2010; Chi et al. 2011; Shinde, Fernandes, and Patravale 2012).

1.7 Infectious Diseases: Nanobiosensing Interface

Morrison et al. (2008) concisely defined a biosensor as, "an analytical device that uses a biological recognition system to target molecules or macromolecules. Biosensors can be coupled to a physicochemical transducer that converts this recognition into a detectable output signal." While neither a new term, nor one that is unique to nanoscience, biosensing is emerging as an extraordinarily fertile area of applied nanobiotechnology, with numerous applications relevant to the diagnosis of IDs. Creative and ingeniously innovative methods utilizing a variety of nanoparticles and nanodevices, along with integrated micro- and nanoscale photoelectric, electromagnetic and microelectromechanical sensors (MEMS), are being developed to enhance cellular and molecular target (analyte) signal detection and transduction.

An ideal biosensing platform strives to achieve the following:

- Detection of biomolecules and biomarkers that may be present in exceedingly low concentrations
- High analytical sensitivity, specificity, and reproducibility
- Quantitation of receptor–ligand binding kinetics
- Multiplex (simultaneous) detection of different varieties of analytes from one sample
- Detection of analytes within environmental or complex biological samples (air, water, foods, tissue, blood, urine, sputum, intra and extracellular fluids)
- Minimal sample processing and preparation
- Rapid analysis with high throughput capacity
- Use of small sample and reagent volumes
- Miniaturization, portability, and ease of use with point-of-care diagnostic capability
- Use of nontoxic reagents
- Low cost

In their characterization of the ideal biosensing platform for use in resource-limited settings (Mabey et al. 2004; Hauck et al. 2010), the World Health Organization has devised the acronym, ASSURED:

- Affordable by those at risk of infection
- Sensitive (few false negatives)
- Specific (few false positives)
- User friendly (simple to perform and requiring minimal training)
- Rapid (to enable treatment at first visit) and Robust (does not require refrigerated storage)
- Equipment free
- Delivered to those who need it

Although a biosensing platform possessing all these characteristics does not yet exist, nanobiotechnology is bringing us closer to achieving these goals. Indeed, Kewal K. Jain (2003) coined the word "nanodiagnostics" to imply very promising applications of nanoscience to biomolecular detection strategies. The size, surface properties, optical, electromagnetic, and electromechanical characteristics of nanoparticles and nanodevices are being exploited to create biorecognition assays that are capable of detecting microorganisms, virulence factors, and host immune markers associated with IDs (Vo-Dinh 2007; Morrison et al. 2008; Hauck et al. 2010; Kaittanis, Santra, and Perez 2010; Tibbals 2010; Chi et al. 2011; Shinde, Fernandes, and Patravale 2012).

Many of the fundamental principles of nanodiagnostics are derived from conventional biorecognition techniques such as immunoassays, fluorescent spectroscopy, fluorescent microscopy, and nucleic acid hybridization, whereby nanomaterials are incorporated into novel biosensing strategies.

Although the field of nanobiotechnology is in its infancy, there already exists a large number of promising ID–nanobiosensing methods that have been applied to the detection of a variety of microorganisms and biomarkers of infection. These techniques are being explored clinically in in *vitro* and in *vivo* settings, as well as in areas of environmental, food, and water safety. For the purposes of this chapter, several nanodiagnostic methods having direct application to IDs will be reviewed. Underlying technical and physicochemical concepts will be described in ways that will hopefully be comprehensible and enlightening to the ID practitioner who might otherwise be rightfully intimidated by terms such as semiconductor nanoparticles, nanocrystal fluorescence resonance energy transfer (FRET), magnetic relaxation nano-switch, surface plasmon nanoprobes, nanoparticle-based bio-barcodes, microcantilever-based nanodetectors, and nanobiosensing integrated MEMS. Indeed, all of that is a mouthful for any self-respecting ID practitioner, but once digested, the principles and concepts essential to nanobiosensing strategies will provide a foundation for appreciating their enormous potential in the diagnosis of IDs.

1.7.1 Photonic Biosensing: Semiconductors

In general, a great deal of nanotechnology's focus is concentrated on the properties of semiconductors, unique materials whose electrical conductivity can be precisely regulated. From the nano-ID standpoint, detailed knowledge of semiconductor quantum physics is not a prerequisite for understanding their potential applications in IDs nanodiagnostics. It is important to recognize, however, that applications of semiconductor nanotechnology extend beyond information age electronics, transistors, and microprocessors. The application of semiconductors to the development of biosensing strategies is an example of the convergence of nanotechnology with biotechnology, and on that basis, a brief discussion of semiconductor principles is in order (Jaiswal and Simon 2004; Alivisatos, Gu, and Larabell 2005; Klostranec and Chan 2006; Gill, Zayats, and Willner 2008).

Semiconductors are unique, because the atoms that make up their crystalline lattice contain valence band electrons that can transition into the higher energy level conduction band. The excitation stimulus for the transition comes from an external electric source (electrical energy), heat (thermal energy), or electromagnetic radiation (EMR) applied to the semiconductor. Once in the conduction band, electrons can either travel, thereby creating electric current, or decay back into the lower energy valence band. This electron decay can be accompanied by the release of energy, and under certain conditions, well exploited by nanoengineers, the energy released can be emitted as light (Jaiswal and Simon 2004; Klostranec and Chan 2006). The difference between valence and conduction band levels, known as the bandgap, is a term often encountered in nanotechnology literature. It is a measure of the amount of energy that is required to excite electrons to transition from the valence band to the conduction band (Chan et al. 2002; Klostranec and Chan 2006). The bandgap energy of semiconductors lies somewhere between that of a metallic conductor, such as copper (valence and conduction bands overlap), and that of an insulator, such as glass (very large bandgap).

The materials used to fabricate semiconductors typically come from column IV of the periodic table: carbon, silicon, and germanium, for example. Alternatively, they can come from combining elements equidistant from column IV. For instance, semiconductors such as cadmium selenide (CdSe) come from the combination of columns II and VI elements. In pure crystalline form, the atoms that comprise semiconductors share an average of four valence electrons. Through the controlled and strategic placement of atoms containing either three or five valence electrons within the semiconductor crystal, a process known as doping, the flow and direction of current can be regulated. This property establishes semiconductors as the fundamental subunit of transistors and microprocessors. Moreover, the potential light-emitting characteristics of semiconductor nanoparticles constitute another powerful property that has great potential in the photonic nanobiosensing realm.

1.7.2 Quantum Dot Semiconductors and Infectious Diseases Biosensing

QDs are semiconductor nanocrystals possessing robust fluorescent properties that are being exploited in a variety of bioanalytical formats (Jaiswal and Simon 2004; Alivisatos, Gu, and Larabell 2005; Arya et al. 2005; Klostranec and Chan 2006; Gill, Zayats, and Willner 2008; Rousserie et al. 2010). They are made from the combination of columns II and VI elements: cadmium selenide, and zinc sulfide, for example. QDs can be synthesized in enumerable nanoscale sizes, and it is this size tunability that imparts one of their most important and unique properties. QD size directly influences their bandgap energy that, in turn, determines fluorescence color emission, as shown in Figure 1.4. QDs ranging in size from 1 to 10 nm emit visible light within the blue to red spectral range. Those smaller than 1 nm and larger than 10 nm emit light in the ultraviolet and near-infrared spectra, respectively.

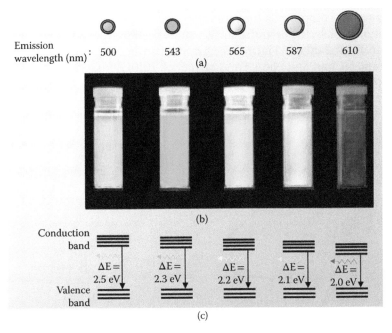

FIGURE 1.4
Tunable optical properties of quantum dots (QDs). (a) Schematic of the size-dependent optical properties of QDs (not drawn to scale). As the QD becomes larger, the optical emission shifts from blue to red. The size range of the QDs is 2–10 nm. (b) Corresponding to (a), real-color emission of vials filled with different-sized QDs suspended in chloroform, excited by a handheld UV lamp. (c) The corresponding bandgap energies for the QDs shown in (a) and (b). (From Klostranec, J.M. and Chan, W.C.W., *Adv. Mater*, 18, 1953–1964, 2006. With permission.)

In addition to size-tunable emission wavelength spectra, QDs possess other unique properties that overcome many of the limitations associated with conventional fluorescent dyes. QDs have very broad excitation (absorption) wavelengths; thus a single light source can simultaneously excite a variety of different-sized QDs. Moreover, the wavelength of light emitted from QDs is characteristically narrow and symmetrical, allowing easy spectral discrimination (deconvolution) when different colored QDs are used simultaneously in multiplex analysis. QD excitation and emission wavelengths can be widely separated (a large Stoke's shift), preventing overlap interference. In contrast to conventional fluorescent dyes, QDs are efficient at producing bright light due to their strong absorption characteristics (high extinction coefficients) and high quantum yield (the ratio of the number of photons emitted to the number of photons absorbed). The luminescence emitted by QDs resists rapid fading (photobleaching or photodegradation). Furthermore, the fluorescence lifetime tends to be significantly longer than that of both organic fluorophores and autofluorescing proteins, resulting in long detection times and reduced background noise inherent in biologically complex specimens. These unique properties have stimulated enormous interest in the potential for QD-based in *vitro* and in *vivo* analysis of dynamic biological processes.

Before QDs can be used as biosensing probes, they must first be encapsulated (passivated) to prevent photo-oxidation and to enhance their fluorescent properties. Passivation may avert QD cytotoxicity by preventing the core elements from leaching into cells (Derfus, Chan, and Bhatia 2004). The composition of the capping agent can modulate QD emission properties; thus, passivation in combination with size-tunable color emission can produce a very large variety of QDs capable of participating in high-throughput, multiplex biosensing platforms (Goldman et al. 2004; Agrawal et al. 2005; Hoshino et al. 2005; Yang and Li 2005; Rousserie et al. 2010; Giri et al. 2011).

QDs are inorganic compounds that are inherently hydrophobic, and in order to be used as biorecognition probes, they require surface modifications that render them water soluble. Most importantly, as shown in Figure 1.5, QDs are readily functionalized by the attachment of affinity ligands (Chan and Nie 1998; Agasti et al. 2010; Rousserie et al. 2010; Sekhon and Kamboj 2010; Cui et al. 2011). QDs have shown great potential in a variety of biosensing formats related to the detection of microorganisms, surrogate markers of infection, and analysis of host–pathogen interactions. These applications will now be explored in greater detail.

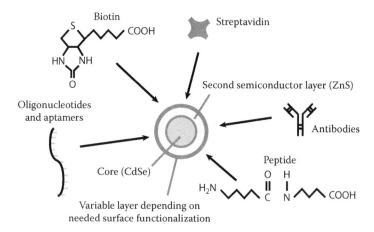

FIGURE 1.5
Schematic representation of ligands that can be conjugated with quantum dots. (Adapted from Rousserie, G. et al., *Crit. Rev. Oncol. Hematol.*, 74, 1, 2010. With permission.)

1.7.2.1 Quantum Dots and Virus Detection

QD-based biosensing strategies have succeeded in the detection of numerous clinically important viruses, including HIV, respiratory syncytial virus (RSV), influenza viruses, human T-cell leukemia virus type 1 (HTLV-1), hepatitis viruses, HPV, and Dengue virus. In addition, analyses of virus–target cell interactions, measurement of surrogate markers of infection, and screening of potentially active antiviral compounds have been developed. Several examples of virus detection strategies are as follows.

1.7.2.2 Quantum Dot Biosensing and Human Immunodeficiency Virus

HIV can be classified as an emerging ID since its discovery in 1983. Although great strides have been made in treatment and prevention, it remains a preeminent global health-care problem. Approximately 35 million people are currently HIV infected, and there are 3 million new cases occurring annually (WHO 2011a). As a result of HIV/AIDS, we have gained immense insight into host–pathogen interactions, immunology, anti-viral pharmacology, and obstacles to health-care delivery (Fauci and Folkers 2012). Several QD-based applications related to HIV detection, target cell interaction, and anti-viral activity will be described.

QD-based detection of HIV, as well as in *vitro* analysis of viral trafficking within target cells, was accomplished by Joo et al. (2008). Streptavidin–conjugated QDs were incubated with HIV-infected cells that had biotin incorporated in their cell membrane. Using confocal microscopy, the luminescence from QDs bound to HIV was detected as viruses budded from the infected cell surfaces. When compared with conventional dye labels, the functionalized QDs were significantly more photostable and provided a means to monitor dynamic events between HIV and their target cells. This in *vitro* method of detection and virus–host cell interaction are applicable to the study of a wide range of membrane-enveloped viruses.

Kim et al. (2009) developed a microfluidic device containing channels functionalized with immobilized anti-HIV antibodies. HIV was captured within the channels when as little as 10 μL of unprocessed whole blood from infected patients was added. The immobilized HIV was then labeled by two differently colored QDs functionalized with anti-HIV antibodies, and their luminescence was detected by standard fluorescent microscopy. The microfluidic immunocapture device was capable of detecting HIV within 10 minutes. It has the potential for rapid point-of-care viral detection. Validation of this promising nano-diagnostic technique under field-testing conditions awaits further investigation.

CD4+ T cells are the major immunologic targets of HIV, and as a result of infection, this cell population declines over time. Monitoring of this decline has a role in determining when anti-retroviral drugs and anti-microbial prophylaxis are initiated (When to Start Consortium 2009). The conventional approach to measure CD4+ T cells utilizes flow cytometry, an expensive and complex technology that is limited, almost exclusively, to technologically advanced regions of the world. Thus, measurement of this biomarker of disease is frequently unavailable in those parts of the world that have the highest prevalence of HIV infection (Cheng et al. 2007b). The need for improved access to CD4+ T cell counts has led to development of techniques having point-of-care capabilities (Yager et al. 2006; Cheng et al. 2007b; Willyard 2007; Jokerst et al. 2008).

The fabrication of inexpensive microfluidic devices capable of selectively trapping and measuring CD4+ T cells has been accomplished by Cheng et al. (2007b). This label-free method requires manual counting of cells by a trained microscopist and possesses many of the features of an ideal biosensor. In a variation of the technique, Jokerst et al. (2008)

designed a similar device and integrated it with an automated miniaturized analyzer. Their assay incorporates a QD-labeling step and produces levels of sensitivity and specificity comparable to flow cytometry. The portable device is simple to use and inexpensive. Moreover, by using QDs functionalized with specific antibodies, it has the potential to detect other classes of T cells.

QDs have been incorporated into FRET-based analyses that have been applied to the detection of HIV-1 protease activity (Choi et al. 2010; Biswas et al. 2011). FRET, a photophysical phenomenon first described over 50 years ago (Förster 1948), has been increasingly applied to the analysis of biological systems. A brief description of FRET is provided, and the reader is referred to detailed reviews on the topic (Selvin 2000; Raymo and Yildiz 2007; Gill, Zayats, and Willner 2008; Sahoo 2011).

In its simplest configuration, FRET is a proximity-dependent interaction, whereby a fluorescent donor molecule transfers energy to a neighboring acceptor molecule. This interaction results in a reduction (quenching) in donor fluorescence intensity and a corresponding excitation of the acceptor. If the acceptor happens to be a fluorophore (it need not be), the energy transfer will result in an increase in its fluorescence intensity. For FRET to occur, in addition to other physical constraints, the donor–acceptor pair must come into close proximity (<10 nm). Conversely, FRET can cease when donor–acceptor pairs, already in close proximity, are made to separate. In this case, the separation of the FRET pair results in a restoration of donor fluorescence. FRET donor–acceptor pairs can be conformationally linked to ligand–receptor biorecognition events. FRET can be thought of as a luminescent switch and is a sensitive tool for analysis of dynamic receptor–ligand binding interactions, spatial distribution, conformation of proteins and nucleic acids, cellular membrane potential, and enzyme activity. Given the nanoscale distances involved, QDs are a very useful addition to FRET-based biosensing platforms, and specific examples of the format will be provided (Algar and Krull 2008).

Cell culture FRET-based analysis of HIV-1 protease has been developed using QDs (Choi et al. 2010; Biswas et al. 2011). Peptides containing an HIV-1 protease recognition site were conjugated to FRET donor QDs. This was followed by the conjugation of a FRET acceptor fluorophore whose close proximity produced quenching of QD emission. In the presence of HIV-1 protease, the donor–acceptor pairs separated, resulting in increased fluorescent intensity as quenching ceased. With this sensing strategy, the effects of protease inhibitor (PI) drugs added to the cell cultures could be quantitatively measured. Rapid, sensitive, and economical real-time quantitation of intracellular HIV-1 protease, as well as analysis of PIs, was accomplished. In addition, QDs were found to be superior to organic dyes as FRET donors.

1.7.2.3 Quantum Dot Biosensing and Respiratory Syncytial Virus

RSV is the leading cause of life-threatening pneumonia in infants and young children and a major cause of respiratory tract infection in both the elderly and immunocompromised (Hall 2004; Halfpenny and Wright 2010). Several promising QD-based nanobiosensing approaches have been applied to the detection of RSV.

After infection by RSV, cells are directed to produce fusion (F) and attachment (G) cell surface proteins. Bentzen et al. (2005) developed a biosensing method designed to detect F and G proteins in an in *vitro* cell culture model that utilized a QD-based immunoassay. At various time points after RSV infection of Hep-2 cells, surface proteins were labeled with biotinylated anti-F and anti-G monoclonal antibodies. Subsequently, streptavidin–conjugated QDs were added to the cultures and bound specifically to the biotinylated antibodies. Confocal laser scanning microscopy detected the fluorescently tagged cells.

This construct permitted measurements of the progression of viral infection and expression of the F and G proteins over time. When compared with experiments utilizing traditional fluorophores, these QD labeling experiments were significantly more sensitive at the earliest time points and provided robust, nonphotobleaching images. In addition, when compared with Western Blot and quantitative polymerase chain reaction (PCR) assays, the QD immunoassay was more sensitive at detecting RSV F protein at earlier time points. The authors suggest that the use of QD probes in this fashion could provide a sensitive means to analyze cell trafficking of viral proteins.

Agrawal et al. (2005) detected RSV using a microcapillary flow system integrated with a confocal microscope. They detected virus that was simultaneously bound to two differently colored QD labels functionalized with RSV anti-F and anti-G monoclonal antibodies. This analysis permitted detection of very low concentrations of virus particles (possibly as low as a single virus) and relative estimation of virus surface protein expression. The authors point out that their experimental setup has potential for multiplexed application for the detection of a variety of ID agents.

Tripp et al. (2007) were able to detect RSV both in *vitro* and in *vivo* using cadmium telluride QDs functionalized with anti-F monoclonal antibodies. In their in *vitro* analysis, the single-step QD labeling experiments were significantly more rapid and produced less background noise when compared with conventional multistep immunostaining techniques. Furthermore, the QD labels detected RSV-infected cells 2–3 days earlier than conventional stains. Their study investigated viral detection in *vivo* by intravenously administering the functionalized QD labels into RSV-infected mice. Sections of lung tissue were visualized by fluorescent microscopy and compared with those stained by conventional immunohistochemical methods. Background interference by tissue autofluorescence was significantly reduced in the QD-stained tissue. This proof-of-concept study suggests that functionalized QDs have robust immunostaining properties in both in *vitro* and in *vivo* formats.

1.7.2.4 Quantum Dot Biosensing and Influenza Virus

Influenza is an emerging and a reemerging ID that results in considerable morbidity and mortality throughout the world (Bartlett and Hayden 2004; Thompson et al. 2010). The potential for future pandemics remains an ongoing threat (Murray et al. 2006; Taubenberger and Morens 2006). Novel QD-based nanobiosensing methods to detect influenza virus have been developed.

Yun et al. (2007) exploited pH-dependent changes in QD fluorescence intensity to detect H9 avian influenza virus. In an in *vitro* system, they capitalized on the ability of a host cell transmembrane ATPase biological motor to pump hydrogen ions out of a cell and into the surrounding media (Yuanbo, Fan, and Jiachang 2005). Anti-H9 virus antibody was attached to the ATPase, and hydrogen ion flux increased when virus bound to the antibody. The biorecognition event was sensed by a pH-dependent increase of fluorescent intensity by QDs bound to the cell membrane. The investigators postulate that the increase in hydrogen ion flux results from a conformational change in the motor resulting from attachment of virus. Incorporating two differently colored QDs to simultaneously sense both H9 influenza and herpes viruses extended the capability of their viral detection technique (Deng et al. 2007).

In a sandwich immunofluorescence capture method devised by Chen et al. (2010), H5N1 avian influenza virus was readily detected using QDs functionalized with rabbit anti-H5N1 antibody. Their strategy used varying dilutions of virus added to 96-well microtiter plates coated with anti-H5N1 monoclonal antibody. A conventional microtiter plate fluorescent reader was able to detect virus at concentrations as low as $1.5 \times 10^{-4}\ \mu g\ mL^{-1}$ of viral protein.

This was an order of magnitude more sensitive than conventional enzyme-linked immunosorbent assay (ELISA). The QD-based method detected virus obtained from chickens, a natural reservoir of avian influenza virus, and was highly sensitive and specific. The authors point out that this QD-based platform is applicable to the detection of a diversity of pathogens.

In a proof-of-concept study, Cui et al. (2011) were the first to develop aptamer-functionalized QDs capable of virus detection. Aptamers are synthesized small oligonucleotides that bind with high specificity to an almost infinite variety of targets (Bunka and Stockley 2006). In this study, aptamers that recognize the hemagglutinin of influenza A virus were constructed, biotinylated, and attached to QDs containing streptavidin on their surface. The viruses bound specifically to the QD probes and were detected by electron microscopy. This study is particularly noteworthy having combined two emerging biotechnology schemes that offer great potential for sensitive and specific detection of a variety of pathogens and biomarkers.

1.7.2.5 Quantum Dot Biosensing and Hepatitis Viruses

Liver disease due to hepatitis B virus (HBV) and hepatitis C virus (HCV) is a global health problem. Worldwide, approximately 500 million people, including 4 million in the United States, are chronically infected (Ly et al. 2012; WHO 2012a). Most individuals with chronic hepatitis remain asymptomatic for many years, are unaware that they are infected, and unknowingly transmit virus to others. In recognition of the large pool of undiagnosed adults with chronic HCV, a recommendation to test all Americans born between 1945 and 1965 has recently been made (Smith et al. 2012). HBV is acquired through parenteral and sexual routes of exposure, and chronic hepatitis due to HBV and HCV can silently progress to cirrhosis and hepatocellular carcinoma. Vaccination is highly effective in preventing HBV infection, and on that basis, it can be considered the first successful anti-cancer vaccine (Chang et al. 1997). Transmission of HCV occurs predominantly by the parenteral route, and access to a safe blood supply from screened donors is central to prevention strategies. Unfortunately, this public health need is often unmet in resource-limited parts of the world (WHO 2011c). The QD-based detection of hepatitis viruses have been accomplished in *vitro*.

Alivisatos et al. were the first to report on the use of QDs for the simultaneous detection of multiple viruses using a microarray approach (Gerion et al. 2003). They employed QDs of three different sizes (three differently colored emission wavelengths), each functionalized with single-stranded DNA (ssDNA) from HBV, HCV, or a sequence specific to the human p53 oncogene. The QD probes were allowed to hybridize with a microarray containing complementary oligonucleotides immobilized on glass slides, and hybridization was achieved with high specificity and selectivity. Moreover, by modifying the stringency of the hybridization conditions, the investigators were able to detect single nucleotide polymorphisms. Remarkably, hybridization could be achieved within 10 minutes at room temperature. In addition, the DNA functionalized QD probes remained stable for more than 6 months when stored at 4°C. This study established the feasibility for multiplex biosensing of microorganisms using QD labels in a microarray format. In addition, the ability to detect single nucleotide polymorphisms will have great value in diagnostics, pharmacogenomics, and the epidemiology of IDs.

1.7.2.6 Quantum Dot Biosensing and Other Viruses

Currently, QD-based nanobiosensing assays have been applied to the detection and study of other clinically relevant viruses, including HTLV-1, Dengue virus, and HPV

(Kampani et al. 2007; Wang, Hsu, and Peng 2008; Yu-Hong, Rui, and Ding 2011). It is likely that virus-detection strategies that incorporate QD labels will continue to proliferate and advance our understanding of virus–host interactions.

1.7.3 Quantum Dots, Bacteria, and Bacterial Toxin Detection

The ability to detect bacteria and their toxins in clinical specimens, environmental samples, and food products represents an essential diagnostic and preventive health-care function. The rapid, sensitive, and specific detection of bacteria and their toxins is of utmost importance in optimizing health care on a global scale, and the application of QDs to methods of detection has generated a great deal of interest. This section will explore several examples of this emerging biosensing field.

1.7.3.1 Quantum Dot Biosensing and Bacteria: General Principles

The first report on QD labeling of bacteria was given by Kloepfer et al. (2003). They investigated conditions by which functionalized QDs might succeed as both extracellular and intracellular fluorescent labels. Their systematic approach investigated the ability of lectin-functionalized CdSe QDs to label the surface of Gram-positive bacteria and of transferrin-functionalized CdSe QDs to gain entry into bacterial cells. Using bacterial cell cultures, they characterized and contrasted features of functionalized QD stability, solubility, binding specificity, and optimal fluorescent excitation and emission spectra.

After being attached to QDs, both lectin and transferrin retained their binding specificity and function. The lectin-QD probes labeled a variety of Gram-positive bacteria. A mixture of three different-sized lectin-QD probes bound to a population of bacteria could be distinguished individually on the basis of emission wavelength spectra. The transferrin QDs were capable of promoting growth of transferrin receptor-positive bacteria in iron-deprived growth media, a characteristic that may be useful in distinguishing virulent from relatively nonvirulent microorganisms (Modun, Morrissey, and Williams 2000). Over a period of several hours, the transferrin QDs appeared to decompose into elemental Cd and Se within the cell. This resulted in brighter fluorescent intensity and a shift in wavelength, suggesting that these intracellular QD labels had undergone photo-oxidation. The mechanism, by which the transferrin QDs gained access into the bacterial cell, and whether they did so intact, requires further investigation.

The study by Kloepfer raises questions regarding the potential role of QDs as bacterial labels. Can other affinity ligands attached to QDs retain their specificity and functionality? Can different functionalized QDs gain entry into bacterial cells, and if so, by what mechanism? Do QDs alter normal cellular function, and are they cytotoxic? The study suggests a possible role for siderophore-functionalized QDs to identify virulent strains of bacteria and to explore the important role of iron binding, uptake, and trafficking within clinically relevant microorganisms.

Another investigation into the ability of QDs to gain entry into bacterial cells was performed by Wenhua et al. (2004). Specifically, they capitalized on the role of supraphysiologic concentrations of divalent Ca^{2+} cations to induce competence (the ability of bacteria to accept and be transformed by a genetic element) in an established strain of *Escherichia coli*. Competence was identified by the ability of bacteria to take up and be transformed by an antibiotic-resistance plasmid. The *E. coli* were incubated with QDs composed of CsSe and CdS, which were 3–4 nm in diameter. Using fluorescence and atomic force microscopy, it was determined that cellular fluorescence occurred predominantly in the Ca^{2+}-treated competent bacteria. Moreover, the cell membrane of the competent bacteria contained holes measuring between 50 and 300 nm. The investigators postulate that the mechanism underlying bacterial cell competence, heretofore unknown, results from a physicochemical, as opposed to a physiological process that enhances cell permeability.

In a study by Dwarakanath et al. (2004), two differently colored QDs were functionalized by antibodies directed against laboratory strains of *E. coli*, *Bacillus subtilis* spores, and *Salmonella typhimurium*. In addition, aptamer DNA-functionalized QDs were constructed and directed against the *E. coli* strain. Fluorescent microscopy detected surface labeling, and spectroscopy detected emission wavelength shifts upon QD binding to their bacterial targets. The intensity of this shift appeared to grow in direct proportion to the number of bacteria in the incubation mixtures. The investigators postulate that the wavelength shift may result from changes in QD shape upon binding to their target or changes in local environmental factors such as pH, electric charge, or hydrophobicity resulting from the binding event. It remains to be determined whether QDs composed of different elements will yield similar changes in emission spectra and whether wavelength shifts can be applied to bacterial detection in a quantitative way.

QD nanobiosensing has been applied to bacteriophage-based detection of bacteria by Edgar et al. (2006). In an effort to overcome the low signal-to-noise ratio and photobleaching limitations of conventional fluorophores, these investigators designed a QD-based labeling strategy that exploited the highly specific bacterial recognition characteristics of bacteriophage. They engineered a phage whose capsid displayed a peptide that could be readily biotinylated by the host bacterial biotin-ligase protein (BLP). After infecting wild-type *E. coli*, phage progeny were biotinylated within the bacterial cell. Upon bacterial cell lysis, the biotinylated phage was detected by QD–streptavidin conjugates. QDs bound to phage did not alter infective titers or cause phage aggregation. Labeled phage was detected by flow cytometry and transmission electron microscopy (TEM), as shown in Figure 1.6. Remarkably, using fluorescent microscopy, one could visualize individual bacteria branded by as few as one QD-labeled phage. Furthermore, since progeny phage acted as surrogate

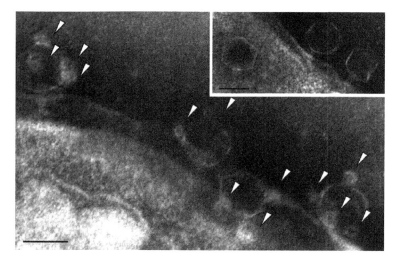

FIGURE 1.6
T7-bio phage bound to streptavidin-functionalized quantum dots (QDs). Transmission electron microscope images of phage–QD targeted bacteria are shown. The arrowheads point to QDs conjugated to the phage head. *(Inset)* Control T7-myc phage that are not biotinylated and therefore have no conjugated QDs (scale bars: 50 nm). (From Edgar, R. et al., *Proc. Natl. Acad. Sci. U S A*, 103, 4841–4845, 2006. With permission.)

amplifiers of the target cell, as few as 10 *E. coli* could be detected within a mixture of several different bacterial species. The investigators were able to detect as few as 20 *E. coli* in 1 mL of river water, and the analysis was performed in approximately 1 hour. A conventional water sampling method took 24 hours to complete and was significantly less sensitive. This QD-based detection strategy may have broad applicability, given the highly conserved nature of BLP in bacterial species. Moreover, by incorporating different-sized QDs and a portable detection device, the technique has the potential for multiplexed use in the field.

1.7.3.2 Quantum Dot Biosensing and Biofilms

Bacterial biofilms, having evolved the ability to colonize both inanimate and animate structures, are fascinating assemblies that make up a large portion of the world's biomass (White 2009). Biofilms play a major role in IDs, a fact that has only recently been appreciated (Costerton et al. 2003; Richards and Melander 2009; Francolini and Donelli 2010; Donlan 2011). Biofilms are composed of either single- or multispecies bacterial and fungal communities that establish themselves on the surfaces of diverse, environmentally compatible materials, including rocks, wood, metals, ceramics, plastics, plumbing, and plant and animal tissues.

As depicted in Figure 1.7, biofilms have a complex architecture composed of strata of microorganisms that range from metabolically inactive and stationary to fully active and dividing. The bacteria are encased in a protective glycoprotein matrix containing water-filled channels that transport nutrients and messenger biomolecules that allow the bacterial populations to communicate with the other (Molin and Tolker-Nielsen 2003). In animal tissues and in the bloodstream, the metabolically active bacteria that reside within the upper biofilm layers can break off and establish infections in adjacent or remote anatomic locations. Bacteria residing within a biofilm are genetically adapted to resist antimicrobial therapy and are a major source of chronic infections, HAIs, and implantable device-related infections (Stickler 2008; Yang, Haagensen et al. 2008; Falagas et al. 2009; Hoa et al. 2010; Hoiby et al. 2010; Foreman, Jervis-Bardy, and Wormald 2011; Koh et al. 2012).

Our understanding of the very complex nature of bacterial biofilms is rudimentary, and tools are needed to help explore these dynamic, multicellular, sophisticated bacterial communities. Nanobiotechnology can aid in analysis of biofilm structure, genomic and proteomic characteristics, and potentially contribute to strategies that prevent bacterial adherence and lead to improved treatment modalities. An example of QD-based biosensing with respect to biofilms will be described.

Kolenbrander and coworkers (Chalmers et al. 2007) succeeded in conjugating primary monoclonal antibodies to QDs and detected a variety of bacterial species within a biofilm model. Their use of primary antibody–QD conjugates as direct labeling probes circumvented limitations associated with indirect immunofluorescence that would otherwise require the use of both primary and secondary antibodies in the assay. They were able to detect planktonic bacteria in *vitro* as well as a variety of bacterial species growing within their in *vivo* model of dental plaque biofilm. The monoclonal antibody-functionalized QD labels produced fluorescent intensity that was equivalent, initially, to traditional fluorophore conjugates. However, the photostability of the QDs resulted in significantly longer periods of bright fluorescence. This property may enable time-lapsed tracking of specific bacteria within the biofilm matrix and enhance our currently limited understanding of the very complex chemical signaling that occurs within these bacterial communities. Furthermore, the QD probes were able to penetrate the biofilm matrix, label bacteria, and achieve single-cell resolution. The simultaneous use of multiple QD probes provided a means of exploring spatial relationships between mixed species within the bacterial communities.

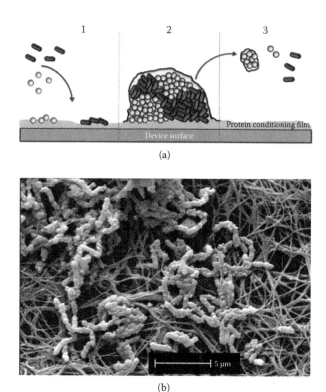

(a)

(b)

FIGURE 1.7
The complex architecture of biofilms. (a) Schematic representation of biofilm formation. Stage 1: microbial adhesion to the surface. Stage 2: exopolysaccharide production and three-dimensional biofilm development. Stage 3: detachment from biofilm of single and clustered cells. (From Francolini, I. and Donelli, G. 2010. *FEMS Immunol Med Microbiol* 59 (3):227–238. With permission.) (b) Biofilm of *Alcaligenes xylosoxidans* in a fibrin-like matrix on the surface of an explanted intravascular catheter (scanning electron microscopic image). (From Donlan, R.M., *Clin. Infect. Dis.*, 52, 1038–1045, 2011. With permission.)

1.7.3.3 Quantum Dot Biosensing and Food-Borne Bacteria

Foodborne diseases are a significant public health problem throughout the world (WHO 2007b). In resource-scarce regions, it is a major cause of life-threatening diarrheal illness in infants, young children, the elderly, and the immunocompromised. In the United States, approximately 9 million people acquire foodborne illness, resulting in 1300 deaths annually (Scallan et al. 2011).

A QD biosensing immunoassay was developed for simultaneous detection of *E. coli* O157:H7 and *S. typhimurium*, common bacterial causes of food-borne illness (Yang and Li 2005). This technique incorporated two differently colored QDs functionalized through a streptavidin linker with biotinylated anti-*E. coli* and anti-*Salmonella* antibodies, respectively. Fluorescence was detected by a portable spectrophotometer. TEM and fluorescent microscopy identified QD probes on the surface of the bacteria. The different-colored QDs fluoresced at distinct, nonoverlapping wavelengths that were readily identified. The limit of detection was approximately 10^4 CFU mL^{-1} and was comparable to a variety of conventional techniques employed in food safety analysis. The investigators subsequently expanded the multiplexing capabilities and were able to simultaneously detect *E. coli*, *Salmonella*, and *Listeria monocytogenes* from a variety of commercial food samples (Wang et al. 2011). Assay modification resulted in detection limits of 20 to 50 CFU mL^{-1} and was performed in less than 2 hours. This represents a significant enhancement over conventional methods.

1.7.3.4 Quantum Dot Biosensing and Syphilis

Approximately 340 million new cases of STDs, including 12 million cases of syphilis, occur each year (WHO 2007a). Rates of syphilis are increasing in both developing and industrialized parts of the world. STD prevention, diagnosis, and treatment remain one of the greatest challenges facing the health-care community, and the availability of and access to diagnostic tests represents a significant obstacle in STD management (Peeling 2006). Applications of nanotechnology may well address many of these public health challenges.

Using lateral flow immunochromatography (LFIC), Yang et al. (2010) developed a rapid, sensitive, and inexpensive screening test for syphilis. In their study of 50 serum samples from known syphilis-positive patients, and an equal number of negative control sera, the QD-based test strips achieved 100% sensitivity and specificity. The investigators point out that this pilot study requires further clinical testing to assess diagnostic accuracy. If additional testing validates the sensitivity and specificity of this technique, it will offer great promise as a point-of-care biosensing strategy. LFIC methods will be discussed in more detail later in Section 1.8.3.

1.7.3.5 Quantum Dot Biosensing and Mycobacteria

Mycobacteria, including *Mycobacteria tuberculosis* (mTb) and the very large group of nontuberculous mycobacterial species, are highly prevalent causes of ID throughout the world. While the absolute number of cases of mTb has been declining since 2006, it is estimated that approximately one-third of the world's population is infected, with the highest prevalence occurring in resource-limited parts of the world (WHO 2011e).

The diagnosis of mycobacterial infections remains a major challenge (Dorman 2010). The organisms tend to grow slowly in culture, and decisions regarding the initiation of empiric antibiotic therapy are always difficult when confronted with a patient in whom mycobacterial infection is suspected. Techniques designed to detect and rapidly identify mycobacteria in clinical specimens represent an area of great interest in clinical microbiology.

In a proof-of-concept in *vitro* study, Otsuka et al. (2004) designed a QD-based immunofluorescent technique that was used to detect a *Mycobacterium bovis* BCG strain. The microorganisms contained a plasmid expressing green fluorescent protein (GFP) and were visualized using confocal fluorescent microscopy. The *M. bovis* were grown on coverslips and treated with primary rabbit anti-BCG antibody. This was followed by treatment with secondary goat anti-rabbit antibody–conjugated QDs having a red-emission wavelength. As shown in Figure 1.8, bright red fluorescence due to surface labeling of *M. bovis* and intracellular green fluorescence were observed. The specificity of the QD label was demonstrated by the absence of staining of *Mycobacterium smegmatis* negative controls. This study is noteworthy given the current limitations in rapid identification of mycobacterial species. The results suggest the possibility that similar QD-labeling strategies may be applied to common human mycobacterial pathogens such as *M. tuberculosis*. Whether QD-based detection of mycobacteria can be accomplished in clinical specimens such as sputum and tissue and whether it will increase the sensitivity of conventional staining methods remains to be determined.

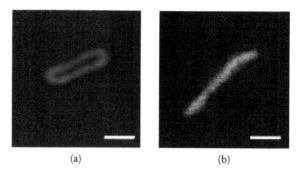

(a)   (b)

FIGURE 1.8
Labeling of BCG (a) and GFP-expressed BCG (b) with anti-BCG antiserum and quantum dot–conjugated anti-rabbit IgG (scale bars: 1 µm). (From Otsuka, Y. et al., *Jpn. J. Infect. Dis.*, 57, 183–184, 2004. With permission.)

1.7.3.6 Quantum Dot Biosensing and Bacterial Toxins

Many species of microorganisms have evolved means to elaborate potent toxins capable of interfering with a variety of cellular and immunologic functions (Hewlett and Hughes 2009). In evolutionary terms, these virulence factors are thought to have developed as a means to gain competitive advantage over host immune defenses. Bacterial toxins have held an important place in the annals of human disease and continue to play a major role as a cause of life-threatening illness throughout the world (Foex 2003; Todar 2006; Schwartz 2009; Kopera 2011). Research on bacterial toxins and their specific mechanisms of action have produced important insights in cell and molecular biology (Schiavo and van der Goot 2001), and interest in toxin detection has recently been heightened by their unfortunate use as biological weapons (Inglesby et al. 2002). Rapid and sensitive methods of toxin detection in clinical specimens and environmental samples are essential ID, public health, and national security goals. Nanobiotechnology may offer promising opportunities as research tools used to explore toxin–host cell interactions and as a practical means of toxin detection.

Goldman et al. (2004) were the first to develop a QD-based multiplex sandwich immunoassay, and they did so in an effort to detect several toxins. Their technique simultaneously detected Shiga-like toxin 1, staphylococcal enterotoxin B, Cholera toxin, and ricin. They utilized four different-sized QDs, each functionalized with specific antitoxin antibodies. Samples containing single and mixtures of toxin at various concentrations were tested. Emission spectra were analyzed spectroscopically, and the signals were deconvoluted and quantitated using a simple mathematical algorithm. Depending on the individual toxin, the limit of detection ranged from 3 to 300 ng mL^{-1} in the single analyte assays. All four toxins could be simultaneously detected when present in a mixture at 30 ng mL^{-1}. The investigators acknowledge that the use of QD labels for simultaneous detection of four toxins overcomes significant limitations encountered with conventional fluorescent dyes. They postulate that the unique fluorescent properties of QDs in combination with relatively simple deconvolution algorithms would allow simultaneous detection of a large number of analytes. Furthermore, this type of biosensing scheme may be amenable to analysis under field testing conditions.

Utilizing total internal reflection fluorescent microscopy (TIRFM), Hoshino et al. (2005) simultaneously detected diphtheria and tetanus toxins by fluorescent immunoassay with QDs conjugated to primary antitoxin antibodies. When compared to an ELISA technique that utilized both primary and secondary antibodies, the TIRFM method was less sensitive. This result was anticipated due to the absence of nonspecific binding when only primary antibody is used in an immunoassay. However, the decreased sensitivity relative to ELISA did not detract from the method's value, since it proved to be quite sensitive, rapid, and capable of simultaneous detection of two important bacterial toxins.

1.7.4 Quantum Dot Biosensing and Fungi

Fungi, including yeast and molds, are environmentally ubiquitous. They inhabit our skin, mucous membranes, lungs, and gastrointestinal tract, where, under normal circumstances, they exist as harmless commensals. They can become opportunistic pathogens capable of exploiting abnormal host immunity and have the potential to cause chronic indolent and acute life-threatening infections (Horn et al. 2009; Neofytos et al. 2009; Pappas et al. 2010). Because of their relatively slow and sometimes fastidious in *vitro* growth characteristics, fungal infections are often challenging to diagnose in a timely fashion. This is particularly problematic when the clinical scenario involves immunocompromised patients—as it often does—in whom rapid microbiologic diagnosis prompting directed antimicrobial therapy is of utmost importance. Thus, nanobiosensing strategies applied to the detection of fungi are an area of great interest. In addition to the importance of fungal detection strategies in clinical medicine, the application of nanobiotechnology to the study of fungal cell biology has great appeal.

Kattke et al. (2011) incorporated QDs into an FRET-based biosensing immunoassay for the detection of aspergillus spores. As depicted in Figure 1.9, their ingenious strategy exploited restoration of fluorescence emission by the displacement of a low-affinity quencher by a high-affinity analyte. In this example, the quenching complex consisted of *Aspergillus fumigatus* spores conjugated to an FRET acceptor (quencher), and the energy donor consisted of QDs conjugated to anti-aspergillus antibody. The assay was designed to identify *Aspergillus amstelodami* spores as the high-affinity analyte.

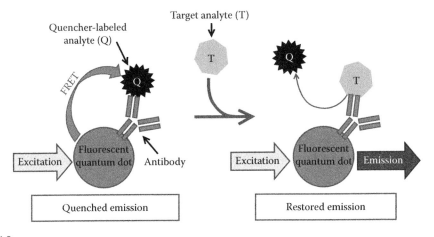

FIGURE 1.9
Mechanism of mold detection. The initial biosensor complex is formed when a quencher-labeled analyte is bound by the antigen-binding site of the quantum dot (QD)–conjugated antibody; when excited, the QD will transfer its energy through fluorescence resonance energy transfer (FRET) to the quencher molecules due to their close proximity. Upon addition of the target analytes, displacement of the quencher-labeled analyte causes disruption of FRET, which translates to an increased QD donor emission signal. (From Kattke, M.D. et al., *Sensors (Basel)*, 11, 6396–6410, 2011. With permission.)

In order to achieve detection of the analyte, the low-binding affinity *A. fumigatus* quenching complex was prebound to the antibody-functionalized QDs. In the presence of the analyte spores of *A. amstelodami*, the quenching complex was displaced from the antibody, FRET ceased, and QD fluorescence intensity increased. The assay was performed in less than 5 minutes and had a limit of detection of 10^3 spores mL^{-1}. This proof-of-concept study suggests that QD-incorporated FRET-based biosensing has the potential for rapid, sensitive, and specific analysis of environmental samples. Furthermore, it establishes an approach for optimizing similar FRET-based assay parameters that can be applied to the detection of a variety of microorganisms. Whether it can be adapted to detect fungi in clinical specimens remains an area of great interest.

Walling et al. (2010) utilized QD conjugates to label both the surface and the interior of three different laboratory strains of the yeast *Saccharomyces cerevisiae*. Using a live cell microarray platform, the QD labels were employed to encode mixtures of yeast that randomly assorted onto a grid of microwells fabricated within a disposable microscope slide. Surface labeling was performed by first incubating the microorganisms with a concanavalan-A biotin conjugate that binds to cell wall glycoproteins, followed by the addition of streptavidin-coated QDs. Intracellular labeling was accomplished by endocytosis of a QD mixture applied to the yeast suspension. To encode the three different yeast strains, differently colored emitting QDs were used. The labeled yeast residing in the microwells were visualized by fluorescent microscopy. Cellular function was not altered by dispersement into microwell arrays or by the presence of QD labels. Both surface and internalized QD labels were readily detected. The biosensing technique proved to be economical and easy to perform compared to flow cytometry. The integration of QD biosensing with live cell arrays is a novel research tool for analysis of dynamic cell processes occurring in mixed cell populations.

1.7.5 Quantum Dot Biosensing and Parasites

In broad terms, all ID agents that derive benefit by causing illness and harming their hosts are, by definition, parasites. In clinical practice, however, the convention is to categorize parasites as higher order organisms, often multicellular, and frequently of tropical origin. This working definition is not entirely accurate, since many parasitic infections are due to unicellular organisms and can occur in temperate climates. For the purposes of this chapter, parasites will be classified as those pathogens that cause protozoan and helminthic diseases.

Parasitic infections, most notably malaria, and as shown in Figure 1.1, those characterized as neglected tropical diseases are responsible for immeasurable disability and over 1.5 million deaths per year (Hotez et al. 2009). Given the diversity of parasites and their prevalence, it is somewhat surprising that relatively few QD-based biosensing strategies have been applied to their detection thus far.

The genus *Cryptosporidium* consists of several species of protozoan parasites that are responsible for widespread waterborne outbreaks of diarrhea throughout the world (D'Antonio et al. 1985; MacKenzie et al. 1995). Cryptosporidiosis tends to cause self-limited diarrhea in immunocompetent adults, but is a frequent cause of prolonged illness in young children and chronic, debilitating diarrhea in immunocompromised hosts. Contamination of food and water supplies by excrement of domestic and wild animals is the main source of infection. Therefore, the detection of *Cryptosporidium* as well as other parasites in the water supply represents an essential public health need. Examples of QD-based detection of *Cryptosporidium parvum* will be described.

Biotinylated monoclonal antibody bound to *C. parvum* oocysts was detected by streptavidin-coated QDs and visualized by fluorescent microscopy (Lee et al. 2004; Zhu, Ang, and Liu 2004). In addition, both *C. parvum* and *Gardia lamblia*, another common water-borne parasite, were concurrently detected using two differently colored QD labels. The sensitivity of detection was found to be equivalent to that of the conventional immuno-fluorescent assay. However, the QDs were significantly more photostable and permitted simultaneous detection of two different parasites.

In contrast to the above studies, Ferrari and Bergquist (2007) found QD labels to be unsuitable for detection of *C. parvum* by flow cytometry. Filtered water samples from a treatment plant were seeded with *C. parvum* oocysts to which various dilutions of either QD–antibody conjugates or conventional fluorophore–antibody conjugates were added. In this study, immunofluorescent intensities of QD labels were significantly lower than that of the fluorophore. In addition, the QD labels appeared to form aggregates and bound nonspecifically to particulate matter in the water. The investigators postulate that the QDs used in this format may have deteriorated as a result of pH and salt concentration in the water samples. This suggests that surface modification of the nanoparticles may be required to enhance their stability in complicated aqueous environments.

Malaria is a mosquito-borne infection caused by one of several species of the proto-zoan parasite *Plasmodium*. Malaria causes approximately 250 million infections and 650,000 deaths annually, with the highest mortality rates occurring in children below the age of 5 years (WHO 2011d). Malaria is mostly confined to regions where the mosquito vectors are endemic. These include sub-Saharan Africa, Asia, and South and Central America. Travelers to these endemic regions are at risk of becoming infected and may not develop signs of illness until after returning from their travels. During the early stages of infection, the symptoms of malaria are nonspecific, and the differential diagnosis always poses a particular challenge to the clinician. Since malaria can rapidly progress to severe, life-threatening illness, timely diagnosis is imperative. Unfortunately, rapid and accurate laboratory confirmation of malaria infection is frequently difficult, and access to diagnostic tests is not always available. This is particularly true in the regions having the highest incidence of infection. Misdiagnosis frequently leads to empiric therapies that contribute to the overuse of anti-malarial drugs and promotes the emergence of malaria resistance. Recent advances in simple, rapid detection tests can improve diagnosis but require training of personnel for proper test interpretation and a health-care infrastruc-ture that is able to respond appropriately to the test results (Bell, Wongsrichanalai, and Barnwell 2006). Nanobiotechnology applied to malaria detection can have a significant impact on the challenges associated with the diagnosis of this major ID.

Red blood cells (RBCs) at varying stages of malaria infection were labeled with polyeth-ylene glycol (PEG)-coated QDs, and the effects of an anti-malarial drug were analyzed by flow cytometry (Ku et al. 2011). The QDs were found to specifically label RBCs containing mature stage parasites, and the effects of the anti-malarial drug were readily detected. The study suggests that QD labels can potentially provide a sensitive and efficient means of anti-malarial drug screening. QD-based malaria detection methods should be pursued in the hopes of establishing sensitive, rapid, and accessible diagnostic schemes.

1.7.6 Quantum Dot Biosensing: Conclusions

QD nanobiosensing strategies are conceptually complex (now, hopefully less so) and offer a diversity of potential applications in ID. QDs can be synthesized in a variety of ways and are also commercially available in kits designed for ease of functionalization. For the ID community, the adage "turn about is fair play" takes on special meaning when one considers that microorganisms have now been engineered to fabricate QDs (Mao et al. 2003; Kumar et al. 2007). QD-based biosensing can be integrated with MEMS devices for multiplexed, high-throughput assays (Vannoy et al. 2011). The ingenious QD-based techniques for detection and study of microbial pathogens have great potential for clinical diagnostics and as research tools. It will be interesting to see what variations and modifications in assay performance will bring QDs closer to fulfill the characteristics of an ideal biosensor.

1.8 Plasmonic Nanobiosensing: Emphasis on Infectious Diseases

Nanobiosensing techniques based on surface plasmon resonance (SPR) and surface-enhanced Raman scattering (SERS) spectroscopy offer a variety of novel, sensitive, and potentially powerful applications for the detection and characterization of microbial pathogens and biomarkers of infection. SPR and SERS are interrelated phenomena that fundamentally depend on the interaction of electron clouds (plasma) present on the surface of metals, particularly gold and silver, with incident light waves (photons). The discoveries of plasmons and Raman scattering date to the early twentieth century (Mie 1908; Landsberg and Mandelstam 1928; Raman and Krishnan 1928), although many centuries earlier, the remarkable interaction of light with nanosized metal colloids was appreciated by artisans who incorporated their vivid and stable colors into pottery and stained glass (Colomban 2009). The recent integration of nanobiotechnology with the physical science of plasmonics has produced a wealth of biosensing capabilities. This section will provide a very simplified description of the physical principles underlying plasmonics and will highlight examples of SPR and SERS biosensing methods that are particularly relevant to ID.

The physical principles underlying plasmonics are complex, not fully elucidated, and definitely challenging to describe. Fortunately, a detailed understanding of the scientific theory is not required in order to gain a working knowledge of, and an appreciation for, the techniques applied to nanobiosensing. It is hoped that the descriptions provided herein will enlighten and stimulate further interest in exploring this fascinating science, and several excellent and detailed reviews are provided (Rich and Myszka 2003; Haes and Van Duyne 2004; Hutter and Fendler 2004; Eustis and EL-Sayed 2006; Hoa, Kirk, and Tabrizian 2007; Homola 2008; Peng and Miller 2010; Garcia 2011; Bedford et al. 2012; Liang et al. 2012).

1.8.1 Plasmonics: Basic Principles

Noble metals have a strong interaction with visible light (Faraday 1857). At the nanoscale, this interaction behaves very differently from that of the bulk metal. For instance, due to SPR phenomena, GNPs and silver nanoparticles (SNPs) appear red and yellow, respectively.

Oscillating light waves impinging on a metallic nanoparticle excite conduction-level electrons that are induced to cluster on the particle's surface and create a dipole electric field. The nanoparticle electrons, governed by quantum confinement effects, oscillate in resonance with the frequency of the light wave and produce the phenomenon of SPR. Due to vibrational energy states of the quantum confined resonating electron clouds that are not fully elucidated and mathematically predicted, SPR results in very large enhancements of light absorption and light scattering (large extinction coefficient cross sections), and in the development of strong, localized electromagnetic fields (Mie 1908; Haes and Van Duyne 2004). These optoelectronic events form the basis of plasmonic biosensing. The magnitude of the events is greatly influenced by the shape (anisotropy), aggregation, or dispersion of the nanostructure (nanospheres, nanorods, nanostars, nanofilms, etc.) in association with its surrounding dielectric environment (materials that are poorly conducting but efficient at supporting an electrostatic field). In the case of GNPs, SPR causes intense absorption of light wavelengths residing in the green to blue spectrum, resulting in the reflection of red light. It should be noted that the properties of gold at the nanoscale confer superior plasmonic biosensing characteristics that even stained glass "nano-artisans" of the fourth century could not have imagined (Eustis and EL-Sayed 2006; Baptista et al. 2008; Bedford et al. 2012).

The surfaces of noble metal nanostructures are readily functionalized with affinity ligand probes. When analytes bind to the functionalized nanostructure, they change the electron density on the metallic surface, resulting in alterations of SPR. Alternatively, probe-target recognition can lead to aggregation or dispersion of nanoparticles, resulting in alterations of SPR. In the case of aggregation, the SPR absorption band shifts toward the red spectrum, resulting in reflection of blue light. The SPR events are readily transduced into a variety of signals that can be sensed by the following general schemes: (1) detection of light scattering, (2) detection of color change, (3) detection of surface plasmon wave (SPW) phenomena, and (4) detection of enhanced Raman scattering signals. These signal transduction motifs form the basis of the categories used herein to describe relevant SPR-based ID-nanobiosensing strategies. This schematic approach is an arbitrary one, and other categorical schemes (genomic, proteomic, immunochemical, organismal) would be equally valid. Although a comprehensive discussion of each of these schemes goes beyond the scope of this chapter, several examples illustrating their application to ID-biosensing will be provided. Two excellent and detailed reviews of plasmonic biosensing with particular relevance to IDs are recommended (Rich and Myszka 2003; Homola 2008). The review by Rich and Myszka cites over 100 articles devoted to SPR-based biosensing of HIV. The authors aptly conclude, "To date, no single biological system has been studied more with this technology. The extensive application of SPR in HIV research eloquently illustrates how biosensors can be used in qualitative modes to detect binding interactions, as well as in quantitative modes to obtain kinetic and thermodynamic parameters for high and low affinity or transient interactions" (Rich and Myszka 2003).

1.8.2 Surface Plasmon Resonance Biosensing Scheme 1: Detection of Light Scattering

By way of background, GNPs are typically capped with negatively charged citrate ions in order to prevent their aggregation in aqueous solution (Sato, Hosokawa, and Maeda 2003; Li and Rothberg 2004). Raising the salt concentration of the solution screens the negative surface charge and induces GNPs to aggregate. ssDNA has been shown to electrostatically adsorb onto the surface of GNPs and prevent particle aggregation. In contrast, double-stranded DNA (dsDNA), with its relatively inflexible double-helix backbone, cannot electrostatically adsorb or prevent salt-induced aggregation. These electrostatic properties provide a means to develop highly sensitive and specific nanobiosensing techniques that exploit the selectivity of DNA hybridization with the SPR signal transduction properties of metal nanostructures (Li and Rothberg 2004; Sato, Hosokawa, and Maeda 2005, 2003; Ray 2006; Darbha et al. 2008; Griffin et al. 2009; Singh et al. 2009).

Ray and colleagues, working extensively with light-scattering properties of GNPs, have developed novel nanobiosensing strategies that they applied to the study of a variety of

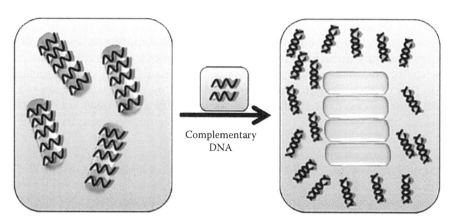

FIGURE 1.10
Schematic representation of the gold-nanorod-based DNA hybridization process. The single-stranded DNA that electrostatically attach to the dispersed gold-nanorods are released in the presence of complementary DNA to which they hybridize. The resulting bare gold-nanorods aggregate and are detected by a change in their light-scattering properties. (From Darbha, G.K. et al., *Chemistry*, 14, 3896–3903, 2008. With permission.)

microorganisms, including HIV, HCV, and *E. coli* O157:H7 (Ray 2006; Darbha et al. 2008; Griffin et al. 2009; Singh et al. 2009). In general, they measure the enhanced light-scattering effects when dispersed GNPs are made to aggregate in solution, as shown in Figure 1.10.

When target viral ssDNA sequences were added to a solution containing GNPs adsorbed with complementary ssDNA probes, hybridization occurred. The resulting dsDNA desorbed from the GNPs and promoted salt-induced aggregation that in turn enhanced SPR-based light-scattering signals. Spectroscopic measurements demonstrated the presence of HIV sequences at a detection limit of 100 pM (10^{-12} M) (Darbha et al. 2008). Moreover, the sequence specificity was capable of identifying single-nucleotide polymorphisms. A similar approach was used to detect HCV RNA with a detection limit of 60 pM (Griffin et al. 2009).

In a variation of this biosensing scheme, the enteric pathogen *E. coli* O157:H7 was detected using GNPs conjugated with anti-*E. coli* antibodies (Singh et al. 2009). Aggregation of the antibody-functionalized GNPs occurred when as few as 50 CFU mL^{-1} of the bacteria were added to the reaction solution, and the intensity of transduced light-scattering signals increased 40 fold. The assay took less than 15 minutes to perform and was highly selective for the O157:H7 strain. The investigators point out that these SPR biosensing strategies do not require the use of fluorescent labels, are highly sensitive and specific, rapid to perform, and potentially able to detect a constellation of microorganisms. Applicability of the method to detection of ID agents in clinical samples and to point-of-care diagnostics awaits further testing.

1.8.3 Surface Plasmon Resonance Biosensing Scheme 2: Colorimetric Detection

Methods that utilize colorimetric detection schemes are appealing, since they do not require the use of expensive and complex optical sensing equipment. Biosensing assays based on SPR as the means of signal transduction are well suited for the design of colorimetric detection platforms. Several assay formats have been applied to ID nanobiosensing, and illustrative examples will be described.

SPR nanobiosensing assays that exploit the sensitivity and specificity of nucleic acid-functionalized GNPs were pioneered by Mirkin and colleagues and merit special emphasis (Mirkin et al. 1996; Taton, Mirkin, and Letsinger 2000; Thaxton, Georganopoulou,

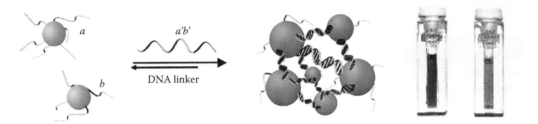

FIGURE 1.11
A mixture of gold nanoparticles with surface-immobilized noncomplementary DNA sequences (a, b) appears red in color and has a strong absorbance at 520 nm. When a complementary DNA sequence (a', b') is added to the solution, the particles are reversibly aggregated, causing a red shift in the surface plasmon absorbance to 574 nm, thus appearing purple in color. (From Thaxton, C.S. et al., *Clin. Chim. Acta.*, 363, 120–126, 2006. With permission.)

and Mirkin 2006). In contrast to the electrostatic adsorption–desorption interactions described earlier, Mirkin developed nanobiosensing strategies that utilize oligonucleotides covalently bound to the surface of GNPs. As shown in Figure 1.11, the assays rely on the ability to cross-link two types of GNPs that differ in their bound ssDNA probes. In this configuration, the GNP probes cross-link by hybridizing at each end of their DNA targets. The cross-linked aggregates produce SPR signals that result in a change in color of the reaction mixture from red to blue. Mirkin and colleagues also developed a scanometric detection format comprised of both oligonucleotide-functionalized GNP probes and capture ssDNA immobilized on a chip substrate (Taton, Mirkin, and Letsinger 2000). Sandwich hybridization of target DNA links the GNP probes to the capture chip. The detection signal is amplified in a step that reduces silver ions onto the GNP surface. The method was 100-fold more sensitive than fluorometric assay and capable of detecting a synthetic anthrax gene sequence at a concentration as low as 50 FM (10^{-15} M). The chip array signals could be visualized by the naked eye or read by a conventional flatbed scanner. A variety of nanobiosensing techniques have evolved from this original approach.

Storhoff et al. (2004) used a variation of the GNP cross-linking method to detect DNA sequences from *mecA*, the gene that confers antibiotic resistance to methicillin-resistant *Staphylococcus aureus* (MRSA). Using both colorimetric and light scattering transduction signals, they were able to detect target sequences from samples containing as little as 10 FM of MRSA total genomic DNA. The technique was rapid, quantitative, and did not require additional signal amplification steps. Because it utilizes a visual detection scheme, it has the potential to become a rapid, sensitive, specific, and relatively inexpensive point-of-care diagnostic assay.

Using a non-cross-linking strategy, Baptista et al. (2006) developed a colorimetric assay to detect *M. tuberculosis* DNA. Their DNA-functionalized GNP technique detected PCR-derived mTb amplicons obtained from a variety of clinical specimens. The SPR-based colorimetric assay demonstrated excellent concordance when compared with conventional acid fast stains and with a commercially available gene probe. In addition, the assay was capable of detecting single-nucleotide polymorphisms within the RNA polymerase locus, a region of the genome that confers resistance to the anti-mycobacterial drug rifampin. The method was sensitive, rapid, and significantly less expensive to perform when compared to other molecular techniques. These preliminary results are promising, although validation of the method will require additional testing of a large number of clinical samples.

GNP-based colorimetric assays that do not rely on nucleic acid probes have been used to detect microorganisms and biomarkers of ID. A colorimetric assay to detect cholera toxin using lactose-functionalized GNPs had a detection limit between 50 and 100 nM (Schofield, Robert, and Russell 2007). The GNP probes could be freeze-dried, stored, and subsequently reconstituted without loss of functionality, suggesting that the assay has promising point-of-care capability. Exploiting unique cell wall constituents of *S. aureus*, Wang et al. (2012) developed a sensitive and rapid colorimetric detection assay using phenylboronic acid-functionalized GNP probes. They could detect *S. aureus* from a variety of spiked samples, including water, milk, and sputum. Furthermore, the sensitivity of the assay was not altered by exposure of the bacteria to vancomycin, a cell wall active antibiotic. Hepatitis B surface antigen (HBsAG) was detected by a biorecognition probe composed of GNPs functionalized with monoclonal HBV surface antibody (Wang et al. 2010). The investigators obtained quantitative measurements of analyte in buffer solution and detected HBsAG concentrations at 0.01 IU mL^{-1}. Of particular importance, this is the first study to demonstrate that this type of biosensing platform can be used to detect HBsAg in the biologically complex milieu of serum and plasma.

LFIC using GNPs has long been applied to the detection of a variety of microorganisms and biomarkers of ID in the form of handheld rapid antigen detection tests (Posthuma-Trumpie, Korf, and van Amerongen 2009). This method relies on the ability to visually detect the red color on a region of a chromatographic strip as GNP probes accumulate in response to a biorecognition event. It is a simple design that is both rapid and inexpensive. The format is most recognizable as the commonly used home pregnancy test. LFIC requires small sample and reagent volumes, and its storage and portability features make it particularly attractive for point-of-care use. LFIC tends to be quite specific but less sensitive than other immunoassays (Nagatani et al. 2006; Carter and Cary 2007; Parolo, de la Escosura-Muñiz, and Merkoçi 2013). This limitation, however, may be offset by its utility as a diagnostic screening tool. LFIC platforms for detection of malaria and STDs are becoming more available in resource-limited regions. However, concerns related to manufacturing quality controls and to the interpretation of test results warrant careful consideration (Laderman et al. 2008; Posthuma-Trumpie, Korf, and van Amerongen 2009; Peeling, Smith, and Bossuyt 2010; WHO 2011d).

1.8.4 Surface Plasmon Resonance Biosensing Scheme 3: Surface Plasmon Wave-Based Detection

In contrast to the locally confined resonant dipole oscillations that occur with noble metal nanoparticles, incident light at a critical angle on a metal nano-thin film induces an SPW that propagates along the film–dielectric interface (Haes and Van Duyne 2004; Hoa, Kirk, and Tabrizian 2007; Homola 2008; Bedford et al. 2012; Ho and Wu 2012). As shown in Figure 1.12, the SPW coupling results in a loss of reflected light energy that can be detected as a dip in reflectance. The surface of a nano-thin metal film can be functionalized with affinity ligands. Biorecognition events, including analyte binding and dissociation, alter the refractive index at the metal–dielectric interface, resulting in a shift in the wavelength of reflected light. Wavelength shifts can be plotted as a function of time, yielding precise measurements of association and dissociation rate constants. Furthermore, indirect competitive binding inhibition experiments can be performed for rapid screening of potentially active drugs. This SPR analytical configuration generates a powerfully sensitive nanobiosensing platform. Several commercially available devices have integrated all of the components (reusable functionalized nano-thin film sensing chips, microfluidic analyte flow channels, spectroscopic detectors, and recording devices) into automated SPR-based biosensors (Gauglitz 2010). Several illustrative examples will briefly be described.

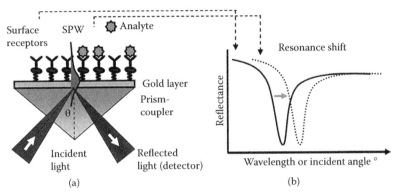

FIGURE 1.12
Tracking surface absorption by surface plasmon resonance, (a) a prism-coupled configuration and (b) resonance shift in the reflected light spectrum. SPR biosensing is accomplished when target analyte binding to the surface bioreceptor results in a change in the conditions of resonance coupling of incident light to the propagating surface plasmon wave (SPW). (From Hoa, X.D. et al., *Biosens. Bioelectron.*, 23, 151–160, 2007. With permission.)

Kumbhat et al. (2010) have developed an SPW biosensing method for detection of Dengue virus infection. Dengue, a reemerging viral infection, is the most prevalent mosquito-borne ID, affecting approximately 50 million people annually (Kroeger, Nathan, and Hombach 2004). Infection occurs predominantly in tropical regions of the world, although the mosquito vectors are capable of spreading to previously nonendemic areas. Dengue infection has an incubation period of 3–14 days, and travelers returning from endemic regions may develop symptoms of illness after they have returned home. The clinical manifestations are nonspecific and can mimic a wide variety of other IDs that comprise a broad differential diagnosis, including malaria, HIV, CMV, EBV, hepatitis, measles, hemorrhagic fever, meningococcemia, typhoid fever, and rickettsiosis.

Currently, there are no highly sensitive, specific, and widely available real-time diagnostic tests for Dengue infection. Viral cultures are not routinely available. Molecular techniques are sensitive and specific but can take several hours to perform and are limited to research laboratories. Immunoassays for the detection of either antigen or specific antibody may lack sensitivity or may not provide results in a meaningful time frame. Hence, a person from an endemic region of the world who develops signs and symptoms of an illness compatible with Dengue is a challenge to the physician—a paradigm not unlike many other patients encountered by ID clinicians. The SPW-based biosensor, developed by Kumbhat, was able to detect Dengue antigen as well as anti-virus antibody in serum with high sensitivity. This proof-of-concept technique illustrates the potential value of this real-time, label-free, and inexpensive ID-nanobiosensing method.

Alterman et al. (2001) developed an efficient SPW-based biosensor for screening of HIV protease inhibitor activity. They compared two different approaches using either protease or PIs immobilized on the nano-thin metal film of a commercial SPW biosensor. On the basis of binding competition assays, they found both approaches to be highly sensitive and to offer an automated and rapid screening alternative to conventional assays.

Naimushin et al. (2002) detected *S. aureus* enterotoxin to demonstrate the utility of their prototype SPW-based biosensor. They were able to detect enterotoxin with high sensitivity in complex substrates, including urine, milk, and seawater. Limits of detection with and without the use of antibody amplification labels were in the femtomolar and nanomolar ranges, respectively. The SPW-based device was powered by a portable battery and used inexpensive, disposable, prefunctionalized sensor modules. This prototype suggests that

SPW-based nanobiosensing can be performed at the point-of-care and can have diverse applications in clinical, environmental, and bio-warfare settings.

1.8.5 Surface Plasmon Resonance Biosensing Scheme 4: Surface-Enhanced Raman Scattering-Based Detection

Raman light scattering refers to an electromagnetic phenomenon that occurs when the energy of incident light is transferred to the vibrating electron cloud that surrounds individual molecules. The light scattering, when analyzed spectroscopically, provides a fingerprint of the molecule's chemical composition that is commonly referred to as Raman spectra (Kneipp and Kneipp 2006; Smith 2008; Hossain et al. 2009; Vo-Dinh, Wang, and Scaffidi 2010; Bantz et al. 2011). RLS spectroscopy has been a fundamental tool of analytical chemistry since its discovery in the early part of the twentieth century (Raman and Krishnan 1928). Unfortunately, the signal intensities generated are very small, and for practical applications, it requires the use of powerful lasers and long data acquisition times. The discovery that enormous signal amplification can be achieved by the adsorption of analyte molecules onto noble metal nanostructures resulted in a major technological advance in RLS spectroscopy (Fleischmann, Hendra, and McQuillan 1974; Nie and Emory 1997). The amplified signals are due, in part, to the local SPR electromagnetic energy conveyed to the analyte by the adjacent nanostructures. This discovery has led to the development of SERS spectroscopy, a powerful analytical tool that can be performed with relatively low-power lasers (battery operated and portable) and short (often less than 10 seconds) data acquisition times.

SERS spectroscopy can acquire Raman fingerprints of the chemical composition of single molecules as well as single microorganisms. Since low-power lasers are used to excite the substrate, SERS spectroscopy neither destroys nor damages the target being probed. Moreover, since water has a negligible Raman scattering fingerprint, spectroscopic analysis is readily accomplished in aqueous environments—a particularly important feature for the detection and analysis of biomolecules. It is well suited for the detection of unique Raman fingerprints produced by nucleic acids, proteins, carbohydrates, and lipids. These characteristics have made SERS spectroscopy a popular biosensing modality having diverse applications in ID (Cao, Jin, and Mirkin 2002; Maquelin et al. 2002, 2003; Wood and McNaughton 2006; Liu et al. 2007; Buijtels et al. 2008; Han, Zhao, and Ozaki 2009; Sayin et al. 2009; Kumbhat et al. 2010; Huang et al. 2010b; Vo-Dinh, Wang, and Scaffidi 2010; Bantz et al. 2011; McNay et al. 2011). Several examples of SERS-based ID nanobiosensing assays will be introduced.

SERS spectroscopy has been applied to the detection and analysis of viruses including HIV, influenza, RSV, West Nile, and a variety of food- and waterborne viruses (Isola, Stokes, and Vo-Dinh 1998; Wabuyele and Vo-Dinh 2005; Shanmukh et al. 2006, 2008; Driskell et al. 2010; Fan et al. 2010; Zhang et al. 2011). The studies demonstrate how SERS technology can work in tandem with the specificity of PCR amplification, distinguish enveloped and nonenveloped viral strains and genotypes, and rapidly generate results with limits of detection considerably lower than conventional assays.

SERS spectroscopy has succeeded in detecting bacterial contaminants in food and water samples (Liu et al. 2007; Dutta, Sharma, and Pandey 2009; Fan et al. 2011; Tay et al. 2012). The technique is capable of detection down to the level of a single bacterium and can distinguish individual species in an aqueous solution containing mixed populations of bacteria. SERS-based techniques detected surrogate anthrax spore biomarkers more rapidly and with equivalent sensitivity when compared to PCR and immunoassays

(Zhang et al. 2005; Cheng et al. 2011). Microcolonies of bacteria cultured from bloodstream isolates were detected and identified within several hours by SERS spectroscopy, compared to 1–2 days by conventional culture methods (Maquelin et al. 2003; Liu et al. 2011). Bacteria in spiked blood samples were captured onto vancomycin-coated SNP arrays and detected by SERS spectroscopy, as depicted in Figure 1.13. This method of detection was also capable of distinguishing antibiotic resistant from antibiotic sensitive bacterial species on the basis of their different Raman spectra. Should this methodology stand up to rigorous clinical testing, it would have an immense impact on the timely diagnosis of life-threatening IDs.

The chemical composition of bacterial biofilms has been analyzed using SERS spectroscopy (Ivleva et al. 2008; Chao and Zhang 2012). Raman spectra of protozoan proteins, polysaccharides, and nucleic acids were attained within the extracellular polymeric matrix. In addition, identification and spatial representation of matrix constituents were possible. The ability to analyze the composition of biofilms will enhance our understanding of the dynamics of these complex structures and may eventually provide insight into ways of preventing biofilm formation on the surfaces of implantable medical devices.

Analysis of SERS spectra from the cell wall of living yeast has been accomplished (Sujith et al. 2009). A dark-field microscopic image of a yeast cell coated with SNPs serving as the plasmon substrate and the corresponding SERS spectra of the cell wall constituents are shown in Figure 1.14. SERS spectroscopy was used to characterize several yeast species (Sayin et al. 2009). The investigators demonstrated the batch-to-batch reproducibility of the SERS spectra as well as differences in cell wall composition between different yeast species. Since the cell wall of fungi is the main target of anti-fungal chemotherapy, it is possible that detailed SERS analysis may lead to the discovery of active anti-fungal compounds.

SERS spectroscopy has been applied to the study of malaria interactions with erythrocytes as well as analysis of the mechanism of action of anti-malarial drugs. Enhanced SERS signals permitted sensitive detection of β-hematin, a surrogate biomarker of early stage parasitic infection (Yuen and Liu 2012). The ability to monitor drug interactions in living cells and to rapidly screen for potentially active anti-malarial drugs make SERS an attractive biosensing modality (Wood and McNaughton 2006).

1.8.6 Plasmonic Nanobiosensing: Conclusions

Plasmonic biosensing offers a variety of label-free, rapid, and sensitive formats for the detection and analysis of microorganisms and biomarkers of ID. The methods utilize an array of devices, ranging from simple, handheld, and economical to complex, bulky, and expensive. Remarkably, these very different formats all depend on the interplay of

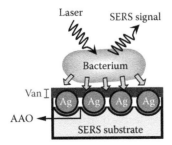

FIGURE 1.13
Bacteria captured on vancomycin-coated silver/anodic aluminum oxide (Van-coated Ag/AAO) SERS substrate. (From Liu, T.Y. et al., *Nat. Commun.*, 2, 538, 2011. With permission.)

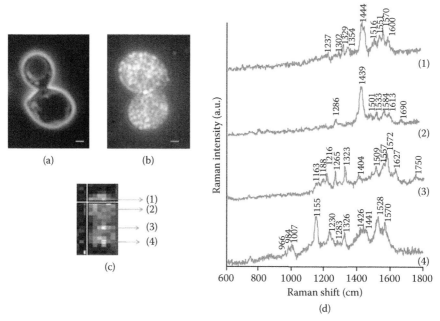

FIGURE 1.14
Dark-field image of (a) a yeast cell and (b) silver adsorbed on yeast cell surface (×100; scale, 1 µm). Raman spectroscopic imaging results for a yeast cell wall with adsorbed silver nanoparticles. The spectra (1) to (4) in (d) have been measured from the positions indicated in (c). (From Sujith, A. et al., *Anal. Bioanal. Chem.*, 394, 1803–1809, 2009. With permission.)

light energy with nanosized noble metals. The technology is capable of detecting single microorganisms and the constituent chemicals that they are made of. Depending on the method utilized, plasmonic biosensing can provide quantitative measurements that reveal ligand–receptor binding affinity. Automated, commercially available plasmonic biosensing instruments exist and have yielded important information pertaining to many agents of ID. The potential ID-nanobiosensing applications are vast, and we await results of rigorous testing to determine which SPR-based methods will be put to clinical use.

1.9 Bio-Barcodes and Infectious Diseases–Nanobiosensing

Bio-barcode is a fitting albeit confusing term that requires explanation for those of us who conjure up black and white lines scanned at the grocery counter. Bio-barcodes are encoded tags that signify the occurrence of a particular biorecognition event. The term "bio-barcode" has been applied to a diverse group of biosensing formats that include (1) a polymeric microbead that encapsulates a precise mixture of QDs that differ in color and intensity (Han et al. 2001; Klostranec et al. 2007; Lee et al. 2007; Giri et al. 2011), (2) nanorods and nanowires containing encoded patterns of different metals (Nicewarner-Pena et al. 2001; Tok et al. 2006), and (3) many short strands of identical DNA that are attached to GNPs to which a biorecognition probe (antibody, DNA complementary to the analyte) is also attached (Nam, Thaxton, and Mirkin 2003; Nam, Stoeva, and Mirkin 2004). By virtue of their encoded signatures, bio-barcodes are well suited for multiplexed biosensing.

Nie et al. first developed the concept of encoding polymeric microbeads by encapsulating precise proportions of multicolored QDs (Han et al. 2001). The ability to functionalize the microbeads with affinity ligands and to tune the colors and fluorescent intensities of the incorporated QDs made it possible to generate unique optical codes that can be used for multiplexed biosensing. Although it is theoretically possible to create millions of uniquely fluorescing codes, the practical limit of discriminatory detection in a multiplexed assay is significantly less than that (Lee et al. 2007).

The potential of the approach is illustrated by a microbead-QD-bio-barcode method that was integrated with an MEMS device to perform multiplexed detection of HIV, HBV, and HCV biomarkers (Klostranec et al. 2007). Using small volumes of spiked serum samples, the analytes were rapidly detected. This proof-of-concept technique was 50 times more sensitive than conventional assay methods. The investigators further expanded the multiplexing capabilities of the QD bio-barcodes by rapidly and simultaneously detecting gene fragments of HIV, malaria, HBV, HCV, and syphilis in spiked buffer solution (Giri et al. 2011).

Nanorods and nanowires, composed of a variety of different metal segments, have been fabricated and employed as bio-barcodes (Nicewarner-Pena et al. 2001; Tok et al. 2006). The metals can be electrodeposited in precise arrangements that produce encoded striped patterns when viewed by an optical microscope. The nanostructures are easily functionalized with specific capture ligands. Analytes bound to the metallic barcodes are then labeled with fluorescent antibodies, resulting in a sandwich configuration, as illustrated in Figure 1.15. The fluorescent signal is keyed to the optical image of the barcode that designates the presence of a particular analyte. The ability to engineer metallic bio-barcodes with nearly infinite permutations of encoded striping patterns lends itself to multiplexed biosensing. In a proof-of-concept study, Tok et al. (2006) were able to rapidly and simultaneously detect three different surrogate toxins with limits of detection comparable to conventional assays.

Mirkin and colleagues developed multistep bio-barcode assays that use short strands of DNA as the barcode element (Nam, Thaxton, and Mirkin 2003; Nam, Stoeva, and Mirkin 2004; Hill and Mirkin 2006; Stoeva et al. 2006). GNPs functionalized with affinity ligands (either DNA or antibody) are also coated with many identical strands of barcode DNA. After binding to analyte, the barcode–GNP–analyte complexes are separated from the reaction mixture, and the barcode DNA are released by heat dehybridization. The barcode DNA are separated from the reaction mixture and allowed to hybridize with capture ssDNA immobilized on a microchip array. Detection of the hybridized strands is accomplished visually or spectroscopically, using either functionalized GNPs or fluorescent dye labels, respectively. The large number of barcode DNA relative to target analyte provides a built-in amplification feature. In a proof-of-concept study, anthrax gene sequences were detected in the attamolar (aM = 10^{-18} M) range (Nam, Stoeva, and Mirkin 2004). Furthermore, the assay was capable of single nucleotide polymorphism selectivity. The technique was used in a multiplex platform to simultaneously detect gene sequences of HBV, smallpox virus, Ebola virus, and HIV, all at femtomolar (fM = 10^{-15} M) concentrations (Stoeva et al. 2006).

Zhang, Huarng, and Alocilja (2010)simultaneously detected salmonella and anthrax gene sequences using a novel modification of the DNA barcode method. The DNA barcodes were linked to either lead or cadmium compounds that are termed nanoparticle tracers (NTs). The NTs are electrochemically detected and obviate the need for the microchip hybridization step of the original bio-barcode assay.

Tang and Hewlett (2010) used highly fluorescent europium (Eu$^+$) nanoparticles in place of GNPs in the DNA barcode assay and were able to detect HIV-1 p24 antigen at a detection

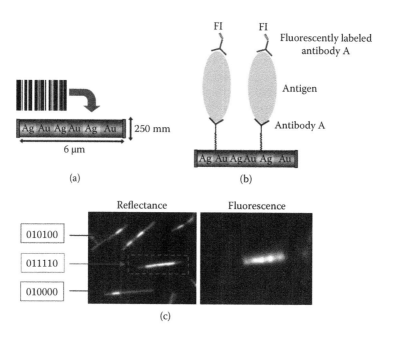

FIGURE 1.15

(a) Analogy between a conventional barcode and a metallic stripe-encoded nanowire (diameter ~ 250 nm; length ~ 6 μm). (b) Schematic of the sandwich immunoassay performed on a nanowire. (c) Postassay reflectance and fluorescence readout of the nanowires. The identity of the antigen present can be easily identified from the stripe pattern of the nanowires; for example, the fluorescently lit nanowire to which a specific anti-analyte antibody was attached has a stripe pattern of 011110 (0 = Au, 1 = Ag). This indicates that this particular analyte is present in the test sample mixture. (From Tok, J.B.H. et al., *Angew Chem. Int. Ed.*, 45, 6900–6904, 2006. With permission.)

limit of 0.5 pg mL^{-1}. Their method reduced the number of steps in the barcode assay and was significantly more sensitive than conventional ELISA.

1.9.1 Bio-Barcode Infectious Diseases–Nanobiosensing: Conclusions

Bio-barcode-based nanobiosensing offers exquisite sensitivity and specificity, along with multiplexing capabilities. The assays are currently in the proof-of-concept stage of development, and it is unclear at this time whether commercial fabrication of the various barcode schemes is feasible. Several of the assays require multiple steps, and performance may ultimately become more practical if integrated automation can be achieved. The methods are applicable to the detection of virtually all clinically relevant agents of ID, and their multiplexing capabilities offer great potential.

1.10 Magnetic Nanoparticles and Infectious Diseases Biosensing: Introduction

The use of MNPs as affinity-labeled probes represents a very promising nanobiosensing platform. Several formats having particular relevance to ID are being explored and appear to offer unique and attractive features with respect to pathogen and biomarker detection

(Grossman et al. 2004; Lee, Yoon, and Weissleder 2009; Haun et al. 2010; Kaittanis et al. 2011). Unlike optical-based assays that must be performed in a clear sample matrix in order to detect biorecognition signals, magnetic-based biosensing may be carried out in complex, turbid samples where the acquired signal is based on magnetic field perturbations. This is one of many attributes of magnetic-based nanobiosensing, and others will be discussed in the following sections.

As is true of most of the nanobiosensing platforms already presented, those utilizing MNPs are fundamentally based on affinity biorecognition events. Functionalized MNPs have been used to capture and separate target molecules from suspension in straightforward magnetic separation schemes. Other assay formats measure either direct change in the magnetization of MNP probes or their induced changes in the atomic spins of neighboring water molecules. Several examples of these novel ID-nanobiosensing assays are given in the following sections.

1.10.1 Magnetic Capture and Separation of Microorganisms

Chen et al. used functionalized iron oxide nanoparticles as affinity probes to detect bacterial pathogens, including *S. aureus* and the uropathogens *E. coli* and *S. saprophyticus* (Ho et al. 2004; Liu et al. 2008). In aqueous suspensions, the nanoparticle-bound microorganisms were captured by the application of an external magnetic field and separated from the rest of the reaction mixture by multiple washing steps. The bacteria were subsequently identified by mass spectrometry. While the technique proved to be rapid and sensitive in aqueous media, it could not be duplicated in urine samples, probably as a result of interfering impurities.

Anti-*E. coli* 0157:H7 antibody-functionalized iron oxide nanoparticles were used to detect and separate the microorganism from ground beef specimens (Varshney et al. 2005). The immunomagnetic separation method was proved to be rapid, sensitive, and more efficient than a conventional assay technique. In proof-of-concept studies, vancomycin functionalized iron oxide nanoparticles (Van-NPs) were used to capture bacteria in aqueous solution as well as in contaminated blood transfusion products (Gu et al. 2003; Gao et al. 2006; Kell et al. 2008).

1.10.2 Magnetic Remanence and Infectious Diseases Nanobiosensing

The size and core metallic components of MNPs can be fabricated in ways that impart superparamagnetic behavior. When subjected to an applied magnetic field, they develop strong magnetic moments (vectors of magnetization). Their small size (approximately 20 nm or less for iron oxide) accommodates a single magnetic domain that aligns with the applied magnetic field lines. When the magnetic field is removed, the magnetic moments return to a randomized state, and in this configuration, an ensemble of MNPs has no net magnetization. The short time it takes for the MNPs to lose their magnetization after the applied field has been removed is designated as magnetic remanence. Superparamagnetic behavior of MNPs has been used to great advantage in biorecognition formats (Tamanaha et al. 2008; Haun et al. 2010).

Using a superconducting quantum interference device (SQUID), Grossman et al. (2004) measured magnetic remanence of antibody-functionalized MNPs bound to *Listeria monocytogenes*. This rather complex device utilizes coils kept at very low temperatures (superconducting magnets) to create a strong, locally applied magnetic field to the sample being analyzed. In their assay, the magnetic field was applied to a suspension containing microorganisms mixed with MNP probes. When the field was removed, the magnetic remanence value of MNPs bound to bacteria was significantly longer than that of unbound MNPs. The remanence value was titrated to organism concentration, resulting in a detection limit of approximately 6×10^6 bacteria in a 20 µL sample. In a separate study using a SQUID sensor, H5N1 influenza virus was detected at a limit of 5 pg mL $^{-1}$ (Yang, Chieh et al. 2008). This technique is quite sensitive and capable of determining association rates of ligand–receptor interactions. Since remanence signals are acquired only from bound MNPs, washing steps to remove unbound particles are not needed. However, given the complexity of the SQUID, the technique, as currently configured, is limited to use in research laboratory settings.

1.10.3 Magnetic Relaxation Nanoswitches and Infectious Diseases Nanobiosensing

Weissleder et al. pioneered nanobiosensing methods that measure the effects of MNPs on neighboring water molecules (Perez et al. 2002; Lee et al. 2008; Haun et al. 2010). Specifically, the technology is based on the ability to detect changes in T2 nuclear magnetic resonance (NMR) signals that occur when MNP probes interact with analytes in solution (Demas and Lowery 2011). While the underlying physics is quite complex, the technology offers a platform worthy of great attention, given the breadth and extent of its biosensing potential. A simplified description of the technology and its fundamental principles will be presented.

Superparamagnetic MNP probes can transition (switch) from a dispersed to an aggregated state upon binding to multivalent analytes. In the presence of an external homogeneous magnetic field, each MNP aggregate produces a localized magnetic field inhomogeneity. When surrounding water molecules diffuse into the regions of inhomogeneity, the hydrogen nuclei lose their synchronous spins, resulting in a decrease of T2 (spin–spin relaxation) signal. This process is shown in Figure 1.16. The change in T2 signal can be quantified and correlated to the concentration of analyte present in the sample. It should be noted that depending on the nature of the analyte being studied—for instance, an enzyme—the assay can be run in the reverse direction, from aggregated to dispersed states. In this case, T2 signal will increase. This biosensing platform has been referred to by several names, although the two most often used are magnetic relaxation nanoswitch (MRnS) and diagnostic magnetic resonance (Perez et al. 2002; Lee et al. 2008).

MRnS assays have been used to detect a variety of microorganisms, including viruses, bacteria, and mycobacteria (Perez et al. 2002, 2003; Lee et al. 2008; Lee, Yoon, and Weissleder 2009; Chung et al. 2011). In aqueous solution containing 25% serum, adenovirus and HSV were detected at limits of 100 viral particles in 100 µL and five viral particles in 10 µL, respectively (Perez et al. 2003). There was a moderate decrease in sensitivity when HSV was detected in spiked serum samples. A variety of Gram-positive bacteria, including staphylococci, streptococci, and enterococci, were detected by highly sensitive and rapid MRnS assays using Van-NPs (Lee et al. 2008; Lee, Yoon, and Weissleder 2009; Chung et al. 2011). Chung et al. (2011) used the MRnS assay to obtain measurement of vancomycin-binding kinetics and were able to detect intracellular *S. aureus* in macrophages. In addition, they functionalized MNPs with the antibiotic daptomycin and showed that it could be used to detect bacteria, albeit with decreased sensitivity relative to vancomycin. The BCG strain of TB was detected in less than 30 minutes at a detection limit of 20 CFU in 1 mL of spiked sputum (Lee, Yoon, and Weissleder 2009). The sensitivity of the assay exceeded that of acid fast stains and was comparable to culture-based detection that required an incubation time of 2 weeks.

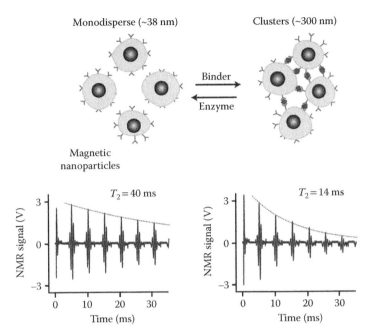

FIGURE 1.16
Principle of the magnetic relaxation nanoswitch (MRnS) assay. When monodisperse magnetic nanoparticles (MNPs) cluster upon binding to targets, the self-assembled clusters become more efficient at dephasing nuclear spins of many surrounding water protons, leading to a decrease in spin–spin relaxation time (T2). The bottom panel shows an example of the proximity assay measured by an MRnS system. Avidin was added to a solution of biotinylated MNPs, causing T2 to decrease from 40 to 14 ms. (From Lee, H. et al., *Nat. Med.*, 14, 869–874, 2008. With permission.)

Kaittanis et al. (2011) discovered that T2 signals initially increase as analyte binds to MNPs and prior to MNP aggregation. They postulate that the analyte–ligand complexes on the surface of the nanoparticles block the nearby water molecules from diffusing close to the local magnetic field inhomogeneity. Utilizing this scheme, the investigators were able to rapidly detect cholera toxin, anthrax DNA, and *Mycobacterium paratuberculosis* DNA with high sensitivity (Kaittanis et al. 2011; Kaittanis et al. 2012). This MRnS assay configuration increases the rapidity of biosensing without a loss of sensitivity. Furthermore, this strategy permits detection of monovalent analytes that would not otherwise lead to MNP probe aggregation.

MRnS biosensing has many appealing features. Billions of water molecules diffusing in close proximity to the MNPs act as built-in signal amplifiers. The NMR signal can be acquired in complex, turbid samples (blood, serum, sputum, urine, cell, and tissue homogenates) that do not require extensive sample preparation. A single reagent (functionalized MNPs) is usually all that is necessary to perform the assay. Small sample volumes are used given the high levels of assay sensitivity. Signal acquisition develops quickly, often yielding results within several minutes. Although NMR detection conjures up the need for very large equipment, MRnS technology has been greatly enhanced by the integration of miniaturized components, including magnets, radiofrequency microcoils, and microfluidic sample concentration channels. Remarkably, miniaturization has led to the development of benchtop and handheld prototypes (Shao et al. 2010; Issadore et al. 2011). For a detailed description of the fundamentals of MRnS-based biosensing, the reader is directed to excellent reviews (Gossuin et al. 2009; Haun et al. 2010; Shao et al. 2010; Demas and Lowery 2011).

A notable and intriguing application of NMR-based ID nanobiosensing was performed by Renshaw and colleagues (Lee et al. 2012). Using superparamagnetic iron oxide (SPIO) nanoparticles functionalized with anti-TB antibodies (SPIO-TBsAb), the investigators succeeded in detecting mTb in infected monocyte cell cultures and within extra-pulmonary granuloma in an in *vivo* mouse model. In the latter case, mice bearing mTb subcutaneous granuloma were injected via tail vein with SPIO-TBsAB probes. The mice underwent magnetic resonance imaging (MRI) that demonstrated a 14-fold reduction in T2 signal intensity at the site of the Tb granuloma. There are several important aspects of this proof-of-concept study, they are (1) SPIO nanoparticles are currently in clinical use as conventional MRI contrast agents, (2) functionalized SPIO nanoparticles may be capable of specifically accumulating at a variety of anatomic sites where they can be detected by MRI, and (3) since extrapulmonary TB is frequently difficult to diagnosis, a highly specific, noninvasive imaging technology would be most welcome. The investigators point out that much additional research is required to further evaluate this innovative technology and to assess its applicability to other IDs.

1.10.4 Magnetic Nanoparticles and Infectious Diseases Biosensing: Miscellanea and Conclusions

MNP-based assays offer great promise in areas of clinical IDs diagnostics, microbiology research, and environmental-, food- and water-related public health. MNPs are relatively inexpensive to manufacture, chemically stable, and biocompatible. Biorecognition signals can be detected rapidly from small volumes of complex samples that require minimal preparation. Since biological samples lack superparamagnetic properties of their own, background interference is minimized. Sensitivity of several MNP-based bioassay methods approaches that of PCR technology. Miniaturization of complex equipment suggests that MRnS formats could achieve point-of-care capabilities.

Not to be outdone by their QD-fabricating microbial distant relatives, magnetotactic bacteria that biomineralize ferromagnetic material into nanoscale magnetosomes have been discovered in nature (Schüler and Frankel 1999). As shown in Figure 1.17, these phylogenetically diverse bacteria contain intracellular nanomagnets that they use for navigation (magnetotaxis). A number of potential, commercial, and biomedical applications for these magnetosomes have been proposed (Schüler and Frankel 1999).

Before leaving the topic of magnetic-based ID-nanobiosensing, it is worth noting that the Gram stain—the work horse of ID detection techniques—has recently undergone a nanomagnetic modification (Budin et al. 2012). Weissleder and colleagues modified crystal violet, the initial component of the stain, by chemically linking it to MNPs. T2 NMR signals obtained from the labeled bacteria correlated very well with semiquantitative analysis from light microscopy. Furthermore, prior to the decolorization step, both Gram-positive and Gram-negative bacteria were labeled, the latter as expected, losing the magnetic label after decolorization. In addition, when GNPs were used in place of MNPs (GNP Gram stain), they were visualized throughout the bacterial surface by TEM. The investigators suggest that the "magnetic Gram stain" could potentially be incorporated into an automated point-of-care diagnostic tool, have a role in detecting bacteria that would otherwise require enrichment steps, and be used to label bacteria in *vivo*. They also postulate that similar MNP modification strategies have the potential to be applied to other conventional microbiology staining techniques.

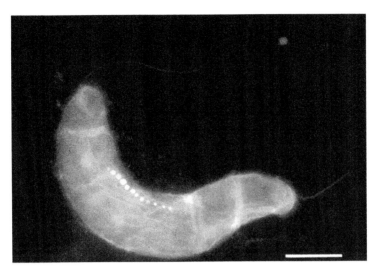

FIGURE 1.17
Electron micrograph of a *Magnetospirillum gryphiswaldense* cell exhibiting the characteristic morphology of magnetic spirilla. The helical cells are bipolarly flagellated and contain up to 60 intracellular magnetite particles in magnetosomes, which are arranged in a chain (scale bar: 0.5 µm). (From Schüler, D. and Frankel, R.B., *Appl. Microbiol. Biotechnol.*, 52, 464–473, 1999. With permission.)

1.11 Microcantilever-Based Infectious Diseases Biosensing

Microcantilevers are unique biosensing tools that are being used to detect a variety of biomolecules as well as microorganisms and biomarkers of ID. The cantilevers can be thought of as microscopic diving board-like structures that register biorecognition events by distortional stress placed on their surface (Hansen and Thundat 2005; Alvarez et al. 2009; Johnson and Mutharasan 2012). They are typically fabricated from crystalline silicon derivatives, metals, glass, or polymers. One surface is coated with a nano-thin layer of material, often gold, that can be readily functionalized with affinity ligands or other media to which analytes adhere. There are two different modes of signal transduction, static and dynamic, that can produce highly sensitive and real-time biosensing results. The static mode, illustrated in Figure 1.18, is a nanomechanical biosensor that measures deflection of the cantilever due to analyte binding. In this configuration, laser light reflecting off the gold surface is deflected when the cantilever bends. The dynamic mode, illustrated in Figure 1.19, measures fluctuations in resonance frequency due to mass change or stiffness induced by analyte binding events.

Several groups have utilized microcantilevers to detect common food contaminants, including *Salmonella*, *E. coli*, *Cryptosporidium*, and *Giardia* (Weeks et al. 2003; Campbell and Mutharasan 2005, 2008; Gfeller, Nugaeva, and Hegner 2005a,b; Campbell et al. 2007; Xu and Mutharasan 2010). In each instance, the cantilever assays were significantly more sensitive and rapid than conventional techniques.

Hegner and coworkers (Gfeller, Nugaeva, and Hegner 2005a,b) functionalized a linear array of microcantilevers with a layer of agarose and exposed them to broth cultures containing *E. coli* in the presence or absence of different antibiotics. Cantilever resonance frequency as a function of bacterial growth on the agarose layer was measured. Active

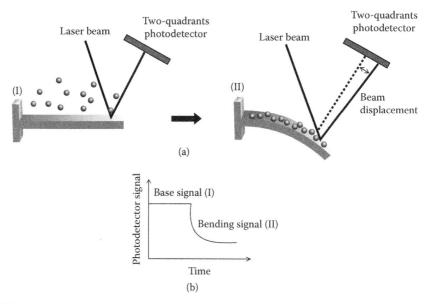

FIGURE 1.18
(a) Scheme of the cantilever bending method and the laser beam displacement readout. The reflected laser beam experiences a displacement with respect to its original position (reflection over the flat cantilever) due to the adsorption of molecules over the upper surface. (b) Photodetector signal read in real time before and after the adsorption of molecules over the cantilever upper surface. (From Alvarez, M. et al., *Methods Mol. Biol.*, 504, 51–71, 2009. With permission.)

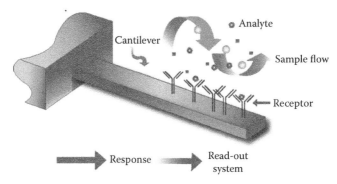

FIGURE 1.19
General schematic of dynamic-mode cantilever sensing principle. (From Johnson, B.N. and R. Mutharasan, *Biosens. Bioelectron.*, 32, 1–18, 2012. With permission.)

growth was detected within 1 hour, comparing quite favorably to the 24 hours needed for standard culture techniques. The mass loading of the microorganisms on the cantilevers correlated with resonance frequency and corresponded closely to lag, exponential, and stationary growth phases of the bacteria. Moreover, selective growth effects on the bacteria by antibiotics were observed in less than 2 hours. This brief time period of antibiotic-resistance testing is most impressive when compared to conventional methods that take between 16 and 48 hours.

Antibody-functionalized resonant cantilevers were used to detect hepatitis A and C viruses in spiked serum (Timurdogan et al. 2011). The assay had a detection limit of 1.6 pM, which was comparable to ELISA. Mutharasan and coworkers (Xu, Sharma, and

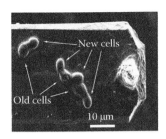

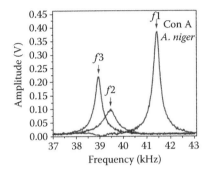

FIGURE 1.20
SEM images of immobilized fungal spores and a corresponding frequency spectra of the cantilever after spore immobilization and spore growth on protein functionalized surface. Upper left panel shows yeast *Saccharomyces cerevisiae* on concanavalin A (Con A) coating: old (three wrinkled cells) and new (four smooth cells) generations. Upper right panel shows mycelial fungus *Aspergillus niger* on Con A coating, mycelia growing from spores. Lower panel resonance frequencies: fl, unloaded cantilever; f2, cantilever with immobilized spores; and f3, cantilever with growing immobilized spores. (From Nugaeva, N. et al., *Biosens. Bioelectron.*, 21, 849–856, 2005. With permission.)

Mutharasan 2010) used cantilevers to detect a mycoplasma species that commonly contaminates cell cultures used for biomedical research. Their assay had four orders higher sensitivity compared with a standard ELISA and took less than 1 hour to perform. Hegner et al. (Nugaeva et al. 2005) detected fungal spores using microcantilevers functionalized with concanavalin A, fibronectin, or anti-aspergillus IgG antibody. The yeast, *S. cerevisiae*, and the mold, *Aspergillus niger*, attached to the microcantilever, germinated, and were detected within a few hours, when compared to several days by conventional methods. As shown in Figure 1.20, mass changes on the cantilever correlated with attachment and growth of the fungi. This proof-of-concept study demonstrates that the microcantilever assay can provide real-time measurement of fungal growth.

1.11.1 Microcantilever-Based Infectious Diseases Biosensing: Conclusions

Nanomechanical cantilever sensors offer sensitive, selective, and real-time ID detection strategies. The ability to functionalize the sensors with agarose and peptides that promote attachment and growth of microorganisms offers a unique biosensing feature. The highly sensitive mechanical characteristics of the microcantilevers permit early detection of microorganism growth and a means of rapid screening for antimicrobial resistance. These features could have a significant impact on early diagnosis of ID agents and timely optimization of antimicrobial therapy.

1.12 Miscellaneous Infectious Diseases–Nanobiosensing Strategies

There is a diversity of biosensing strategies that do not easily fit into the categorical scheme presented thus far. With respect to IDs, several strategies are noteworthy, and all are deserving of mention, given their uniqueness and potential as diagnostic tools. An arbitrary classification scheme and brief description of relevant biosensing platforms are as follows.

1.12.1 Fluorescent Europium Nanoparticle Biosensors

Fluorescent europium chelate nanoparticles (EuNPs) embedded in either silica or polystyrene shells possess high quantum yields, long fluorescent lifetimes, and are capable of producing intense luminescence (Harma, Soukka, and Lovgren 2001). To date, they have been used in immunoassay formats to detect adenovirus and bacterial toxins (Valanne et al. 2005; Ai, Zhang, and Lu 2009; Tang et al. 2009). In each case, the use of EuNP probes displayed higher sensitivity than that of conventional immunoassays.

1.12.2 Silica Nanoparticle Biosensors

Silica nanoparticles (SiNPs) are easily fabricated, chemically and thermally stable, dispersible in aqueous media, biocompatibile, and readily functionalized (Tan et al. 2004; Knopp, Tang, and Niessner 2009). They have been used to encapsulate thousands of fluorophore dye molecules (dye-doped SiNPs) to create brightly amplified biosensing tags capable of detecting microorganisms and biomarkers of ID.

Tan and coworkers (Zhao et al. 2004) detected *E coli* O157:H7 with antibody-functionalized fluorescent dye-doped SiNPs using several formats, including flow cytometry, spectroscopy, and high throughput microtiter assay. In this proof-of-concept study, the detection limit was as low as one bacterial cell per sample, and results were obtained in 20 minutes. In addition, they were able to perform the analysis on spiked ground beef samples. In a separate study, the investigators exploited the versatility of SiNP fluorescent probes and developed a multiplex FRET-based assay that simultaneously detected *E. coli*, *Salmonella*, and *S. aurues* (Wang et al. 2007). They succeeded in detecting *M. tuberculosis* in spiked urine and sputum samples by immunofluorescence and flow cytometry (Qin et al. 2007, 2008). Staphylococcal enterotoxin was detected, using a sandwich fluorescent dye-doped SiNP immunoassay (Hun and Zhang 2007). The method was rapid, sensitive, and capable of detecting toxin in spiked water and milk samples. Antibody-functionalized SiNP probes were used to simultaneously detect CD4+ and CD45+ T-cell subsets from blood samples, utilizing disposable microchips in a portable flow cytometer (Yun et al. 2010). The fluorescent SiNPs were 100 times brighter than conventional immunofluorescent probes. This prototypical assay has potential point-of-care capability.

1.12.3 Liposome Nanobiosensors

Many biorecognition events occur at the cell membrane and lead to transduction of biomolecular signals that control a multitude of cell functions. Capitalizing on this fundamental feature of cell biology, liposomes have been synthesized to resemble the structure of cell membranes and enlisted as nanobiosensors. Liposomes expressing sialic acid residues were used in a colorimetric assay to detect influenza virus in a proof-of-concept study

(Reichert et al. 1995). The strategy is unique in that the sialic acid molecules, synthesized with chemical side chains that turn color upon influenza binding, act as both sensing and transducing elements. Fluorescent dyes were encapsulated by liposomes that were functionalized to express ganglioside receptors on the lipid bilayer (Singh, Harrison, and Schoeniger 2000). The liposome constructs were used to detect tetanus, botulinum, and cholera toxins. The large surface area and the internal volume of liposomes allow incorporation of many biorecognition and reporter molecules capable of achieving high specific activity and signal amplification, respectively.

1.12.4 Carbon Nanotube and Nanowire Biosensors

Carbon nanotubes (CNTs) are remarkable structures made of hollow graphene sheets having diameters ranging from 1 to 100 nm and lengths as long as several micrometers (Valcarcel et al. 2005; Rivas et al. 2007). Their structure, comprised entirely of surface carbon atoms, confers enormous ratios of surface area to weight. They can exist in either single or multiwall CNT configurations (SWNT and MWNT, respectively) and possess unique electrical, electrochemical, and mechanical properties. Their surfaces can be readily functionalized with a variety of biomolecules and used as highly sensitive and selective biosensors.

SWNTs functionalized with anti-*E. coli* O157-H7 antibodies were used as monovalent probes to selectively bind to the surface of the bacteria in a proof-of-concept study (Elkin et al. 2005). In a separate study, the SWNTs were functionalized with galactose derivatives and used as multivalent probes to bind *E. coli* O157-H7 (Gu et al. 2005). In the latter case, the bacteria were found to clump in the presence of the SWNT probes as a result of the multivalent nature of the binding event. In conjunction with their affinity biorecognition potential, CNTs have been used as electrochemical biosensors capable of measuring enzyme–substrate interactions, nucleic acid hybridization, and aptamer–ligand interactions (Besteman et al. 2003; He and Dai 2004; So et al. 2008; Zelada-Guillen et al. 2009).

Nanowires can be composed of a variety of metallic, semiconducting, and organic materials. Their very large aspect ratios (length-to-width) in conjunction with the quantum confinement effects on nanowire electrons create unique electrochemical properties that impart impressive biosensing proficiency (Cui et al. 2001). Analyte binding to affinity ligand-functionalized nanowires results in instantaneous transduction of electric signals. In an assay that incorporated antibody functionalized silicon nanowires, Lieber and coworkers demonstrated single virus particle detection sensitivity (Patolsky et al. 2004). In addition, the assay was able to simultaneously detect influenza and adenovirus in a multiplexed nanowire array. This platform has the potential for exquisitely sensitive, rapid, multiplex detection of ID-relevant targets, as well as the ability to quantify ligand–receptor binding kinetics (Patolsky, Zheng, and Lieber 2006).

The electrochemical nanobiosensing platform offers many attractive features, including ultra-high sensitivity, selectivity, and real-time detection. Specimen samples do not require labeling or amplification steps. Point-of-care capability of electrochemical biosensing is likely to develop as integration with MEMS devices is accomplished.

1.12.5 Microbial-Based Biosensing

In keeping with the microbiologic bias of this chapter, delight is taken in briefly describing potential nanobiosensing strategies that exploit the incredible biology and architecture of microorganisms. The bacteria *Geobacter sulfurreducens* possess nanowire filaments composed of pilin proteins capable of conducting electrons along long distances (Malvankar and Lovley 2012). These nonpathogenic, biofilm-forming, anaerobic bacteria inhabit soil sediments, where they oxidize organic matter and metals into environmentally safe compounds. The potential to exploit their electric conducting properties for a variety of purposes, including biosensing, represents an exciting scientific frontier. For an in-depth discussion of the fascinating field of electromicrobiology, including a description of microbial nanowires and their potential uses, the reader is directed to the review by Lovley (2012).

Protein nanocages derived from the plant pathogen cowpea chlorotic mottle virus (CCMV) were used to target *S. aureus* cell surface protein-A and evaluate biofilm architecture produced by the bacteria (Suci et al. 2007). These well-characterized, icosahedral, noninfectious virus nanoparticles were dual functionalized with (1) anti-protein-A antibody and (2) fluorescent tags. The multifunctional CCMV nanoparticles coated the surface of *S. aureus* at very high density and penetrated well into the in *vitro* biofilm model used in the study. In addition, as proof-of-concept, the investigators substituted the fluorescent tags with the MRI contrast agent gadolinium. They postulate that the gadolinium-functionalized CCMV probes could potentially allow MRI of biofilms within tissue samples.

1.13 Infectious Diseases Nanobiosensing: Summary and Conclusions

A great deal of emphasis has been placed on ID nanobiosensing in this chapter. The underlying physicochemical principles are complex, and the potential to favorably impact ID diagnosis, research, and public safety is great. Although the nanobiosensing platforms are diverse in function, they share many fundamental similarities. Their ultra-high sensitivity and specificity depend, in large part, on the ability to attach affinity ligands to nanostructured probes. Indeed, the breadth and variety of the ligands capable of functionalizing nanoprobes is extraordinary. Many of the platforms do not require the use of fluorescent dye labels or nucleic acid amplification; several permit biosensing in complex media, requiring minimal preparation. A number of nanobiosensing platforms are configured as bioorthogonal schemes that make it possible to analyze form and function of intact microorganisms and host–pathogen interactions. Many of the assay formats are being integrated with MEMS and appear capable of point-of-care deployment. Several platforms offer the potential for in *vivo* biosensing.

At this early stage in their evolution, many ID relevant nanobiosensors appear to possess features that come close to resembling that of an "ideal biosensor." However, there is much to be accomplished in moving forward from proof-of-concept stages to mainstream diagnostic assays. Rigorous quality control standards must be established for the engineering and fabrication of the nanostructures and devices. Equally rigorous preclinical testing is required to establish the sensitivity, specificity, and field applicability of the biosensor and to demonstrate that this new technology adds to, or surpasses, the characteristics of currently available methods. Assessments of nanoparticle potential toxicities need to be addressed. Proceeding through the regulatory process will take a great deal of time, money, and effort. Hopefully, the process will be thorough, expeditious, and fruitful in advancing the cause for sensitive, specific, rapid, inexpensive, easily performed, and widely accessible IDs detection techniques.

1.14 Application of Nanotechnology to Antimicrobial Therapy and Prevention of Infectious Diseases: Introduction

This chapter, while placing an emphasis on the ID–nanobiosensing interface, should not be viewed as minimizing the equally important relationships between nanotechnology and both antimicrobial therapy and infection prevention. There is a critical need for new and novel antimicrobials, environmental biocides, water treatment strategies, defouling agents, and vaccines. A great deal of nanotechnology research is being devoted to applications in these areas of IDs and public health. The framework of nanoscience principles introduced in earlier sections is equally applicable here. Indeed, we will revisit several nanotechnology concepts and nanomaterials as they are applied to antimicrobials and infection prevention strategies.

1.14.1 Perspectives on Antimicrobial Therapy and Prevention of Infectious Diseases

In the 70 years since the discovery of penicillin, medical science has witnessed a dramatic change in research and development of antimicrobial pharmaceuticals; a 50-year span of initial exuberance has given way to a recent, and potentially calamitous, dearth of new anti-infective agents (Boucher et al. 2009; Infectious Diseases Society of America 2011; White 2011; Wise 2011). At the same time, the emergence and spread of resistant microorganisms has accelerated at an alarming rate, crossing all global boundaries and occurring in clinically relevant viruses, bacteria, mycobacteria, fungi, and parasites. Resistance to antimicrobials is particularly prevalent in health-care settings, where the substantial use and frequent overuse of antibiotics foster the emergence of highly resistant microorganisms (Rice 2009; Millar 2012). It is also problematic in the community where antimicrobial resistance in HIV, *S. aureus*, TB, and malaria continue to spread. There is a critical need for innovative approaches to the development of anti-infective agents and a replenishment of the antimicrobial pipeline.

Prevention of IDs can be considered in two broad, interrelated contexts: (1) preventing the growth and transmission of pathogens from animate and inanimate sources before they have an opportunity to cause infection and (2) stimulating innate host immunity that will either attenuate the severity of infection or completely prevent infection from occurring in the first place. In the first instance, disinfection is linked to the development of active and safe biocides, as well as water and environmental decontamination (Lambert 2004; Shannon et al. 2008). In the latter instance, the development and administration of safe and effective vaccines are perhaps the most important public health strategies (Keegan and Bilous 2004; Plotkin and Plotkin 2004).

Nanotechnology research and development in areas of antimicrobial therapy, disinfection, and immunization have generated intense interest. Numerous applications have proven to be quite successful and are in current use, while many others are in various phases of testing. The remainder of this chapter will explore several examples of these important aspects of the ID–nanotechnology interface.

1.14.2 Overview of Antimicrobial Pharmacology Principles

For an antimicrobial to work in *vivo*, it must first reach the site of infection and do so in a bioactive form. Antimicrobial uptake, distribution, metabolism, and elimination are considered PK variables, whereas the interplay of antimicrobial concentration and time-dependent activities on pathogen survival are deemed PD variables (Craig 1998). PK–PD characteristics guide the development of antimicrobial agents and help to define optimal dosing parameters (Ambrose et al. 2007; Owens Jr and Shorr 2009). Antimicrobial activity also depends on the nature of the infecting microorganism as well as on its inoculum size, phase of growth (actively dividing vs. stationary phases), resistance traits, and the cellular and anatomic locations in which the microorganisms reside. Equally relevant aspects of pharmacology include the evaluation of potential toxicities and the interaction between antimicrobials and other classes of drugs. A frequently overlooked consideration of antimicrobial use is the profound effects it can have on the ecology of normal microbial flora and the complications that can occur as a result.

The evaluation of antimicrobial activity takes on a somewhat different set of principles when applied to animate and inanimate surfaces as well as to water and environmental disinfection. A brief perspective is in order. Globally, approximately 1 billion people lack access to safe water supplies, resulting in widespread outbreaks of potentially lethal infectious diarrheal diseases that become self-perpetuating (Shannon et al. 2008; WHO 2012b). Colonized and contaminated surfaces, both animate and inanimate, which would benefit from active and sustained disinfection are another major source of transmission of microorganisms in both community and health-care settings (Lichter, Van Vliet, and Rubner 2009). Resisting microbial adhesion, killing microbes on contact, leaching of the anti-infective compounds into the immediate surrounding environment, and safety are the parameters most often considered in biofilm prevention (Li et al. 2008; Lichter, Van Vliet, and Rubner 2009).

A diverse group of nanoparticles can be broadly categorized as possessing inherent antimicrobial activity or acting as drug delivery vehicles in targeted antimicrobial therapy. In the latter case, they may enhance antimicrobial PK–PD characteristics, reduce toxicity, and deliver antimicrobials to intracellular and anatomic sites at concentrations, and for durations of time, that would not be achieved otherwise. Several examples of nanotechnology applications to anti-infective therapy and IDs prevention will be presented, and the reader is referred to excellent and detailed reviews of the topic (Li et al. 2008; Lichter, Van Vliet, and Rubner 2009; Zhang et al. 2010; Huh and Kwon 2011; Seil and Webster 2012).

1.14.3 Nanoparticles Having Inherent Antimicrobial Activity

As summarized in Table 1.4, a number of nanoscale metals, metal oxides, naturally occurring and synthetic polymers, and engineered nanoparticles possess antimicrobial properties (Li et al. 2008; Huh and Kwon 2011; Seil and Webster 2012). Although there is general agreement on the probable mechanisms of antimicrobial action, as summarized in Figure 1.21, precise means by which nanoparticles interact with microorganisms have not been fully established. Cationic nanoparticles interacting with the anionic cell membrane of microorganisms appear to cause mechanical disruption or interfere with electron transport. Semiconductor nanoparticles—for instance, TiO_2—can be activated by ultraviolet light and catalyze reactions that produce reactive oxygen species (ROS) that damage proteins and nucleic acids. This photocatalytic property offers great potential as a means of water disinfection (Li et al. 2008). Metal ions released from nanoparticles have been postulated to damage microbial cell membranes, mitochondria, and DNA. Nanoparticles, such as SiO_2, and the naturally occurring polycationic biopolymer, chitosan, induce cell agglomeration (flocculation) and membrane damage (Rabea et al. 2003). The SPR properties of SNPs, GNPs, and CNTs are exploited in photothermal killing of microorganisms to which they attach (Zharov et al. 2006; Kim et al. 2007; Mamouni et al. 2011). Similarly, GNPs

TABLE 1.4

Summary of Select Studies Concerning the Antimicrobial Effects of Nanoparticles

Chemistry	Organisms	Proposed Mechanism	Reference
ZnO	*Staphylococcus aureus, Escherichia coli*	ROS inhibition, membrane disruption	Reddy et al. 2007, Jones et al. 2007, Nair et al. 2009, Liu et al. 2009b, Padmavathy and Vijayaraghavan 2008
ZnO ions	*Pseudomonas aeruginosa, S. aureus, Candida albicans*	ROS inhibition	McCarthy, Zeelie, and Krause 1992
SNP	*E. coli, Vibrio cholera, Salmonella typhi, P. aeruginosa*	Membrane disruption, Ag ion interference with DNA replication	Morones et al. 2005, Pal, Tak, and Song 2007, Zhou et al. 2012
SNP	HIV-1	Binds to gp 120, inhibits binding to target cell	Elechiguerra et al. 2005
SNP	TB strains, *Candida* sp.	Membrane disruption, ROS inhibition	Seth et al. 2011, Hwang et al. 2012, Silva et al. 2012
SNP	*Leishmania* sp., *Microfilaria* sp.	ROS inhibition	Allahverdiyev et al. 2011, Singh et al. 2012
Cu	*E. coli, Bacillus subtilis*	Protein inactivation	Yoon et al. 2007
CuO	*S. aureus, MRSA, P. aeruginosa, Proteus* sp., *E. coli*	Membrane disruption, enzyme dysfunction	Ren et al. 2009
Fe₃O₄	*S. aureus, Staphylococcus Epidermidis*	ROS inhibition, membrane disruption	Tran et al. 2010, Taylor and Webster 2009
Al₂O₃	*E. coli, B. subtilis, Pseudomonas fluorescens*	ROS inhibition, Cell penetration, flocculation	Simon-Deckers et al. 2009, Jiang, Mashayekhi, and Xing 2009
TiO₂	*E. coli*	Membrane disruption	Simon-Deckers et al. 2009
SiO₂	*E. coli, B. subtilis, P. fluorescens*	Flocculation, membrane disruption	Jiang, Mashayekhi, and Xing 2009
Chitosan	*E. coli, S. aureus, Enterococcus* sp., *P. aeruginosa, C. albicans*	Flocculation, membrane disruption, chelation, enzyme dysfunction,	Qi et al. 2004, EL-Sharif and Hussain 2011
Chitosan–Ag composite	*S. aureus*	Synergistic activity	Potara et al. 2011
GNP	*Toxoplasma gondii, S. aureus, P. aeruginosa*	Plasmon-induced photodynamic and photothermal-targeted cell destruction	Pissuwan et al. 2007, Zharov et al. 2006, Norman et al. 2008
GNP	*E. coli*, TB strain	Membrane disruption	Zhou et al. 2012
MgO	*S. aureus, B. subtilis*	ROS inhibition	Huang et al. 2005
CNT	*E. coli, S. typhi, S. aureus, B. subilis, P. aeruginosa*	Membrane disruption, ROS inhibition	Kang et al. 2007, Arias and Yang 2009; Liu et al. 2009a, Vecitis et al. 2010
CNT	*E. coli, Salmonella* sp.	Photothermal targeted cell destruction	Kim et al. 2007; Mamouni et al. 2011
Fullerenes	*S. aureus, E. coli, P. aeruginosa, C. albicans*	Photodynamic targeted cell destruction	Huang et al. 2010
Fullerenes	Gram-positive bacteria	Membrane disruption	Tsao et al. 2002
Biodegradable nanopolymers	*S. aureus, B. subtilis, Enterococcus* sp., *Cryptococcus* sp.	Membrane disruption	Nederberg et al. 2011

Source: Adapted from Seil, J.T., and T.J. Webster, *Int. J. Nanomedicine*, 7, 2767–2781, 2012.

Abbreviations: HIV, human immunodeficiency virus; ZnO, zinc oxide; ROS, reactive oxygen species; SNP, silver nanoparticle; Cu, copper; CuO, copper oxide; Fe₃O₄, iron oxide; Al₂O₃, aluminum oxide; TiO₂, titanium dioxide; SiO₂, silicon dioxide; Ag, silver; GNP, gold nanoparticle; MgO, magnesium oxide; CNT, carbon nanotube; MRSA, methicillin resistant *S. aureus*; sp, species.

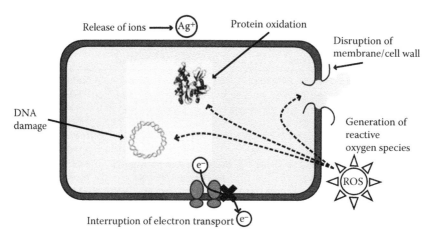

FIGURE 1.21

Various mechanisms of antimicrobial activities exerted by nanomaterials. *Abbreviation:* ROS, reactive oxygen species. (From Li, Q. et al., *Water Res.*, 42, 4591–4602, 2008. With permission.)

and pure carbon fullerene nanoparticles can generate antimicrobial ROS using photodynamic activation strategies (Maisch 2009; Perni et al. 2011).

The mechanisms by which nanoparticles inhibit or kill microorganisms are uniquely different from those of conventional antibiotics, and it is this feature that makes them particularly attractive. Pathogens that have developed resistance to antibiotics in current use are likely to be sensitive, at least initially, to nanoparticle antimicrobial mechanisms. Moreover, the ability of microorganisms to develop resistance may prove particularly difficult, given the multiple antimicrobial mechanisms and cellular sites of action of the nanoparticles. It must be emphasized, however, that the majority of studies that have evaluated the antimicrobial actions of nanoparticles to date have utilized standard *in vitro* techniques; there is a paucity of data on PK–PD parameters, *in vivo* toxicity, and drug interaction characteristics for most of these compounds (Hagens et al. 2007). Furthermore, the potential effect of nanoparticles on normal microbial flora in humans, animals, and the environment raises concerns that warrant a great deal of research. SNPs will be presented herein to illustrate the potential opportunities as well as the challenges associated with the development of nanoparticles possessing inherent antimicrobial activity.

1.14.4 Silver Nanoparticles: Antimicrobial Activities and Anxieties

Silver, in several configurations, has been used for centuries to treat a variety of ailments having both infectious and non-ID origins (Klasen 2000a,b; Alexander 2009). From as early as the fourth century BCE, silver chalices and containers were used to store water and food in order to prevent them from spoiling. Silver colloids were the predominant antimicrobials until the 1940s, when they were largely replaced by the newly discovered antibiotics. Topical applications of silver compounds to wounds are a common practice, and deposition of antimicrobial SNPs on textile fibers, bandages, and articles of clothing is an area of scientific and commercial interest (Strohal et al. 2005; Dubas, Kumlangdudsana, and Potiyaraj 2006; Rai, Yadav, and Gade 2009). SNPs have been used for water purification and to coat or impregnate medical devices in an effort to prevent biofilm-associated infections (Furno et al. 2004; Roe et al. 2008; Zodrow et al. 2009; Cao and Liu 2010).

Numerous studies investigating the mechanisms that underlie antimicrobial properties of silver are detailed in excellent reviews (Rai, Yadav, and Gade 2009; Cao and Liu 2010; Huh and Kwon 2011). Although the precise nature of its antimicrobial actions remains a matter of speculation, catalysis of reactions that increase oxidative stress appear to be fundamentally involved. In addition, it is unclear whether silver ions, particulates, or a combination of both are primarily responsible for antimicrobial activity. Dissolved silver ions bind to protein sulfhydryl groups, and this interaction results in the production of ROS that damage microbial cell membranes and nucleic acids. SNPs appear to cause microbial cell damage by similar mechanisms. Moreover, the high surface area and surface reactivity of SNPs are thought to enhance the inherent biocidal activity of silver. There is evidence suggesting that antimicrobial activity is dependent on the size and shape of the SNP, with smaller and more angulated particles displaying greater potency (Pal, Tak, and Song 2007). Direct binding and inactivation of cell membrane components, mitochondria, and DNA by SNPs have also been described. Furthermore, in the presence of activating wavelengths of light, plasmonic phenomena appear to enhance SNP antimicrobial activity (photocatalysis).

Silver is a broad-spectrum antimicrobial; its biocidal effects have been observed in viruses, bacteria, fungi, and parasites, as outlined in Table 1.4. In health-care settings, silver formulations have seen their widest application in the management of wounds where both antimicrobial and wound-healing properties have been observed. An interest in silver-based wound dressings, particularly for patients suffering from burns, increased in the 1960s, when bacterial resistance to topically applied conventional antibiotics began to emerge (Klasen 2000a, b). Since then, a variety of silver-based compounds, alone or in combination with antibiotics, has become available for wound management. Several types of silver-coated or -impregnated medical devices, including bladder and vascular catheters and endotracheal tubes, have been studied (Furno et al. 2004; Roe et al. 2008; Afessa et al. 2010). Results of these investigations have met with mixed success, and it is clear that additional research of the many different types of silver-deposition strategies used in health-care appliances is needed. Consumer products containing silver-based compounds and SNPs have proliferated (Rai, Yadav, and Gade 2009; Lem et al. 2012).

Silver-containing cosmetics, toothpastes, textiles, paints, food packaging, water filters, and countless other products are being manufactured. An Internet search will yield a plethora of nano-silver items that can be purchased from around the world. This raises questions about the rigor of standards applied to the efficacy of these products and also raises concerns regarding their potential to promote the emergence of microbial resistance, toxicities, and environmental harm from exposure to formulations of nano-silver and other silver-based compounds (Chopra 2007; Ahamed, Alsalhi, and Siddiqui 2010; Johnston et al. 2010).

The ability of microorganisms to develop resistance to silver ions and SNPs remains an area of debate. Much of the uncertainty stems from a lack of well-established standards of testing for silver and SNP microbial resistance, toxicokinetics, and toxicodynamics (Chopra 2007; Johnston et al. 2010). In a review by Chopra (2007), detailed discussion is made of reports describing inducible plasmid and chromosomally mediated bacterial resistance as well as an efflux-mediated resistance mechanism identified in an isolate of candida. Given the multiple mutations that would be required by microorganisms to overcome the diverse array of silver's mechanisms of action, rapid emergence of resistance within a given species is unlikely. However, the presence of resistance plasmids, the ease with which they can be transmitted, and the impressive resistance-acquisition track record of microorganisms suggest that concern, vigilance, and generally accepted standards of testing are needed.

Similar concerns exist regarding potential toxicities (Rai, Yadav, and Gade 2009; Ahamed, Alsalhi, and Siddiqui 2010; Johnston et al. 2010). As pointed out by Hagens et al. (2007), there are gaps in our understanding of absorption, distribution metabolism, and excretion of nanoparticles, including SNPs. Furthermore, given the diverse formulations that exist for silver compounds, it is difficult to extrapolate the results from individual studies and apply them to the general class of agents. Toxicity due to silver ions and SNPs has been described in in *vitro* mammalian cell cultures and in in *vivo* animal models (Ahamed, Alsalhi, and Siddiqui 2010; Johnston et al. 2010). Inhalation, ingestion, cutaneous absorption, and where and for how long silver compounds accumulate in organs require close analysis. Oxidative stress appears to be the mechanism underlying the toxicities induced by silver exposure, and at the present time, there is a paucity of data on dose–response relationships. In addition to potential cytotoxic effects, there is evidence to suggest that SNPs can modify immune function by altering the production of cytokines (Klippstein et al. 2010). While the currently available studies of silver toxicity are a cause for concern, it is premature to become discouraged about its potential therapeutic and preventive value. Indeed, virtually all antibiotics have associated potential toxicities, and a fundamental tenet of ID practice dictates that consideration will be given, on an individual patient basis, to the benefits versus risks of their administration. One hopes that similar considerations will be made with respect to SNPs—and all nanoparticles, for that matter—and that administration of these agents will be conducted sensibly and with knowledge of detailed toxicological and environmental impact data.

The theme of this chapter in concert with the amazing characteristics of microbes dictate that there is a mention of SNP biosynthesis by the enzymatic reduction of silver nitrate to form nano-silver colloids (Ingle et al. 2008; Balaji et al. 2009; Birla et al. 2009; Nanda and Saravanan 2009; Saravanan and Nanda 2010). The reactions are performed using culture supernatants that contain extracellular enzymes produced by *S. aureus*, as well as a variety of fungi. The SNPs produced by this process possess antimicrobial activity, and the investigators suggest that this technology represents a simple, inexpensive, and eco-friendly way to produce SNPs.

1.14.5 Nanostructures Having Inherent Antibiofilm Activity

A unique construct of antimicrobial nanomaterials having particular relevance to biofilm prevention can be found in the design of surface structures coated with polyelectrolyte multilayers (PEMs) (Lichter, Van Vliet, and Rubner 2009). As the name implies, these diverse, charged, polymeric nano-thin films are coated, layer-by-layer, onto substrates (e.g., metal alloys, food packaging, glass, paper, and microelectronics) and possess antibacterial properties. PEM coatings are being studied for their bacterial adhesion resistance, contact killing, and biocide leaching capabilities. Their physical and chemical properties can be tuned, and a variety of inherent antimicrobial nanoparticles—for example chitosan, TiO_2, CNTs, and SNPs—can be incorporated into their structure (Podsiadlo et al. 2005; Fu et al. 2006; Yuan et al. 2007). PEM coatings applied to implantable medical devices are being evaluated for their anti-biofilm forming properties (Fu et al. 2006; Chua et al. 2008). Anti-adhesive and contact killing properties, in the absence of biocide leaching, could confer a theoretically inexhaustible framework of anti-biofilm activity. Furthermore, the relatively low fabrication costs, and the ability to withstand harsh cleaning and sterilization procedures, make PEM coatings an attractive nanobiotechnology platform. As discussed by Lichter, Van Vliet, and Rubner (2009), several in *vitro* models of PEM coatings have shown excellent antimicrobial activity in low, but not in high nutrient environments; the reason for this observed phenomena has not yet been elucidated. The investigators point out that significant challenges in anti-biofilm research exist, and a great deal of work is required to optimize the functional and chemical properties of these promising antimicrobial nanostructures.

Strategies that incorporate antimicrobial nanoparticles into synthetic biomaterials that will be used for tissue regeneration and medical device implantation are an area of great interest (Taylor and Webster 2009, 2011). The design of nanotextured surfaces that impede microorganism attachment is being explored (Colon, Ward, and Webster 2006; Puckett et al. 2010). Paints containing photocatalyzable TiO_2 as well as SNPs are being tested for their antimicrobial and antibiofilm activities (Kumar et al. 2008; Caballero et al. 2010). Although all are very intriguing from an infection control standpoint, the potential environmental risks, as mentioned earlier, remain a concern and warrant detailed analysis.

1.14.6 Nanostructures as Antimicrobial Carriers

The small size, large surface area, and surface functionality of nanoparticles have attracted great interest in terms of their potential as carriers of conventional antibiotics and immunomodulators. The novel PK–PD characteristics of the nanoparticle-bound antibiotics may enhance anti-infective strategies in several advantageous ways: (1) antibiotic administration and distribution to anatomic and intracellular sites that would have been inaccessible otherwise; (2) increased antibiotic solubility; (3) enhanced microbicidal activity; (4) specific targeting to localized areas of infection; (5) reduced rates of antibiotic metabolism and elimination, with concomitant prolongation of antibiotic exposure to microorganisms; (6) controlled drug release; (7) reduced systemic antibiotic concentrations; and (8) reduced antibiotic toxicity. The reader is referred to the reviews by Zhang et al. (2010) and Huh and Kwon (2011) for excellent and detailed presentations of this topic. In addition, the outstanding review by Nowacek and Gendelman (2009) focuses attention on the challenges of treating HIV infection of the central nervous system and on nanocarrier delivery of anti-retroviral agents across the blood-brain barrier (BBB). Table 1.5 provides a summary of selected nanoparticle antimicrobial carriers and identifies their unique characteristics. A discussion of liposomes as being representative of the class of nanoparticle antimicrobial carrier is as follows.

1.14.7 Liposomes: Approved Antimicrobial Nanocarriers

Thirty years after their discovery in the 1960s, liposomes were approved by the U.S. Food and Drug Administration (FDA) as nanoparticles capable of carrying the antifungal agent amphotericin B (Lopez-Berestein et al. 1989; Lian and Ho 2001). Shortly thereafter, liposomal carriage of the chemotherapeutic agent doxorubicin was approved for the treatment of AIDS-related Kaposi's sarcoma (Bogner et al. 1994). Liposomes are capable of carrying both water soluble and lipid soluble drugs, and currently, research on liposomal carriage of every class of antibiotic is in various phases of development (Drulis-Kawa and Dorotkiewicz-Jach 2010).

Several formulations varying in liposome size and structure exist, and they are engineered to provide carrying capacity that is optimized relative to the parent drug. Liposome surfaces can be modified by the attachment of compounds, for instance, PEG, that reduce their uptake by the macrophage phagocyte system (MPS), thereby increasing circulation time and sustaining antimicrobial release. Surface functionalization of liposomes with immunoglobulins, peptides, and oligosaccharides is being investigated as a means of

TABLE 1.5

Summary of Select Nanoparticle Antimicrobial Carriers

Nanoparticle	Salient Pharmacologic Characteristics	Parent Antimicrobial	References
Liposomes	Biocompatible, biodegradable formulations varying in natural and synthetic phospholipid and cholesterol composition; intravenous, topical, and inhaled; drug loading of hydrophobic and hydrophilic parent antimicrobials; prolonged circulation and tissue retention times; MPS intracellular targeting; microbial membrane fusion, yielding large drug payload; ease of surface functionalization; potential pH-dependent controlled release; potential anti-biofilm activity; reduced toxicity of parent compound; complex and expensive fabrication; short shelf-life	Antivirals, antibacterials, antimycobacterials, antifungals, antiparasitics	Pinto-Alphandary, Andremont, and Couvreur 2000; Drulis-Kawa and Dorotkiewicz-Jach 2010; Walsh et al. 1998; Gelperina et al. 2005; Beaulac et al. 1996; Omri, Suntres, and Shek 2002; Sanderson and Jones 1996
SLNPs	Biocompatible, biodegradable formulations with high safety profile derived from oil/water emulsions; intravenous, topical, inhaled, and excellent topical occlusives; high drug-loading capacity of hydrophobic and hydrophilic parent antimicrobials; prolonged topical, circulation, and tissue retention times; MPS intracellular targeting; reduced toxicity; simple and inexpensive fabrication; long shelf-life	Antibacterials, antimycobacterials, antifungals	Müller, Mäder, and Gohla 2000; Gelperina et al. 2005; Pandey and Khuller 2005; Cavalli et al. 2002, 2003; Souto et al. 2004; Bhalekar et al. 2009
Polymeric NPs[a]	Biocompatible, biodegradable formulations; linear polymers and amphiphilic block copolymers that encapsulate parent antimicrobials; precise tuning of properties during fabrication; intravenous, oral, and inhaled; high drug-loading capacity of hydrophobic and hydrophilic parent antimicrobials; enhanced water solubility; prolonged circulation and tissue retention times; MPS intracellular and microbial targeting; ease of surface functionalization; long shelf-life	Antivirals, antibacterials, antimycobacterials, antifungals, antiparasitics	Soppimath et al. 2001; Cheng et al. 2007; Shah and Amiji 2006; Fattal et al. 1989; Forestier et al. 1992; Balland et al. 1996; Fontana et al. 1998; Turos et al. 2007; Ohashi et al. 2009; Espuelas et al. 2002

Dendrimers	Several formulations of precise globular, highly branched polymers containing a central core and internal voids; biocompatibility being investigated; very high drug-loading capacity of hydrophobic and hydrophilic parent antimicrobials, including silver; enhanced water solubility; ease of surface functionalization; size and surface modification-dependent circulation and tissue retention times; intravenous, inhaled, oral, topical, subcutaneous, intravaginal, rectal, condom coating	Antiviral (STDs), antibacterials, antimycobacterials, antifungals, antiparasitics	Wijagkanalan, Kawakami, and Hashida 2011; Svenson 2009; Patton et al. 2006; Balogh et al. 2001; Cheng et al. 2007; Kumar et al. 2006; Ma et al. 2007; Winnicka et al. 2012; Bhadra, Bhadra and Jain 2005
Chitosan	Biocompatible chitin-derived polysaccharide biopolymer; biodegradable; inherent antimicrobial activity; antimicrobial encapsulation with enhanced activity; prolonged release of parent antimicrobial; prolonged tissue retention time; pH-dependent controlled release; topical, oral; ease of fabrication; long shelf-life	Antibacterials, antimycobacterials, antifungals, silver	Rabea et al. 2003; Jain and Banerjee 2008; Risbud et al. 2000; Pandey et al. 2005; Lucinda-Silva and Evangelista 2003; Potara et al. 2011; Saravanan et al. 2011
CNT	Biocompatibility unknown; nonbiodegradable; inherent antimicrobial activity; ease of functionalization; high drug-loading capacity; intracellular targeting; enhanced amphotericin activity; oral administration in a rodent model; ease of fabrication; long shelf-life	Antifungal, antiparasitic	Klumpp et al. 2006; Wu et al. 2005; Prajapati et al. 2012
Nitric oxide releasing NPs	Broad-spectrum antimicrobial activity; encapsulation by silica NPs, dendrimers, and hydrogel/chitosan composites; sustained NO release; biocompatibility unknown; potential topical, inhalation; targeted drug delivery; anti-biofilm activity	Antivirals, antibacterials, antifungals	Jones et al. 2010; Hetrick et al. 2008; Sun et al. 2012; Friedman et al. 2008

Abbreviations: MPS, macrophage phagocytic system; SLNPs, solid lipid nanoparticles; NPs, nanoparticles; STDs, sexually transmitted diseases; CNT, carbon nanotube; NO, nitric oxide.

[a] A variety of self-assembled copolymers and linear polymers, including poly(D,L-lactide) (PLA); poly lactic-co-glycolic acid (PLGA); Poloxamer 188-coated poly(epsilon-caprolactone) (PCL); polyethylene glycol (PEG)-PLA, polyalkyl acrylates; polymethyl methacrylate; and others.

targeting cells, tissues, microorganisms, and biofilms. Intracellular pathogens that would otherwise escape the effects of conventional antibiotics become targets of the antimicrobials delivered by liposomes into the cells of the MPS. Large payloads of antibiotic can be delivered to microorganisms when liposomes fuse with their cell membranes. The membrane fusion could potentially bypass membrane-bound antimicrobial resistance mechanisms.

Distribution and deposition of liposomes vary on the basis of size, charge, surface modification, and route of administration. It has been suggested that increased vascular permeability resulting from the inflammatory response due to infection increases local liposome accumulation and antimicrobial deposition (Jain 1987). Accumulation of liposomes in lymph nodes after cutaneous and intramuscular administration has been proposed as a possible means of targeting HIV that sequester in these locations (Désormeaux and Bergeron 1998). Gendelman and coworkers have pioneered research that has investigated MPS-mediated delivery of anti-retroviral containing nanoparticles, including liposomes, across the BBB (Nowacek and Gendelman 2009).

Liposomes are biocompatible, biodegradable, and have an established record of safety. When encapsulated by liposomes, the notable toxicities of amphotericin B are significantly reduced (Walsh et al. 1998). Liposomal amphotericin B has been widely used to treat a broad variety of fungal infections and is also being used to treat visceral infections due to the parasite *Leishmania donovani* (Davidson et al. 1991). At the current time, liposomal encapsulation of different classes of antimicrobial agents is being studied for a variety of acute and chronic IDs, including those associated with biofilms (Jones 2005), and hopefully the arduous, expensive, and long preclinical testing phase will yield clinically valuable applications (Pinto-Alphandary, Andremont, and Couvreur 2000). Clinical trials are being conducted on the treatment of HPV infection using multilamellar liposomes that encapsulate the anti-viral cytokine interferon alpha (Foldvari and Moreland 1997). The advantages of this approach to treatment include topical application, reduced systemic side effects, and avoidance of painful injections. Problematic aspects of liposomal therapy include complexity, cost of manufacturing, and relatively short shelf-life. Given their versatility and potentially broad applicability to the treatment of a variety of IDs, it is hoped that these shortcomings can be minimized.

1.14.8 Nanoparticles, Vaccines, and Vaccine Delivery

It has been posited that after the provision of safe water, the most beneficial impact on public health comes from the use of vaccines (Plotkin and Plotkin 2004). The cell biology underlying the deliberate stimulation of an immune response by immunization is complex and not fully elucidated. Equally complex are the interplay between population dynamics, environmental reservoirs of pathogens, and the logistics of disease eradication programs. With regard to these logistics, the ability to get effective vaccines—those that can withstand nonrefrigerated storage and transport and that can provide lasting immunity to populations irrespective of their location—is a public health goal of daunting magnitude. In broad terms, vaccination seeks to provoke durable mucosal, cellular, and humoral immune responses. To do so safely has proven to be one of the most challenging aspects of medicine. Nanotechnology is being applied to vaccinology in an effort to potentiate the immune response to antigens and to provide novel ways of encapsulating and delivering vaccines. This active area of investigation is discussed in several excellent reviews (Singh, Chakrapani, and O'Hagan 2007; Peek, Middaugh, and Berkland 2008; Nandedkar 2009; Look et al. 2010).

Nanotechnology provides an opportunity to tailor vaccine strategies that take into account specific aspects of the host–pathogen interaction, including the initial site of infection (nasal, oral, lung, gastrointestinal, urogenital, and cutaneous) and the mechanisms of immuno-modulation. Nanoparticles that encapsulate and deliver antigens and adjuvants may serve to protect their cargo in *vivo* from proteolytic, mechanical, and chemical damage, thereby enhancing vaccine biodistribution. With respect to the site and mechanism of naturally occurring infections, nanoscale delivery systems can be designed to administer vaccines via several routes (cutaneous, inhaled, oral, parenteral, ocular, vaginal, rectal) that facilitate a robust immune response. Moreover, immune enhancement may be achieved through sustained vaccine release by nanocarrier design (Shahiwala, Vyas, and Amiji 2007; Nandedkar 2009). Surface modification of the nanoparticle can promote specific tissue uptake and facilitate interactions with antigen presenting cells (APCs) (Cui, Hsu, and Mumper 2003; Kwon et al. 2005; des Rieux et al. 2006). Nanoparticle delivery systems may allow prolonged storage of vaccine components at ambient temperatures (Abdelwahed et al. 2006; Makidon et al. 2010). This would provide a means to transport these agents to those parts of the world that are currently unable to receive them due to refrigeration storage requirements.

Encapsulation strategies can combine both antigen and adjuvant in an effort to enhance immune responses. Recently, Kasturi and colleagues (Kasturi et al. 2011) encapsulated two FDA-approved adjuvants along with antigen into a nanoparticle polymer and elicited synergistic increases in levels of neutralizing antibody. This strategy induced complete protection against influenza virus infections in animal models. Important features of their approach are the potential to induce lasting immunity and to stimulate both cellular and humoral arms of the immune response. The investigators postulate that this strategy could produce a robust and universal adjuvant that may potentiate immunity of vaccines against a diverse array of IDs, agents.

Nanoparticles themselves are being explored as adjuvants to potentiate the immune response to antigens (Singh, Chakrapani, and O'Hagan 2007). Immunogenicity studies demonstrate conflicting results that appear to depend on nanoparticle size, surface charge, mode of delivery, and type of antigen being utilized. The characteristics of size and surface properties likely influence the uptake and processing of the particles by mucosal and lymphoid tissues as well as by APCs. Nanoadjuvant delivery systems also include viral-vectors and virosomes that are capable of stimulating both cellular and humoral immunity (Mischler and Metcalfe 2002; Van Kampen et al. 2005). Virosome-based vaccines for influenza virus have been licensed for use in many parts of the world (Peek, Middaugh, and Berkland 2008). A virus-like particle nanovaccine has been FDA approved for immunization against HPV (Foldvari 2011). Liposomes can deliver both antigens and adjuvants to APCs, thereby stimulating both a cellular and humoral immune response. A lipsome-based hepatitis A vaccine has been licensed for use in many parts of the world (Bovier 2008). Other nanoparticle-based vaccines in varying phases of development are presented in detail in the excellent review by Peek, Middaugh, and Berkland (2008).

1.15 Infectious Diseases—Nano Therapeutic and Prevention Strategies: Summary

In an effort to reduce the burden of IDs, advances must be made in the development of novel antimicrobial agents, safe water technology, and durable, safe, and available vaccines.

Ongoing progress in any one of these areas will have an extraordinarily beneficial impact on human health, and applications of nanotechnology to address therapy and prevention of infections represent an emerging and very exciting phase of biomedical research. Microorganisms, both good and bad, have demonstrated impressive ways of causing illness, developing antimicrobial resistance, and evading immune defenses. In tackling IDs, therapeutics and prevention through nanotechnology, we must proceed with caution. Harm to the host, the environment, or the ecology of commensal microorganisms are an unacceptable trade-off. Enormous gains in public health are already being realized, and there is much more to come from this nascent technology.

1.16 Nanotechnology–Infectious Diseases Interface: Conclusions

All difficult things have their origin in that which is easy, and great things in that which is small.

Lao-Tzu (sixth century BCE)

The intent of this chapter was to provide a means for the ID practitioner and the nanoscientist to meet on common grounds and recognize that they have a great deal to discuss. The argument has been posited that the nanotechnology–ID interface is expansive and robust but not yet fully realized by either party. Indeed, one has only to look at the reference list that follows to appreciate that just a small portion of the bibliography comes from "mainstream" clinical ID sources. Most of the references exist in "technology-based" journals that are not likely to be routinely accessed by the average ID practitioner, and that should come as no surprise, given the new and rapidly emerging nature of the technology. The convergence of the disciplines of ID and nanoscience has begun to generate benefits that impact IDs prevention, diagnosis, and therapy. The interface will provide extraordinary opportunities for basic science research, clinical care, and public health. It is hoped that this chapter will, even in a small way, promote a partnership capable of combating the major ID challenges that confront all of us.

Acknowledgments

The author thanks Dr. David Kuehler and Dr. Clemente Montero for their very helpful suggestions in the preparation of this chapter. The author thanks Dr. Chrystal Reed for her assistance in preparing the tables, Elizabeth Sweeney for her graphic arts expertise, and Ann Marie L'Hommedieu for her help with manuscript preparation. A very special acknowledgement to Dr. Ann Carey Tobin for helpful suggestions, a great deal of love and encouragement, and a nudge down the path of nanoscience. This chapter is dedicated to the memory of Dr. John Nicholas Vecchio, forever—young friend, and the finest ID doctor I have ever known.

References

Abdelwahed, W., G. Degobert, S. Stainmesse, and H. Fessi. 2006. Freeze-drying of nanoparticles: formulation, process and storage considerations. *Adv Drug Deliv Rev* 58 (15):1688–1713.

Afessa, B., A.F. Shorr, A.R. Anzueto, D.E. Craven, R. Schinner, and M.H. Kollef. 2010. Association between a silver-coated endotracheal tube and reduced mortality in patients with ventilator-associated pneumonia. *Chest* 137 (5):1015–1021.

Agasti, S.S., S. Rana, M.H. Park, C.K. Kim, C.C. You, and V.M. Rotello. 2010. Nanoparticles for detection and diagnosis. *Adv Drug Deliv Rev* 62 (3):316–328.

Agrawal, A., R.A. Tripp, L.J. Anderson, and S. Nie. 2005. Real-time detection of virus particles and viral protein expression with two-color nanoparticle probes. *J Virol* 79 (13):8625–8628.

Ahamed, M., M.S. Alsalhi, and M.K. Siddiqui. 2010. Silver nanoparticle applications and human health. *Clin Chim Acta* 411 (23–24):1841–1848.

Ai, K., B. Zhang and L. Lu. 2009. Europium-based fluorescence nanoparticle sensor for rapid and ultrasensitive detection of an anthrax biomarker. *Angew Chem Int Ed Engl* 48(2):304–308.

Alexander, J.W. 2009. History of the medical use of silver. *Surg Infect* 10 (3):289–292.

Algar, W.R. and U.J. Krull. 2008. Quantum dots as donors in fluorescence resonance energy transfer for the bioanalysis of nucleic acids, proteins, and other biological molecules. *Anal Bioanal Chem* 391 (5):1609–1618.

Alivisatos, A.P., W. Gu, and C. Larabell. 2005. Quantum dots as cellular probes. *Annu Rev Biomed Eng* 7:55–76.

Allahverdiyev, A.M., E.S. Abamor, M. Bagirova, C.B. Ustundag, C. Kaya, F. Kaya, and M. Rafailovich. 2011. Antileishmanial effect of silver nanoparticles and their enhanced antiparasitic activity under ultraviolet light. *Int J Nanomedicine* 6:2705.

Alterman, M., H. Sjobom, P. Safsten, P.O. Markgren, U.H. Danielson, M. Hamalainen, S. Lofas et al. 2001. P1/P1′ modified HIV protease inhibitors as tools in two new sensitive surface plasmon resonance biosensor screening assays. *Eur J Pharm Sci* 13 (2):203–212.

Alvarez, M., L.G. Carrascosa, K. Zinoviev, J.A. Plaza, and L.M. Lechuga. 2009. Biosensors based on cantilevers. *Methods Mol Biol* 504:51–71.

Ambrose, P.G., S.M. Bhavnani, C.M. Rubino, A. Louie, T. Gumbo, A. Forrest, and G.L. Drusano. 2007. Pharmacokinetics-pharmacodynamics of antimicrobial therapy: it's not just for mice anymore. *Clin Infect Dis* 44(1):79–86.

Archibald, L.K. and L.B. Reller. 2001. Clinical microbiology in developing countries. *Emerg Infect Dis* 7:302–305.

Arias, L.R. and L. Yang. 2009. Inactivation of bacterial pathogens by carbon nanotubes in suspensions. *Langmuir* 25 (5):3003–3012.

Arya, H., Z. Kaul, R. Wadhwa, K. Taira, T. Hirano, and S.C. Kaul. 2005. Quantum dots in bio-imaging: revolution by the small. *Biochem Biophys Res Commun* 329 (4):1173–1177.

Aylward, B., K.A. Hennessey, N. Zagaria, J.M. Olive, and S. Cochi. 2000. When is a disease eradicable? 100 years of lessons learned. *Am J Public Health* 90 (10):1515.

Baddour, L.M., A.E. Epstein, C.C. Erickson, B.P. Knight, M.E. Levison, P.B. Lockhart, F.A. Masoudi, E.J. Okum, W.R. Wilson, and L.B. Beerman. 2010. Update on cardiovascular implantable electronic device infections and their management a scientific statement from the American Heart Association. *Circulation* 121 (3):458–477.

Balaji, D.S., S. Basavaraja, R. Deshpande, D. Mahesh, B.K Prabhakar, and A. Venkataraman. 2009. Extracellular biosynthesis of functionalized silver nanoparticles by strains of Cladosporium cladosporioides fungus. *Colloids Surf B: Biointerfaces* 68 (1):88–92.

Balland, O., H. Pinto-Alphandary, A. Viron, E. Puvion, A. Andremont, and P. Couvreur. 1996. Intracellular distribution of ampicillin in murine macrophages infected with *Salmonella typhimurium* and treated with (3H) ampicillin-loaded nanoparticles. *J Antimicrob Chemother* 37 (1):105–115.

Balogh, L., D.R. Swanson, D.A. Tomalia, G.L. Hagnauer, and A.T. McManus. 2001. Dendrimer–silver complexes and nanocomposites as antimicrobial agents. *Nano Lett* 1 (1):18–21.

Bantz, K.C., A.F. Meyer, N.J. Wittenberg, H. Im, O. Kurtulus, S.H. Lee, N.C. Lindquist, S.H. Oh, and C.L. Haynes. 2011. Recent progress in SERS biosensing. *Phys Chem Chem Phys* 13 (24):11551–11567.

Baptista, P., E. Pereira, P. Eaton, G. Doria, A. Miranda, I. Gomes, P. Quaresma, and R. Franco. 2008. Gold nanoparticles for the development of clinical diagnosis methods. *Anal Bioanal Chem* 391 (3):943–950.

Baptista, P.V., M. Koziol-Montewka, J. Paluch-Oles, G. Doria, and R. Franco. 2006. Gold-nanoparticle-probe–based assay for rapid and direct detection of *Mycobacterium tuberculosis* DNA in clinical samples. *Clin Chem* 52 (7):1433–1434.

Barker, I., J. Brownlie, C. Peckham, J. Pickett, W. Stewart, J. Waage, P. Wilson, and M. Woodhouse. 2006. Foresight: *Infectious Diseases: Preparing for the Future. A Vision of Future Detection, Identification and Monitoring Systems*. London: Office of Science and Innovation.

Barshes, N.R., S.E. Goodpastor, and J.A. Goss. 2004. Pharmacologic immunosuppression. *Front Biosci* 9:411–420.

Bartlett, J.G. and F.G. Hayden. 2004. Influenza A (H5N1): will it be the next pandemic influenza? *Ann Intern Med* 143:460–461.

Beaulac, C., S. Clement-Major, J. Hawari, and J. Lagacé. 1996. Eradication of mucoid *Pseudomonas aeruginosa* with fluid liposome-encapsulated tobramycin in an animal model of chronic pulmonary infection. *Antimicrob Agents Chemother* 40 (3):665–669.

Bedford, E.E., J. Spadavecchia, C.M. Pradier, and F.X. Gu. 2012. Surface plasmon resonance biosensors incorporating gold nanoparticles. *Macromol Biosci* 12:724–739.

Bell, D., C. Wongsrichanalai, and J.W. Barnwell. 2006. Ensuring quality and access for malaria diagnosis: how can it be achieved? *Nat Rev Microbiol* 4 (9 Suppl):S7–S20.

Bentzen, E.L., F. House, T.J. Utley, J.E. Crowe Jr, and D.W. Wright. 2005. Progression of respiratory syncytial virus infection monitored by fluorescent quantum dot probes. *Nano Lett* 5 (4):591–595.

Besteman, K., J.O. Lee, F.G.M. Wiertz, H.A. Heering, and C. Dekker. 2003. Enzyme-coated carbon nanotubes as single-molecule biosensors. *Nano Lett* 3 (6):727–730.

Bhadra, D., S. Bhadra, and N.K Jain. 2005. Pegylated lysine based copolymeric dendritic micelles for solubilization and delivery of artemether. *J Pharm Pharm Sci* 8 (3):467–482.

Bhalekar, M.R., V. Pokharkar, A. Madgulkar, N. Patil, and N. Patil. 2009. Preparation and evaluation of miconazole nitrate-loaded solid lipid nanoparticles for topical delivery. *AAPS PharmSciTech* 10 (1):289–296.

Birla, S.S., V.V. Tiwari, A.K. Gade, A.P. Ingle, A.P. Yadav, and M.K. Rai. 2009. Fabrication of silver nanoparticles by *Phoma glomerata* and its combined effect against *Escherichia coli*, *Pseudomonas aeruginosa* and *Staphylococcus aureus*. *Lett Appl Microbiol* 48 (2):173–179.

Biswas, P., L.N. Cella, S.H. Kang, A. Mulchandani, M.V. Yates, and W. Chen. 2011. A quantum-dot based protein module for in vivo monitoring of protease activity through fluorescence resonance energy transfer. *Chem Commun* 47 (18):5259–5261.

Black, R.E., S.S. Morris, and J. Bryce. 2003. Where and why are 10 million children dying every year? *Lancet* 361 (9376):2226–2234.

Bogner, J.R., U. Kronawitter, B. Rolinski, K. Truebenbach, and F.D. Goebel. 1994. Liposomal doxorubicin in the treatment of advanced AIDS-related Kaposi sarcoma. *J Acquir Immune Defic Syndr* 7 (5):463–468.

Boucher, H.W., G.H. Talbot, J.S. Bradley, J.E. Edwards, D. Gilbert, L.B. Rice, M. Scheld, B. Spellberg, and J. Bartlett. 2009. Bad bugs, no drugs: no ESKAPE! An update from the Infectious Diseases Society of America. *Clin Infect Dis* 48 (1):1–12.

Bovier, P.A. 2008. Epaxal®: a virosomal vaccine to prevent hepatitis A infection. *Expert Rev Vaccines* 7 (8):1141–1150.

Budin, G., H.J. Chung, H. Lee, and R. Weissleder. 2012. A magnetic Gram stain for bacterial detection. *Angew Chem Int Ed Engl* 51 (31):7752–7755.

Buijtels, P., H.F.M. Willemse-Erix, P.L.C. Petit, H.P. Endtz, G.J. Puppels, H.A. Verbrugh, A. van Belkum, D. van Soolingen, and K. Maquelin. 2008. Rapid identification of mycobacteria by Raman spectroscopy. *J Clin Microbiol* 46 (3):961–965.

Bunka, D.H.J. and P.G. Stockley. 2006. Aptamers come of age—at last. *Nat Rev Microbiol* 4 (8):588–596.

Caballero, L., K.A. Whitehead, N.S. Allen, and J. Verran. 2010. Photoinactivation of *Escherichia coli* on acrylic paint formulations using fluorescent light. *Dyes Pigm* 86 (1):56–62.

Campbell, G.A. and R. Mutharasan. 2005. Detection of pathogen *Escherichia coli* O157:H7 using self-excited PZT-glass microcantilevers. *Biosens Bioelectron* 21 (3):462–473.

Campbell, G.A. and R. Mutharasan. 2008. Near real-time detection of *Cryptosporidium parvum* oocyst by IgM-functionalized piezoelectric-excited millimeter-sized cantilever biosensor. *Biosens Bioelectron* 23 (7):1039–1045.

Campbell, G.A., J. Uknalis, S.I. Tu, and R. Mutharasan. 2007. Detection of *Escherichia coli* O157:H7 in ground beef samples using piezoelectric excited millimeter-sized cantilever (PEMC) sensors. *Biosens Bioelectron* 22 (7):1296–1302.

Cao, H. and X. Liu. 2010. Silver nanoparticles-modified films versus biomedical device-associated infections. *Wiley Interdiscip Rev Nanomed Nanobiotechnol* 2 (6):670–684.

Cao, Y.C., R. Jin, and C.A. Mirkin. 2002. Nanoparticles with Raman spectroscopic fingerprints for DNA and RNA detection. *Science* 297 (5586):1536–1540.

Carter, D.J. and R.B. Cary. 2007. Lateral flow microarrays: a novel platform for rapid nucleic acid detection based on miniaturized lateral flow chromatography. *Nucleic Acids Res* 35 (10):1–11.

Cavalli, R., A. Bargoni, V. Podio, E. Muntoni, G.P. Zara, and M.R. Gasco. 2003. Duodenal administration of solid lipid nanoparticles loaded with different percentages of tobramycin. *J Pharm Sci* 92 (5):1085–1094.

Cavalli, R., M.R. Gasco, P. Chetoni, S. Burgalassi, and M.F. Saettone. 2002. Solid lipid nanoparticles (SLN) as ocular delivery system for tobramycin. *Int J Pharm* 238 (1):241–245.

Chalmers, N.I., R.J. Palmer, L. Du-Thumm, R. Sullivan, W. Shi, and P.E. Kolenbrander. 2007. Use of quantum dot luminescent probes to achieve single-cell resolution of human oral bacteria in biofilms. *Appl Environ Microbiol* 73 (2):630–636.

Chan, W.C.W. and S. Nie. 1998. Quantum dot bioconjugates for ultrasensitive nonisotopic detection. *Science* 281 (5385):2016–2018.

Chan, W.C.W., D.J. Maxwell, X. Gao, R.E. Bailey, M. Han, and S. Nie. 2002. Luminescent quantum dots for multiplexed biological detection and imaging. *Curr Opin Biotechnol* 13 (1):40–46.

Chang, M.H., C.J. Chen, M.S. Lai, H.M. Hsu, T.C. Wu, M.S. Kong, D.C. Liang, W.Y. Shau, and D.S. Chen. 1997. Universal hepatitis B vaccination in Taiwan and the incidence of hepatocellular carcinoma in children. *N Eng J Med* 336 (26):1855–1859.

Chao, Y. and T. Zhang. 2012. Surface-enhanced Raman scattering (SERS) revealing chemical variation during biofilm formation: from initial attachment to mature biofilm. *Anal Bioanal Chem* 404 (5):1465–1475.

Chen, L., Z. Sheng, A. Zhang, X. Guo, J. Li, H. Han, and M. Jin. 2010. Quantum-dots-based fluoro-immunoassay for the rapid and sensitive detection of avian influenza virus subtype H5N1. *Luminescence* 25 (6):419–423.

Cheng, H.W., Y.Y. Chen, X.X. Lin, S.Y. Huan, H.L. Wu, G.L. Shen, and R.Q. Yu. 2011. Surface-enhanced Raman spectroscopic detection of *Bacillus subtilis* spores using gold nanoparticle based substrates. *Anal Chim Acta* 707 (1):155–163.

Cheng, J., B.A. Teply, I. Sherifi, J. Sung, G. Luther, F.X. Gu, E. Levy-Nissenbaum, A.F. Radovic-Moreno, R. Langer, and O.C. Farokhzad. 2007a. Formulation of functionalized PLGA–PEG nanoparticles for in vivo targeted drug delivery. *Biomaterials* 28 (5):869–876.

Cheng, X., D. Irimia, M. Dixon, J.C. Ziperstein, U. Demirci, L. Zamir, R.G. Tompkins, M. Toner, and W.R. Rodriguez. 2007b. A microchip approach for practical label-free CD4+ T-cell counting of HIV-infected subjects in resource-poor settings. *J Acquir Immune Defic Syndr* 45 (3):257–261.

Cheng, Y., H. Qu, M. Ma, Z. Xu, P. Xu, Y. Fang, and T. Xu. 2007c. Polyamidoamine (PAMAM) dendrimers as biocompatible carriers of quinolone antimicrobials: an in vitro study. *Eur J Med Chem* 42 (7):1032–1038.

Chi, X., D. Huang, Z. Zhao, Z. Zhou, Z. Yin, and J. Gao. 2011. Nanoprobes for in vitro diagnostics of cancer and infectious diseases. *Biomaterials* 33 (1):189–206.

Chin, C.D., V. Linder, and S.K. Sia. 2006. Lab-on-a-chip devices for global health: past studies and future opportunities. *Lab Chip* 7 (1):41–57.

Choi, Y., J. Lee, K. Kim, H. Kim, P. Sommer, and R. Song. 2010. Fluorogenic assay and live cell imaging of HIV-1 protease activity using acid-stable quantum dot–peptide complex. *Chem Commun* 46 (48):9146–9148.

Chopra, I. 2007. The increasing use of silver-based products as antimicrobial agents: a useful development or a cause for concern? *J Antimicrob Chemother* 59 (4):587–590.

Chua, P.H., K.G. Neoh, E.T. Kang, and W. Wang. 2008. Surface functionalization of titanium with hyaluronic acid/chitosan polyelectrolyte multilayers and RGD for promoting osteoblast functions and inhibiting bacterial adhesion. *Biomaterials* 29 (10):1412–1421.

Chung, H.J., T. Reiner, G. Budin, C. Min, M. Liong, D. Issadore, H. Lee, and R. Weissleder. 2011. Ubiquitous detection of gram-positive bacteria with bioorthogonal magnetofluorescent nanoparticles. *ACS Nano* 5 (11):8834–8841.

Collins, S. 2011. Report from the First Workshop on Nanomedicine for Infectious Diseases of Poverty, at Magaliesberg, South Africa on March 27–31, 2011, http://i-base.info/htb/14934.

Colomban, P.J. 2009. The use of metal nanoparticles to produce yellow, red and iridescent colour from bronz age to present times in lustre pottery and glass: solid state chemistry, spectroscopy and nanostructure. *J Nano Res* 8:109–132.

Colon, G., B.C. Ward, and T.J. Webster. 2006. Increased osteoblast and decreased *Staphylococcus epidermidis* functions on nanophase ZnO and TiO$_2$. *J Biomed Mater Res A* 78 (3):595–604.

Costerton, W., R. Veeh, M. Shirtliff, M. Pasmore, C. Post, and G. Ehrlich. 2003. The application of biofilm science to the study and control of chronic bacterial infections. *J Clin Invest* 112 (10):1466–1477.

Craig, W.A. 1998. Pharmacokinetic/pharmacodynamic parameters: rationale for antibacterial dosing of mice and men. *Clin Infect Dis* 26 (1):1–10.

Cui, Y., Q. Wei, H. Park, and C.M. Lieber. 2001. Nanowire nanosensors for highly sensitive and selective detection of biological and chemical species. *Science* 293 (5533):1289–1292.

Cui, Z., C.H. Hsu, and R.J. Mumper. 2003. Physical characterization and macrophage cell uptake of mannan-coated nanoparticles. *Drug Dev Ind Pharm* 29 (6):689–700.

Cui, Z.Q., Q. Ren, H.P. Wei, Z. Chen, J.Y. Deng, Z.P. Zhang, and X.E. Zhang. 2011. Quantum dot–aptamer nanoprobes for recognizing and labeling influenza A virus particles. *Nanoscale* 3 (6):2454–2457.

D'Antonio, R.G., R.E. Winn, J.P. Taylor, T.L. Gustafson, W.L. Current, M.M. Rhodes, G.W. Gary Jr., and R.A. Zajac. 1985. A waterborne outbreak of cryptosporidiosis in normal hosts. *Ann Intern Med* 103 (6 (Pt 1)):886–888.

Darbha, G.K., U.S. Rai, A.K. Singh, and P.C. Ray. 2008. Gold-nanorod-based sensing of sequence specific HIV-1 virus DNA by using hyper-Rayleigh scattering spectroscopy. *Chemistry* 14 (13):3896–3903.

Davidson, R.N., A. Scott, M. Maini, A.D.M. Bryceson, and S.L. Croft. 1991. Liposomal amphotericin B in drug-resistant visceral leishmaniasis. *Lancet* 337 (8749):1061–1062.

de Martel, C., J. Ferlay, S. Franceschi, J. Vignat, F. Bray, D. Forman, and M. Plummer. 2012. Global burden of cancers attributable to infections in 2008: a review and synthetic analysis. *Lancet Oncol* 13:607–615.

Demas, V. and T.J. Lowery. 2011. Magnetic resonance for in vitro medical diagnostics: superparamagnetic nanoparticle-based magnetic relaxation switches. *New J Phys* 13 025005.

Deng, Z., Y. Zhang, J. Yue, F. Tang, and Q. Wei. 2007. Green and orange CdTe quantum dots as effective pH-sensitive fluorescent probes for dual simultaneous and independent detection of viruses. *J Phys Chem* 111 (41):12024–12031.

Derfus, A.M., W.C.W. Chan, and S.N. Bhatia. 2004. Probing the cytotoxicity of semiconductor quantum dots. *Nano Lett* 4 (1):11–18.

des Rieux, A., V. Fievez, M. Garinot, Y.J. Schneider, and V. Préat. 2006. Nanoparticles as potential oral delivery systems of proteins and vaccines: a mechanistic approach. *J Control Release* 116 (1):1–27.

Désormeaux, A. and M.G. Bergeron. 1998. Liposomes as drug delivery system: a strategic approach for the treatment of HIV infection. *J Drug Target* 6 (1):1–15.

DeWitt, L.M. 1918. The use of gold salts in the treatment of experimental tuberculosis in Guinea-Pigs: XVIII. Studies on the biochemistry and chemotherapy of tuberculosis. *J Infect Dis* 23(5):426–437.

Donlan, R.M. 2011. Biofilm elimination on intravascular catheters: important considerations for the infectious disease practitioner. *Clin Infect Dis* 52 (8):1038–1045.

Donlan, R.M. and J.W. Costerton. 2002. Biofilms: survival mechanisms of clinically relevant microorganisms. *Clin Microbiol Rev* 15 (2):167–193.

Dorman, S.E. 2010. New diagnostic tests for tuberculosis: bench, bedside, and beyond. *Clin Infect Dis* 50 (Suppl 3):S173–S177.

Driskell, J.D., Y. Zhu, C.D. Kirkwood, Y. Zhao, R.A. Dluhy, and R.A. Tripp. 2010. Rapid and sensitive detection of rotavirus molecular signatures using surface enhanced Raman spectroscopy. *PloS one* 5(4):e10222.

Drulis-Kawa, Z. and A. Dorotkiewicz-Jach. 2010. Liposomes as delivery systems for antibiotics. *Int J Pharm* 387 (1):187–198.

Drusano, G.L. 2007. Pharmacokinetics and pharmacodynamics of antimicrobials. *Clin Infect Dis* 45 (Suppl 1):S89–S95.

Dubas, S.T., P. Kumlangdudsana, and P. Potiyaraj. 2006. Layer-by-layer deposition of antimicrobial silver nanoparticles on textile fibers. *Colloids Surf A Physicochem Eng Asp* 289 (1):105–109.

Dutta, R.K., P.K. Sharma, and A.C. Pandey. 2009. Surface enhanced Raman spectra of *Escherichia Coli* cell using ZnO nanoparticles. *Dig J Nanomater Biostruct* 4:83–87.

Dwarakanath, S., J.G. Bruno, A. Shastry, T. Phillips, A. John, A. Kumar, and L.D. Stephenson. 2004. Quantum dot-antibody and aptamer conjugates shift fluorescence upon binding bacteria. *Biochem Biophys Res Commun* 325 (3):739–743.

Dykman, L.A. and N.G. Khlebtsov. 2010. Gold nanoparticles in biology and medicine: recent advances and prospects. *Acta Naturae* 4:2595–2606.

Edgar, R., M. McKinstry, J. Hwang, A.B. Oppenheim, R.A. Fekete, G. Giulian, C. Merril, K. Nagashima, and S. Adhya. 2006. High-sensitivity bacterial detection using biotin-tagged phage and quantum-dot nanocomplexes. *Proc Natl Acad Sci U S A* 103 (13):4841–4845.

Elechiguerra, J.L., J.L. Burt, J.R. Morones, A. Camacho-Bragado, X. Gao, H.H. Lara, and M.J. Yacaman. 2005. Interaction of silver nanoparticles with HIV-1. *J Nanobiotechnology* 3 (6):1–10.

Elkin, T., X. Jiang, S. Taylor, Y. Lin, L. Gu, H. Yang, J. Brown, S. Collins, and Y.P. Sun. 2005. Immuno-carbon nanotubes and recognition of pathogens. *Chembiochem* 6 (4):640–643.

El-Sharif, A.A. and M.H.M. Hussain. 2011. Chitosan–EDTA new combination is a promising candidate for treatment of bacterial and fungal infections. *Curr Microbiol* 62 (3):739–745.

Ermert, V., A.H. Fink, A.P. Morse, and H. Paeth. 2012. The impact of regional climate change on malaria risk due to greenhouse forcing and land-use changes in tropical Africa. *Environ Health Perspect* 120 (1):77.

Ervin, F.R. 1986. The bell tolls for the infectious diseases clinician [with comments]. *J Infect Dis* 153 (2):183–188.

Espuelas, M.S., P. Legrand, P.M. Loiseau, C. Bories, G. Barratt, and J.M. Irache. 2002. In vitro anti-leishmanial activity of amphotericin B loaded in poly(epsilon-caprolactone) nanospheres. *J Drug Target* 10 (8):593–599.

Eustis, S. and M.A. EL-Sayed. 2006. Why gold nanoparticles are more precious than pretty gold: noble metal surface plasmon resonance and its enhancement of the radiative and nonradiative properties of nanocrystals of different shapes. *Chem Soc Rev* 35 (3):209–217.

Falagas, M.E., A.M. Kapaskelis, V.D. Kouranos, O.K. Kakisi, Z. Athanassa, and D.E. Karageorgopoulos. 2009. Outcome of antimicrobial therapy in documented biofilm-associated infections: a review of the available clinical evidence. *Drugs* 69 (10):1351–1361.

Fan, C., Z. Hu, A. Mustapha, and M. Lin. 2011. Rapid detection of food- and waterborne bacteria using surface-enhanced Raman spectroscopy coupled with silver nanosubstrates. *Appl Microbiol Biotechnol* 92 (5):1053–1061.

Fan, C., Z. Hu, L.K. Riley, G.A. Purdy, A. Mustapha, and M. Lin. 2010. Detecting food- and water-borne viruses by surface-enhanced Raman spectroscopy. *J Food Sci* 75 (5):M302–M307.

Faraday, M. 1857. The Bakerian lecture: experimental relations of gold (and other metals) to light. *Phil Trans R Soc Lond* 147:145–181.

Fattal, E., M. Youssef, P. Couvreur, and A. Andremont. 1989. Treatment of experimental salmonellosis in mice with ampicillin-bound nanoparticles. *Antimicrob Agents Chemother* 33 (9):1540–1543.

Fauci, A.S. 2001. Infectious diseases: considerations for the 21st century. *Clin Infect Dis* 32 (5):675–685.

Fauci, A.S. and G.K. Folkers. 2012. Toward an AIDS-free generation. *JAMA* 308 (4):343–344.

Faulk, W.P. and G.M. Taylor. 1971. An immunocolloid method for the electron microscope. *Immunochemistry* 8 (11):1081–1083.

Ferrari, B.C. and P.L. Bergquist. 2007. Quantum dots as alternatives to organic fluorophores for *Cryptosporidium* detection using conventional flow cytometry and specific monoclonal antibodies: lessons learned. *Cytometry A* 71 (4):265–271.

Fleischmann, M., P.J. Hendra, and A.J. McQuillan. 1974. Raman spectra of pyridine adsorbed at a silver electrode. *Chem Phys Lett* 26 (2):163–166.

Foex, B.A. 2003. How the cholera epidemic of 1831 resulted in a new technique for fluid resuscitation. *Emerg Med J* 20 (4):316–318.

Foldvari, M. 2011. HPV infections: can it be eradicated using nanotechnology? *Nanomedicine* 8 (2):131–135.

Foldvari, M. and A. Moreland. 1997. Clinical observations with topical liposome-encapsulated interferon alpha for the treatment of genital papillomavirus infections. *J Liposome Res* 7 (1):115–126.

Fonkwo, P.N. 2008. Pricing infectious disease. *EMBO Rep* 9:S13–S17.

Fontana, G., G. Pitarresi, V. Tomarchio, B. Carlisi, and P.L. San Biagio. 1998. Preparation, characterization and in vitro antimicrobial activity of ampicillin-loaded polyethylcyanoacrylate nanoparticles. *Biomaterials* 19 (11):1009–1017.

Foreman, A., J. Jervis-Bardy, and P.J. Wormald. 2011. Do biofilms contribute to the initiation and recalcitrance of chronic rhinosinusitis? *Laryngoscope* 121 (5):1085–1091.

Forestier, F., P. Gerrier, C. Chaumanrd, A.M. Quero, P. Couvreur, and C. Labarre. 1992. Effect of nanoparticle-bound ampicillin on the survival of Listeria monocytogenes in mouse peritoneal macrophages. *J Antimicrob Chemother* 30 (2):173–179.

Förster, T. 1948. Zwischenmolekulare energiewanderung und fluoreszenz. *Annalen der Physik* 437 (1–2):55–75.

Francolini, I. and G. Donelli. 2010. Prevention and control of biofilm-based medical-device-related infections. *FEMS Immunol Med Microbiol* 59 (3):227–238.

Friedman, A.J., G. Han, M.S. Navati, M. Chacko, L. Gunther, A. Alfieri, and J.M. Friedman. 2008. Sustained release nitric oxide releasing nanoparticles: characterization of a novel delivery platform based on nitrite containing hydrogel/glass composites. *Nitric Oxide* 19 (1):12–20.

Fu, J., J. Ji, D. Fan, and J. Shen. 2006. Construction of antibacterial multilayer films containing nanosilver via layer-by-layer assembly of heparin and chitosan-silver ions complex. *J Biomed Mater Res A* 79 (3):665–674.

Furno, F., K.S. Morley, B. Wong, B.L. Sharp, P.L. Arnold, S.M. Howdle, R. Bayston, P.D. Brown, P.D. Winship, and H.J. Reid. 2004. Silver nanoparticles and polymeric medical devices: a new approach to prevention of infection? *J Antimicrob Chemother* 54 (6):1019–1024.

Gao, J., L. Li, P.L. Ho, G.C. Mak, H. Gu, and B. Xu. 2006. Combining fluorescent probes and biofunctional magnetic nanoparticles for rapid detection of bacteria in human blood. *Adv Mater* 18 (23):3145–3148.

Garcia, M.A. 2011. Surface plasmons in metallic nanoparticles: fundamentals and applications. *J Phys D Appl Phys* 44 (28):1–20.

Gauglitz, G. 2010. Direct optical detection in bioanalysis: an update. *Anal Bioanal Chem* 398 (6):2363–2372.

Gelperina, S., K. Kisich, M.D. Iseman, and L. Heifets. 2005. The potential advantages of nanoparticle drug delivery systems in chemotherapy of tuberculosis. *Am J Respir Crit Care Med* 172 (12):1487–1490.

Gerion, D., F. Chen, B. Kannan, A. Fu, W.J. Parak, D.J. Chen, A. Majumdar, and A.P. Alivisatos. 2003. Room-temperature single-nucleotide polymorphism and multiallele DNA detection using fluorescent nanocrystals and microarrays. *Anal Chem* 75 (18):4766–4772.

Gfeller, K.Y., N. Nugaeva, and M. Hegner. 2005a. Micromechanical oscillators as rapid biosensor for the detection of active growth of *Escherichia coli*. *Biosens Bioelectron* 21 (3):528–533.

Gfeller, K.Y., N. Nugaeva, and M. Hegner. 2005b. Rapid biosensor for detection of antibiotic-selective growth of *Escherichia coli*. *Appl Environ Microbiol* 71 (5):2626–2631.

Gill, R., M. Zayats, and I. Willner. 2008. Semiconductor quantum dots for bioanalysis. *Angew Chem Int Ed* 47 (40):7602–7625.

Gill, V.J, Fedorko, D.P., Witebsky, F.G. 2005. The clinical and microbiological laboratory. In *Principles and Practices of Infectious Diseases*, edited by G.L. Mandell, J.E. Bennett, and R. Dolin. Philadelphia: Elsevier Churchill Livingstone.

Giri, S., E.A. Sykes, T.L. Jennings, and W.C.W. Chan. 2011. Rapid screening of genetic biomarkers of infectious agents using quantum dot barcodes. *ACS Nano* 5 (3):1580–1587.

Goldman, E.R., A.R. Clapp, G.P. Anderson, H.T. Uyeda, J.M. Mauro, I.L. Medintz, and H. Mattoussi. 2004. Multiplexed toxin analysis using four colors of quantum dot fluororeagents. *Anal Chem* 76 (3):684–688.

Gossuin, Y., P. Gillis, A. Hocq, Q.L. Vuong, and A. Roch. 2009. Magnetic resonance relaxation properties of superparamagnetic particles. *Wiley Interdiscip Rev: Nanomed Nanobiotechnol* 1 (3):299–310.

Grace, D., F. Mutua, P. Ochungo, R. Kruska, K. Jones, L. Brierley, L. Lapar et al. 2012. Mapping of poverty and likely zoonoses hotspots. Report to the Department for International Development. Nairobi, Kenya: ILRI: 1–119. http://hdl.handle.net/10568/21161.

Griffin, J., A.K. Singh, D. Senapati, E. Lee, K. Gaylor, J. Jones-Boone, and P.C. Ray. 2009. Sequence-specific HCV RNA quantification using the size-dependent nonlinear optical properties of gold nanoparticles. *Small* 5 (7):839–845.

Grossman, H.L., W.R. Myers, V.J. Vreeland, R. Bruehl, M.D. Alper, C.R. Bertozzi, and J. Clarke. 2004. Detection of bacteria in suspension by using a superconducting quantum interference device. *Proc Natl Acad Sci U S A* 101 (1):129–134.

Gu, H., P.L. Ho, K.W.T. Tsang, L. Wang, and B. Xu. 2003. Using biofunctional magnetic nanoparticles to capture vancomycin-resistant enterococci and other gram-positive bacteria at ultralow concentration. *J Am Chem Soc* 125 (51):15702–15703.

Gu, L., T. Elkin, X. Jiang, H. Li, Y. Lin, L. Qu, T.R. Tzeng, R. Joseph, and Y.P. Sun. 2005. Single-walled carbon nanotubes displaying multivalent ligands for capturing pathogens. *Chem Commun (Camb)* (7):874–876.

Gyssens, I.C. 2011. Antibiotic policy. *Int J Antimicrob Agents* 38 (Suppl):11–20.

Haes, A.J. and R.P. Van Duyne. 2004. A unified view of propagating and localized surface plasmon resonance biosensors. *Anal Bioanal Chem* 379 (7):920–930.

Hagens, W.I., A.G. Oomen, W.H. de Jong, F.R. Cassee, and A.J.A.M. Sips. 2007. What do we (need to) know about the kinetic properties of nanoparticles in the body? *Regul Toxicol Pharmacol* 49 (3):217–229.

Halfpenny, K.C. and D.W. Wright. 2010. Nanoparticle detection of respiratory infection. *Wiley Interdiscip Rev: Nanomed Nanobiotechnol* 2 (3):277–290.

Hall, C.B. 2004. Respiratory syncytial virus. In *Principles and Practice of Clinical Virology, Fifth Edition*, edited by A.J. Zuckerman, J.E. Banatvala, J.R. Pattison, P.D. Griffiths, and B.D. Shoub. Chichester, England: John Wiley & Sons.

Han, M., X. Gao, J.Z. Su, and S. Nie. 2001. Quantum-dot-tagged microbeads for multiplexed optical coding of biomolecules. *Nat Biotechnol* 19 (7):631–635.

Han, X.X., B. Zhao, and Y. Ozaki. 2009. Surface-enhanced Raman scattering for protein detection. *Anal Bioanal Chem* 394 (7):1719–1727.

Hannu, T. 2011. Reactive arthritis. *Best Pract Res Clin Rheumatol* 25 (3):347–357.

Hansen, K.M. and T. Thundat. 2005. Microcantilever biosensors. *Methods* 37 (1):57–64.

Harma, H., T. Soukka, and T. Lovgren. 2001. Europium nanoparticles and time-resolved fluorescence for ultrasensitive detection of prostate-specific antigen. *Clin Chem* 47 (3):561–568.

Hauck, T.S., S. Giri, Y. Gao, and W.C.W. Chan. 2010. Nanotechnology diagnostics for infectious diseases prevalent in developing countries. *Adv Drug Deliv Rev* 62 (4):438–448.

Haun, J.B., T.J. Yoon, H. Lee, and R. Weissleder. 2010. Magnetic nanoparticle biosensors. *Wiley Interdiscip Rev: Nanomed Nanobiotechnol* 2 (3):291–304.

He, P. and L. Dai. 2004. Aligned carbon nanotube-DNA electrochemical sensors. *Chem Commun (Camb)* (3):348–349.

Hetrick, E.M., J.H. Shin, N.A. Stasko, C.B. Johnson, D.A. Wespe, E. Holmuhamedov, and M.H. Schoenfisch. 2008. Bactericidal efficacy of nitric oxide-releasing silica nanoparticles. *ACS Nano* 2 (2):235–246.

Hewlett, E.L. and M.A. Hughes. 2009. Toxins. In *Principles and Practice of Infectious Diseases*, edited by G.L. Mandell, J.E. Bennett, and R. Dolin. Philadelphia, PA: Elsevier Churchill Livingstone.

Hill, H.D. and C.A. Mirkin. 2006. The bio-barcode assay for the detection of protein and nucleic acid targets using DTT-induced ligand exchange. *Nat Protoc* 1 (1):324–336.

Ho, H.P. and S.Y. Wu. 2012. Biomedical sensing using surface plasmon resonance. In *Nanotechnology in Biology and Medicine*, edited by T. Vo-Dinh. New York: CRC Press.

Ho, K.C., P.J. Tsai, Y.S. Lin, and Y.C. Chen. 2004. Using biofunctionalized nanoparticles to probe pathogenic bacteria. *Anal Chem* 76 (24):7162–7168.

Hoa, M., M. Syamal, M.A. Schaeffer, L. Sachdeva, R. Berk, and J. Coticchia. 2010. Biofilms and chronic otitis media: an initial exploration into the role of biofilms in the pathogenesis of chronic otitis media. *Am J Otolaryngol* 31 (4):241–245.

Hoa, X.D., A.G. Kirk, and M. Tabrizian. 2007. Towards integrated and sensitive surface plasmon resonance biosensors: a review of recent progress. *Biosens Bioelectron* 23 (2):151–160.

Hodgkiss-Harlow, K.D. and D.F. Bandyk. 2011. Antibiotic Therapy of aortic graft infection: treatment and prevention recommendations. *Semin Vasc Surg* 24 (4):191–198.

Hoiby, N., T. Bjarnsholt, M. Givskov, S. Molin, and O. Ciofu. 2010. Antibiotic resistance of bacterial biofilms. *Int J Antimicrob Agents* 35 (4):322–332.

Homola, J. 2008. Surface plasmon resonance sensors for detection of chemical and biological species. *Chem Rev* 108 (2):462–493.

Hooton, T.M., S.F. Bradley, D.D. Cardenas, R. Colgan, S.E. Geerlings, J.C. Rice, S. Saint, A.J. Schaeffer, P.A. Tambayh, and P. Tenke. 2010. Diagnosis, prevention, and treatment of catheter-associated urinary tract infection in adults: 2009 International Clinical Practice Guidelines from the Infectious Diseases Society of America. *Clin Infect Dis* 50 (5):625–663.

Horn, D.L., D. Neofytos, E.J. Anaissie, J.A. Fishman, W.J. Steinbach, A.J. Olyaei, K. A. Marr, M.A. Pfaller, C.H. Chang, and K.M. Webster. 2009. Epidemiology and outcomes of candidemia in 2019 patients: data from the prospective antifungal therapy alliance registry. *Clin Infect Dis* 48 (12):1695–1703.

Hoshino, A., K. Fujioka, N. Manabe, S. Yamaya, Y. Goto, M. Yasuhara, and K. Yamamoto. 2005. Simultaneous multicolor detection system of the single-molecular microbial antigen with total internal reflection fluorescence microscopy. *Microbiol Immunol* 49 (5):461–470.

Hossain, M.K., Y. Kitahama, G.G. Huang, X. Han, and Y. Ozaki. 2009. Surface-enhanced Raman scattering: realization of localized surface plasmon resonance using unique substrates and methods. *Anal Bioanal Chem* 394 (7):1747–1760.

Hotez, P.J., A. Fenwick, L. Savioli, and D.H. Molyneux. 2009. Rescuing the bottom billion through control of neglected tropical diseases. *Lancet* 373 (9674):1570–1575.

Huang, L., D.Q. Li, Y.J. Lin, M. Wei, D.G. Evans, and X. Duan. 2005. Controllable preparation of Nano-MgO and investigation of its bactericidal properties. *J Inorg Biochem* 99 (5):986–993.

Huang, L., M. Terakawa, T. Zhiyentayev, Y.Y. Huang, Y. Sawayama, A. Jahnke, G.P. Tegos, T. Wharton, and M.R. Hamblin. 2010a. Innovative cationic fullerenes as broad-spectrum light-activated antimicrobials. *Nanomedicine* 6 (3):442–452.

Huang, W.E., M. Li, R.M. Jarvis, R. Goodacre, and S.A. Banwart. 2010b. Shining light on the microbial world the application of Raman microspectroscopy. *Adv Appl Microbiol* 70:153–186.

Huh, A.J. and Y.J. Kwon. 2011. "Nanoantibiotics": a new paradigm for treating infectious diseases using nanomaterials in the antibiotics resistant era. *J Control Release* 156 (2):128–145.

Hun, X. and Z. Zhang. 2007. A novel sensitive staphylococcal enterotoxin C 1 fluoroimmunoassay based on functionalized fluorescent core-shell nanoparticle labels. *Food Chem* 105 (4):1623–1629.

Hunt, P. 2006. The human right to the highest attainable standard of health: new opportunities and challenges. *Trans R Soc Trop Med Hyg* 100:603–607.

Hutter, E. and J.H. Fendler. 2004. Exploitation of localized surface plasmon resonance. *Adv Mater* 16 (19):1685–1706.

Hwang, I.S., J. Lee, J.H. Hwang, K.J. Kim, and D.G. Lee. 2012. Silver nanoparticles induce apoptotic cell death in *Candida albicans* through the increase of hydroxyl radicals. *FEBS J* 279 (7):1327–1338.

Infectious Diseases Society of America. 2011. Combating antimicrobial resistance: policy recommendations to save lives. *Clin Infect Dis* 52 (Suppl 5):S397–S428.

Ingle, A., A. Gade, S. Pierrat, C. Sonnichsen, and M. Rai. 2008. Mycosynthesis of silver nanoparticles using the fungus *Fusarium acuminatum* and its activity against some human pathogenic bacteria. *Curr Nanosci* 4 (2):141–144.

Inglesby, T.V., T. O'Toole, D.A. Henderson, J.G. Bartlett, M.S. Ascher, E. Eitzen, A.M. Friedlander et al. 2002. Anthrax as a biological weapon, 2002: updated recommendations for management. *JAMA* 287 (17):2236–2252.

Isola, N.R., D.L. Stokes, and T. Vo-Dinh. 1998. Surface-enhanced Raman gene probe for HIV detection. *Anal Chem* 70 (7):1352–1356.

Issadore, D., C. Min, M. Liong, J. Chung, R. Weissleder, and H. Lee. 2011. Miniature magnetic resonance system for point-of-care diagnostics. *Lab Chip* 11 (13):2282–2287.

Ivleva, N.P., M. Wagner, H. Horn, R. Niessner, and C. Haisch. 2008. In situ surface-enhanced Raman scattering analysis of biofilm. *Anal Chem* 80 (22):8538–8544.

Jabes, D. 2011. The antibiotic R&D pipeline: an update. *Curr Opin Microbiol* 14 (5):564–569.

Jain, D. and R. Banerjee. 2008. Comparison of ciprofloxacin hydrochloride-loaded protein, lipid, and chitosan nanoparticles for drug delivery. *J Biomed Mater Res B: Appl Biomater* 86 (1):105–112.

Jain, K.K. 2003. Nanodiagnostics: application of nanotechnology in molecular diagnostics. *Expert Rev Mol Diagn* 3 (2):153–161.

Jain, K.K. 2005. Nanotechnology in clinical laboratory diagnostics. *Clinica Chimica Acta* 358 (1):37–54.

Jain, K.K. 2007. Applications of nanobiotechnology in clinical diagnostics. *Clin Chem* 53 (11):2002–2009.

Jain, R.K. 1987. Transport of molecules in the tumor interstitium: a review. *Cancer Res* 47 (12):3039–3051.

Jaiswal, J.K. and S.M. Simon. 2004. Potentials and pitfalls of fluorescent quantum dots for biological imaging. *Trends Cell Biol* 14 (9):497.

Jiang, W., H. Mashayekhi, and B. Xing. 2009. Bacterial toxicity comparison between nano-and microscaled oxide particles. *Environ Pollut* 157 (5):1619–1625.

Johnson, B.N. and R. Mutharasan. 2012. Biosensing using dynamic-mode cantilever sensors: a review. *Biosens Bioelectron* 32:1–18.

Johnston, H.J., G. Hutchison, F.M. Christensen, S. Peters, S. Hankin, and V. Stone. 2010. A review of the in vivo and in vitro toxicity of silver and gold particulates: particle attributes and biological mechanisms responsible for the observed toxicity. *Crit Rev Toxicol* 40 (4):328–346.

Jokerst, J.V., P.N. Floriano, N. Christodoulides, G.W. Simmons, and J.T. McDevitt. 2008. Integration of semiconductor quantum dots into nano-bio-chip systems for enumeration of CD4+ T cell counts at the point-of-need. *Lab Chip* 8 (12):2079–2090.

Jones, G., R.W. Steketee, R.E. Black, Z.A. Bhutta, and S.S. Morris. 2003. How many child deaths can we prevent this year? *Lancet* 362 (9377):65–71.

Jones, K.E., N.G. Patel, M.A. Levy, A. Storeygard, D. Balk, J.L. Gittleman, and P. Daszak. 2008. Global trends in emerging infectious diseases. *Nature* 451 (7181):990–993.

Jones, M.L., J.G. Ganopolsky, A. Labbé, C. Wahl, and S. Prakash. 2010. Antimicrobial properties of nitric oxide and its application in antimicrobial formulations and medical devices. *Appl Microbiol Biotechnol* 88 (2):401–407.

Jones, M.N. 2005. Use of liposomes to deliver bactericides to bacterial biofilms. *Methods Enzymol* 391:211–228.

Jones, N., B. Ray, K.T. Ranjit, and A.C. Manna. 2007. Antibacterial activity of ZnO nanoparticle suspensions on a broad spectrum of microorganisms. *FEMS Microbiol Lett* 279 (1):71–76.

Joo, K.I., Y. Lei, C.L. Lee, J. Lo, J. Xie, S.F. Hamm-Alvarez, and P. Wang. 2008. Site-specific labeling of enveloped viruses with quantum dots for single virus tracking. *ACS Nano* 2 (8):1553–1562.

Juma, C. and L Yee-Cheong. 2005. UN Millenium Project 2005, Innovtion: Applying Knowledge in Development, Task Force on Science, Technology, and Innovation: CTA. http://www.unmillenniumproject.org/documents/Science-complete.pdf.

Kaittanis, C., H. Boukhriss, S. Santra, S.A. Naser, and J.M. Perez. 2012. Rapid and sensitive detection of an intracellular pathogen in human peripheral leukocytes with hybridizing magnetic relaxation nanosensors. *PloS One* 7(4): e35326.

Kaittanis, C., S. Santra, and J.M. Perez. 2010. Emerging nanotechnology-based strategies for the identification of microbial pathogenesis. *Adv Drug Deliv Rev* 62 (4):408–423.

Kaittanis, C., S. Santra, O.J. Santiesteban, T.J. Henderson, and J.M. Perez. 2011. The assembly state between magnetic nanosensors and their targets orchestrates their magnetic relaxation response. *J Am Chem Soc* 133 (10):3668–3676.

Kampani, K., K. Quann, J. Ahuja, B. Wigdahl, Z.K. Khan, and P. Jain. 2007. A novel high throughput quantum dot-based fluorescence assay for quantitation of virus binding and attachment. *J Virol Methods* 141 (2):125–132.

Kang, S., M. Pinault, L.D. Pfefferle, and M. Elimelech. 2007. Single-walled carbon nanotubes exhibit strong antimicrobial activity. *Langmuir* 23 (17):8670–8673.

Kass, E.H. 1987. History of the specialty of infectious diseases in the United States. *Ann Intern Med* 106 (5):745–756.

Kasturi, S.P., I. Skountzou, R.A. Albrecht, D. Koutsonanos, T. Hua, H.I. Nakaya, R. Ravindran, S. Stewart, M. Alam, and M. Kwissa. 2011. Programming the magnitude and persistence of antibody responses with innate immunity. *Nature* 470 (7335):543–547.

Kattke, M.D., E.J. Gao, K.E. Sapsford, L.D. Stephenson, and A. Kumar. 2011. FRET-based quantum dot immunoassay for rapid and sensitive detection of *Aspergillus amstelodami*. *Sensors (Basel)* 11 (6):6396–6410.

Keegan, R. and J. Bilous. 2004. Current issues in global immunizations. *Semin Pediatr Infect Dis* 15 (3):130–136.

Kell, A.J., G. Stewart, S. Ryan, R. Peytavi, M. Boissinot, A. Huletsky, M. G. Bergeron, and B. Simard. 2008. Vancomycin-modified nanoparticles for efficient targeting and preconcentration of Gram-positive and Gram-negative bacteria. *ACS Nano* 2 (9):1777–1788.

Kiechle, F.L. and C.A. Holland. 2009. Point-of-care testing and molecular diagnostics: miniaturization required. *Clin Lab Med* 29 (3):555–560.

Kim, B.Y.S., J.T. Rutka, and W.C.W. Chan. 2010. Nanomedicine. *N Engl J Med* 363:2434–2443.

Kim, J.W., E.V. Shashkov, E.I. Galanzha, N. Kotagiri, and V.P. Zharov. 2007. Photothermal antimicrobial nanotherapy and nanodiagnostics with self-assembling carbon nanotube clusters. *Lasers Surg Med* 39 (7):622–634.

Kim, Y.G., S. Moon, D.R. Kuritzkes, and U. Demirci. 2009. Quantum dot-based HIV capture and imaging in a microfluidic channel. *Biosens Bioelectron* 25 (1):253–258.

Klasen, H.J. 2000a. A historical review of the use of silver in the treatment of burns. II. Renewed interest for silver. *Burns* 26 (2):131–138.

Klasen, H.J. 2000b. Historical review of the use of silver in the treatment of burns. I. Early uses. *Burns* 26 (2):117–130.

Klevens, R.M., J.R. Edwards, C.L. Richards, Jr., T.C. Horan, R.P. Gaynes, D.A. Pollock, and D.M. Cardo. 2007. Estimating health care-associated infections and deaths in U.S. hospitals, 2002. *Public Health Rep* 122 (2):160–166.

Klippstein, R., R. Fernandez-Montesinos, P.M. Castillo, A.P. Zaderenko, and D. Pozo. 2010. Silver nanoparticles interactions with the immune system: implications for health and disease. In *Silver Nanoparticles*, edited by D.P. Perez. Rijeka, Croatia: In Tech Europe.

Kloepfer, J.A., R.E. Mielke, M.S. Wong, K.H. Nealson, G. Stucky, and J.L. Nadeau. 2003. Quantum dots as strain-and metabolism-specific microbiological labels. *Appl Environ Microbiol* 69 (7):4205–4213.

Klostranec, J.M. and W.C.W. Chan. 2006. Quantum dots in biological and biomedical research: recent progress and present challenges. *Adv Mater* 18 (15):1953–1964.

Klostranec, J.M., Q. Xiang, G.A. Farcas, J.A. Lee, A. Rhee, E.I. Lafferty, S.D. Perrault, K.C. Kain, and W.C. Chan. 2007. Convergence of quantum dot barcodes with microfluidics and signal processing for multiplexed high-throughput infectious disease diagnostics. *Nano Lett* 7 (9):2812–2818.

Klumpp, C., K. Kostarelos, M. Prato, and A. Bianco. 2006. Functionalized carbon nanotubes as emerging nanovectors for the delivery of therapeutics. *Biochim Biophys Acta (BBA)-Biomembranes* 1758 (3):404–412.

Kneipp, K. and H. Kneipp. 2006. Single molecule Raman scattering. *Appl Spectrosc* 60 (12):322A–334A.

Knobler, S.L., T. Burroughs, A. Mahmoud, and S.M. Lemon. 2006. *Forum on Microbial Threats, Ensuring an Infectious Disease Workforce: Education and Training Needs for the 21st Century—Workshop Summary*, edited by S.L. Knobler, T. Burroughs, A. Mahmoud, and S.M. Lemon. Washington, D.C.: The National Academies Press. http://www.nap.edu/openbook.php?record_id=11563.

Knopp, D., D. Tang, and R. Niessner. 2009. Review: bioanalytical applications of biomolecule-functionalized nanometer-sized doped silica particles. *Anal Chim Acta* 647 (1):14–30.

Koh, D.B.C., I.K. Robertson, M. Watts, and A.N. Davies. 2012. Density of microbial colonization on external and internal surfaces of concurrently placed intravascular devices. *Am J Crit Care* 21 (3):162–171.

Kopera, D. 2011. Botulinum toxin historical aspects: from food poisoning to pharmaceutical. *Int J Dermatol* 50 (8):976–980.

Kroeger, A., M. Nathan, and J. Hombach. 2004. Dengue. *Nat Rev Microbiol* 2 (5):360–361.

Ku, M.J., F.M. Dossin, Y. Choi, C.B. Moraes, J. Ryu, R. Song, and L.H. Freitas-Junior. 2011. Quantum dots: a new tool for anti-malarial drug assays. *Malar J* 10:118.

Kumar, A., P.K. Vemula, P.M. Ajayan, and G. John. 2008. Silver-nanoparticle-embedded antimicrobial paints based on vegetable oil. *Nat Mater* 7 (3):236–241.

Kumar, P.V., A. Asthana, T. Dutta, and N.K. Jain. 2006. Intracellular macrophage uptake of rifampicin loaded mannosylated dendrimers. *J Drug Target* 14 (8):546–556.

Kumar, S.A., A.A. Ansary, A. Ahmad, and M.I. Khan. 2007. Extracellular biosynthesis of CdSe quantum dots by the fungus, *Fusarium oxysporum*. *J Biomed Nanotechnol* 3 (2):190–194.

Kumbhat, S., K. Sharma, R. Gehlot, A. Solanki, and V. Joshi. 2010. Surface plasmon resonance based immunosensor for serological diagnosis of dengue virus infection. *J Pharm Biomed Anal* 52 (2):255–259.

Kunin, C.M. 1993. Resistance to antimicrobial drugs—a worldwide calamity. *Ann Intern Med* 118 (7):557.

Kwon, Y.J., E. James, N. Shastri, and J.M.J. Fréchet. 2005. In vivo targeting of dendritic cells for activation of cellular immunity using vaccine carriers based on pH-responsive microparticles. *Proc Natl Acad Sci U S A* 102 (51):18264–18268.

Laderman, E.I., E. Whitworth, E. Dumaual, M. Jones, A. Hudak, W. Hogrefe, J. Carney, and J. Groen. 2008. Rapid, sensitive, and specific lateral-flow immunochromatographic point-of-care device for detection of herpes simplex virus type 2-specific immunoglobulin G antibodies in serum and whole blood. *Clin Vaccine Immunol* 15 (1):159–163.

Lambert, P.A. 2004. Mechanisms of action of biocides. In *Principles and Practice of Disinfection, Preservation and Sterilization*, edited by A.P. Fraise, P.A. Lambert, and J-Y. Maillard. Boston, MA: Blackwell Publishing Ltd.

Landsberg, G. and L. Mandelstam. 1928. A novel effect of light scattering in crystals. *Naturwissenschaften* 16:557.

Lee, C.N., Y.M. Wang, W.F. Lai, T.J. Chen, M.C. Yu, C.L. Fang, F.L. Yu et al. 2012. Super-paramagnetic iron oxide nanoparticles for use in extrapulmonary tuberculosis diagnosis. *Clin Microbiol Infect* 18 (6):E149–E157.

Lee, H., E. Sun, D. Ham, and R. Weissleder. 2008. Chip–NMR biosensor for detection and molecular analysis of cells. *Nat Med* 14 (8):869–874.

Lee, H., T.J. Yoon, and R. Weissleder. 2009. Ultrasensitive detection of bacteria using core-shell nanoparticles and an NMR-filter system. *Angew Chem Int Ed Engl* 48 (31):5657–5660.

Lee, J.A., A. Hung, S. Mardyani, A. Rhee, J. Klostranec, Y. Mu, D. Li, and W.C.W. Chan. 2007. Toward the accurate read-out of quantum dot barcodes: design of deconvolution algorithms and assessment of fluorescence signals in buffer. *Adv Mater* 19 (20):3113–3118.

Lee, L.Y., S.L. Ong, J.Y. Hu, W.J. Ng, Y. Feng, X. Tan, and S.W. Wong. 2004. Use of semiconductor quantum dots for photostable immunofluorescence labeling of *Cryptosporidium parvum*. *Appl Environ Microbiol* 70 (10):5732–5736.

Lee, W.G., Y.G. Kim, B.G. Chung, U. Demirci, and A. Khademhosseini. 2010. Nano/microfluidics for diagnosis of infectious diseases in developing countries. *Adv Drug Deliv Rev* 62 (4):449–457.

Lem, K., A. Choudhury, A.A. Lakhani, P. Kuyate, J.R. Haw, D.S. Lee, Z. Iqbal, and C.J. Brumlik. 2012. Use of nanosilver in consumer products. *Recent Pat Nanotechnol* 6 (1):60–72.

Li, H. and L. Rothberg. 2004. Colorimetric detection of DNA sequences based on electrostatic interactions with unmodified gold nanoparticles. *Proc Natl Acad Sci U S A* 101 (39):14036–14039.

Li, Q., S. Mahendra, D.Y. Lyon, L. Brunet, M.V. Liga, D. Li, and P.J. Alvarez. 2008. Antimicrobial nanomaterials for water disinfection and microbial control: potential applications and implications. *Water Res* 42 (18):4591–4602.

Lian, T. and R.J.Y. Ho. 2001. Trends and developments in liposome drug delivery systems. *J Pharm Sci* 90 (6):667–680.

Liang, A., Q. Liu, G. Wen, and Z. Jiang. 2012. The surface-plasmon-resonance effect of nanogold/silver and its analytical applications. *TrAC Trends Analyt Chem* 37:32–47.

Lichter, J.A., K.J. Van Vliet, and M.F. Rubner. 2009. Design of antibacterial surfaces and interfaces: polyelectrolyte multilayers as a multifunctional platform. *Macromolecules* 42 (22):8573–8586.

Liu, J.C., P.J. Tsai, Y.C. Lee, and Y.C. Chen. 2008. Affinity capture of uropathogenic *Escherichia coli* using pigeon ovalbumin-bound $Fe_3O_4@Al_2O_3$ magnetic nanoparticles. *Anal Chem* 80 (14):5425–5432.

Liu, S., L. Wei, L. Hao, N. Fang, M.W. Chang, R. Xu, Y. Yang, and Y. Chen. 2009a. Sharper and faster "nano darts" kill more bacteria: a study of antibacterial activity of individually dispersed pristine single-walled carbon nanotube. *ACS Nano* 3 (12):3891–3902.

Liu, T.Y., K.T. Tsai, H.H. Wang, Y. Chen, Y.H. Chen, Y.C. Chao, H.H. Chang, C.H. Lin, J.K. Wang, and Y.L. Wang. 2011. Functionalized arrays of Raman-enhancing nanoparticles for capture and culture-free analysis of bacteria in human blood. *Nat Commun* 2:538.

Liu, Y., L. He, A. Mustapha, H. Li, ZQ Hu, and M. Lin. 2009b. Antibacterial activities of zinc oxide nanoparticles against *Escherichia coli* O157: H7. *J Appl Microbiol* 107 (4):1193–1201.

Liu, Y., Y.R. Chen, X. Nou, and K. Chao. 2007. Potential of surface-enhanced Raman spectroscopy for the rapid identification of *Escherichia coli* and *Listeria monocytogenes* cultures on silver colloidal nanoparticles. *Appl Spectrosc* 61 (8):824–831.

Look, M., A. Bandyopadhyay, J.S. Blum, and T.M. Fahmy. 2010. Application of nanotechnologies for improved immune response against infectious diseases in the developing world. *Adv Drug Deliv Rev* 62 (4–5):378–393.

Lopez-Berestein, G., G.P. Bodey, V. Fainstein, M. Keating, L.S. Frankel, B. Zeluff, L. Gentry, and K. Mehta. 1989. Treatment of systemic fungal infections with liposomal amphotericin B. *Arch Intern Med* 149 (11):2533–2536.

Lovley, D.R. 2012. Electromicrobiology. *Annu Rev Microbiol* 66:391–409.

Lucinda-Silva, R.M. and R.C. Evangelista. 2003. Microspheres of alginate-chitosan containing isoniazid. *J Microencapsul* 20 (2):145–152.

Ly, K.N., J. Xing, R.M. Klevens, R.B. Jiles, J.W. Ward, and S.D. Holmberg. 2012. The increasing burden of mortality from viral hepatitis in the United States between 1999 and 2007. *Ann Intern Med* 156 (4):271–278.

Ma, M., Y. Cheng, Z. Xu, P. Xu, H. Qu, Y. Fang, T. Xu, and L. Wen. 2007. Evaluation of polyamidoamine (PAMAM) dendrimers as drug carriers of anti-bacterial drugs using sulfamethoxazole (SMZ) as a model drug. *Eur J Med Chem* 42 (1):93–98.

Mabey, D., R.W. Peeling, A. Ustianowski, and M.D. Perkins. 2004. Tropical infectious diseases: diagnostics for the developing world. *Nat Rev Microbiol* 2 (3):231–240.

MacKenzie, W.R., W.L. Schell, K.A. Blair, D.G. Addiss, D.E. Peterson, N.J. Hoxie, J.J. Kazmierczak, and J.P. Davis. 1995. Massive outbreak of waterborne *Cryptosporidium* infection in Milwaukee, Wisconsin: recurrence of illness and risk of secondary transmission. *Clin Infect Dis* 21 (1):57–62.

Maisch, T. 2009. A new strategy to destroy antibiotic resistant microorganisms: antimicrobial photodynamic treatment. *Mini Rev Med Chem* 9 (8):974–983.

Makidon, P.E., S.S. Nigavekar, A.U. Bielinska, N. Mank, A.M. Shetty, J. Suman, J. Knowlton, A. Myc, T. Rook, and J.R. Baker Jr. 2010. Characterization of stability and nasal delivery systems for immunization with nanoemulsion-based vaccines. *J Aerosol Med Pulm Drug Deliv* 23 (2):77–89.

Malvankar, N.S. and D.R. Lovley. 2012. Microbial nanowires: a new paradigm for biological electron transfer and bioelectronics. *ChemSusChem* 5 (6):1039–1046.

Mamouni, J., Y. Tang, M. Wu, B. Vlahovic, and L. Yang. 2011. Single-walled carbon nanotubes coupled with near-infrared laser for inactivation of bacterial cells. *J Nanosci Nanotechnol* 11 (6):4708–4716.

Mao, C., C.E. Flynn, A. Hayhurst, R. Sweeney, J. Qi, G. Georgiou, B. Iverson, and A.M. Belcher. 2003. Viral assembly of oriented quantum dot nanowires. *Proc Natl Acad Sci U S A* 100 (12):6946–6951.

Maquelin, K., C. Kirschner, L.P. Choo-Smith, N.A. Ngo-Thi, T. van Vreeswijk, M. Stammler, H.P. Endtz, H.A. Bruining, D. Naumann, and G.J. Puppels. 2003. Prospective study of the performance of vibrational spectroscopies for rapid identification of bacterial and fungal pathogens recovered from blood cultures. *J Clin Microbiol* 41 (1):324–329.

Maquelin, K., L.P. Choo-Smith, H.P. Endtz, H.A. Bruining, and G.J. Puppels. 2002. Rapid identification of Candida species by confocal Raman microspectroscopy. *J Clin Microbiol* 40 (2):594–600.

McCarthy, T.J., J.J. Zeelie, and D.J. Krause. 1992. The antimicrobial action of zinc ion/antioxidant combinations. *J Clin Pharm Ther* 17 (1):51–54.

McNay, G., D. Eustace, W.E. Smith, K. Faulds, and D. Graham. 2011. Surface-enhanced Raman scattering (SERS) and surface-enhanced resonance Raman scattering (SERRS): a review of applications. *Appl Spectrosc* 65 (8):825–837.

Mie, G. 1908. Contributions to the optics of turbid media particularly of colloidal metal solutions. *Ann Phys* 25:377–445.

Millar, M. 2012. Constraining the use of antibiotics: applying Scanlon's contractualism. *J Med Ethics* 38 (8):465–469.

Mirkin, C.A., R.L. Letsinger, R.C. Mucic, and J.J. Storhoff. 1996. A DNA-based method for rationally assembling nanoparticles into macroscopic materials. *Nature* 382 (6592):607–609.

Mischler, R. and I.C. Metcalfe. 2002. Inflexal® V a trivalent virosome subunit influenza vaccine: production. *Vaccine* 20:B17–B23.

Modun, B., J. Morrissey, and P. Williams. 2000. The staphylococcal transferrin receptor: a glycolytic enzyme with novel functions. *Trends Microbiol* 8 (5):231–237.

Molin, S. and T. Tolker-Nielsen. 2003. Gene transfer occurs with enhanced efficiency in biofilms and induces enhanced stabilisation of the biofilm structure. *Curr Opin Biotechnol* 14 (3):255–261.

Morens, D.M., G.K. Folkers, and A.S. Fauci. 2008. Emerging infections: a perpetual challenge. *Lancet Infect Dis* 8 (11):710.

Morones, J.R., J.L. Elechiguerra, A. Camacho, K. Holt, J.B. Kouri, J.T. Ramírez, and M.J. Yacaman. 2005. The bactericidal effect of silver nanoparticles. *Nanotechnology* 16 (10):2346.

Morrison, D.W.G., M.R. Dokmeci, U. Demirci, and A. Khademhosseini. 2008. Clinical applications of micro- and nanoscale biosensors. In *Biomedical Nanostructures*, edited by K.E. Gonsalves, C.L. Laurencin, C.R. Halberstadt, and L.S. Nair. Hoboken, NJ: John Wiley & Sons, Inc.

Müller, R.H., K. Mäder, and S. Gohla. 2000. Solid lipid nanoparticles (SLN) for controlled drug delivery—a review of the state of the art. *Eur J Pharm Biopharm* 50 (1):161–177.

Murray, C.J.L., A.D. Lopez, B. Chin, D. Feehan, and K.H. Hill. 2006. Estimation of potential global pandemic influenza mortality on the basis of vital registry data from the 1918–20 pandemic: a quantitative analysis. *Lancet* 368 (9554):2211–2218.

Nagatani, N., R. Tanaka, T. Yuhi, T. Endo, K. Kerman, Y. Takamura, and E. Tamiya. 2006. Gold nanoparticle-based novel enhancement method for the development of highly sensitive immunochromatographic test strips. *Sci Technol Adv Mater* 7 (3):270–275.

Naimushin, A.N., S.D. Soelberg, D.K. Nguyen, L. Dunlap, D. Bartholomew, J. Elkind, J. Melendez, and C.E. Furlong. 2002. Detection of *Staphylococcus aureus* enterotoxin B at femtomolar levels with a miniature integrated two-channel surface plasmon resonance (SPR) sensor. *Biosens Bioelectron* 17 (6–7):573–584.

Nair, S., A. Sasidharan, V.V. Divya Rani, D. Menon, S. Nair, K. Manzoor, and S. Raina. 2009. Role of size scale of ZnO nanoparticles and microparticles on toxicity toward bacteria and osteoblast cancer cells. *J Mater Sci: Mater Med* 20:235–241.

Nam, J.M., C.S. Thaxton, and C.A. Mirkin. 2003. Nanoparticle-based bio-bar codes for the ultrasensitive detection of proteins. *Science* 301 (5641):1884–1886.

Nam, J.M., S.I. Stoeva, and C.A. Mirkin. 2004. Bio-bar-code-based DNA detection with PCR-like sensitivity. *J Am Chem Soc* 126 (19):5932–5933.

Nanda, A. and M. Saravanan. 2009. Biosynthesis of silver nanoparticles from *Staphylococcus aureus* and its antimicrobial activity against MRSA and MRSE. *Nanomedicine* 5 (4):452–456.

Nandedkar, T.D. 2009. Nanovaccines: recent developments in vaccination. *J Biosci* 34 (6):995–1003.

Nederberg, F., Y. Zhang, J.P.K. Tan, K. Xu, H. Wang, C. Yang, S. Gao, X.D. Guo, K. Fukushima, and L. Li. 2011. Biodegradable nanostructures with selective lysis of microbial membranes. *Nat Chem* 3 (5):409–414.

Neofytos, D., D. Horn, E. Anaissie, W. Steinbach, A. Olyaei, J. Fishman, M. Pfaller, C. Chang, K. Webster, and K. Marr. 2009. Epidemiology and outcome of invasive fungal infection in adult hematopoietic stem cell transplant recipients: analysis of Multicenter Prospective Antifungal Therapy (PATH) Alliance registry. *Clin Infect Dis* 48 (3):265–273.

Nicewarner-Pena, S.R., R.G. Freeman, B.D. Reiss, L. He, D.J. Pena, I.D. Walton, R. Cromer, C.D. Keating, and M.J. Natan. 2001. Submicrometer metallic barcodes. *Science* 294 (5540):137–141.

Nie, S. and S.R. Emory. 1997. Probing single molecules and single nanoparticles by surface-enhanced Raman scattering. *Science* 275 (5303):1102–1106.

Norman, R.S. J.W. Stone, A. Gole, C.J. Murphy, and T.L. Sabo-Attwood. 2008. Targeted photothermal lysis of the pathogenic bacteria, *Pseudomonas aeruginosa*, with gold nanorods. *Nano Lett* 8 (1):302–306.

Nowacek, A. and H.E. Gendelman. 2009. NanoART, neuroAIDS and CNS drug delivery. *Nanomedicine* 4 (5):557–574.

Nugaeva, N., K.Y. Gfeller, N. Backmann, H.P. Lang, M. Duggelin, and M. Hegner. 2005. Micromechanical cantilever array sensors for selective fungal immobilization and fast growth detection. *Biosens Bioelectron* 21 (6):849–856.

Ohashi, K., T. Kabasawa, T. Ozeki, and H. Okada. 2009. One-step preparation of rifampicin/poly (lactic-co-glycolic acid) nanoparticle-containing mannitol microspheres using a four-fluid nozzle spray drier for inhalation therapy of tuberculosis. *J Control Release* 135 (1):19–24.

Olano, J.P. and D.H. Walker. 2011. Diagnosing emerging and reemerging infectious diseases: the pivotal role of the pathologist. *Arch Pathol Lab Med* 135 (1):83–91.

Omri, A., Z.E. Suntres, and P.N. Shek. 2002. Enhanced activity of liposomal polymyxin B against *Pseudomonas aeruginosa* in a rat model of lung infection. *Biochem Pharmacol* 64 (9):1407–1414.

Otsuka, Y., K. Hanaki, J. Zhao, R. Ohtsuki, K. Toyooka, H. Yoshikura, T. Kuratsuji, K. Yamamoto, and T. Kirikae. 2004. Detection of *Mycobacterium bovis* Bacillus Calmette-Guerin using quantum dot immuno-conjugates. *Jpn J Infect Dis* 57 (4):183–184.

Owens Jr, R.C. and A.F. Shorr. 2009. Rational dosing of antimicrobial agents: pharmacokinetic and pharmacodynamic strategies. *Am J Health-Syst Pharm* 66 (12 Suppl 4):S23–S30.

Padmavathy, N., and R. Vijayaraghavan. 2008. Enhanced bioactivity of ZnO nanoparticles—an antimicrobial study. *Sci Technol Adv Mat* 9 (3):035004.

Pal, S., Y.K. Tak, and J.M. Song. 2007. Does the antibacterial activity of silver nanoparticles depend on the shape of the nanoparticle? A study of the gram-negative bacterium *Escherichia coli*. *Appl Environ Microbiol* 73 (6):1712–1720.

Pandey, R. and G.K. Khuller. 2005. Solid lipid particle-based inhalable sustained drug delivery system against experimental tuberculosis. *Tuberculosis* 85 (4):227–234.

Pandey, R., Z. Ahmad, S. Sharma, and G.K. Khuller. 2005. Nano-encapsulation of azole antifungals: potential applications to improve oral drug delivery. *Int J Pharm* 301 (1):268–276.

Pappas, P.G., B.D. Alexander, D.R. Andes, S. Hadley, C.A. Kauffman, A. Freifeld, E.J. Anaissie et al. 2010. Invasive fungal infections among organ transplant recipients: results of the Transplant-Associated Infection Surveillance Network (TRANSNET). *Clin Infect Dis* 50 (8):1101–1111.

Parija, S.C. and I. Praharaj. 2011. Drug resistance in malaria. *Indian J Med Microbiol* 29 (3):243.

Parolo, C., A. de la Escosura-Muñiz, and A. Merkoçi. 2013. Enhanced lateral flow immunoassay using gold nanoparticles loaded with enzymes. *Biosens Bioelectron* 40:412–416.

Patolsky, F., G. Zheng, and C.M. Lieber. 2006. Nanowire sensors for medicine and the life sciences. *Nanomedicine (Lond)* 1 (1):51–65.

Patolsky, F., G. Zheng, O. Hayden, M. Lakadamyali, X. Zhuang, and C.M. Lieber. 2004. Electrical detection of single viruses. *Proc Natl Acad Sci U S A* 101 (39):14017–14022.

Patton, D.L., Y.T. Cosgrove Sweeney, T.D. McCarthy, and S.L. Hillier. 2006. Preclinical safety and efficacy assessments of dendrimer-based (SPL7013) microbicide gel formulations in a nonhuman primate model. *Antimicrob Agents Chemother* 50 (5):1696–700.

Peek, L.J., C.R. Middaugh, and C. Berkland. 2008. Nanotechnology in vaccine delivery. *Adv Drug Deliv Rev* 60 (8):915–928.

Peeling, R.W. 2006. Testing for sexually transmitted infections: a brave new world? *Sex Transm Infect* 82 (6):425–430.

Peeling, R.W., P.G. Smith, and P.M. Bossuyt. 2010. A guide for diagnostic evaluations. *Nat Rev Microbiol* 8 (12 Suppl):S2–S6.

Peng, H.I. and B.L. Miller. 2010. Recent advancements in optical DNA biosensors: exploiting the plasmonic effects of metal nanoparticles. *Analyst* 136 (3):436–447.

Perez, J.M., F.J. Simeone, Y. Saeki, L. Josephson, and R. Weissleder. 2003. Viral-induced self-assembly of magnetic nanoparticles allows the detection of viral particles in biological media. *J Am Chem Soc* 125 (34):10192–10193.

Perez, J.M., L. Josephson, T. O'Loughlin, D. Högemann, and R. Weissleder. 2002. Magnetic relaxation switches capable of sensing molecular interactions. *Nat Biotechnol* 20 (8):816–820.

Perni, S., P. Prokopovich, J. Pratten, I.P. Parkin, and M. Wilson. 2011. Nanoparticles: their potential use in antibacterial photodynamic therapy. *Photochem Photobiol Sci* 10 (5):712–720.

Petersdorf, R.G. 1986. Whither infectious diseases? Memories, manpower, and money. *J Infect Dis* 153 (2):189–195.

Petti, C.A., C.R. Polage, T.C. Quinn, A.R. Ronald, and M.A. Sande. 2006. Laboratory medicine in Africa: a barrier to effective health care. *Clin Infect Dis* 42 (3):377–382.

Pinto-Alphandary, H., A. Andremont, and P. Couvreur. 2000. Targeted delivery of antibiotics using liposomes and nanoparticles: research and applications. *Int J Antimicrob Agents* 13 (3):155–168.

Pissuwan, D., S.M. Valenzuela, C.M. Miller, and M.B. Cortie. 2007. A golden bullet? Selective targeting of Toxoplasma gondii tachyzoites using antibody-functionalized gold nanorods. *Nano Lett* 7 (12):3808–3812.

Planinsic, G., A. Lindell, and M. Remskar. 2009. Themes of nanoscience for the introductory physics course. *Eur J Phys* 30 (4):S17.

Plotkin, S.L. and S.A. Plotkin. 2004. A short history of vaccination. *Vaccines* 5:1–16.

Podsiadlo, P., S. Paternel, J.M. Rouillard, Z. Zhang, J. Lee, J.W. Lee, E. Gulari, and N.A. Kotov. 2005. Layer-by-layer assembly of nacre-like nanostructured composites with antimicrobial properties. *Langmuir* 21 (25):11915–11921.

Posthuma-Trumpie, G.A., J. Korf, and A. van Amerongen. 2009. Lateral flow (immuno) assay: its strengths, weaknesses, opportunities and threats. A literature survey. *Anal Bioanal Chem* 393 (2):569–582.

Potara, M., E. Jakab, A. Damert, O. Popescu, V. Canpean, and S. Astilean. 2011. Synergistic antibacterial activity of chitosan–silver nanocomposites on *Staphylococcus aureus*. *Nanotechnology* 22 (13):135101.

Prajapati, V.K., K. Awasthi, T.P. Yadav, M. Rai, O.N. Srivastava, and S. Sundar. 2012. An oral formulation of amphotericin B attached to functionalized carbon nanotubes is an effective treatment for experimental visceral leishmaniasis. *J Infect Dis* 205 (2):333–336.

Puckett, S.D., E. Taylor, T. Raimondo, and T.J. Webster. 2010. The relationship between the nanostructure of titanium surfaces and bacterial attachment. *Biomaterials* 31 (4):706–713.

Qi, L., Z. Xu, X. Jiang, C. Hu, and X. Zou. 2004. Preparation and antibacterial activity of chitosan nanoparticles. *Carbohydr Res* 339 (16):2693–2700.

Qin, D., X. He, K. Wang, and W. Tan. 2008. Using fluorescent nanoparticles and SYBR Green I based two-color flow cytometry to determine *Mycobacterium tuberculosis* avoiding false positives. *Biosens Bioelectron* 24 (4):626–631.

Qin, D., X. He, K. Wang, X.J. Zhao, W. Tan, and J. Chen. 2007. Fluorescent nanoparticle-based indirect immunofluorescence microscopy for detection of *Mycobacterium tuberculosis. J Biomed Biotechnol* 2007 (7):89364.

Rabea, E.I., M.E.T. Badawy, C.V. Stevens, G. Smagghe, and W. Steurbaut. 2003. Chitosan as antimicrobial agent: applications and mode of action. *Biomacromolecules* 4 (6):1457–1465.

Rai, M., A. Yadav, and A. Gade. 2009. Silver nanoparticles as a new generation of antimicrobials. *Biotechnol Adv* 27 (1):76–83.

Raman, C.V. and K.S. Krishnan. 1928. A new type of secondary radiation. *Nature* 121:501–502.

Ray, P.C. 2006. Diagnostics of single base-mismatch DNA hybridization on gold nanoparticles by using the hyper-Rayleigh scattering technique. *Angew Chem Int Ed* 45 (7):1151–1154.

Raymo, F.M. and I. Yildiz. 2007. Luminescent chemosensors based on semiconductor quantum dots. *Phys Chem Chem Phys* 9 (17):2036–2043.

Reddy, K.M., K. Feris, J. Bell, D.G. Wingett, C. Hanley, and A. Punnoose. 2007. Selective toxicity of zinc oxide nanoparticles to prokaryotic and eukaryotic systems. *App Phys Lett* 90 (21):213902-1–213902-3.

Reichert, A., J.O. Nagy, W. Spevak, and D. Charych. 1995. Polydiacetylene liposomes functionalized with sialic acid bind and colorimetrically detect influenza virus. *J Am Chem Soc* 117 (2):829–830.

Ren, G., D. Hu, E.W.C. Cheng, M.A. Vargas-Reus, P. Reip, and R.P. Allaker. 2009. Characterisation of copper oxide nanoparticles for antimicrobial applications. *Int J Antimicrob Agents* 33 (6):587–590.

Rice, L.B. 2009. The clinical consequences of antimicrobial resistance. *Curr Opin Microbiol* 12 (5):476–485.

Rich, R.L. and D.G. Myszka. 2003. Spying on HIV with SPR. *Trends Microbiol* 11 (3):124–133.

Richards, J.J. and C. Melander. 2009. Controlling bacterial biofilms. *Chembiochem* 10 (14):2287–2294.

Risbud, M.V., A.A. Hardikar, S.V. Bhat, and R.R. Bhonde. 2000. pH-sensitive freeze-dried chitosan–polyvinyl pyrrolidone hydrogels as controlled release system for antibiotic delivery. *J Control Release* 68 (1):23–30.

Rivas, G.A., M.D. Rubianes, M.C. Rodriguez, N.F. Ferreyra, G.L. Luque, M.L. Pedano, S.A. Miscoria, and C. Parrado. 2007. Carbon nanotubes for electrochemical biosensing. *Talanta* 74 (3):291–307.

Roduner, E. 2006. Size matters: why nanomaterials are different. *Chem Soc Rev* 35 (7):583–592.

Roe, D., B. Karandikar, N. Bonn-Savage, B. Gibbins, and J.B. Roullet. 2008. Antimicrobial surface functionalization of plastic catheters by silver nanoparticles. *J Antimicrob Chemother* 61 (4):869–876.

Rosi, N.L. and C.A. Mirkin. 2005. Nanostructures in biodiagnostics. *Chem Rev* 105 (4):1547–1562.

Rousserie, G., A. Sukhanova, K. Even-Desrumeaux, F. Fleury, P. Chames, D. Baty, V. Oleinikov, M. Pluot, J.H.M. Cohen, and I. Nabiev. 2010. Semiconductor quantum dots for multiplexed biodetection on solid-state microarrays. *Crit Rev Oncol Hematol* 74 (1):1.

Rudan, I., S. El Arifeen, R.E. Black, and H. Campbell. 2007. Childhood pneumonia and diarrhoea: setting our priorities right. *Lancet Infect Dis* 7 (1):56–61.

Sahoo, H. 2011. Förster resonance energy transfer—a spectroscopic nanoruler: principle and applications. *J Photochem Photobiol* 12 (1):20–30.

Sanderson, N. and M. Jones. 1996. Encapsulation of vancomycin and gentamicin within cationic liposomes for inhibition of growth of *Staphylococcus epidermidis. J Drug Target* 4 (3):181–189.

Saravanan, M. and A. Nanda. 2010. Extracellular synthesis of silver bionanoparticles from *Aspergillus clavatus* and its antimicrobial activity against MRSA and MRSE. *Colloids Surf B: Biointerfaces* 77 (2):214–218.

Saravanan, S., S. Nethala, S. Pattnaik, A. Tripathi, A. Moorthi, and N. Selvamurugan. 2011. Preparation, characterization and antimicrobial activity of a bio-composite scaffold containing chitosan/nano-hydroxyapatite/nano-silver for bone tissue engineering. *Int J Biol Macromol* 49 (2):188–193.

Sato, K., K. Hosokawa, and M. Maeda. 2003. Rapid aggregation of gold nanoparticles induced by non-cross-linking DNA hybridization. *J Am Chem Soc* 125 (27):8102–8103.

Sato, K., K. Hosokawa, and M. Maeda. 2005. Non-cross-linking gold nanoparticle aggregation as a detection method for single-base substitutions. *Nucleic Acids Res* 33 (1):e4.

Sayin, I., M. Kahraman, F. Sahin, D. Yurdakul, and M. Culha. 2009. Characterization of yeast species using surface-enhanced Raman scattering. *Appl Spectrosc* 63 (11):1276–1282.

Scallan, E., R.M. Hoekstra, F.J. Angulo, R.V. Tauxe, M.A. Widdowson, S.L. Roy, J.L. Jones, and P.M. Griffin. 2011. Foodborne illness acquired in the United States—major pathogens. *Emerg Infect Dis* 17 (1):7–17.

Schiavo, G. and F.G. van der Goot. 2001. The bacterial toxin toolkit. *Nat Rev Mol Cell Biol* 2 (7):530–537.

Schofield, C.L., A. Robert, and D.A. Russell. 2007. Glyconanoparticles for the colorimetric detection of cholera toxin. *Anal Chem* 79 (4):1356–1361.

Schüler, D. and R.B. Frankel. 1999. Bacterial magnetosomes: microbiology, biomineralization and biotechnological applications. *Appl Microbiol Biotechnol* 52 (4):464–473.

Schwartz, M. 2009. Dr. Jekyll and Mr. Hyde: a short history of anthrax. *Mol Aspects Med* 30 (6):347–355.

Scott, R.D. 2009. The Direct Medical Costs of Healthcare-Associated Infections in U.S. Hospitals and the Benefits of Prevention. Atlanta, GA: Centers for Disease Control and Prevention. http://www.cdc.gov/hai/burden.html.

Seil, J.T. and T.J. Webster. 2012. Antimicrobial applications of nanotechnology: methods and literature. *Int J Nanomedicine* 7:2767–2781.

Sekhon, B.S. and S.R. Kamboj. 2010. Inorganic nanomedicine—part 1. *Nanomedicine* 6 (4):516–522.

Selvin, P.R. 2000. The renaissance of fluorescence resonance energy transfer. *Nat Struct Biol* 7 (9):730–734.

Seth, D. S.R. Choudhury, S. Pradhan, S. Gupta, D. Palit, S. Das, N. Debnath, and A. Goswami. 2011. Nature-inspired novel drug design paradigm using nanosilver: efficacy on multi-drug-resistant clinical isolates of tuberculosis. *Curr Microbiol* 62 (3):715–726.

Shah, L.K. and M.M. Amiji. 2006. Intracellular delivery of saquinavir in biodegradable polymeric nanoparticles for HIV/AIDS. *Pharm Res* 23 (11):2638–2645.

Shahiwala, A., T.K. Vyas, and M.M. Amiji. 2007. Nanocarriers for systemic and mucosal vaccine delivery. *Recent Pat Drug Deliv Formul* 1 (1):1–9.

Shanmukh, S., L. Jones, J. Driskell, Y. Zhao, R. Dluhy, and R.A. Tripp. 2006. Rapid and sensitive detection of respiratory virus molecular signatures using a silver nanorod array SERS substrate. *Nano Lett* 6 (11):2630–2636.

Shanmukh, S., L. Jones, Y.P. Zhao, J.D. Driskell, R.A. Tripp, and R.A. Dluhy. 2008. Identification and classification of respiratory syncytial virus (RSV) strains by surface-enhanced Raman spectroscopy and multivariate statistical techniques. *Anal Bioanal Chem* 390 (6):1551–1555.

Shannon, M.A., P.W. Bohn, M. Elimelech, J.G. Georgiadis, B.J. Marinas, and A.M. Mayes. 2008. Science and technology for water purification in the coming decades. *Nature* 452 (7185):301–310.

Shao, H., T.J. Yoon, M. Liong, R. Weissleder, and H. Lee. 2010. Magnetic nanoparticles for biomedical NMR-based diagnostics. *Beilstein J Nanotechnol* 1:142–154.

Shinde, S.B., C.B. Fernandes, and V.B. Patravale. 2012. Recent trends in in vitro nanodiagnostics for detection of pathogens. *J Control Release* 159:164–180.

Shuman, E.K. 2010. Global climate change and infectious diseases. *N Engl J Med* 362 (12):1061–1063.

Silva, S., P. Pires, D.R. Monteiro, M. Negri, L.F. Gorup, E.R. Camargo, D.B. Barbosa et al. 2012. The effect of silver nanoparticles and nystatin on mixed biofilms of *Candida glabrata* and *Candida albicans* on acrylic. *Med Mycol* 51 (2):178–184.

Simon-Deckers, A., S. Loo, M. Mayne-L'hermite, N. Herlin-Boime, N. Menguy, C. Reynaud, B. Gouget, and M. Carrière. 2009. Size-, composition-and shape-dependent toxicological impact of metal oxide nanoparticles and carbon nanotubes toward bacteria. *Environ Sci Technol* 43 (21):8423–8429.

Singh, A.K., D. Senapati, S. Wang, J. Griffin, A. Neely, P. Candice, K.M. Naylor, B. Varisli, J.R. Kalluri, and P.C. Ray. 2009. Gold nanorod based selective identification of *Escherichia coli* bacteria using two-photon Rayleigh scattering spectroscopy. *ACS Nano* 3 (7):1906.

Singh, A.K., S.H. Harrison, and J.S. Schoeniger. 2000. Gangliosides as receptors for biological toxins: development of sensitive fluoroimmunoassays using ganglioside-bearing liposomes. *Anal Chem* 72 (24):6019–6024.

Singh, M., A. Chakrapani, and D. O'Hagan. 2007. Nanoparticles and microparticles as vaccine-delivery systems. *Expert Rev Vaccines* 6 (5):797–808.

Singh, S.K., K. Goswami, R.D. Sharma, M.V.R. Reddy, and D. Dash. 2012. Novel microfilaricidal activity of nanosilver. *Int J Nanomedicine* 7:1023.

Small, P.M. 2009. Tuberculosis: a new vision for the 21st century. *Kekkaku* 84 (11):721–726.

Smith, B.D., R.L. Morgan, G.A. Beckett, Y. Falck-Ytter, D. Holtzman, C.G. Teo, A. Jewett et al. 2012. Recommendations for the identification of chronic hepatitis C virus infection among persons born during 1945–1965. *MMWR Morb Mortal Wkly Rep* 61 (RR-4):1–32.

Smith, W.E. 2008. Practical understanding and use of surface enhanced Raman scattering/surface enhanced resonance Raman scattering in chemical and biological analysis. *Chem Soc Rev* 37 (5):955–964.

So, H.M., D.W. Park, E.K. Jeon, Y.H. Kim, B.S. Kim, C.K. Lee, S.Y. Choi, S.C. Kim, H. Chang, and J.O. Lee. 2008. Detection and titer estimation of *Escherichia coli* using aptamer-functionalized single-walled carbon-nanotube field-effect transistors. *Small* 4(2):197–201.

Soppimath, K.S., T.M. Aminabhavi, A.R. Kulkarni, and W.E. Rudzinski. 2001. Biodegradable polymeric nanoparticles as drug delivery devices. *J Control Release* 70 (1):1–20.

Souto, E.B., S.A. Wissing, C.M. Barbosa, and R.H. Müller. 2004. Development of a controlled release formulation based on SLN and NLC for topical clotrimazole delivery. *Int J Pharm* 278 (1):71–77.

Spellberg, B., R. Guidos, D. Gilbert, J. Bradley, H.W. Boucher, W.M. Scheld, J.G. Bartlett, and J. Edwards Jr. 2008. The epidemic of antibiotic-resistant infections: a call to action for the medical community from the Infectious Diseases Society of America. *Clin Infect Dis* 46 (2):155–164.

Stickler, D.J. 2008. Bacterial biofilms in patients with indwelling urinary catheters. *Nat Clin Pract Urol* 5 (11):598–608.

Stoeva, S.I., J.S. Lee, C.S. Thaxton, and C.A. Mirkin. 2006. Multiplexed DNA detection with biobarcoded nanoparticle probes. *Angew Chem* 118 (20):3381–3384.

Storhoff, J.J., A.D. Lucas, V. Garimella, Y.P. Bao, and U.R. Muller. 2004. Homogeneous detection of unamplified genomic DNA sequences based on colorimetric scatter of gold nanoparticle probes. *Nat Biotechnol* 22 (7):883–887.

Strohal, R., M. Schelling, M. Takacs, W. Jurecka, U. Gruber, and F. Offner. 2005. Nanocrystalline silver dressings as an efficient anti-MRSA barrier: a new solution to an increasing problem. *J Hosp Infect* 60 (3):226–230.

Suci, P.A., D.L. Berglund, L. Liepold, S. Brumfield, B. Pitts, W. Davison, L. Oltrogge et al. 2007. High-density targeting of a viral multifunctional nanoplatform to a pathogenic, biofilm-forming bacterium. *Chem Biol* 14 (4):387–398.

Sujith, A., T. Itoh, H. Abe, K. Yoshida, M.S. Kiran, V. Biju, and M. Ishikawa. 2009. Imaging the cell wall of living single yeast cells using surface-enhanced Raman spectroscopy. *Anal Bioanal Chem* 394 (7):1803–1809.

Sun, B., D.L. Slomberg, S.L. Chudasama, Y. Lu, and M.H. Schoenfisch. 2012. Nitric oxide-releasing dendrimers as antibacterial agents. *Biomacromolecules* 13:3343–3354.

Suzuki, H., Y. Saito, and T. Hibi. 2009. Helicobacter pylori and gastric mucosa-associated lymphoid tissue (MALT) lymphoma: updated review of clinical outcomes and the molecular pathogenesis. *Gut Liver* 3 (2):81–87.

Svenson, S. 2009. Dendrimers as versatile platform in drug delivery applications. *Eur J Pharm Biopharm* 71 (3):445–462.

Tallury, P., A. Malhotra, L.M. Byrne, and S. Santra. 2010. Nanobioimaging and sensing of infectious diseases. *Adv Drug Deliv Rev* 62 (4):424–437.

Tamanaha, C.R., S.P. Mulvaney, J.C. Rife, and L.J. Whitman. 2008. Magnetic labeling, detection, and system integration. *Biosens Bioelectron* 24 (1):1–13.

Tan, W., K. Wang, X. He, X.J. Zhao, T. Drake, L. Wang, and R.P. Bagwe. 2004. Bionanotechnology based on silica nanoparticles. *Med Res Rev* 24 (5):621–638.

Tang, S. and I. Hewlett. 2010. Nanoparticle-based immunoassays for sensitive and early detection of HIV-1 capsid (p24) antigen. *J Infect Dis* 201 (Suppl 1):S59–S64.

Tang, S., M. Moayeri, Z. Chen, H. Harma, J. Zhao, H. Hu, R.H. Purcell, S.H. Leppla, and I.K. Hewlett. 2009. Detection of anthrax toxin by an ultrasensitive immunoassay using europium nanoparticles. *Clin Vaccine Immunol* 16 (3):408–413.

Taton, T.A., C.A. Mirkin, and R.L. Letsinger. 2000. Scanometric DNA array detection with nanoparticle probes. *Science* 289 (5485):1757–1760.

Taubenberger, J.K. and D.M. Morens. 2006. 1918 Influenza: the mother of all pandemics. *Rev Biomed* 17:69–79.

Tay, L.L., P.J. Huang, J. Tanha, S. Ryan, X. Wu, J. Hulse, and L.K. Chau. 2012. Silica encapsulated SERS nanoprobe conjugated to the bacteriophage tailspike protein for targeted detection of Salmonella. *Chem Commun* 48 (7):1024–1026.

Taylor, E. and T.J. Webster. 2011. Reducing infections through nanotechnology and nanoparticles. *Int J Nanomedicine* 6:1463–1473.

Taylor, E.N. and T.J. Webster. 2009. The use of superparamagnetic nanoparticles for prosthetic biofilm prevention. *Int J Nanomedicine* 4:145–152.

Thaxton, C.S., D.G. Georganopoulou, and C.A. Mirkin. 2006. Gold nanoparticle probes for the detection of nucleic acid targets. *Clin Chim Acta* 363 (1):120–126.

Theron, J., T. Eugene Cloete, and M. de Kwaadsteniet. 2010. Current molecular and emerging nanobiotechnology approaches for the detection of microbial pathogens. *Crit Rev Microbiol* 36 (4):318–339.

Thompson, M.G., D.K. Shay, H. Zhou, C. Bridges, P.Y. Cheng, E. Burns, J.S. Bresee, and N.J. Cox. 2010. Estimates of deaths associated with seasonal influenza—United States, 1976–2007. *MMWR Morb Mortal Wkly Rep* 59 (33):1057–1062.

Tibbals, H.F. 2010. *Medical Nanotechnology and Nanomedicine*. Boca Raton, FL: CRC Press.

Timurdogan, E., B.E. Alaca, I.H. Kavakli, and H. Urey. 2011. MEMS biosensor for detection of Hepatitis A and C viruses in serum. *Biosens Bioelectron* 28 (1):189–194.

Todar, K. 2006. *Bacterial Protein Toxins*, ed K. Todar. http://www.textbookofbacteriology.net.

Tok, J.B.H., F. Chuang, M.C. Kao, K.A. Rose, S.S. Pannu, M.Y. Sha, G. Chakarova, S.G. Penn, and G.M. Dougherty. 2006. Metallic striped nanowires as multiplexed immunoassay platforms for pathogen detection. *Angew Chem Int Ed* 45 (41):6900–6904.

Tran, N., A. Mir, D. Mallik, A. Sinha, S. Nayar, and T.J. Webster. 2010. Bactericidal effect of iron oxide nanoparticles on *Staphylococcus aureus*. *Int J Nanomedicine* 5:277.

Tripp, R.A., R. Alvarez, B. Anderson, L. Jones, C. Weeks, and W. Chen. 2007. Bioconjugated nanoparticle detection of respiratory syncytial virus infection. *Int J Nanomedicine* 2 (1):117–124.

Tsao, N., T.Y. Luh, C.K. Chou, T.Y. Chang, J.J. Wu, C.C. Liu, and H.Y. Lei. 2002. In vitro action of carboxyfullerene. *J Antimicrob Chemother* 49 (4):641–649.

Turos, E., J.Y. Shim, Y. Wang, K. Greenhalgh, G. Reddy, S. Dickey, and D.V. Lim. 2007. Antibiotic-conjugated polyacrylate nanoparticles: new opportunities for development of anti-MRSA agents. *Bioorg Med Chem Lett* 17 (1):53–56.

U.S. Department of Health and Human Services, Administration on Aging. 2011. *A profile of older Americans: 2011*. Washington, D.C.: U.S. Department of Health and Human Services.

U.S. Department of State. 2000. The global infectious disease threat and its implications for the United States, NIE 99-17D. In *National Intelligence Estimate*. National Intelligence Council.

United Nations General Assembly, 56th session. 2001. Road map towards the implementation of the United Nations Millennium Declaration. Report of the Secretary-General, UN document no. A/56/326.

Valanne, A., S. Huopalahti, T. Soukka, R. Vainionpaa, T. Lovgren, and H. Harma. 2005. A sensitive adenovirus immunoassay as a model for using nanoparticle label technology in virus diagnostics. *J Clin Virol* 33 (3):217–223.

Valcarcel, M., B.M. Simonet, S. Cardenas, and B. Suarez. 2005. Present and future applications of carbon nanotubes to analytical science. *Anal Bioanal Chem* 382 (8):1783–1790.

Van Kampen, K.R., Z. Shi, P. Gao, J. Zhang, K.W. Foster, D.T. Chen, D. Marks, C.A. Elmets, and D.C. Tang. 2005. Safety and immunogenicity of adenovirus-vectored nasal and epicutaneous influenza vaccines in humans. *Vaccine* 23 (8):1029–1036.

Vannoy, C.H., A.J. Tavares, M.O. Noor, U. Uddayasankar, and U.J. Krull. 2011. Biosensing with quantum dots: a microfluidic approach. *Sensors* 11 (10):9732–9763.

Varshney, M., L. Yang, X.L. Su, and Y. Li. 2005. Magnetic nanoparticle-antibody conjugates for the separation of *Escherichia coli* O157:H7 in ground beef. *J Food Prot* 68 (9):1804–1811.

Vecitis, C.D., K.R. Zodrow, S. Kang, and M. Elimelech. 2010. Electronic-structure-dependent bacterial cytotoxicity of single-walled carbon nanotubes. *ACS Nano* 4 (9):5471–5479.

Vo-Dinh, T. 2007. Nanotechnology in biology and medicine: the new frontier. In *Nanotechnology in Biology and Medicine: Methods, Devices, and Applications*, edited by T. Vo-Dinh. Boca Raton, FL: CRC Press.

Vo-Dinh, T., H.N. Wang, and J. Scaffidi. 2010. Plasmonic nanoprobes for SERS biosensing and bioimaging. *J Biophotonics* 3 (1–2):89–102.

Wabuyele, M.B. and T. Vo-Dinh. 2005. Detection of human immunodeficiency virus type 1 DNA sequence using plasmonics nanoprobes. *Anal Chem* 77 (23):7810–7815.

Wahrenberger, A. 1995. Pharmacologic immunosuppression: cure or curse? *Crit Care Nurs Q* 17 (4):27–36.

Walling, M.A., S. Wang, H. Shi, and J.R. Shepard. 2010. Quantum dots for positional registration in live cell-based arrays. *Anal Bioanal Chem* 398 (3):1263–1271.

Walsh, T.J., V. Yeldandi, M. McEvoy, C. Gonzalez, S. Chanock, A. Freifeld, N.I. Seibel et al. 1998. Safety, tolerance, and pharmacokinetics of a small unilamellar liposomal formulation of amphotericin B (AmBisome) in neutropenic patients. *Antimicrob Agents Chemother* 42 (9):2391–2398.

Wang, C.H., Y.S. Hsu, and C.A. Peng. 2008. Quantum dots encapsulated with amphiphilic alginate as bioprobe for fast screening anti-dengue virus agents. *Biosens Bioelectron* 24 (4):1012–1019.

Wang, H., Y. Li, A. Wang, and A. Slavik. 2011. Rapid, sensitive, and simultaneous detection of three foodborne pathogens using magnetic nanobeadbased immunoseparation and quantum dot-based multiplex immunoassay. *J Food Prot* 74 (12):2039–2047.

Wang, J., J. Gao, D. Liu, D. Han, and Z. Wang. 2012. Phenylboronic acid functionalized gold nanoparticles for highly sensitive detection of *Staphylococcus aureus*. *Nanoscale* 4 (2):451–454.

Wang, L., W. Zhao, M.B. O'Donoghue, and W. Tan. 2007. Fluorescent nanoparticles for multiplexed bacteria monitoring. *Bioconjug Chem* 18 (2):297–301.

Wang, X., Y. Li, H. Wang, Q. Fu, J. Peng, Y. Wang, J. Du, Y. Zhou, and L. Zhan. 2010. Gold nanorod-based localized surface plasmon resonance biosensor for sensitive detection of hepatitis B virus in buffer, blood serum and plasma. *Biosens Bioelectron* 26 (2):404–410.

Warsinke, A. 2009. Point-of-care testing of proteins. *Anal Bioanal Chem* 393 (5):1393–1405.

Weeks, B.L., J. Camarero, A. Noy, A.E. Miller, L. Stanker, and J.J. De Yoreo. 2003. A microcantilever-based pathogen detector. *Scanning* 25 (6):297–299.

Wenhua, L., X. Haiyan, X. Zhixiong, L. Zhexue, O. Jianhong, C. Xiangdong, and S. Ping. 2004. Exploring the mechanism of competence development in *Escherichia coli* using quantum dots as fluorescent probes. *J Biochem Biophys Methods* 58:59–66.

When To Start Consortium. 2009. Timing of initiation of antiretroviral therapy in AIDS-free-infected patients: a collaborative analysis of 18 HIV cohort studies. *Lancet* 373:1352–1363.

White, A.R. 2011. Effective antibacterials: at what cost? The economics of antibacterial resistance and its control. *J Antimicrob Chemother* 66 (9):1948–1953.

White, W.C. 2009. Biofims before biofilms. In *Handbook of Applied Biomedical Microbiology: a Biofilm Approach*, edited by D.S. Paulson. Boca Raton, FL: CRC Press.

WHO, Geneva. 2007a. *Global Strategy for the Prevention and Control of Sexually Transmitted Infections: 2006–2015*. http://www.who.int/entity/reproductivehealth/publications/rtis/.../index.html.

WHO, Geneva. 2007b. *WHO Initiative to Estimate the Global Burden of Foodborne Diseases*. http://www.who.int/foodsafety/publictions/foodborne_disease/burden_nov07/en.

WHO, Geneva. 2008. *The Global Burden of Disease: 2004 Update*. http://www.who.int/healthinfo/global_burden_disease/en/.

WHO, Geneva. 2009. *Climate Change and Human Health*. http://www.who.int/globalchange/en.

WHO, Geneva. 2011a. *Global HIV/AIDS Response, Epidemic Update and Health Sector Progress towards Universal Access*: Progress Report 2011. http://www.who.int/hiv/pub/progress_report2011/en/index.html.

WHO, Geneva. 2011b. *Report on the Burden of Endemic Health Care-Associated Infection Worldwide*. http://www.who.int/gpsc/country_work/burden_hcai/en/.

WHO, Geneva. 2011c. *WHO Global Database on Blood Safety*, Summary Report 2011. http://who.int/bloodsafety/global_database/en/.

WHO, Geneva. 2011d. *WHO Global Malaria Programme: Good Practices for Selecting and Procuring Rapid Diagnostic Tests for Malaria*. http://who.int/publications/2011/9789241501125_eng.pdf.

WHO, Geneva. 2011e. *Global Tuberculosis Control*. WHO Report 2011. http://www.who.int/tb/publications/global_report/en/.

WHO, Geneva. 2011f. *WHO World Malaria Report: 2011.* http://www.who.int/malaria/world_malaria_report_2011/en/.

WHO, Geneva. 2012a. *Prevention and Control of Viral Hepatitis Infection: Framework for Global Action, 2012.* http://www.who.int/csr/disease/hepatitis/Framework/en/index.html.

WHO, Geneva. 2012b. *WHO UN—WaterGlobal Annual Assessment of Sanitation and DrinkingWater (GLAAS) 2012 Report: The Challenge of Extending and Sustaining Services*. http://www.who.int/water_sanitation_health/en/.

Wijagkanalan, W., S. Kawakami, and M. Hashida. 2011. Designing dendrimers for drug delivery and imaging: pharmacokinetic considerations. *Pharm Res* 28 (7):1500–1519.

Willyard, C. 2007. Simpler tests for immune cells could transform AIDS care in Africa. *Nat Med* 13 (10):1131–1131.

Winnicka, K., M. Wroblewska, P. Wieczorek, P.T. Sacha, and E. Tryniszewska. 2012. Hydrogel of ketoconazole and PAMAM dendrimers: formulation and antifungal activity. *Molecules* 17 (4):4612–4624.

Wise, R. 2011. The urgent need for new antibacterial agents. *J Antimicrob Chemother* 66 (9):1939–1940.

Wolk, D., S. Mitchell, and R. Patel. 2001. Principles of molecular microbiology testing methods. *Infect Dis Clin North Am* 15:1157–1204.

Wood, B.R. and D. McNaughton. 2006. Resonance Raman spectroscopy in malaria research. *Expert Rev Proteomics* 3 (5):525–544.

Wu, W., S. Wieckowski, G. Pastorin, M. Benincasa, C. Klumpp, J.P. Briand, R. Gennaro, M. Prato, and A. Bianco. 2005. Targeted delivery of amphotericin B to cells by using functionalized carbon nanotubes. *Angew Chem Int Ed* 44 (39):6358–6362.

Xu, S. and R. Mutharasan. 2010. Rapid and sensitive detection of *Giardia lamblia* using a piezoelectric cantilever biosensor in finished and source waters. *Environ Sci Technol* 44 (5):1736–1741.

Xu, S., H. Sharma, and R. Mutharasan. 2010. Sensitive and selective detection of mycoplasma in cell culture samples using cantilever sensors. *Biotechnol Bioeng* 105 (6):1069–1077.

Yager, P., T. Edwards, E. Fu, K. Helton, K. Nelson, M.R. Tam, and B.H. Weigl. 2006. Microfluidic diagnostic technologies for global public health. *Nature* 442 (7101):412–418.

Yang, H., D. Li, R. He, Q. Guo, K. Wang, X. Zhang, P. Huang, and D. Cui. 2010. A Novel quantum dots–based point of care test for syphilis. *Nanoscale Res Lett* 5 (5):875–881.

Yang, L. and Y. Li. 2005. Simultaneous detection of *Escherichia coli* O157: H7 and *Salmonella typhimurium* using quantum dots as fluorescence labels. *Analyst* 131 (3):394–401.

Yang, L., J.A.J. Haagensen, L. Jelsbak, H.K. Johansen, C. Sternberg, N. Høiby, and S. Molin. 2008. In situ growth rates and biofilm development of *Pseudomonas aeruginosa* populations in chronic lung infections. *J Bacteriol* 190 (8):2767–2776.

Yang, S.Y., J.J. Chieh, W.C. Wang, C.Y. Yu, C.B. Lan, J.H. Chen, H.E. Horng, C.Y. Hong, H.C. Yang, and W. Huang. 2008. Ultra-highly sensitive and wash-free bio-detection of H5N1 virus by immuno-magnetic reduction assays. *J Virol Methods* 153 (2):250–252.

Yoon, K.Y., J. Hoon Byeon, J.H. Park, and J. Hwang. 2007. Susceptibility constants of *Escherichia coli* and *Bacillus subtilis* to silver and copper nanoparticles. *Sci Total Environ* 373 (2):572–575.

Yuan, W., J. Ji, J. Fu, and J. Shen. 2007. A facile method to construct hybrid multilayered films as a strong and multifunctional antibacterial coating. *J Biomed Mater Res B: Appl Biomater* 85 (2):556–563.

Yuanbo, C., Z. Fan, and Y. Jiachang. 2005. Detecting proton flux across chromatophores driven by F0 F1 ATPase using *N*-(fluorescein-5-thiocarbamoyl)-1, 2-dihexadecanoyl-*sn*-glycero-3-phosphoethanolamine, triethylammonium salt. *Anal Biochem* 344 (1):102–107.

Yuen, C. and Q. Liu. 2012. Magnetic field enriched surface enhanced resonance Raman spectroscopy for early malaria diagnosis. *J Biomed Opt* 17 (1):17005-1–17005-7.

Yu-Hong, W., C. Rui, and L. Ding. 2011. A quantum dots and superparamagnetic nanoparticle-based method for the detection of HPV DNA. *Nanoscale Res Lett* 6 (1):1–9.

Yun, H., H. Bang, J. Min, C. Chung, J.K. Chang, and D.C. Han. 2010. Simultaneous counting of two subsets of leukocytes using fluorescent silica nanoparticles in a sheathless microchip flow cytometer. *Lab Chip* 10 (23):3243–3254.

Yun, Z., D. Zhengtao, Y. Jiachang, T. Fangqiong, and W. Qun. 2007. Using cadmium telluride quantum dots as a proton flux sensor and applying to detect H9 avian influenza virus. *Anal Biochem* 364 (2):122–127.

Zaghloul, M.S. and I. Gouda. 2012. Schistosomiasis and bladder cancer: similarities and differences from urothelial cancer. *Expert Rev Anticancer Ther* 12 (6):753–763.

Zelada-Guillen, G.A., J. Riu, A. Duzgun, and F.X. Rius. 2009. Immediate detection of living bacteria at ultralow concentrations using a carbon nanotube based potentiometric aptasensor. *Angew Chem Int Ed Engl* 48 (40):7334–7337.

Zeuzem, S., S.V. Feinman, J. Rasenack, E.J. Heathcote, M.Y. Lai, E. Gane, J. O'Grady et al. 2000. Peginterferon alfa-2a in patients with chronic hepatitis C. *N Engl J Med* 343 (23):1666–1672.

Zhang, D., M.C. Huarng, and E.C. Alocilja. 2010. A multiplex nanoparticle-based bio-barcoded DNA sensor for the simultaneous detection of multiple pathogens. *Biosens Bioelectron* 26 (4):1736–1742.

Zhang, H., M.H. Harpster, H.J. Park, P.A. Johnson, and W.C. Wilson. 2011. Surface-enhanced Raman scattering detection of DNA derived from the west nile virus genome using magnetic capture of Raman-active gold nanoparticles. *Anal Chem* 83 (1):254–260.

Zhang, L., D. Pornpattananangku, C.M. Hu, and C.M. Huang. 2010. Development of nanoparticles for antimicrobial drug delivery. *Curr Med Chem* 17 (6):585–594.

Zhang, X., M.A. Young, O. Lyandres, and R.P. Van Duyne. 2005. Rapid detection of an anthrax biomarker by surface-enhanced Raman spectroscopy. *J Am Chem Soc* 127 (12):4484–4489.

Zhao, X., L.R. Hilliard, S.J. Mechery, Y. Wang, R.P. Bagwe, S. Jin, and W. Tan. 2004. A rapid bioassay for single bacterial cell quantitation using bioconjugated nanoparticles. *Proc Natl Acad Sci U S A* 101 (42):15027–15032.

Zharov, V.P., K.E. Mercer, E.N. Galitovskaya, and M.S. Smeltzer. 2006. Photothermal nanotherapeutics and nanodiagnostics for selective killing of bacteria targeted with gold nanoparticles. *Biophys J* 90 (2):619–627.

Zhou, Y., Y. Kong, S. Kundu, J.D. Cirillo, and H. Liang. 2012. Antibacterial activities of gold and silver nanoparticles against *Escherichia coli* and bacillus Calmette-Guerin. *J Nanobiotechnology* 10:19.

Zhu, L., S. Ang, and W.T. Liu. 2004. Quantum dots as a novel immunofluorescent detection system for *Cryptosporidium parvum* and *Giardia lamblia*. *Appl Environ Microbiol* 70 (1):597–598.

Zimmerli, W., A. Trampuz, and P.E. Ochsner. 2004. Prosthetic-joint infections. *N Engl J Med* 351 (16):1645–1654.

Zodrow, K., L. Brunet, S. Mahendra, D. Li, A. Zhang, Q. Li, and P.J. Alvarez. 2009. Polysulfone ultrafiltration membranes impregnated with silver nanoparticles show improved biofouling resistance and virus removal. *Water Res* 43 (3):715–723.

2

Nanotechnology Applications in Dermatology

David Schairer, MD; Jason Chouake, MD; Adnan Nasir, MD; and Adam Friedman, MD

CONTENTS

2.1 Introduction to Dermatology

Dermatology is the branch of medicine that specializes in the diagnosis, treatment, and prevention of diseases of the skin, hair, nails, and mucosa. The skin is not only the largest organ of the human body but is also unique, as assessment can easily be accomplished by direct visual examination and local tissue sampling. The dermatologist has the unusual advantage of being able to visually examine the skin and correlate it with the pathology present below the surface.

To become a dermatologist, one receives the general training by completing medical school and at least 1 year of internship in general medicine, surgery, or pediatrics. This is then followed by 3 years of dermatology residency, which can be followed by subspecialty fellowships in dermatopathology, pediatric dermatology, or procedural dermatology. Today's dermatologists are trained in the basic science of skin, skin diseases, dermatopathology, treatment of skin diseases, and skin surgery. Phototherapy, laser treatment, cosmetic surgery, immunopathology, and patch testing are also part of the training. Lab or clinical research is an integral part of the dermatologist's experience and most publish in scientific and medical journals during this period of residency. Because of his or her training in scientific method, knowledge of cutaneous physiology and pathophysiology, and experience in treating skin diseases, a dermatologist is able to independently evaluate new ideas and products and choose whether or not to recommend or use them in practice. Education and public health have always been a major part of dermatology. In the past, infections were a big problem, and may be again in the future due to bioterrorism. The last 10 years of education has focused on cancer and aging prevention by promoting sun protection and early diagnosis of skin cancer.

From a therapeutic standpoint, cutaneous drug delivery offers many advantages over alternative routes of administration with regard to target-specific impact, decreased systemic toxicity, avoidance of first-pass metabolism, variable dosing schedules, and broadened utility to diverse patient populations. A complicating factor is that the skin has evolved mechanisms to impede exogenous molecules, especially hydrophilic ones, from safe passage. The horny layer of the stratum corneum (the topmost layer of the skin) is tightly bonded to an intercellular lipid matrix, making the passage of therapeutics a serious challenge.[1] This strong barrier to molecular activity is quite effective at blocking large drugs (molecular mass > 500 Da), which of course make up the majority of active therapeutics.[2] Mechanical abraders and microneedles can open a limited number of relatively wide ($\geq 10^3$ nm) pores in the skin barrier, which can allow for transient passage of small and even large molecules (or even bacteria).[3] Disruption with either ultrasound (phonopheresis) or high-voltage electrical pulsing (electroporation) has been used to force larger materials through this complex barrier. Chemical penetration enhancers are also used to perturb the epidermal barrier, though safety concerns have limited their efficacy.[4–6] Furthermore, many substances that could, in theory, be used as topical therapeutics have several disadvantages in that they are (1) weakly soluble or insoluble in water, (2) degraded or inactivated before reaching the appropriate target, and (3) nonspecifically distributed to tissues and organs, resulting in undue adverse side effects and limited efficacy at the target site.

With this in mind, it is no surprise that dermatology is in need of new topical therapies. Many believe that progress in drug delivery has tapered, and some investigators are concerned that development of new treatments in dermatology has reached a plateau. In fact, the Food and Drug Administration (FDA) has approved only a handful of patents for new drugs and treatments related to dermatology. Therefore, upcoming breakthroughs will likely rely on improvements or better use of existing agents. Although the hope is that novel molecules and therapies will be developed, it is obviously more cost-effective for manufacturers to repurpose a well-established, FDA-approved product in a superior vehicle than to invent a new product altogether. Novel delivery vehicles generated through nanotechnology are raising the exciting prospect for controlled and sustained drug delivery across the impenetrable skin barrier. Particles of size 500 nm and smaller exhibit a host of unique properties that are superior to their bulk material counterparts.[7–9] Small size is a necessary feature, but other properties are needed for nanomaterials to achieve efficacy as a topical delivery vehicle. Optimally, these nanoparticles should (1) carry drugs through cutaneous pores in the primary skin barrier, (2) release the transported drug spontaneously once penetration is achieved, and (3) exhibit low rates of cutaneous drug clearance allowing for deep/targeted deposition and prolonged action of the carrier-transported drugs. In addition, these products should be able to adjust to relevant physiologic variations as part of their design and targeting.

In this chapter, the following 10 most common dermatologic diseases encountered by physicians will be reviewed:

1. Acne vulgaris
2. Dermatitis
3. Actinic keratosis
4. Skin cancer (nonmelanoma/melanoma)
5. Viral warts
6. Benign nevi
7. Psoriasis
8. Epidermoid cyst
9. Seborrheic keratosis
10. Dermatophytosis

Their epidemiology, pathophysiology, clinical features, and current therapeutic modalities will be discussed to provide a basis for the utilization of nanotechnology for the diagnosis and treatment of these conditions.

2.2 Acne Vulgaris

2.2.1 Epidemiology

Acne vulgaris is one of the most common dermatologic diseases. It affects between 40 and 50 million people each year.[10] Although it peaks in adolescence, where it can be seen in up to 85% of individuals, 12% of adult women and 3% of adult men suffer from acne.[11] Acne is most often an isolated problem; however, it may also be associated with underlying disease or exogenous influences such as polycystic ovarian syndrome, hyperandrogenism, hypercortisolism, precocious puberty, and corticosteroid therapy.[10]

2.2.2 Pathophysiology

Acne results from a disorder of the pilosebaceous apparatus. Comedo formation, sebum production, colonization by *Propionibacterium acnes*, and inflammation are the key steps in the development of acne.[12] Comedones are accumulations of keratinocytes and sebum within a follicle.[10] Microcomedones, which enlarge to become comedones, are the result of increased cohesion and production of keratinocytes at the follicular infundibulum and ostium. Continued production of keratinocytes and sebum behind the occlusion causes the comedo to grow and eventually rupture into the surrounding tissue. Comedo formation is driven by subclinical levels of inflammation.[13] However, when they rupture, keratin and sebum are released into the surrounding tissue, where they cause a clinically observable inflammatory response characterized by erythema and swelling and at the cellular level, recruitment of lymphocytes and neutrophils.[12] As the inflammatory process progresses, neutrophils predominate and may result in the formation of pustules, nodules, and cysts.

The Gram-positive bacteria *P. acnes* and to a lesser extent *P. granulosum* and *P. parvum* are present and contribute to acne lesions.[10] They produce enzymes and lipases that promote comedo rupture, comedogenic and proinflammatory factors, and glycocalyx polymers. The glycocalyx acts as an adhesive that further binds keratinocytes together.[14]

Sebum secretion and therefore acne is sensitive to steroid hormones such as androgens and corticosteroids.[15] Examples of androgen-induced acne include neonatal acne, acne of adolescence, and polycystic ovarian syndrome. On the other hand, estrogens decrease sebum production either by direct action on the sebaceous unit or by inhibiting androgen production.[16]

2.2.3 Clinical Features

Acne vulgaris falls on a clinical spectrum of severity and can be subdivided into mild, moderate, and severe categories.[17] The mild category encompasses noninflammatory comedones. These small 1 mm papules can present with two morphologies: closed comedones, better known by the lay term whiteheads, and open comedones, better known as blackheads. Unlike closed comedones that lack an apparent follicular opening, open comedones have a dilated follicular opening. Open comedones owe their black appearance to exposed melanin deposits and oxidized lipids.[12]

When comedones become enflamed and enlarge into papules, pustules, nodules, or cysts, they enter the moderate category.[17] The inflammation is characterized by erythema, induration, and tenderness. Pustules, nodules, and cysts are filled with a mixture of sterile

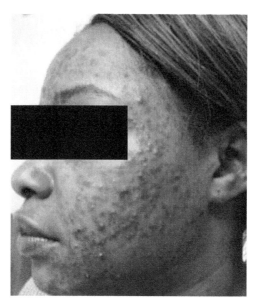

FIGURE 2.1
Acne vulgaris. Severe cases of acne (as pictured above) are characterized by extensive inflammatory and pustular lesions. Such cases are challenging to treat and decrease the quality of life of those affected.

pus and serosanguineous fluid.[12] In the most severe forms, such as nodular or conglobate
acne, the nodules and cysts can coalesce and interconnect to form sinus tracts.

The lesions of acne vulgaris are predominantly found on the face but are also commonly
seen on the back and chest (Figure 2.1).[11] When the inflammatory lesions resolve, they can
leave behind scars and hyperpigmentation.

2.2.4 Current Treatment

The primary goals of acne treatment are resolution of inflammatory lesions, prevention of
future comedo formation, and prevention of inflammation.[17] Retinoids, vitamin A derivatives, that bind to retinoic acid receptor (RAR) and subsequently activate nuclear transcription have been shown to act on all three fronts.[18] They normalize follicular keratinization,
cause expulsion of existing comedones, have anti-inflammatory effects on existing lesions,
and prevent formation of new comedones. Retinoids also bind to the retinoic x receptor
(RXR) that induce apoptosis. This property of RXR underlies the use of some retinoids in
the treatment of mycosis fungoides and Kaposi's sarcoma.

Currently, first, second, and third generation retinoids are available.[19] The first generation retinoids are retinol (alcohol form of vitamin A), tretinoin (all-*trans*-retinoic acid, vitamin A acid), isotretinoin (13-*cis*-retinoic acid), and alitretinoin (9-*cis*-retinoic acid). While
the first-generation retinoids have cyclic end groups, the second-generation retinoids, such
as acitretin and etretinate, have aromatized rings. The third-generation retinoids have
improved receptor selectivity as compared to the first- and second-generation retinoids.
Tazarotene and adapalene bind more readily to RAR, and bexarotene to RXR. The specificity of third-generation retinoids allows for selective activation of RAR and or RXR and for
reduction of off-target effects.

The use of topical retinoids is limited by erythema, dryness, scaling, and photosensitiv-
ity as well as restrictions in its storage and coadministration with benzoyl peroxide (BP).[17]

Some of these limitations have been overcome by novel delivery systems, which will be discussed in detail, and by later generation retinoids. Adapalene, for example, is unlike other retinoids, as it is not degraded by light and is not oxidized by BP and thus can be used in the daytime and in combination with BP.[18] Because of their efficacy, safety, and ease of use, topical retinoids are key components of first-line therapy for mild and moderate acne vulgaris and are the preferred maintenance therapy for all forms of acne.[20]

Oral retinoids in the form of isotretinoin are used when topical retinoids and antibiotics fail. Isotretinoin is thought to act through one of its metabolites and is notable for its potential to induce a remission of acne in 70% of cases.[18] Because isotretinoin is given orally, it causes side effects both in the skin and systemically, ranging from cheilitis, xerosis, skin fragility, xerophthalmia, hyperostosis, and myalgias with elevated levels of creatine phosphokinase, both in the skin and in other organs. Elevation of triglycerides and cholesterol is common and requires laboratory monitoring. Teratogenicity evidenced by craniofacial, cardiac, thymus, and central nervous system defects has led to the formation of the iPledge program for physicians who prescribe oral retinoids.[21] The program requires women taking isotretinoin to use two forms of birth control during use and be followed monthly with serial pregnancy tests to prevent the teratogenic sequelae that could ensue if an unexpected pregnancy occurs.

The second key component of acne therapy is antimicrobial agents.[20] Topical agents include clindamycin, erythromycin, BP, azelaic acid, and sulfacetamide. These agents inhibit the growth of *P. acnes*, preventing the inflammation caused by the bacteria.[10] Azelaic acid also shows comedolytic properties. In moderate and severe cases of acne, oral antibiotics are first-line treatment.[20] Just like their topical counterparts, the oral antibiotics act by inhibiting *P. acnes* growth. Erythromycin, tetracycline, doxycycline, and minocycline are the most commonly used agents. However, resistance has been reported in all four. Salicylic acid is a topically applied agent with comedolytic and mild anti-inflammatory properties that does not have antimicrobial properties.

Hormonal interventions have also been shown to be effective in the treatment of acne.[20] They act by decreasing hormone-driven sebum production. Oral contraceptives are second-line treatment for women with acne and are most effective in adult women with persistent inflammatory lesions, especially those that fluctuate with the menstrual cycle or those that are unresponsive to oral antibiotics. The majority of oral contraceptives are formulated with a progesterone component to reduce the risk of endometrial cancer associated with unopposed estrogens.[19] However, progesterone has androgenic activity.[22] Thus, oral contraceptive formulations with low androgenic progesterone are preferred.[23] The oral antiandrogens such as spironolactone and cyproterone are most effective in patients with elevated androgens.[20] They are often used in combination with oral contraceptives.

In addition to daily topical or oral therapies, office procedures such as intralesional corticosteroids, comedo extraction, and electrocautery and optical therapies are available for the treatment of acne. Intralesional triamcinolone is effective in nodulocystic acne lesions but can cause atrophy if excess triamcinolone is injected.[24] Manual expression of a comedo can provide rapid improvement of lesions and reduce the number of future inflammatory lesions.[25] Laser- and ultraviolet (UV) light-based optical therapies for acne have shown promising results in small randomized control trials.[26] Optical therapies are thought to act by heating sebaceous glands and photochemical inactivation of *P. acnes*. Adjuvants, such as aminolevulinic acid (ALA), are frequently used to potentiate the effects of optical therapies.

Current acne therapies are effective in controlling inflammation but struggle to reduce the initiating events of acne, such as comedo formation.[10] Conversely, topical retinoids, which do reduce comedo formation, also cause irritation. Thus, multiple medications must be used in combination for a patient to gain some control over their acne. These regimens quickly become complex and expensive. It is not uncommon for patients to use oral antibiotics along with one topical medication in the morning and another at night.[18] On average, such a regimen costs 1722 dollars a month, not including any over-the-counter products a patient may use.[27] Oral retinoids may be simpler and less costly for patients. However, many physicians are wary of prescribing an oral retinoid to a patient with mild or moderate acne because of the drug's toxicities.

2.2.5 Applications in Nanotechnology

The efficacy of topical drugs in the treatment of acne is well established. In reviewing the various options above, it is evident that there are two large categories consisting of agents with a primarily keratolytic or keratoregulatory action (retinoids and retinoid-like drugs, BP, salicylic acid, and azelaic acid) and antibiotics (clindamycin, erythromycin, and dapsone).[28–30] In general, many of the aforementioned topical therapies suffer from various related side effects, including irritation, erythema, xerosis, peeling and scaling, bacterial resistance, and resulting dyschromia from the associated irritation in patients of darker skin types—adverse events that continue to be major limiting factors influencing patient compliance and ultimately impacting efficacy.[28,31–34] To overcome these adverse events, novel vesicular and nanoparticulate drug delivery systems have been proposed to reduce these side effects commonly experienced by patients using topical treatments.[35–40] As the greatest strides in nanotechnology have been made in the acne arena, a brief review of topical drug delivery and various nanomaterials will be presented.

One approach to reduce these side effects is to use lower concentrations; however, conventional formulations containing lower concentrations of active ingredients often fall short, resulting in less significant results. Novel delivery systems, on the other hand, present the potential to reduce the side effects without compromising efficacy. Vesicular and particulate drug delivery systems such as liposomes, niosomes, micro/nanospheres or sponges, solid lipid nanoparticles (SLNs), and nanoemulsions all have the potential for improved penetration and controlled release of active substances.[29,37,41–46] The application of these systems to the skin distributes the topical agents gradually and, in some cases, has shown the ability to reduce the irritancy of some drugs while maintaining a similar efficacy when compared with conventional formulations.

In addition to the controlled release of the drug into the epidermis, these systems can also promote follicular targeting, creating high local concentrations of the active compounds in the pilosebaceous unit.[39,47–49] Briefly, a topically applied material has three possibilities to penetrate into the skin: transcellular, intercellular, and follicular. The hair follicle is a skin appendage with a complex structure containing many cell types that produce highly specialized proteins. The hair follicle is in a continuous cycle of growth: anagen is the hair growth phase, catagen the involution phase, and telogen is the resting phase. The hair follicle represents a great target for skin treatment owing to its deep extension into the dermis and thus provides much deeper penetration and absorption of compounds beneath the skin than seen with the transdermal route. In the case of skin diseases and of cosmetic products, delivery to sweat glands or to the pilosebaceous unit is essential for the effectiveness of certain drugs, specifically in acne. Increased accumulation in the pilosebaceous unit could result in improved outcomes related to topical therapy. The vesicular and particulate drug delivery systems have the advantage of penetrating more efficiently into the hair follicles than do nonparticulate systems, such as conventional formulations.[48] This provides a high local concentration over a prolonged period, allowing for increased efficacy (Figures 2.2 and 2.3).

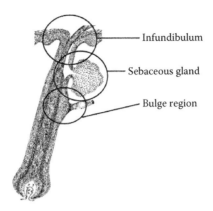

FIGURE 2.2
Diagram of the hair follicle. The hair follicle plays an essential role in the pathophysiology of acne. However, these same structures can also be targeted by nanodelivery systems. Targeted delivery allows for therapeutic doses to involve tissue while minimizing toxicity to uninvolved cells. (From Vogt, A., *J. Investig. Dermatol. Symp. Proc.*, 10, 252, 2005.[50] With permission.)

FIGURE 2.3
Three-dimensional case of a hair follicle done in cyanoacrylate. The follicular ostea is on the left side of the image and the bulb is in the bottom right. A glandular structure in the top right is connected to the hair follicle by a duct. (From Lademann, J., *J. Investig. Dermatol. Symp. Proc.*, 10, 301, 2005. With permission.)

2.2.5.1 Liposomes

Liposomes are microscopic spherical vesicles that are formed of phospholipids that have associated themselves spontaneously in a bilayer containing a centralized aqueous cavity.[47] Bilayers provide permeability barriers between exterior and interior compartments. A large group of biological membrane lipids that spontaneously form bilayers in water are the phospholipids, and in fact, liposomes are most commonly formed by natural origins, though synthetic or semisynthetic phospholipids may be used.[47] The ability of phospholipids to form a bilayer structure is because of their amphipathic character, resulting from the presence of a polar or hydrophilic (water attracting) head-group region and a nonpolar, lipophilic (water repellent) tail. The hydrophilic head groups orient toward the aqueous phase, and the lipophilic tails orient to each other in the presence of water. Therefore, liposomes contain a lipophilic compartment within the bilayer membranes and hydrophilic compartments between the membranes. Under the right conditions, water-soluble substances can be stored in the water phase and lipophilic substances into the lipid phase. Compounds such as methotrexate (MTX), which cannot penetrate the epidermis, may be used for topical application when encapsulated in penetrating liposomes.[51] It has also been shown that liposomes may also improve stabilization of unstable drugs through encapsulation.[52,53]

Not all liposomes are created equal. It has been shown that the penetration of liposomes through the stratum corneum decreases with increasing diameters.[37,54,55] Therefore, the preferred structures for drug delivery are those that fall in the nanoscale of 50–100 nm in diameter. Next to the preferred sizes, an essential characteristic of liposomes for penetration through the stratum corneum is their state in a liquid crystal phase. The lipid bilayer passes from a gel into a liquid crystal phase at a critical phase transition temperature. This liquid crystal state is essential for liposomes to interact simultaneously with the lipid and the aqueous compartments of the stratum corneum and for delivering entrapped drugs into the skin.

The similarity of lipid composition of liposomes and membranes in the epidermis enables liposomes to penetrate the epidermal barrier to a higher extent as compared with other topical preparations. This may result in an increased drug absorption into the epidermis and a decreased clearance of drug from the epidermis, resulting in a much longer sustained drug release and reduction of drug absorption into the blood. Furthermore, liposomes have the potential to target drugs into the pilosebaceous structures, and therefore can be used for treatment of hair follicle–associated and sebaceous gland–associated disorders.[39,47,49]

Interestingly, decreasing the size of liposomes is not the only feature that allows for improved transcutaneous penetration, but they themselves may serve as penetration enhancers. The use of conventional penetration enhancers (e.g., dimethylsulfoxide or propylene glycol) has rendered side effects not only because of the potential for increased systemic drug level, but also because these agents can induce irritant and allergic reactions. Cutaneous irritation from liposomes is not expected, because liposomes are similar to epidermal lipids, are biodegradable, and are derived from nontoxic lipids.

As acne is one of the most common skin diseases, the pursuit to improve the efficacy and safety profiles of our armament of acne therapeutics through liposomal encapsulation has been rigorous and will be divided by agent.

2.2.5.1.1 Benzoyl Peroxide

BP has both a mild keratolytic and a potent bactericidal effect to which *P. acnes* has yet to show resistance.[56] Although an effective acne therapy, skin irritation is an expected adverse event, often appreciated at efficacious doses. One approach to improving efficacy could be encapsulation in liposomes by reducing the side effects associated with topical application and ultimately improving patient compliance.

In one, small ($n = 30$) clinical trial, liposomal BP gels significantly improved therapeutic responses (about twofold) at all evaluation time points as compared with those of the commercially available BP gel.[57] Furthermore, liposomal BP gels showed significantly less irritation in the first 2 weeks of treatment, with no reports of "burning" as per subjects. Conversely, the BP gel–treated group showed increased irritation through the eighth week of treatment. Stability evaluation of the BP liposomes showed a good stability profile (almost 100% of drug retained, 90 days) at refrigeration temperature (2°C–8°C), room temperature (25°C), and body temperature (37°C).

BP-loaded liposomes have also been shown to impart a significant antibacterial effect in the infundibula against both Propionibacteria and Micrococcaceae in acne patients (20 patients, 4-week treatment) when compared with other BP formulations. These results showed the potential infundibular targeting of liposomal BP.[58]

2.2.5.1.2 Clindamycin

Antibiotics have been used in the management of inflammatory acne vulgaris for decades,[33,59,60] with clindamycin being one of the most commonly used options. In previous studies, lipid compositions have been evaluated in an attempt to improve the effectiveness of clindamycin-encapsulated liposomes in acne treatment.[61–63]

In one study, treatment of acne patients with liposomal clindamycin resulted in better efficacy than nonliposomal lotion forms.[63] Application of a conventional lotion solution, a nonliposomal emulsion lotion, and a liposomal emulsion lotion resulted in decreases of 42.9%, 48.3%, and 62.8%, respectively, in the total number of lesions after a 4-week treatment. In another study, a double-blind clinical trial ($n = 76$ patients, 6-week treatment), the efficacy of clindamycin 1% liposomes to reduce open comedones (33.3%) was greater than that for the free clindamycin 1% solution (8.3%), also appearing therapeutically superior when closed comedones, papules, and pustules were examined.[62]

2.2.5.1.3 Salicylic Acid

Salicylic acid, a keratolytic agent with both comedolytic and antimicrobial activity, is available over the counter as well as is incorporated into peeling agents. It has been efficiently used in the treatment of acne as well as cosmetic repair of postinflammatory pigment alteration and pitted scars.[28,32,33] The majority of work to date on salicylic acid liposomes has been primarily preclinical.[64,65] In one investigation, when compared with free salicylic acid dispersion, salicylic liposomes not only prolonged the release of salicylic acid across the porcine skin but also enhanced retention in the skin by a factor of 10.[64] Furthermore, stability of the liposomal preparation was shown after 12 weeks of storage at refrigeration temperature (4°C–5°C)—liposomes retained their structure and showed minimal (4.01%) salicylic acid leakage.

2.2.5.1.4 Retinoids

Retinoids regulate several critical physiological and pathological processes. These biological response modifiers exert their pleiotropic effects through the interaction with nuclear receptors.[29,30] In acne treatment, topical retinoids are important, because they reverse the abnormal desquamation by affecting follicular epithelial turnover and maturation of cells. In addition, some topical retinoids have an effect on inflammation by modulating the immune response, inflammatory mediators, and migration of inflammatory cells.[66] Tretinoin, or all-*trans*-retinoic acid, is a polymechanistic, effective drug that is frequently used in the topical treatment of mild to moderate acne. Retinoic acid enables comedolysis and normalizes the maturation of the follicular epithelium so as to cease comedone formation, and has been shown to have anti-inflammatory and possibly even antimicrobial activity.[29] Unfortunately, topical application of retinoic acid and its derivatives generally results in local irritation, mild to severe erythema, dryness, peeling, and scaling.[29,30] As mentioned earlier, these adverse events tend to diminish patient compliance, compromising the efficacy of the therapy. It is in fact with retinoic acid that the majority of work on liposomal encapsulation has been pursued both in the in *vitro* and preclinical setting[67–70] as well as clinical studies, which will be the primary focus of this chapter (Figure 2.4).

With respect to preclinical animal studies, there have been a number of investigations suggesting increased skin penetration yet decreased percutaneous absorption of retinoic acid encapsulated in liposomes. In a hairless rat skin model, liposomal retinoic acid showed a significantly higher distribution in the epidermis (mainly stratum corneum) and dermis as compared to tretinoin in an alcohol gel (41% and 13%, respectively, with liposomes, versus 18% and 8% with the gel).[70] Conversely, in *vitro* skin permeation was statically reduced for liposomes (1.6%) when compared with the gel (3.1%) in this same study.

In another in *vivo* study, retinoic acid–loaded liposomes prolonged drug release through hairless rat skin and promoted greater amounts of the drug deposited into different epidermal strata as well as the dermis, increasing the retinoic acid concentration in these layers (about twofold) when compared with nonliposomal systems.[68]

A second strong investigatory focus is on improving the photostability limitation with retinoid-based acne therapies. The encapsulation of retinoic acid in liposomes was shown to improve the drug photolability. When liposomes developed by Brisaert et al.[71] were submitted to photodegradation (xenon lamp, 15 minutes), degradation constants in liposomes

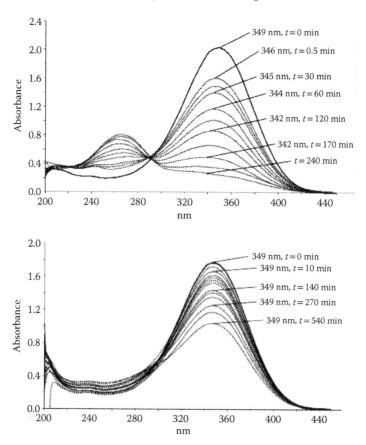

FIGURE 2.4
Photostability of 13 retinoic acid in ethanol and in liposomes. When 13 retinoic acid dissolved in ethanol is exposed to light, it is almost entirely depleted within 4 hours (top). In contrast, 13 retinoic acid packaged into liposomes is significantly more stable, with over half the original concentration intact after 9 hours (bottom). (From Ioele, G., *Int. J. Pharm.* 293, 251, 2005. With permission.)

were approximately 1.8 times slower than in castor oil. These liposomes also presented a high stability for 1 year (about 95% retinoic acid retained), even when stored at 25°C.

There have been a number of clinical studies evaluating the efficacy of liposomal retinoic acid. In a double-blind, split-face clinical trial,[69] 20 patients received retinoic acid liposomal gel (0.01%) on one side of the face and a commercial gel on the other (0.025% or 0.05%) once daily for 10 weeks. The global evaluation of efficacy (comedones, papules, and pustules) indicated no significant differences between the treatment groups, indicating that the retinoic acid–encapsulated liposomes allowed for the reduction of drug concentration without compromising efficacy. Even more importantly, the liposomal preparation induced less burning and erythema than the 0.025% and 0.05% gels.

In a larger, double-blind study (30 patients, mild to moderate acne, 12 weeks), liposomal tretinoin 0.025% gel was compared to commercially available tretinoin 0.025% gel.[72] Liposomal tretinoin showed better efficacy (~1.5-fold) than the nonliposomal preparation, and at every evaluation time point, it was significantly more effective. In addition, it was observed that the mean percent reduction in comedonal lesions was significant between the two groups, with 62.36% for retinoic acid gel and 94.17% for liposomal retinoic acid gel. All adverse events, with the exception of scaling, were remarkably decreased with the use of liposomal retinoic acid gel compared with the plain retinoic acid gel.

2.2.5.1.5 Antiandrogenic Therapy

Although the pathogenesis of acne is multifactorial and highly variable from case to case, the role and impact of androgens is well established. Enhanced dihydrotestosterone formation via 5-α-reductase stimulates sebocyte proliferation and sebum production, one of the key pathogenic factors involved in acne vulgaris. Furthermore, the presence of androgen receptors in pilosebaceous unit has been shown, thereby suggesting a role for androgens in keratinization.[36] Therefore, it is no surprise that antiandrogen therapies have been used in the treatment of acne, especially in the clear setting of androgen excess such as in polycystic ovarian syndrome. In the past, oral antiandrogenic therapies have been limited by their systemic effects, including teratogenicity in women and gynecomastia in men. Topical application through liposomal preparations offers a unique approach to overcoming these potential systemic side effects associated with oral administration. The following is a brief review of ongoing work with topical antiandrogens.

Finasteride is an antiandrogen that blocks the production of dihydrotestosterone from testosterone by competitively and specifically inhibiting the type II 5-α-reductase isozyme, and is approved for the treatment of benign prostatic hypertrophy and androgenic alopecia in men. Both in *vitro* permeation and in *vivo* deposition studies evaluating liposomal finasteride showed the potential of liquid-state liposomes to successfully deliver finasteride to the pilosebaceous unit.[73,74]

Cyproterone acetate is a potent steroidal antiandrogen and can suppress luteinizing hormone, thereby effectively decreasing testosterone levels. A topical liposomal cyproterone acetate lotion (2 mg cyproterone acetate/mL; $n = 12$) was compared to an oral formulation containing 0.035 mg ethinyl estradiol, 2 mg cyproterone acetate ($n = 12$), and a placebo lotion ($n = 16$).[75] After 3 months of topical cyproterone acetate application, lesion count decrease from baseline was comparable to that of the systemically treated group. No adverse events were noted in either group; however, serum cyproterone acetate levels were 10 times lower after topical application, compared with orally administered cyproterone acetate.

Even more recently, a novel nonsteroidal antiandrogen (RU 58841) that showed significant impact on sebaceous gland activity in animal models was evaluated preclinically in a liposomal formulation.[76–78] The in *vivo* cutaneous distribution studies on hairless rat skin showed that the greatest concentrations in the epidermis and dermis were sustained over a shorter period in the alcohol solution groups (3–6 hours), as compared with the liposomal preparation used (24 hours). The liposomal formulation also increased the drug concentration at a depth between 30 and 150 μm, indicating an accumulation in the sebaceous ducts and upper portion of the glands.[77]

2.2.5.1.6 Combination Therapy through Liposomes

Considering that acne is a multifactorial disease, combination therapies using agents with varying mechanisms of action are recognized as an effective treatment strategy. In fact, it is rare to treat acne with monotherapy, especially when using a topical antibiotic, as the likelihood of developing microbial resistance is high. Topical retinoids in combination with topical antimicrobials have been shown to reduce inflammatory and noninflammatory acne lesions faster and to a greater degree than antimicrobial therapy alone. This may be explained by the fact that combination treatments target several areas of acne pathophysiology simultaneously. In addition, topical retinoids may affect skin permeability and facilitate the penetration of the topical antibiotic.[66] Once again, the main limiting factor is the associated irritation, which can be heightened when medications are used concurrently. In addition, drug interactions when using combination therapy can result in inactivation/degradation of the active ingredients: for example, BP and a retinoid. The advantages of liposomal preparations are quite clear.

In one double-blind study with mild to moderate acne patients ($n = 30$; 12-week study period), a comparison of concomitant treatment with BP 2.5% gel in the morning and tretinoin 0.025% gel in the evening to their liposomal counterparts revealed significantly better (1- to 1.5-fold) therapeutic response with the liposomal treated group.[57] There was a significant improvement in both comedonal and inflammatory types of acne lesions as well as a reduction in duration of therapy. These factors in turn improved patient compliance. This is consistent with the biological mechanisms of action, as retinoids can unplug the follicular ostia, allowing BP access to the intrafollicular *P. acnes* to exert its antibacterial action.[60]

2.2.5.2 Niosomes

Niosomes are nonionic surfactant vesicles, formed from the self-assembly of nonionic amphiphiles in aqueous media. Niosomes present a bilayer structure similar to that of liposomes[79] and can be considered the next evolutionary step for liposomes. Niosomes are able to entrap hydrophilic and hydrophobic molecules. The great availability of surfactants, plus their low cost and stability, has led to the investigation of these colloidal carriers as an alternative to conventional liposomes. However, there has been limited work performed with niosomes in the acne arena. Release rates of retinoic acid from various sizes and charges were evaluated in *vitro,* and penetration/drug concentration in pig skin was investigated.[52,80–82] Smaller and negatively charged niosomes showed greater skin penetration than conventional forms of Retin-A along with greater epidermal concentrations. It is likely that this platform will be used in the near future to deliver antiacne agents, as the in *vitro* data are encouraging.

2.2.5.3 Solid Lipid Nanoparticles

SLNs are nanoparticles made from solid lipids, ranging in diameter from approximately 50 to 1000 nm. SLNs were first developed for improving delivery of parenteral nutrition, simply by replacing the liquid lipid (oil) of the emulsion droplets with a solid lipid.[83] In contrast to emulsions for parenteral nutrition, which are generally stabilized with lethicin, SLNs can be stabilized by multiple surfactants or polymers. Compared with vesicular and other particulate systems such as liposomes, SLNs have more advantages for drug delivery, such as good tolerability and biodegradation, high bioavailability, and sustained release due to their solid matrix.[84] Moreover, SLNs are easily scaled up to an industrial size at minimal cost and without the use of organic solvents.[85] Some problems should be noted, however, such as limited encapsulation efficiency and drug expulsion from the carrier due to lipid polymorphic transformations.[86] The release from SLNs depends on the localization of the drug in the solid lipid matrix. If the drug is localized on the outer shell, burst release will be observed and, probably, no controlled release will be achieved. If the drug is homogeneously distributed within the lipid matrix, however, controlled release can be achieved (Figure 2.5).

In dermatology, SLNs have emerged as an alternative to liposomes due to the aforementioned advantages, such as improved physical and photostability, low cost compared to phospholipids, and ease of scale-up and manufacturing. Most importantly, their potential in epidermal targeting[43,83,87] and follicular delivery[39,54,88,89] has been established and is equivalent to that of liposomes. As with liposomes, research evaluating the improved efficacy and tolerability of SLN-encapsulated antiacne drugs has been pursued.

2.2.5.3.1 Retinoids

The preparation and characterization of tretinoin-loaded SLNs have been reported in the literature.[40,43,48,87,90–93] Cutaneous irritation studies have been carried out on rabbits (Draize test) and mice showing that retinoic acid SLN gel was significantly less irritating to skin compared with the marketed retinoic acid cream (Retin-A).[40,43] In addition, *in vitro* permeation studies through rat skin indicated that retinoic acid SLN gel presented a permeation profile comparable to that of the conventional retinoic acid cream. Even more importantly, the SLNs also improved the retinoic acid photostability in comparison to methanolic retinoic acid solution when both formulations were exposed to 180 minutes of natural sunlight.[94]

The loading of isotretinoin has also been investigated as a means of avoiding systemic administration and therefore bypassing the teratogenic risks and side effects associated with its use. Isotretinoin permeation of rat skin was not observed with loaded SLNs, despite a significant accumulation within the tissue, in contrast to the retinoid dissolved in 95% ethanol.[422] The increase in accumulative uptake of isotretinoin in the skin suggests that SLN encapsulation provides for enhanced skin-targeting effect.

2.2.5.3.2 Antiandrogens

2.2.5.3.2.1 Cyproterone Acetate The encapsulation of topical antiandrogens has also been found to promote site-targeted increase in drug concentrations, specifically in the upper skin layers after SLN application. Human skin penetration of cyproterone acetate-loaded SLNs resulted in a fourfold increase in epidermal concentrations as compared with the uptake from creams and nanoemulsions. This formulation was also compared to incorporation into nanostructured lipid carriers and microspheres, which resulted in only a two- to threefold increase.[95]

2.2.5.3.2.2 RU-58841 Myristate The encapsulation of the myristate prodrug form of the antiandrogen RU 58841 in SLNs has been evaluated. Penetration and permeation studies

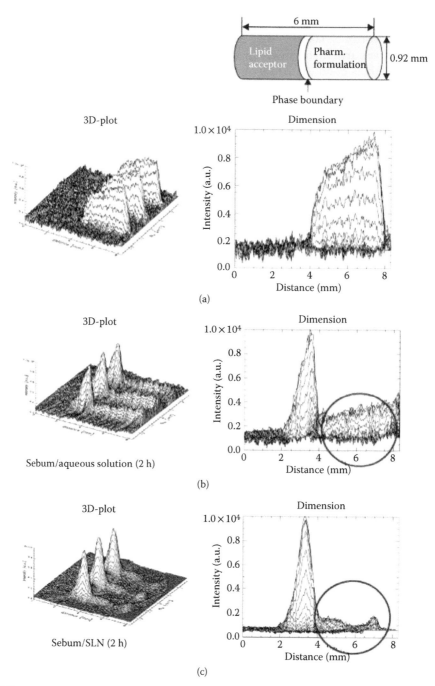

FIGURE 2.5

Electron spin resonance demonstrates that the movement of a probe from solid lipid nanoparticles (SLNs) into sebum occurs more completely than from an aqueous solution. The left side of a 6 mm capillary is loaded with sebum (the lipid acceptor) and the right side is loaded with an electron spin probe either in aqueous solution or within SLN (top diagram). (a) Immediately after loading, high levels of the probe are localized to the right side of the capillary tube. (b) After 2 hours, some, but not all, of the probe has moved from the aqueous solution (on the right) into the sebum-containing portion (on the left). (c) In comparison, when the probe is delivered in SLN, it is almost entirely distributed within the sebum at 2 hours. (From Kuchler, S., *Int. J. Pharm.*, 390, 225, 2010. With permission.)

were conducted using reconstructed human epidermis and excised porcine skin. Minimal skin permeation was observed even with the rapid conversion of the prodrug to its active form by various skin cell lines, and intact SLN were found to target the pilosebaceous unit using electron microscopy.[76]

2.2.5.4 Microspheres/Particles

Microencapsulation is a process through which micronized particles, either solid or liquid, are coated with inert, natural, and synthetic polymeric materials. The resulting composite is a microparticle. When the core of a particle is a polymeric matrix that contains a drug that is homogeneously dispersed throughout this matrix, the system is entitled a microsphere.[96–98] Microspheres offer numerous advantages when compared with the conventional formulations, undescoring their potential in the controlled release of administered drugs through the systemic pathway. Interest in their use for improving topical application has exponentially increased in recent years. Microspheres have showed good stability when applied to the skin, ease of preparation, stability and protection of the encapsulated drug against degradation, and controlled/sustained release.

Microspheres can be considered a drug depot, serving as a reservoir system for the active agent. When applied on the skin, the amount of free active agent in the formulation penetrates into the epidermis, followed by the drug release from the microspheres.[97,99–102] This provides immediate and sustained drug delivery without overloading the epidermis and subsequent increase of transdermal penetration. The use of microspheres in the acne therapeutic arena has been investigated.

2.2.5.4.1 Benzoyl Peroxide

Several investigations on cutaneous irritation, percutaneous absorption in *vitro* and in *vivo*, and the efficacy of the encapsulated BP in microspheres have been pursued.[36,39,99,100,103] In one study, cumulative irritancy of BP microspheres versus BP lotion was evaluated over 21 days in rabbits and humans.[103] The lotion containing free BP at 2.5% and 5% produced significantly greater irritation than that observed in entrapped BP. Encapsulated BP was also associated with a lesser percutaneous absorption in *vitro* and in *vivo* when compared with free BP. Even more importantly, corresponding in *vivo* human antimicrobial efficacy studies showed that application of the formulations containing entrapped BP significantly reduced counts of *P. acnes* and aerobic bacteria and the free fatty acid/triglyceride ratio in skin lipids. Similarly, BP-loaded microsponges have also been evaluated for their ability to reduce noninflammatory and inflammatory acne lesions.[104,105]

2.2.5.4.2 Retinoids

Similarly with liposomes and SLN, attempts to increase the efficacy and tolerability of retinoic acid have been pursued using microspheres and microsponges.[101,106–112] In fact, there are currently commercially available formulations of Retin-A-loaded microspheres (Retin-A, Micro™, Ortho Dermatological). Ex *vivo* mouse skin cutaneous permeation studies showed that the retinoic acid permeation from this microsponge technology was significantly less than that observed with the gel form of native drug. A multicenter, double-blind, placebo-controlled, 12-week study was undertaken, revealing a statistically significant reduction in inflammatory and noninflammatory acne lesions treated using retinoic acid–loaded microsponges as compared to control.

Recently there have been several studies looking at the efficacy and cumulative irritancy of retinoids either in microsphere or in native form, as well as comparing various microsphere formulations. In one study, 483 participants aged 10–14 years with mild to moderate acne were evaluated to compare the efficacy and tolerability of once-daily treatment with micronized tretinoin gel 0.05%, tretinoin gel microsphere 0.1%, and vehicle over 12 weeks.[110] Inflammatory and noninflammatory lesion reduction and treatment success was comparable between micronized tretinoin gel 0.05% and tretinoin gel microsphere 0.1%. Inflammatory (46.3%) and noninflammatory (45.7%) lesion reduction with tretinoin gel 0.05% was significantly greater than vehicle (37.1% and 27.9%, respectively). However, micronized tretinoin gel 0.05% provided a comparable lesion reduction and treatment success versus tretinoin gel microsphere 0.1%, with a better cutaneous tolerability profile.

A recent open-label study was conducted with 40 patients 8–12 years of age with mild/moderate acne to assess the tolerability and safety of tretinoin 0.04% microsphere gel.[101] Patients were treated for 12 weeks and were evaluated at baseline and at weeks 3, 6, and 12. Treatment-associated adverse events were minimal, with mild skin irritation being most commonly recorded in the first 3 weeks of therapy. The tretinoin 0.04% microsphere gel was found to be effective and safe for the treatment of acne vulgaris in this 8- to 12-year-old population, and the treatment was generally well tolerated.

In addition, since the commercialization of new retinoid formulations as well as the development of combination therapy routines, several studies have been published comparing the tolerance and efficacy of microsphere formulations with other retinoic acid derivatives (adapalene and tazorotene)[113] and in the setting of other therapies, such as BP.[101,113]

2.2.5.5 Nanoemulsions

Nanoemulsions are oil-in-water (o/w) emulsions with mean droplet diameters ranging from 50 to 1000 nm, though the average droplet size is generally between 100 and 500 nm. Nanoemulsions have been used in parenteral nutrition for a long time. These nanoemulsions typically contain 10%–20% oil stabilized with 0.5%–2% egg or soybean lecithin.[113,114] Because of their lipophilic interior, nanoemulsions are more suitable for the transport of lipophilic compounds than are liposomes. Similar to liposomes, they support the skin penetration of active ingredients and thus increase their concentration in the skin. Furthermore, nanoemulsions have gained increasing interest due to their own bioactive effects. It has been shown that nanoemulsions may reduce the transepidermal water loss (TEWL), indicating that the barrier function of the skin is strengthened from their application.[115,116]

Nanoemulsions have been pursued in targeting the pilosebaceous unit for the treatment of acne. NB-003 is an antimicrobial o/w emulsion in development for the topical treatment of acne. NB-003 has been shown to concentrate in the pilosebaceous unit and has potent in *vitro* bactericidal activity against multidrug-resistant clinical isolates of *P. acnes*.[117] NB-003 was evaluated in a pig skin model designed to mimic clinical studies that measure the reduction of *P. acnes* in human volunteers. It was found that NB-003 was more effective at reducing surface *P. acnes* on pig skin than were commercial products containing BP.

In summary, acne vulgaris is one of the most common skin conditions worldwide, impacting a broad range of patient populations. Because of the financial and time constraints of novel agent development, the trend in acne therapeutics has been more focused on improving or enhancing established therapeutics. Nanotechnology is a clear means to achieve this goal, as evidenced by the wealth of research and application of various nanomaterials to this arena. It is anticipated that growth in this area will continue to rise exponentially.

2.3 Atopic Dermatitis

2.3.1 Epidemiology

Atopic dermatitis (AD) is predominantly first seen in children,[118] with a prevalence of 6.8%–17.2% of school-aged children being affected in the United States.[119] Only 10% of patients with AD are adults.[118] Asthma and allergic rhinitis, components of the atopic spectrum, are seen in 30% and 35% of patients with AD, respectively.[120] Childhood AD is financially and emotionally taxing on families and is equivalent to caring for a child with type I diabetes mellitus.[121] Scratching and rubbing are common in both children and adults, which can disturb sleep and represent a significant cause of distress.[122] The prevalence of AD has been increasing since World War II, possibly due to the growth in urban populations, as urban populations have a higher prevalence of AD than their rural counterparts.[123] The cause of this association is unknown but has been speculated to be because of decreased early exposure to antigens and infections in the urban setting.[124]

2.3.2 Pathophysiology

AD is a systemic disorder that clinically manifests in the skin. Even clinically unaffected skin is drier, more easily irritated, and has a higher lymphocyte content than skin of patients without AD.[118] Irritation by allergens, scratching, and skin flora activates the immune cells in the skin and initiates inflammatory pathways. Changes in the skin barrier, including alterations in ceramide content and lower levels of antimicrobial peptides, decrease the skin's normal resistance to these irritants.[125]

The immune response in the skin of AD patients is biphasic.[125] It is Th2 mediated with eosinophilia and elevation of IgE in acute lesions and Th1 mediated in chronic lesions. Dysfunctions in regulatory T cells and dendritic cell, both of which can regulate Th1 and Th2 T cells, may be responsible for the immune response seen in AD.

Twin studies have confirmed that genetics play a significant role in the development of AD.[123] Several genes, including the filaggrin gene, have been associated with AD.[126] Filaggrin plays a key role in the differentiation of the epidermis. Linkage studies have also implicated the regions encoding Th2 cytokines and IgE receptors.

Acute lesions are histologically characterized by spongiosis (edema) and perivascular T-cell lymphocytic infiltrates.[118] Chronic lesions show psoriasiform acanthosis or hyperplasia of the epidermis and an increase in dendritic cells and monocytes.

2.3.3 Clinical Features

The manifestation of AD varies by patient age and duration of disease.[118] Acute AD appears as erythematous, edematous papules and plaques with vesicles, and oozing and crusting of the affected skin. Lesions are very pruritic, often covered in excoriations from scratching. As the lesions persist, the edema subsides, and scaling develops. Chronic rubbing, scratching, and picking leads to lichenified (increased skin markings) plaques and prurigo nodularis (pickers nodules). When lesions resolve, hypopigmentation or hyperpigmentation may be present.[123]

In children, the distribution of AD correlates to areas of rubbing.[118] In infants, the face, scalp, and extensor surfaces are involved, while the diaper area tends to be spared. As children start to crawl, knees and forearms become the common sites of involvement. After the age of two, the distribution begins to approach the flexural distribution of adult hood, involving the antecubital and popliteal fossae, wrists, hands, and neck.

A variety of other skin changes can be seen in AD including[118] xerosis, dry skin; keratosis pilaris, horny plugs in hair follicles with or without perifollicular erythema; ichthyosis vulgaris, increased scaling most prominently on the shins; Dennie–Morgan lines which are horizontal skin folds of the lower eyelid; allergic shiners, edema, and vascular congestion of the lower eyelid; increased skin markings on the hands; and cheilitis, characterized by dry, crusting, fissuring skin around the lips.

Concurrent bacterial, viral, and fungal infections are seen in AD. Yellow crusts and pustules suggest that AD has become superinfected with *Staphylococcus aureus*. The painful, vesiculopustular or punched-out lesions of eczema herpeticum (a disseminated herpes simplex viral infection) can resemble bacterial superinfection.[127] Dermatophytes like such as *Trichophyton rubrum* can easily invade the stratum corneum and infection can exacerbate AD.

Allergen testing, food challenges, and IgE levels may be useful in selected cases, but their widespread use is not suggested, because they are expensive and time consuming.[127] Biopsy is of little use in diagnosis (Figures 2.6 and 2.7).[123]

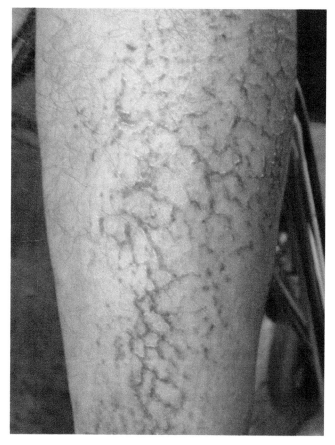

FIGURE 2.6

Asteatotic eczema. This scaling, erythematous eruption, resembling a dry river bed, hints at the underlying inflammation and breakdown in the skin barrier seen in eczema

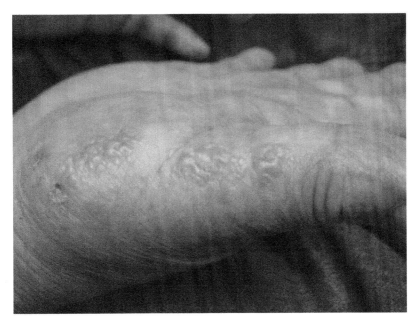

FIGURE 2.7
Dyshidrotic eczema. Characterized by small and itchy blisters on the hands and feet, dyshidrotic eczema represents one of the more difficult-to-treat eczema variants. The palmar and plantar surfaces are resistant to traditional drug-delivery methods because of their thickness and decreased number of follicles and pores.

2.3.4 Current Treatment

Corticosteroids, emollients, wet wraps, and patient education are first-line therapies.[127] There is little difference between short-term and long-term corticosteroid therapy except the risk of developing adverse events.[123] With more severe forms of AD, more potent corticosteroids may be necessary. The mechanism of action of corticosteroids is covered in Section 2.8.5.

Emollients improve the dry skin associated with AD, but do little to improve AD when used alone. When used in combination with a corticosteroid, emollients improve the efficacy of the corticosteroid.[127] Wet wraps are also effective in improving dry skin and the efficacy of corticosteroids.

Topical calcineurin inhibitors are also safe and effective alternatives to steroids. Low-dose preparations are available for use in children 2 to 15 years of age.[123] Phototherapy is also effective for both acute flairs and chronic AD. Although phototherapy is effective and well tolerated, adoption of this therapy has been limited because of the lack of treatment guidelines and poor access to phototherapy centers.[128] For extremely resistant cases, immunosuppression with systemic steroids, azathioprine, cyclosporin, and MTX can be effective.[127] However, use of these medications must be balanced against their known toxicities. The mechanisms of these immunosuppressive agents are covered in Section 2.8 of this chapter.

Antihistamines, both topical and systemic, do little to improve AD, with the exception of systemic first-generation antihistamines that improve sleep because of their sedative properties.[127] Topical preparations run the risk of sensitization. If allergic triggers can be identified, attempts should be made to avoid them. Doxepin, a tricyclic antidepressant, can provide itch relief but is sedating. Antibiotics, antivirals, and antifungals may be necessary when AD becomes super infected.[127]

The current therapies for AD focus on treating disease flares but do little to prevent lesions.[129] Frequent reapplication of topical medication and the use of occlusive dressing are difficult for children. Furthermore, children are often undertreated, because physicians and parents try to avoid the long-term effects of corticosteroids and phototherapy. Serologic monitoring and the threat of opportunistic infections limit the use of immunosuppressants to only the most severe cases.

2.3.5 Application in Nanotechnology

2.3.5.1 Diagnosis

Noninvasive or minimally invasive methods for the diagnosis of AD can offer significant advantages over predecessors.[129–131] Proteomic profiling is one method of assessing AD. In a pilot study of a small subset of patients with AD, the utility of proteomics was demonstrated.[132] Patients with a history of AD and eczema herpeticum and *S. aureus* colonization were tape stripped on lesional and nonlesional skin. Proteins related to the skin barrier were assessed by mass spectrometry. Skin barrier proteins, including filaggrin-2, corneodesmosin, desmoglein-1, desmocollin-1, and transglutaminase-3, and proteins comprising natural moisturizing factors, such as arginase-1, caspase-14, and γ-glutamyl cyclotransferase, were diminished in lesional skin when compared to nonlesional skin in patients with a history of AD and no history of eczema herpeticum. Epidermal fatty acid–binding protein was expressed in higher levels in patients colonized with methicillin-resistant *S. aureus*. This study showed that proteomic profiling might noninvasively give clues about the pathogenesis of AD. It may also be useful in personalized medicine, helping identify patients at risk for serious cutaneous bacterial and viral superinfections.

In *vivo* reflectance confocal microscopy has been shown in case studies to distinguish eczema from Paget's disease.[129] Similar methodologies have been used for the evaluation of allergic contact dermatitis.[130] Fast optical in *vivo* topometry, which combines laser profilometry and confocal microscopy, has been shown to distinguish healthy skin from atopic skin and to monitor improvement with the use of topical therapy.[133] If imaging techniques could be combined with biomarkers using, for example, quantum dots (QDs), the reliability and specificity of diagnosis could be enhanced.

2.3.5.2 Treatment

As discussed earlier, AD may be multifactorial and may be associated with a variety of factors, including deficiencies in the skin barrier; exogenous and endogenous stressors, including climate and stress; diminished antimicrobial defense against bacterial, viral, and fungal pathogens; and enhanced inflammatory responsiveness. Therapy for AD can include topical therapies (baths, wraps, corticosteroids, immunomodulators, antihistamines, emollients, counterirritants, and barrier preparations), systemic therapies (antihistamines, anxiolytics, and immunosuppressive agents), phototherapy, and inpatient therapy.[134,135] Traditional therapies may lack specificity, may be inconvenient, and may be associated with systemic toxicity.

Tacrolimus, an immunomodulator originally used to prevent organ transplant rejection, has been formulated topically in a lipid nanoparticle.[136,137] The nanoparticulate form is more soluble than the unencapsulated counterpart. It has better delivery and retention kinetics in the epidermis. These characteristics predict greater efficacy, a longer duration of activity, and reduced systemic release and toxicity.

UV-induced dermatitis is believed to involve reactive oxygen species (ROS). Platinum has been shown to catalyze the dissipation of ROS. Nanoparticulate platinum is more potent than macroparticulate platinum, because catalysis is a surface phenomenon, and nanoparticulate platinum has a very high comparative surface to volume ratio. Nanoparticulate platinum stabilized with a layer of polyacrylic acid was found to reduce UV-induced ROS production in HaCAT cells. This effect was also noted in UV-irradiated mice pretreated with a nanoparticulate platinum gel, preventing UV-induced inflammation and apoptosis.

AD is generally not associated with a higher incidence of allergic contact dermatitis, including nickel dermatitis. However, the most common cause of contact dermatitis is nickel, and nickel contact allergy can exacerbate AD. Nickel can be found on jewelry and everyday items such as cabinet handles, cell phones, and coins. Nickel must penetrate the epidermis to induce contact allergy. Topical agents to reduce the penetration of nickel into skin have benefited from nanoformulation.[138,139] Calcium nanoparticles capture nickel ions by cation exchange and cause nickel to remain on the skin surface. In *vitro*, and in *vivo*, this capture and ease of washing away nickel was shown with nanoparticles 500 nm and smaller containing calcium carbonate or calcium phosphate. About 90% fewer nanoparticles by mass are needed to match the efficacy of ethylene diamine tetra acetic acid (Figure 2.8).

One of the hallmarks of eczema is inflammation. ROS are generated during inflammation, and fullerenes have been shown to quench ROS.[140–145] In an in *vivo* model of inflammation induced by phorbol 12-myristate 13-acetate, fullerene nanoparticles reduced inflammation and edema. Toxicity concerns regarding fullerenes and fullerols are being revised.

Clobetasol propionate is a potent topical corticosteroid used for the management of AD. A solid lipid nanoparticle containing clobetasol propionate has been developed and characterized.[146] Optimized nanoparticles have an average size of 177 nm and entrap 92% of drugs. The nanoparticulate form of clobetasol had a higher uptake in the epidermis and enhanced retention in the skin compared with unencapsulated clobetasol. In a double-blind pilot study of 16 patients with chronic eczema, both forms of clobetasol improved outcomes, but nanoparticulate clobetasol recipients had a 1.9 times greater reduction in inflammation and a 1.2-fold greater reduction in itching compared with native clobetasol recipients.

A nanoemulsion containing prednicarbate is also being developed for the treatment of AD.[147,148] The emulsion contains phytosphingosine to confer a positive charge for enhanced epidermal binding and penetration. The active ingredient, prednicarbate, is located on the stabilizer layer of the nanoemulsion, and this translocation from the inner oil phase imparts increased prednicarbate stability.

Angiogenesis has been shown in at least one model to be important in the pathogenesis of AD. Nascent capillaries formed by neoangiogenesis contain fenestrae that enhance their permeability to particles in the 150 nm range. Passive targeting of AD with systemic administration of stealth steroids and stealth anti-inflammatory medications may be a useful therapeutic strategy.[149,150]

Some of the side effects related to topical steroids stem from penetration into the dermis, where they can cause atrophy, telangiectasia, and purpura. Betamethasone 17-valerate is a potent class I topical steroid that has been used in the treatment of AD.[151] Monstearin betamethasone 17-valerate SLNs show persistence in the epidermis and slow controlled release. Other SLNs containing beeswax and betamethasone 17-valerate did not remain in the skin or dermal–epidermal junction, and readily diffused into the dermis. By localizing topical steroids to the epidermis and papillary dermis, nanoparticulate preparations of topical steroids may confer therapeutic benefit in AD while minimizing side effects (Figure 2.9).

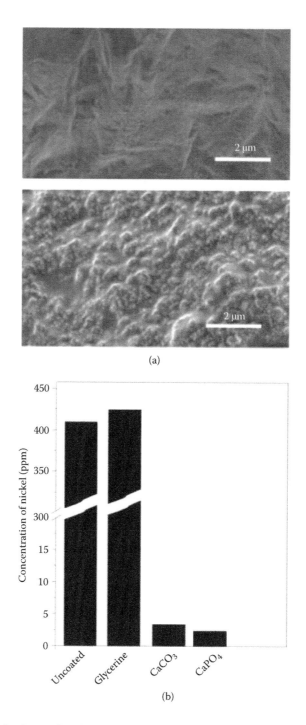

FIGURE 2.8
Calcium carbonate- and calcium phosphate-containing nanoparticles decrease the entry of nickel into skin. (a) Uncoated pig skin (top) and pig skin coated with nanoparticles (bottom) are exposed to nickel for 48 hours. (b) Skin coated with either calcium carbonate (CaCO$_3$) or calcium phosphate (CaPO$_4$) decreases the amount of nickel entering into the skin. (From Vemula, P.K., *Nat. Nanotechnol.*, 6, 291, 2011. With permission.)

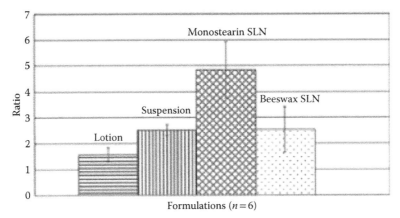

FIGURE 2.9
Betamethasone valerate content ration in human skin. Monostearin solid lipid nanoparticle (SLN) is significantly more effective in delivering betamethasone valerate than is a commercial lotion. However, not all SLNs are made equal. A beeswax SLN performs similarly to a simple suspension of betamethasone but not as well as the monostearine SLN valerate. $N = 6$, error bars represent standard deviation. (From Zhang, J., *J. Pharm. Sci.*, 100, 896, 2011. With permission.)

In models of rheumatoid arthritis, glucocorticoids encapsulated in liposomes penetrated arthritic joints, which have fenestrated capillaries, resulting in passive targeting.[152,153] This type of passive targeting allows for greater efficacy, prolonged duration of therapy, need for less medication, and fewer side effects. Neoangiogenesis in tumors mediated by vascular endothelial growth factor (VEGF) is also associated with increased capillary permeability. Evidence of enhanced capillary permeability by a similar mechanism has been seen in models of AD. Passive targeting using nanoencapsulated liposomal corticosteroids may be beneficial in the therapy of AD.

Because one of the fundamental aspects of AD is impaired skin barrier function, the use of topical preparations that enhance the skin barrier can improve outcomes. Emollients have been shown to reduce the requirements for topical steroids. Replacement emollients that are optimized to correct the barrier defect typically contain physiologic lipids—such as cholesterol, free fatty acids, and ceramide—in appropriate proportions that mimic physiologic cutaneous conditions. Barrier creams containing ceramide-3 lipid nanoparticles in physiologic ratios showed improvement in disease severity and symptomatology for atopic dermatitis, irritant contact dermatitis, and allergic contact dermatitis.[134] Patients treated with balanced lipid barrier creams in combination with topical steroids had a more favorable response than patients treated with barrier cream alone.

Gene silencing techniques may also be useful in the management of AD. One tool for silencing genes is siRNA. Penetration of siRNA into skin, including the skin of patients with AD, is prevented by enzymes, lipids, and cell membranes. Nanoparticles containing Tat analogs combined with AT1002, which increases paracellular transport across the epidermal barrier, increase the skin permeability of water-soluble siRNA.[154,155] In animal models, siRNA that silences RelA, a member of the NF-κB family, was successfully delivered topically into the skin of mice with AD and reduced ear thickness, clinical skin severity score, skin cytokine levels, and serum IgE levels. Gene silencing of CD86—a costimulatory molecule on dendritic cells expressed after antigen uptake—using a cream containing nanoemulsified CD86 siRNA was shown to reduce murine model contact hypersensitivity and AD.[156]

One of the steps in an inflammatory response involves extravasation of leukocytes from the circulation into the interstitium. This process of leukocyte migration is mediated by cell adhesion molecules known as selectins. E-, L-, and P-selectins bind carbohydrate ligands (lectins). Dendrimers are richly arborized nanoparticles, which can contain many ligands per unit mass on their surface. Dendrimers made of polyglycerol sulfate are efficient multivalent inhibitors of L-selectin on leukocytes and P-selectin on endothelia cells. This binding can be manipulated by altering particle size and sulfation of the core polymer. In animal models of contact dermatitis, nanopolymeric dendritic polyglycerol sulfates were efficient inhibitors of leukocyte extravasation, and reduced edema and inflammation on par with glucocorticoids. At least one instance of toxicity from topical application of dendrimers has been reported (Figure 2.10).[157-160]

Colonization with *S. aureus* has been shown to cause secondary infection and exacerbate inflammation in patients with eczema. In one case, *S. aureus* endocarditis was seen as a complication of acupuncture therapy for AD.[161-167] A meta-analysis of randomized controlled trials of patients with AD who were treated to minimize *S. aureus* colonization showed no benefit of oral antibiotics in AD individuals, regardless of infection status.[168] Antibacterial soaps, baths, or antiseptics also did not show any benefit. There was equivocal benefit of adding antibiotics to topical corticosteroids. In one trial, the use of silver textiles did not show any improvement for AD, despite a high use of concomitant topical steroids. The meta-analysis was limited by small study sizes and limited data reported. Textiles treated to reduce friction have been shown to reduce symptoms and TEWL in patients with AD.[169-176] Imaging techniques using nanofiber-impregnated textiles may be useful for monitoring skin lesion size and distribution and for the photodynamic therapy (PDT) of AD.[170,172,175,177-179]

In summary, the diagnosis of AD, traditionally done using clinical criteria will be enhanced with nano-based methodologies. Genome, transcriptome, and proteome analysis will help distinguish AD from other related dermatitides. Advanced imaging techniques

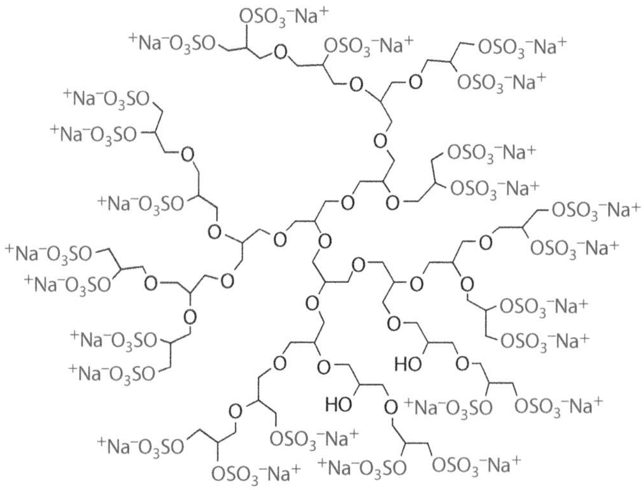

FIGURE 2.10
Dendritic polyglycerol sulfate. Nanoscale polymers of dendritic polyglycerol sulfates inhibit leukocyte extravasation and reduce edema and inflammation. (Reproduced with permission from Dernedde, J., *Proc. Natl. Acad. Sci.*, 107, 19679, 2010.)

may also be used to noninvasively diagnose and monitor AD.[178,180] Therapies using nanotechnology may be developed that have greater efficacy, specificity, and fewer side effects.

2.4 Actinic Keratoses

2.4.1 Epidemiology

Actinic keratoses (AKs) are cutaneous neoplasms that consist of growths of cytologically aberrant epidermal keratinocytes that develop in response to prolonged exposure to UV radiation. Although the true prevalence of AK is not known, it was found to occur in 14% of patients visiting dermatologists in the United States (suggesting the actual prevalence is much higher).[181,182]

2.4.2 Pathophysiology

Although genetic and environmental factors play a role in the development of AKs, it has long been recognized that UV radiation is a key etiologic factor responsible for their development. This is evidenced by the fact that more than 80% of AKs are distributed on habitually sun-exposed areas of the body (scalp, head, neck, forearms, etc).[183] Although there has been some debate regarding their true nature or even if this entity exists, it is generally believed that AKs are the initial lesions in a disease continuum that progresses to squamous cell carcinoma (SCC).[184] Biologically, multiple mutations and ultimately attenuation of cell-cycle control are the key steps in the transition to malignancy.[185–187] Although multiple mutations are involved, such as those in p16 and RAS, it is believed that the primary alteration is consistently in the p53 proto-oncogene. Generally, UV radiation via sun exposure causes mutations in cell-cycle regulatory proteins such as p53.[188] These mutations eventually prevent p53 from playing its role in cell-cycle arrest and DNA repair, leading to unchecked growth and tumor formation.[189]

2.4.3 Clinical Features

There are two important questions to consider when considering and evaluating the AK lesion. First, how frequently does a SCC arise from a preexisting AK, and second, for a typical patient with several AKs, what is the chance of developing an SCC? Fortunately, there are numerous investigations that have pursued these very questions. Marks et al.[190] showed that approximately 60% of SCCs arose from AK lesions, as well as that each AK has a 0.075% risk of transforming into an SCC per year. This value has been extrapolated to suggest that over a 10-year period, a given individual with an average of ~7 AKs has a 10% risk of developing an SCC.[191] Therefore, the "10%" risk refers to a patient with multiple AKs, not the patient with the single lesion. Overall, two points are evident. First, it is clear that the risk of developing an SCC increases considerably with and proportionally to an increasing number of AKs present on a given person per year.[192] Second and even more importantly, in the average patient with AKs, the very low yearly transformation rate for individual AKs translates into a substantial risk over a lifetime.[193,195] However, it has also been shown that up to 25% of all solar keratoses may remit spontaneously within 12 months and with reduced exposure to sunlight.[191] Putting this all together, the

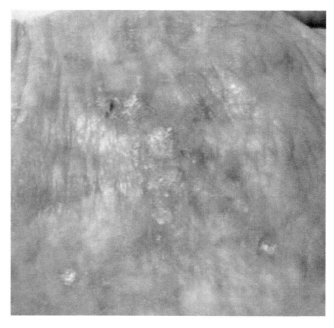

FIGURE 2.11
Actinic keratosis: Characterized by small rough raised erythematous guttate plaques of skin that has been exposed to the sun for long periods.

progression to SCC is probably less than commonly stated, but it is not zero. Therefore, these data suggest that therapy for skin cancer prevention is warranted.

The typical patient with AKs is an older, fair-skinned, light-eyed individual, with a history of sun exposure, who burns and freckles rather than tans, and has significant actinic damage on examination. Common signs and symptoms include pruritus, burning or stinging pain, bleeding, and crusting. The typical AK lesion presents most commonly as a 2–6 mm erythematous, flat, rough, or scaly papule. They are better felt than seen; gently abrading lesions with a fingernail usually induces pain, even in early subtle lesions, which can be a helpful diagnostic finding. They are most commonly found in areas of chronically sun-exposed skin such as the face, scalp, and the dorsa of the hands (Figure 2.11).[196,197]

2.4.4 Current Treatment

The factors influencing how treatment should proceed may be determined by (1) the medical status of the patient; (2) lesion characteristics such as size, location, duration, changes in growth pattern of single, isolated lesions, suspicious lesions, and plaques or diffuse lesions; (3) previous treatment; and (4) anatomic location.[198] Although treatment of symptomatic AKs in patients with pruritic or painful lesions is warranted to minimize symptoms, asymptomatic patients may decide that no treatment is needed after the physician has explained the risks of skin malignancy and the treatment options.[196]

Treatment modalities for AKs can be divided into lesion-targeted therapies and field therapies. Lesion-targeted therapies include liquid nitrogen cryotherapy, curettage with or without electrocautery, and shave excision. Field therapies include topical treatments, 5-fluorouracil (5-FU) cream and solution, 5% and 3.75% imiquimod cream, and 3%

diclofenac gel. Also included are procedural therapies such as cryopeeling, dermabrasion, medium-depth chemical peel, deep chemical peel, laser resurfacing, and PDT.

Liquid nitrogen cryosurgery is the most common destructive procedure, and is usually administered with a spray device or cotton tip applicator.[199] Potential disadvantages of cryotherapy include pain and discomfort to the patient, unsightly blisters and crusted wounds for a week or longer, hypopigmentation, scarring, and possible alopecia in treated areas.

Field-directed therapies, such as topical 5-FU, are another effective form of treatment. Fluorouracil is a cytotoxic agent that destroys dysplastic keratinocytes through interference of DNA and RNA synthesis via blocking methylation of deoxyuridylic acid to thymidylic acid.[200] Many individuals find 5-FU therapy difficult. This is because treatment of lesions commonly causes significant erythema and erosions, swelling, and pain, which may be mild, moderate, or severe, resulting in temporary cosmetic disfigurement.[194,196,198]

Imiquimod is the first in a new class of immune response modifiers that are effective in treating AK. Imiquimod works by activating macrophages and other cell by binding to cell-surface receptors, such as toll receptor 7, and inducing the release of proinflammatory cytokines, including interferon (IFN)-α, tumor necrosis factor (TNF)-α, and interleukin (IL)-12.[201] This results in antineoplastic and antiviral effects. Common side effects found in phase III trials for two and three times weekly applications were severe erythema (17.7%), crusting (8.4%), and ulceration (2.3%) for the twice weekly application[195] and severe erythema (30.6%–33.26%), severe crusting (27.4%–29.9%), severe flaking (8.7%–10.2%), and erosion/ulceration (10.2%) for thrice weekly application.[193] Laser surgery is usually effective for individual lesions, but for extensive facial lesions, facial resurfacing is effective.

PDT is another modality that can be used to effectively treat AKs, but its use is limited by availability and cost. PDT involves the use of a photosensitizing agent, oxygen, and light of a specific wavelength to produce controlled cell death. Typically ALA is applied topically, and then converted to porphyrins in the heme biosynthesis pathway. Dysplastic and neoplastic cells preferentially take up more porphyrins, and are therefore preferentially destroyed when the porphyrins are photoactivated by light between 400 and 600 nm.[194,200] The main adverse effects are pain and stinging during treatment and erythema and crusting after. Studies have found that PDT has improved cosmesis when compared to 5-FU and improved tolerability when compared to cryotherapy.[195,201,202]

2.4.5 Applications in Nanotechnology

2.4.5.1 Diagnosis

Traditional methods for diagnosis of AK have included medical history; however, diagnostic methods such as KOH preparation (to distinguish from tinea corporis), shave biopsy or punch biopsy may be performed. Skin biopsy may have variable reliability. Diagnostic tools using nanotechnology require less tissue, can be atraumatic, can be rapid, and may offer advantages in terms of sensitivity and specificity.

Diagnostic methods for assessing AK in addition to physical examination may include the use of dermoscopy. Dermoscopic features of AK sometimes overlap with other lesions and can sometimes make it difficult to distinguish pigmented AKs from lentigo maligna and other pigmented disorders. Reflectance confocal microscopy can also aid in the diagnosis of AK. Optical coherence tomography (OCT) can aid in the diagnosis of AK as well.[204–216]

Microarray nucleic acids have been tethered to nanocantilevers for the real-time analysis of gene expression. This technology is being adapted for the diagnosis of melanoma and can be modified for AK.[217–220] Techniques that capitalize on nanotechnology evaluate

AKs for molecular features including gene and protein expression.[221–223] Distinguishing AK from SCC is a molecular challenge, because many investigators believe that AK is an early SCC in situ. A few investigators have been able to find differences in gene expression profiles that may separate AKs from SCC. One study compared gene expression profiles in paraffin-embedded sections of normal skin, AK, and SCC using an Affymetrix U133plus2.0 array. They found 382 genes differentially expressed between SCC and normal skin, 423 differentially expressed genes between AK and normal skin and 9 differentially expressed genes between AK and SCC. They suggested that nucleic acid–based microarrays might be useful for the diagnosis, prognosis, or future therapy of AK and SCC.

2.4.5.2 Treatment

Traditional therapy for AK includes observation, and some studies have demonstrated that lesions appear and regress spontaneously. As mentioned above, cryotherapy using liquid nitrogen is successful in eradicating AKs, while thicker lesions may require paring or curetting followed by cryotherapy, or electrodessication and curettage (ED&C). Chemotherapy for AK includes topical therapy with immunomodulators (such as imiquimod), antimetabolites (such as 5-FU or its derivative, capecitabine), retinoids (such as tretinoin and acitretin) or nonsteroidal anti-inflammatory drugs (such as celecoxib).[205,224–248] PDT using ALA has also been successful but is limited by significant pain. Other methods for managing AKs include chemical peels and laser resurfacing.

The delivery of 5-FU to the epidermis using nanoparticles has been studied. Because of its surface characteristics, 5-FU has poor penetration capability in the epidermis.[249] Niosomes (250 nm), transferosomes (150 nm), and liposomes (120 nm) containing 5-FU were developed and characterized. In tissue culture, transferosomes were twice as efficient (82%) at entrapment compared to niosomes and liposomes. Transferosomes were much more cytotoxic than niosomes and liposomes, and all nanoparticles were more cytotoxic than free 5-FU (IC50 = (1.02 μmol/L, 9.91 μmol/L, 6.83 μmol/L, 15.89 μmol/L, respectively).

The use of nanoemulsions for delivery of PhotoFluor in the treatment of AK has also been investigated.[224] ALA is chemically unstable in solution and has poor skin penetration. A stabilized nanoemulsion form of ALA (BF-200 ALA) was developed and studied in a phase III clinical trial in Regensburg, Germany. A total of 122 patients were studied, with 4–8 moderate AK lesions on the face or scalp, in a double-blind, placebo-controlled, multicenter trial. The efficacy of BF-200 ALA was studied after one and two treatments with PDT using two different light sources (Aktilite CL128 and PhotoDyn 750). BF-200 ALA was shown to be superior to a placebo in complete clearance of AK lesions (64% vs. 11%).

PDT using ALA is effective in eradicating atypical cells but is limited by subjective pain. A nanoemulsion of ALA was compared to methyl ALA in the treatment of AK. In an academic medical center in Heidelberg, Germany, 173 patients with 965 treated areas were enrolled in a retrospective study. Subjects all had multiple AKs and had undergone extensive treatment area PDT with 5-aminolevulinic acid methylester (MAL) (424 treated areas) or 5-ALA nanoemulsion (BF-200-ALA) (541 treated areas). Patients were provided with a visual analog scale to rate their pain during therapy, and treatment interruptions due to pain were recorded. Patients treated with MAL reported a lower VAS (5 vs. 5.8), a lower amount of "severe" pain (25% vs. 36%), and fewer treatment interruptions (13.2% vs. 19.9%) compared to patients treated with BF-200-ALA. Methods using nanotechnology to enhance delivery of topical anesthetic theoretically could make PDT more tolerable. However, at least one study has showed that topical anesthesia with morphine 0.3% gel is not effective as a pain reliever for PDT of AK.

Chemoprevention of nonmelanoma skin cancer and AKs has been shown using cele-coxib, capecitabine, retinoids, PDT, and diet.[226,248,250–252] Diet may work through many mechanisms, including antioxidant mechanisms.[254,255] Topical DL-α-tocopherol has shown utility in preventing AK, and formulations using nanostructured lipid carriers and elec-trospun cellulose acetate nanofiber mats are being explored for enhanced skin delivery and stability of tocopherol.[228,241,255–269] Topical resveratrol is being developed in nanopar-ticulate form. Valrubicin and difluoromethylornithine in combination with triamcinolone has been shown to treat AKs.[252,270] Sunscreen has demonstrated cost-effectiveness in the management and prevention of AK.[229,241] The addition of antioxidants such as tocopherol to sunscreen may enhance their efficacy, and nanoencapsulation is allowing for more effec-tive and cosmetically elegant combination products.[257,259,264,265,267–269]

Mutations in the epidermal growth factor receptor (EGFR) pathway have been linked to SCC and AK.[232] The EGFR tyrosine kinase can activate signaling in the following antiapop-totic pathways: PI3K/AKT, JAK-STAT, and ERK/MAPK. Inhibitors of the epidermal growth factor tyrosine kinase include gefitinib and erlotinib. These are engineered antibodies with specific targeting properties. By their size ($5 \times 5 \times 20$ nm), specificity, and the precision engi-neering involved in their design and manufacture, biological antibodies can be classified as nanotherapeutic agents. Erlotinib has been useful in the treatment of some AKs.

Topical cyclooxygenase 2 (COX-2) inhibitors have been used for the treatment of AK, and nanotechnology avenues are being pursued to enhance their efficacy. The mecha-nism of action is unclear but may involve inhibition of angiogenesis, restoration of normal apoptosis, reduced cell proliferation, and reduced inflammation. Recently, diclofenac, a topical COX-2 inhibitor, has been shown to be useful in the management of AKs in the periorbital area.[271] Topical 1% piroxicam gel has also shown efficacy in the treatment of AK. Nanoemulsion-based transdermal delivery of celecoxib and a related compound, ace-clofenac, has been demonstrated in *vitro* and in animal models, respectively.[253]

With appropriate combinations of topical and systemic therapies, innovative strategies using nanotechnology can be developed for the prevention and management of AK.

2.5 Skin Cancer

2.5.1 Nonmelanoma Skin Cancer

2.5.1.1 Epidemiology

Nonmelanoma skin cancer (NMSC) is the most common human cancer—the incidence of skin cancer is greater than that of all other cancers combined.[253,272] Basal cell carcinoma (BCC) and SCC are the most common cutaneous malignancies. Though there are no exact numbers, because these cases are not required to be reported to cancer registries, accord-ing to one report, in 2006 an estimated 3.5 million cases of NMSC occurred, and approxi-mately 2.2 million people were treated for NMSC.[273] An estimated 3190 deaths from NMSC occurred in 2011.[273] Many host and environmental factors contribute to the development of NMSC. Skin color is a major risk factor. Factors that confer the greatest risk of developing NMSC include having light complexion, burning easily, and tanning poorly. Other risk factors include increasing age, male gender, and the presence of precancerous skin lesions.

Cumulative lifetime UV exposure is a risk factor for SCC and BCC. Higher incidence rates are found among those with occupational exposure to sunlight, as well as those

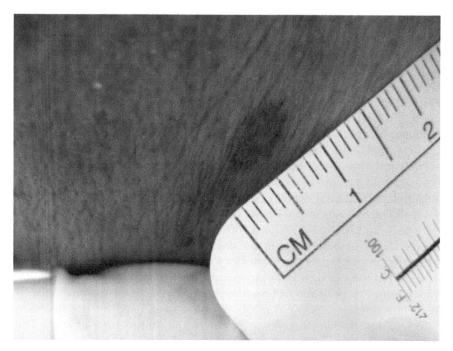

FIGURE 2.12
Pigmented basal cell carcinoma (BCC): Pigmented BCC is an uncommon variant of nodular BCC, which shares all the other clinical and histological features of the common nonpigmented histological subtype. Telangiectasia and ulceration are common features of BCC.

living in latitudes closer to the equator. SCC can also arise from chronic skin ulcers and thermal burn scars. Exposure to chemicals such as arsenic and coal tar products, as well as ionizing radiation, increases the risk of BCC and SCC (Figure 2.12).[195,274,275]

2.5.1.2 Pathophysiology

Similar to other epithelial cancers, such as breast and colon cancer, most of the genes implicated in the development of NMSC are mutations in tumor suppressor genes. Mutations in p16 and p53 genes are commonly found in NMSC as compared to other epithelial malignancies, clearly demarcating the role of UV radiation–induced mutagenesis in the development of NMSC.[195,253,274–277]

UVB radiation causes the formation of dimeric photoproducts between adjacent pyrimidine bases on the same strand of DNA. The number of cyclobutane pyrimidine dimers (CPDS) that formed in a basal cell of human epidermis after exposure to three minimal doses of solar light at about 60 minutes exposure at noon in the summer in Kobe, Japan, was calculated to be approximately 100,000 per cell.[278] The CPDS found to be most mutagenic are thymine–cytosine and cytosine–cytosine dimers. In addition, UVA, UVB, and UVC radiation can result in the formation of ROS causing oxidative damage to the cell and DNA. Oxidative DNA damage is one of the proposed mechanisms for the induction of SCC from arsenic exposure.[276,278,279] UV radiation induces a wide array of DNA damage, including protein DNA linkages and DNA single strand breaks.[280]

One of the main mechanisms of human cells to mitigate the potential damage to DNA by UV radiation is by nucleotide excision repair (NER). The importance of an intact NER system for the prevention of UV-induced skin cancers is emphasized by the high occurrence of skin cancer on the sun-exposed skin of patients with xeroderma pigmentosum (XP). Patients with XP have defective NER and are therefore extremely sensitive to UV-induced DNA damage.[276,278,280] In patients with functional NER, most DNA damage caused by UVB radiation is repaired effectively. However, there are a small percentage of these DNA lesions that are repaired incorrectly, which result in mutations of oncogenes such as *ras*, and tumor suppressor genes, such as *p53* or *PTCH*. Ultimately, these mutations provide the potential for the cells to transform and immortalize, leading to malignant tumor cells.

Mutations in *p53* are found in more than 90% of SCCs and more than 50% of BCCs. Mutations of *p53* are also found in 50% of AKs, suggesting that *p53* mutation is one of the first steps in SCC pathogenesis. Patched (PTCH) gene mutations are found in 50%–60% of BCC. PTCH encodes a transmembrane protein that inhibits smoothened protein (SMO) from translocating to the nucleus and inducing downstream transcription. Sonic hedgehog protein (SHH) regulates this inhibition by binding to PTCH and blocking its inhibition of SMO. Overexpression of SHH has been shown to induce BCC in mice, and mutations of SMO have been identified in sporadic BCC, showing that SMO can function as an oncogene for BCC formation.[281,282] Mutations in PTCH have also been shown to result in BCC formation, best represented by basal cell nevus syndrome, in which an autosomal dominant mutation in PTCH results in the formation of numerous BCCs.

There are no known precursor lesions for development of BCC. Precursor lesions for SCC include AKs, and Bowen's disease (SCC in situ). Marks et al.[190] showed that approximately 60% of SCCs arose from AK lesions as well as that each AK has a 0.075% risk of transforming into an SCC per year. However, because multiple AKs are typically found, the risk of an individual developing SCC may be substantially higher. SCC in situ appears as a well-demarcated erythematous, scaly plaque that is confined to the epidermis histologically.[275]

2.5.1.3 Clinical Features

BCCs can present with a variety of clinical types, which can be broadly categorized on the basis of whether they behave as indolent or aggressive tumors. The early tumors of BCC are commonly small, translucent, or pearly, raised and rounded areas covered by thin epidermis through which a few dilated superficial vessels can be seen.[195] BCCs typically found at this stage include the most common nodular type responsible for 75% of cases and usually are located on the face. The next most common indolent variant is the superficial variant, which typically presents on the trunk or extremities as a slowly expanding erythematous scaly patch. Both nodular and superficial forms may contain melanin, imparting a brown, blue, or black color to these lesions.[283] Aggressive variants include infiltrating, morpheaform, basosquamous, and micronodular types. Infiltrating and morpheaform variants can present as rapidly expanding ill-defined erythematous or whitish indurated plaques that are found on the face. Other manners of presentation of BCC are a small, pearly papule, or as a keratotic or slightly indurated area, or as a small and superficial ulcer.[284] Casual inspection of superficial BCCs may suggest that they are patches of eczema, psoriasis, or Bowen's disease, but when the scale is removed and the edge is stretched, the thread like margin of the lesion will reveal that it is, in fact, more consistent with a BCC. The typical BCC runs a slow progressive course of peripheral expansion. Suspicious lesions occurring in high-risk areas, such as the central portion of the face should undergo prompt biopsy.[195,283,284]

SCC does not often arise from healthy looking skin. Usually there is some evidence of longstanding photodamage: AK, lichenification, irregular pigmentation, and telangiectasia.

FIGURE 2.13
Squamous cell carcinoma (SCC): external photograph of an erythematous scaly papule with ulceration consistent with SCC.

The first clinical evidence of malignancy is induration. The area can be plaque-like, verrucous, or ulcerated. The lesion usually feels firm. The limits of the indurated area are not well demarcated and typically extend beyond the margins of the lesion. The resistance to pressure is much greater than is expected from an inflammatory lesion. In the earlier stages, the better-differentiated tumors are papillomatous and capped by a keratotic crust. Later, this crust can shed to reveal an ulcer or eroded tumor that may have a purulent exuding surface that bleeds easily. The lesion border can be rounded; however, premalignant lesions are often asymmetrical at first. Red flags include fissures, erosions, or ulcers that fail to heal and bleed recurrently. Most commonly, SCCs develop on sites that have the greatest sun exposure. They form on the backs of the hands and forearms, the upper part of the face, and on the lower lip and pinna. SCCs usually evolve faster than BCC (Figure 2.13).[195,275]

2.5.1.4 Histopathology

The diagnosis of BCC or SCC can be confirmed by a punch or shave biopsy. A shave biopsy provides sufficient tissue sample for the diagnosis of BCC, AK, or SCC in situ. Punch biopsies provide a full thickness skin specimen that aids interpretation of the specimen by revealing tumor depth and tissue architecture. Punch biopsy is indicated if invasive SCC or melanoma is included on the differential diagnosis. Histologically, a biopsy positive for BCC will show nests of atypical basaloid cells that are invading the dermis. These cells are characterized by peripheral palisading and mucin deposition in the surrounding stroma. The cells inside the palisade are usually poorly organized, and mitotic figures may be frequent.[195,275] Morpheaform BCC has a dense fibrous stroma and narrow cords of infiltrating tumor cells. Histopathologic findings in SCC are a proliferation of squamous cells into the dermis accompanied by atypical variation in size and shape of cells, enlargement of hyperchromatic nuclei,

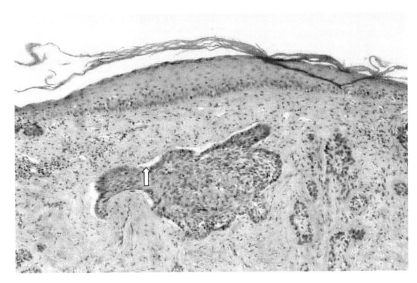

FIGURE 2.14
Basal cell pathology: there is a proliferation of basaloid cells parallel to the long axis of the epidermis. We also see slit-like stromal retraction with mucin deposition in the papillary dermis (arrow).

and atypical mitotic figures. Invasive SCC begins when the atypical keratinocytes breach the dermal basement membrane and invade the dermis. The cells of SCC can vary from large well-differentiated polygonal cells with vesicular nuclei to completely anaplastic cells with basophilic cytoplasm with no evidence of their origin. Well-differentiated tumors show areas of maturation that form parakeratotic horny pearls. The invading tumor stimulates an inflammatory reaction in the dermis. Even in extensively ulcerated tumors, the tumor cells maintain their connection with the epidermis, and the origin of the tumor can be traced to atypical epidermal cells. SCC can be graded cytologically as well-differentiated, moderately differentiated, or poorly differentiated (Figure 2.14).[195,275]

2.5.1.5 Current Treatment

There are many methods of treating biopsy-proven NMSC. The primary method varies depending on the diagnosis, histologic subtype of the cancer, anatomic location, and a variety of clinical factors. Any recurrent and metastatic tumors are associated with high morbidity and mortality and are considered to be high risk. The size and location also determine whether the tumor is high or low risk. Locations associated with high risk of recurrence include the central face, eyelids, nose, chin mandible, pre/postauricular skin, temple, ear, genitalia, hands, and feet. Areas at low risk of recurrence include the trunk and extremities. Treatments include surgical removal, radiotherapy, cryotherapy, photo-therapy, and creams. Factors that would make NMSC a high risk include recurrent tumor, high-risk location, >2 cm diameter, aggressive histopathology, ill-defined borders, rapidly growing and incompletely excised lesion, or >4 mm depth (SCC only). NMSC with these high-risk factors should be treated with Mohs micrographic surgery, or radiation in patients unwilling or unable to undergo surgical treatment. NMSC with low-risk factors—such as nodular or superficial histology (BCC only), well-defined histology, well-defined borders, and a slow-growing tumor—can be treated with 5-FU cream, 5% imiquimod cream, PDT,

ED&C, or excision. 5-FU, 5% imiquimod, and PDT are only considered for SCC in situ, and superficial BCC on the trunk or extremities.[195,253,272,283,285]

For optimal management of NMSCs, a treatment plan should be based on a clear understanding of the pros and cons of the different treatment modalities. The advantages of the topical creams (5-FU, imiquimod) include ease of use, home application, noninvasive procedure, and potential for excellent cosmetic result—and they may help avoid surgery. Disadvantages include less long-term data regarding cure rates, absent margin control, 6- to 16-week treatment course, temporary cosmetic disfigurement, and significant irritation and erythema at the site of application.[283,286] An additional concern is that the persistent tumor may become buried and require extensive surgical treatment for complete excision in the future. Advantages of PDT are that they are simple to perform and noninvasive, give excellent cosmetic results, and have minimal restriction of postprocedural activity. Disadvantages of PDT include the necessity of specialized equipment, absence of histologic margin control, moderate pain, availability of limited data for use in treatment of skin cancer, and the need of patients to stay out of sunlight for 24–48 hours. ED&C is a simple way of treating cancers in the office. The technique is most effective for the destruction of well-defined, superficial cancers. Disadvantages include the absence of histologic margin control, slower healing by secondary intention, and greater potential for suboptimal cosmesis. Surgical excision of clinically uninvolved skin is the mainstay of treatment for NMSC on the trunk and extremities. Margins of 4 mm are desired in BCC and SCC of low-risk character, and 6 mm for high-risk tumors if Mohs surgery is not available or feasible. Disadvantages include subtotal margin control (less than Mohs surgery), requirement of equipment and assistance, lack of tissue conservation, and postoperative activity restriction.[195,283,285,287]

Mohs micrographic surgery is a specialized technique of excision and margin examination that provides the highest cure rates and maximal tissue conservation. Cure rates of BCC and SCC with Mohs surgery approach 99%. The technique is best suited where evidence of clear margins is important (high-risk tumors) or in areas where tissue conservation is important (eyelid, lips, genitalia, etc.). Tissue is removed in stages, and the tissue is histologically examined at each stage for the evidence of clear margins. Once clear histologic margins are obtained, the tumor-free defect is ready for immediate reconstruction. Mohs technique ensures that 100% of the surgical margin is histologically examined, and that only malignant tissue (with minimal normal tissue) is removed. Disadvantages of Mohs surgery include the necessity of specialized training and facilities, increased time for complete margin control, and postoperative activity restriction if the defect is reconstructed.[195,283,285,287]

2.5.2 Melanoma

2.5.2.1 Epidemiology

The current incidence rates of melanoma have continued to increase since 1960 and are highest among the developed countries.[288] This may be the result of a change in behavior, increased screening, or a combination. It has long been known that melanoma occurs at the highest incidence in fair-skinned populations.[289] The U.S. SEER registries note that rates are higher among non-Hispanic whites. The rates for white males are 19.4/100,000, and the rates for white females are 14.4/100,000.[288,290] The rates of melanoma in Hispanic white males and females are 3.0/100,000 and 3.2/100,000, respectively. Melanoma mortality increased to 157% in men at the age of 65 and above between 1969 and 1999.[291] When melanoma is identified in racial groups other than whites, it is more likely to be of a more advanced stage.[288] Melanoma has different incidence patterns based on anatomic site. The age distribution of trunk melanomas peaks at age 54 for males and at age 44 for females. In contrast, the age-specific incidence of face/ear melanomas peaks at age 77 for males and at age 78 for females.[292]

2.5.2.2 Pathophysiology

These age-specific patterns support the "divergent pathway model," which proposes that melanoma may emerge from more than one causal pathway.[293] The first pathway, where melanoma develops on the trunk in the younger age population, occurs in areas with less sun exposure such as the trunk. These younger patients have a higher inherent propensity for melanocyte proliferation. Less sun exposure would be needed to induce the proliferation of melanocytes in these patients. In addition, this proliferation occurs in areas of the body with a higher proportion of unstable melanocytes, such as the trunk. Patients who have a lower propensity for melanocyte proliferation would need greater sun exposure to induce proliferation and would therefore be more likely to develop melanoma at an older age, in a site with chronic sun exposure, such as the face.[293–296] This theory is supported by the fact that nevus-associated melanomas occur far more commonly on the trunk than on the head and neck.[297] De novo melanomas are more likely to arise in older patients, on the head and neck, and are associated with solar elastosis.[289,297,298]

Phenotypic risk factors associated with increased risk of melanoma include a large number of melanocytic nevi on the skin, a family history of melanoma, fair skin that burns and does not tan, hair color (red vs. dark), and a high freckle density. The strongest risk factor is the presence of increased numbers of melanocytic nevi.[289,299] Many of these phenotypes are genetically determined. The *MC1R* gene codes for the melanocortin receptor, which partially determines the presence of red hair and freckling. This gene modulates the ratio of pheomelanin and eumelanin produced by the melanocyte.[299]

Melanoma oncogenes that have been identified include *NRAS* and other oncogenes from the RAS family. NRAS mutations have been found in approximately 15% of melanomas.[289] NRAS mutations have not been definitively associated with specific histopathologic subtypes. In contrast, BRAF mutations have been reproducibly associated with specific clinical and histopathologic subtypes. BRAF-mutated melanomas are more likely to be found in younger patients with melanoma than in older patients. BRAF mutations are also more common in the superficial spreading melanoma (SSM) subtype and are more likely to be found on the trunk and on intermittently rather than on chronically sun-exposed skin.[293,300]

2.5.2.3 Clinical Features

The diagnosis of melanoma is based on recognizing a progressively changing melanocytic lesion that is becoming irregular in shape and color. The ABCDE acronym for melanoma diagnosis applies mainly to the most common variant of melanoma, SSM. The American ABCDE mnemonic is A = asymmetry, B = irregular border, C = irregular color, D = diameter >6 mm, and E = evolving.[299,301] Problems with this acronym include the fact that most seborrheic keratoses (SKs), which are very common in older patients, will often exhibit ABCDE features. In addition, amelanotic melanomas do not usually exhibit these features, and melanomas that develop de novo (not as part of a preexisting nevus) will be smaller than 6 mm at an earlier stage.

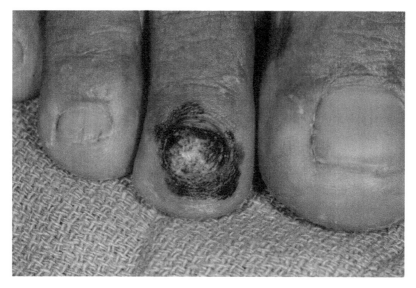

FIGURE 2.15
Acral melanoma: acral lentiginous melanoma occurs on the acral or peripheral portions of the limb, on the plantar palmar surfaces of the feet or hands, or in the subungual areas of the fingers and toes. The lesion is often missed in the early stages and is often attributed to injury.

It is important to recognize that nevi may undergo reversible changes in color and texture that may be incited by chronic rubbing. These types of changes are generally symmetric and uniform, whereas asymmetric changes in color or shape within a changing lesion are more characteristic of melanoma. In addition, in young patients, new nevi are not of concern, unless they appear different from a patient's other nevi (Figure 2.15).

2.5.2.4 Histopathology

The crucial pathological features of SSM are the presence of a focus of malignant melanocytes invading the dermis. In the adjacent epidermis, foci of in situ malignant change should be evident. Cytologically abnormal melanocytes are found in the suprabasal layers of the epidermis singly and in groups. In malignant melanocytic nevi, atypical melanocytes collect in the upper layers of epidermis, proliferate at the dermal–epidermal junction, and invade the underlying dermis. The cells in the deeper epidermis will not show maturation, as is seen in benign nevi. Malignant cells also have a higher nuclear to cytoplasm ratio, vary in size, have denser nuclei, and have abnormal mitotic figures.[299,302,303]

2.5.2.5 Current Treatment

The current recommendations by the American Joint Committee on Cancer (AJCC) to the tumor node metastasis (TNM) melanoma staging system for the *AJCC Cancer Staging Manual*, 7th edition, include the following: (1) melanoma thickness and tumor ulceration continue to define the T category, (2) tumor mitotic rate (as mitosis/mm^2) as an important independent predictor of survival, (3) nodal tumor deposits of any size are included in the staging of nodal disease, and (4) M-category is primarily defined by the site or sites of distant metastases. Increased serum lactate dehydrogenase also remains a powerful adverse predictor of survival, and (5) lymphoscintigraphy and lymph node mapping remain important components of melanoma staging and should be used to identify occult stage II regional nodal disease in patients with clinical stage IB or II melanoma.[304,305]

Melanoma accounts for the majority of skin cancer–related deaths, but treatment is nearly always curative with early detection of the disease. All patients with suspected malignant melanoma should have an excision biopsy of the lesion with a margin of 1–2 mm of normal skin. Incisional or punch biopsies should not be performed in suspected melanoma for three reasons: (1) sampling error, (2) inaccurate tumor thickness be obtained due to biopsy trauma, and (3) the diagnosis of melanocytic lesions depends in part on the overall shape and symmetry of the whole lesion.[299]

In its early stages, melanoma can be surgically cured, leading to 5-year survival rates greater than 90%. Metastatic disease, however, is uniformly fatal, with survival rates less than 2%.[306] Patients with stage I and II melanomas who have had surgery for primary melanoma and have no evidence of spread beyond the primary site require follow-up at varying intervals for a varying number of years. Accepted intervals and duration of follow-up are 3 monthly for the first 3–5 years, depending on the stage, and yearly for life,[307] but these intervals can be interpreted according to the needs of individual patients.

Metastatic melanoma is highly resistant to chemotherapy, hormonal therapy, radiation therapy, and current immunologic therapies. However, there are several promising phase II studies suggesting long-term benefits with immunotherapeutic approaches. Dacarbazine (DTIC) remains the only cytotoxic drug approved by the FDA for treatment of metastatic melanoma. Single-agent chemotherapy produces objective response rates of less than 20%. However, in a small subset of patients with lung metastases and otherwise good prognostic factors, long-term disease control with good quality of life is achievable. Generally, long-term follow-up of patients treated with DTIC alone shows that less than 2% can be expected to survive 6 years.[308] Temozolomide (TMZ) is an oral formulation of the prodrug of DTIC that has improved CNS penetration for potential treatment of brain metastases. TMZ was not approved by the FDA for treatment of malignant melanoma, because results of the study showed only equivalence of TMZ to DTIC, and the study was designed to show its superiority over DTIC. With modern antiemetics, DTIC is now much better tolerated by patients and can be given in an outpatient setting.

Combination chemotherapy, particularly the four-drug Dartmouth regimen, including cisplatin, DTIC, carmustine, and tamoxifen, produced responses in 46% of 141 patients. The median response duration was 7 months.

Melanoma is an immunogenic tumor. Immune-mediated regression is common in primary tumors, and immune-mediated vitiligo in stage IV disease is a good prognosticator for response to chemotherapy. The most widely used adjuvant is IFN-α. Though IFN-α seems to increase disease-free survival in some patients, there has not been shown any significant overall survival benefit. Though there are no well-recognized factors that predict outcome, there does seem to be some host variation in immune response that predicts outcome. In a study of 200 melanoma patients treated with IFN-α 2b, the median rate of survival in patients with signs of autoimmunity after being treated with IFN were significantly higher than those of patients without autoimmunity.[309]

The resistance of melanoma to conventional chemotherapy has lead investigators to pursue novel approaches, such as protein kinase inhibitors. Sorafenib targets the adenosine triphosphate (ATP)–binding site of the BRAF kinase, and has been found to inhibit both wild-type and mutant BRAF in *vitro*. Sorafenib has been found to inhibit a spectrum of kinases, including RAF proto-oncogene serine/threonine protein kinase, platelet-derived growth factor receptor 2, flt-3, and c-kit, among others. Responses to sorafenib have not been correlated with BRAF mutation status to date. One phase III trial, which added sorafenib to carboplatin and paclitaxel, found that it did not improve any of the end points over a placebo.[310] It is believed to have failed because of its nonspecific, broad kinase inhibitor activity and associated toxicity. However, newer, more potent, and more specific inhibitors show promise.[311]

Two human anti-CTLA-4 monoclonal antibodies have been tested in clinical trials: ipilimumab and tremelimumab. Responses have been observed with both antibodies administered as single-agent therapy. However, a randomized phase III trial of 655 melanoma patients failed to show a better overall survival versus chemotherapy.

To date, when compared with standard chemotherapy, no other drug has shown benefits in terms of survival up to now. Interest remains in biological therapies for melanoma, and there are several drugs in clinical trials that may offer hope in the future.

2.5.2.6 Applications in Nanotechnology

2.5.2.6.1 Diagnosis

The most common skin cancers encountered in clinical practice are melanoma, SCC, and BCC. Noninvasive methods of diagnosing skin cancer include NIR reflectance confocal microscopy, OCT, and terahertz pulsed spectroscopy imaging.[128,130,132,207–211,214,312–332] Terahertz pulsed spectroscopy imaging relies on changes in the nanomechanical properties of epidermal cells, the interstitium, and the extracellular matrix during the transition to malignancy. In the IR spectrum (1060 nm), OCT has been used for the imaging of NMSC. Combining OCT with photoacoustic scanning is being used for three-dimensional (3D) morphological imaging of the skin and may have applications for the delineation of tumors, vascular lesions, deeper soft tissue masses, wounds, and inflammatory diseases such as dermatitis and psoriasis.[130,132]

The goals of nanotechnology for the diagnosis of skin cancer are to be faster, more sensitive, more specific, less invasive, and require less tissue than traditional methods.

Melanoma has been shown to differ from benign nevi in a number of ways. Recently, the DNA-methylation profile of genes has been shown in arrays to differ among nevi and melanomas.[333] In one study at the University of North Carolina, Chapel Hill, NC, Conway et al. found 26 CpG sites in 22 genes that had significantly different methylation patterns for melanoma and nevi. Of these, 12 CpG loci were highly predictive of melanoma. A similar approach has been used to identify differential hypermethylation of promoters in SCC of the head and neck.[334]

Circulating tumor markers are also being studied for the diagnosis of skin cancer. Surface plasmon resonance of silver nanoparticles has been used to make a label-free method of detecting p53 in the serum of patients with SCC of the head and neck.[335,336]

Small noncoding RNAs, called microRNAs (miRNA), have been shown to be involved in regulating gene expression during cancer progression. Dicer is an enzyme that cleaves precursor RNAs into miRNAs. Dicer expression was compared in one study in melanocytic nevi, melanoma, sarcoma, and nonmelanoma carcinomas.[337,338] Dicer expression was found to be significantly elevated in melanoma compared to benign nevi. Elevated Dicer expression also correlated with melanoma mitotic index, Breslow depth, nodal metastasis, and AJCC clinical stage.

Gene expression profiling comparing cultured SCC cells with normal keratinocytes identified 435 differentially expressed genes. Of these 435, 154 were differentially expressed in SCC when compared to normal skin. Of the 154, 37 were differentially expressed nonmalignant hyperproliferating psoriatic skin. This final group was designated as SCC specific.

Of the 37 genes, 29 were differentially regulated in *vitro,* and 21 of the 29 were upregulated. The effect of knocking down these 21 genes was tested individually with siRNA. Polo-like kinase-1 (PLK1) and C20orf20 knockdowns consistently interfered with SCC viability. Inhibition of these gene products with a PLK inhibitor or siRNA to C20orf20 reduced explanted tumor volume in *vivo.*[339]

A proprietary array using tape stripping of the stratum corneum has been shown to be useful in gene expression profiling of melanocytic lesions.[340] RNA from tape-stripped skin is amplified and analyzed on a microarray of 312 genes differentially expressed in melanoma, melanocytic nevi, and normal skin specimens. Although many of the genes in the dataset are involved in normal melanocyte development and physiology, some are involved in melanoma growth, cancer, and cell-cycle control. With a heuristic training dataset of 37 melanomas and 37 nevi, a 17-gene classifier was found to distinguish in situ melanoma and invasive melanoma from melanocytic nevi with 100% sensitivity and 88% specificity.

In some families that are at high risk for melanoma, mutations can be seen in the cyclin-dependent kinase inhibitor family (CDKN2A/CDK4). Lang et al.[341] developed a high-throughput multiplexed bead-based PCR assay for detecting 39 different germ-line variants in kindreds prone to melanoma. The assay was correctly able to identify 1540/1603 of susceptible individuals, and 1540/1545 individuals whose variants were included in the probe set.

A small association has been shown between defects in telomere genes and melanoma, particularly four single-nucleotide polymorphisms (SNPs) (rs2853676[T], rs2242652[A], rs2981096[G], and rs401681[C]) that are associated with a higher odds ratio of 1.43, 1.50, 1.87, and 0.73, respectively, for melanoma.[342] No consistent association was noted between telomere SNPs and BCC or SCC.

In a Spanish population, polymorphisms associated with base excision repair, NER, and oxidative stress, particularly the *NOS1* oxidative stress gene, were associated with a susceptibility to malignant melanoma.[343] In animal studies of *Xiphophorous,* the association of defective NER and melanoma was less clear.[344]

Genes governing pigment have also been implicated in increasing the risk of NMSC. These include genes for the melanocortin receptor MC1R, agouti signaling protein (ASIP), tyrosinase (TYR), tyrosinase-related protein 1 (TYRP1), and oculocutaneous albinism II (OCA2). The relationship between pigment, especially hair color, and nevus counts has evolved. In a study of 654 white children in Colorado, Aalborg et al.[345] found that 9-year olds who burned and tanned and had dark hair had higher nevus counts than similar cohorts with light hair.

One of the goals of personalized medicine is to target therapy to the individual. Tailored therapy has at least two components, the status of the host and the susceptibility of the disease. Melanoma typically is poorly responsive to chemotherapy. Traditional assays, such as the ATP-based tumor chemosensitivity array, correlate well with outcomes in melanoma but require fresh tissue and sophisticated lab facilities.[346–350] In one study, the authors compared the traditional ATP assay with a quantitative RT-PCR of 93 chemotherapy resistance genes using RNA from formalin-fixed paraffin-embedded tissue.[349] The data correlated well for resistance to DTIC, treosulfan, and cisplatin. Genes affected in common included *HSP70, EGFR, IAP2, PTEN, ERC1, XPA, XRCC1, XRCC6, Ki67, p21,* and *p27.*

Combining common nonspecific biomarkers such as blood groups with less common more specific biomarkers such as BRAF or CpG may lead to enhanced diagnosis and prognosis of skin cancer. For example, in studies of blood groups, subjects with non-O blood groups had a slightly reduced risk of SCC (14%) and BCC (4%) but no altered risk for melanoma.[351] With the advent of nanoarrays, the number of markers that can be included in assays is virtually unlimited, greatly adding to the specificity and sensitivity of diagnosis as well as personalized medicine.

Mutations associated with susceptibility to chemotherapy have been identified in melanoma.[352,353] BRAF is mutated in a minority of malignant melanomas.[311] BRAF inhibitors such as ipilimumab can dramatically alter the course of disease, even in patients with widespread metastases. Identifying individuals with BRAF mutations may permit rapid early therapy and may obviate unnecessary therapy in patients who lack the mutation, saving cost and needless delay.

An important aspect of tumor staging in melanoma is assessment of sentinel node involvement.[354,355] Typically, sentinel node identification involves injection of a radiolabel and dye. The combined methods are highly sensitive but operator-technique dependent, require surgical incision, and subject patients to radiation exposure. QDs have been used for sentinel node identification in animal models.[356,357] QDs are label free, easy to inject, and allow for real-time visualization of dye using NIR fluorescent light. Concern over the potential toxicity in humans of heavy metal (Cd, Hg, Te, Pb)–containing QDs has limited their widespread adoption. Erogbogbo et al. have designed QDs containing silicon tuned to fluoresce in the NIR range.[358,359] These QDs, when encapsulated, can be coupled to molecules to allow tumor vasculature targeting, sentinel lymph node mapping, and multicolor NIR imaging in *vivo* in murine models. The silicon dots are encapsulated in PEGylated micelles and bioconjugated in nanospheres that show stable fluorescence and long tumor accumulation times. Pons et al.[360] have manufacutred $CuInS2/ZnS$ QDs and successfully used them for NIR sentinel lymph node mapping in murine models. They compared toxicity with $CdTeSe/CdZnS$ QDs and showed 10-fold less toxicity and inflammation with the $CuInS2/ZnS$ QDs (Figure 2.16).

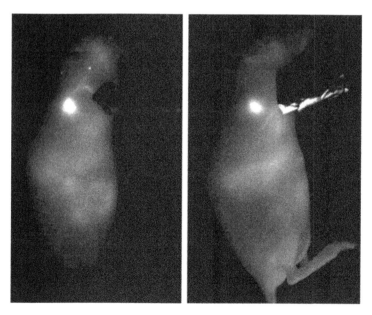

FIGURE 2.16
Near-infrared fluorescence imaging of right anterior lymph node 15 minutes (left) and 7 days (right) after injection of 20 pmol micelle-encapsulated quantum dots into the right anterior paw. (Reprinted with permission from Pons, T., *ACS Nano.*, 4, 2531, 2010. Copyright 2010 American Chemical Society.)

DNA-methylation patterns have been associated with progression of nevi to melanoma. One group examined the DNA-methylation status of paraffin-embedded tissue from patients with melanoma and benign nevi.[333,361–363] Using a 1505 CpG site microarray, 22 CpG sites were identified in 22 genes that differed in methylation status between melanomas and nevi. Microarrays or cantilever arrays may be useful in determining the DNA-methylation status of these genes and may facilitate the diagnosis of melanoma. Assessment of the methylation status of the line-1 element may be useful in predicting melanoma survival.

2.5.2.6.2 Treatment

BCC has been associated with defects in the patched gene and the hedgehog signaling pathway, including smoothened and Gli1.[364] This pathway has also been shown to be aberrant in glioblastoma, medulloblastoma, and rhabdomyosarcoma.[365] Small-molecule inhibitors of smoothened (such as GDC-0449) have shown efficacy in treating advanced and metastatic BCC.[366,367]

The role of angiogenesis in the maintenance and growth of tumors has been well studied. Antibodies specific to the EDB domain of fibronectin-binding, tumor-induced angiogenic vessels.[368,369] When coupled to a porphyrin-based photosensitizer, the antibodies still localize to tumor vasculature but in the presence of photoirradiation, lead to vascular disruption and tumor eradication in mouse models of F9 (teratocarcinoma) and A431 (human epithelial carcinoma).

Biologically active antibodies, so-called biologics, are nanoparticulate (5 × 5 × 20 nm)-specific agents that target receptors.[370] A number of biologics have shown therapeutic benefit in the treatment of cutaneous malignancies, including melanoma, with high rates of efficacy and low toxicity.[353,371] Small-molecule inhibitors of signaling pathways in melanoma and BCC have been developed as well.[365,367,372] Nanoparticulate delivery vehicles may enhance efficacy and decrease toxicity of these agents.

Theragnostics is the combination of simultaneous diagnosis and therapy of a target lesion.[373–377] Nanotechnology is particularly amenable to theragnostics. A targeting ligand is coupled to a macromolecule that is a visible by an imaging modality (e.g., NIR fluorescence). The same imaging modality can activate or deliver a drug that leads to targeted therapy. An example would be a gold nanoshell tuned to a surface plasmon resonance frequency of 800 nm coupled to a targeting ligand, such as a melanocyte-stimulating hormone.[378] These gold nanoparticles that can be home to melanoma, are visible under NIR light, and can toxically heat up melanoma cells while leaving surrounding tissue unharmed. This type of selective photothermolysis of tumor has been shown in murine models of melanoma.

ALA has been shown to highlight margins of SCC, making them visible in fluorescent light.[379] The margins seen in fluorescent light correspond well to margins excised by Mohs micrographic surgery. Real-time visualization of tumor margins may make Mohs micrographic surgery more rapid and convenient and, by reducing tissue processing and analysis steps, might reduce costs.

Proteomics may allow for analysis of serologic markers for tumor presence, tumor burden, and for tracking tumor response to therapy.[380,381] Detection of small quantities of biomarkers accurately and rapidly using, for example, QDs or nanocantilevers, may allow for more rapid characterization of tumor features.

OCT has been studied in the United Kingdom for mapping SCC.[207,208,214,313,323,324,328] The OCT map of SCC is used to guide topical PDT, leading to complete response and no recurrence in at least one case study. While other studies have shown no significant benefit in the use of OCT for margin control in removal of cutaneous malignancies, this technique may show promise when combined with SCC-targeted photoresponsive biomarkers, such as QDs tuned to an NIR wavelength. Tumor delivery of a photodynamic agent is important. In one study, intralesional ALA was shown to result in complete eradication of SCC treated with PDT.[382]

Macromolecular antitumor drugs including photosensitizers such as protoporphyrin IX can be concentrated in tumors with the topical application of nitroglycerin.[383] Nitroglycerin enhances vascular drug delivery two- to threefold. In murine models, nitroglycerin showed enhanced therapeutic effect yet no increased toxicity. Vehicles that enhance targeted delivery of nitroglycerin may be useful adjuncts in PDT.

Photoacoustic tomography using gold nanocages coupled to MSH allows for high-resolution real-time in *vivo* visualization of melanoma.[130,313] This can allow for early diagnosis, accurate staging, and an image-guided resection of tumor.

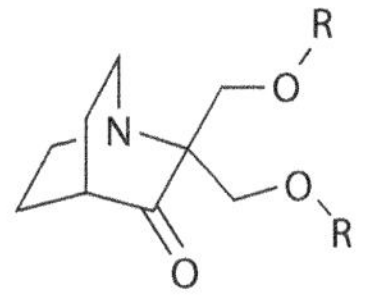

FIGURE 2.17
Chemical structure of 2,2-disubstituted quinuclidinone. (Reproduced with permission from Malki, A. et al., *Bioorg. Med. Chem. Lett.*, 16, 1156, 2006.386.)

The tumor suppressor p53 is not typically mutated in melanoma. It is involved in cell-cycle arrest, senescence, and apoptosis. It is mutated in tumors that are typically resistant to chemotherapy and radiation. Its expression can be induced by integrin αv-mediated signaling. A new class of small molecules known as quinuclidinones reactivate p53 by inducing correct folding and functionality of the protein and triggering apoptosis.[384] Studies of cultured melanoma cells in 3D collagen gels showed that functional restoration of p53 by the quinuclicidinone APR-246 (PRIMA-1Met) (p53 reactivation and induction of massive apoptosis) led to apoptosis, expression of Apaf1, PUMA, caspase-3 and -9, and led to suppression of melanoma xenografts in *vivo* (Figure 2.17).

The induction of antitumor immunity is another strategy for eradication of melanoma. Studies have shown that CpG oligonucleotides enhance Th1 responsiveness to tumors.[362,363,386] The epidermis and the reticular dermis are abundant reservoirs of antigen-presenting cells, including dendritic cells and Langerhans cells. Transdermal delivery of topically applied CpG oligonucleotides can be enhanced by iontophoresis.[386,387] In a murine model, transdermal delivery of CpG oligonucleotides using iontophoresis induced the expression of proinflammatory cytokines and Th1 cytokines in the local skin and draining lymph nodes and induced antitumor activity against B16F1 melanoma in animal models.

Small inhibitor RNAs have showed utility in downregulating gene expression. Chen et al.[388] have used liposomal nanoparticles composed of *N,N*-distearyl-*N*-methyl-*N*-2-(*N*'-arginyl) aminoethyl ammonium chloride (DSAA), a guanidinium-containing cationic lipid, to deliver c-myc siRNA to melanoma cell cytoplasm. Targeting was achieved via the anisamide receptor, a sigma receptor found on the surface of many cancers including melanoma. The nanoparticles were able to induce apoptosis and increase tumor sensitivity to paclitaxel. In addition, siRNAs delivered by intravenously administered polymeric protein nanoparticles have been used to target gene expression in human subjects with melanoma.

The role of miRNAs in the regulation of cancer growth has become increasingly understood over the past 10 years. miRNAs are small noncoding fragments of RNA that target the 3' untranslated portion of genes, leading to inhibition by mRNA translation suppression or mRNA cleavage. miRNAs are more stable and tolerant of RNAses than mRNAs. Melanoma-associated mRNAs have been identified in abundance and have been shown to be involved in the transition from melanocyte to melanoma (let-7a, let-7b, miR-155, miR-145), from primary melanoma to metastasis (miR-126, let-7b, miR-182), and from melanocyte to metastasis (miR-133a, MiR-221/222).[337,338,361,389–394] These miRNAs have been shown to target genes associated with migration, proliferation, cell survival, dedifferentiation, and invasion (ITGB3, CCND1, MET, MNT, MITF, FOXO3, c-Kit, p27, PLZF, RUNX3). Isoflavones such as genistein have been shown to alter miRNA expression (miR-27a and its target ZBTB10) and inhibit the growth of human uveal melanoma in *vivo* and in *vitro.*

miRNAs also circulate in the serum of cancer patients and may be useful as diagnostic markers, markers of prognosis, markers of recurrence, or markers of response to therapy. miRNAs may be actively secreted into the extracellular space to regulate genes in neighboring cells. Analysis of miRNAs may lead to personalized profiling of patient cancer phenotype and more accurately targeted therapy. Screening strategies have identified small-molecule inhibitors of the miRNA pathway, such as estradiol, 5-FU, diazobenzene, and enoxacin.[153,391,393]

The role of COX-2 in colon cancer is well established. Inhibitors of COX-2 have been used as treatments for colon cancers and as preventive agents.[247] COX-2 inhibitors have also shown utility in the prevention of skin cancer. COX-2 inhibition strategies using siRNA have been developed.[395] By altering the flora of the gut using *E. coli* expressing siRNA against COX-2, a long-term preventive strategy for colon cancer has been devised.[395] Similar tools to alter the microbiome of the skin may be helpful in some patients. Similar gene-silencing techniques using siRNA to knock out TSPAN1 in a human SCC line (A431) have resulted in inhibition of cell proliferation, migration, and infiltration.[396] TSPAN1 is a tetraspanin in the family of hydrophobic cell-surface proteins that play a role in cell growth, development, and migration. Overexpression has been shown to lead to colon cancer progression and gastric cancer migration and invasion.

Another novel gene-silencing strategy involves reversing inhibition of dendritic cell activity. Indoleamine 2,3-dioxygenase (IDO) is an enzyme that degrades tryptophan.[397–400] Dendritic cells that express IDO inhibit T-cell responses and may be considered immunosuppressive. Silencing of IDO with siRNA was studied in *vitro* and in *vivo* in a murine model. In *vivo* cutaneous delivery of IDO siRNA resulted in tumor growth inhibition and prolonged animal survival in models of MBT-2 and CT-26 tumors. This effect could be adoptively transferred using CD11c dendritic cells. Furthermore, IDO siRNA enhanced the antitumor efficacy of Her2/Neu DNA vaccination. Similar effects were noted with the homolog, IDO-2. Similar results were obtained in animal models of B16 melanoma, immune responsiveness being enhanced through the silencing of STAT3 in dendritic cells using siRNA to STAT3 in poly-L-glycolic acid nanoparticles.[399,400]

Topical delivery of antineoplastics has been enhanced with packaging in chitosan nanoparticles.[401–403] In one study, nanoparticles ranging in size from 114 to 192 nm and with a positive zeta potential were synthesized. These properties conferred high bioavailability and decreased immunogenicity in an animal model of ocular delivery. SLNs containing 5-FU are being developed for sustained release in the treatment of colon cancer and may be adapted for cutaneous use. Tumor cells may overexpress folate receptors, and functionalizing nanoparticles with folate may control enhanced delivery of 5-FU to nonmelanoma skin cancer cells.[159,404–407] Nanoparticles of 5-FU made with hyaluronic acid and poly(lactide-*co*-glycolide) have enhanced plasma permeation and retention effects, giving them selective passive targeting to solid tumors. 5-FU-loaded *N*-succinyl-chitosan nanoparticles and 5-FU SLNs (182 nm negative zeta potential) showed similar targeting properties and had a long half-life in the circulation of animal models. Nanoparticles of 5-FU containing poly(*N*-isopropylacrylamide-*co*-acrylic acid), a thermosensitive and pH-sensitive polymer, allowed pH- and temperature-dependent control of release of active drug.[408] 5-FU-loaded polybutyl-cyanoacrylate nanoparticles were used in a pilot study of 32 patients (mean age 74, range 56–90) with biopsy-proven superficial BCC.[409] Treatment was daily for 35–40 days. Thirty-one of the 32 patients achieved histologically confirmed tumor resolution and tolerated the treatment well.

MTX has also shown utility in the treatment of head and neck SCC. In animal models, dendrimers containing MTX could be delivered at higher doses than free drug, had higher efficacy than free drug, and had reduced systemic toxicity, targeting folate-expressing tumor populations efficiently. Lipid drug conjugate nanoparticles of MTX have higher oral bioavailability and less gastrointestinal toxicity than free drug. Dendrimers containing poly(ester-amine) groups enhance the solubility of MTX, which is highly hydrophobic, and increase bioavailability.[161,404,410] Nanogels containing copolymerized *N*-isopropylacrylamide (NIPAM) and butylacrylate were loaded with MTX and applied to skin in a porcine model. This nanogel delivered MTX topically, and the delivery was enhanced with Na_2CO_3.[412] Calcium phosphate nanoparticles 262 nm in size were also shown to retain MTX at physiologic pH and to release 90% within 3–4 hours of exposure to endosomal pH (5.5–6.0). MTX nanoparticles conjugated to Fe_3O_4 (magnetite) via a trichloro-s-triazine linker also showed good solubility and stability under physiologic conditions.[413] They also had specificity to a tumor cell line (9L) over a healthy cell line (cultured pulmonary artery endothelium) and localized intracellularly (Figure 2.18).

Iron dioxide nanoparticles coated with (3-aminopropyl) trimethoxysilane monolayers can be conjugated to MTX through amidation.[414] MTX is cleaved and released from the nanoparticles under low pH conditions, which simulate the interior of endosomes. In

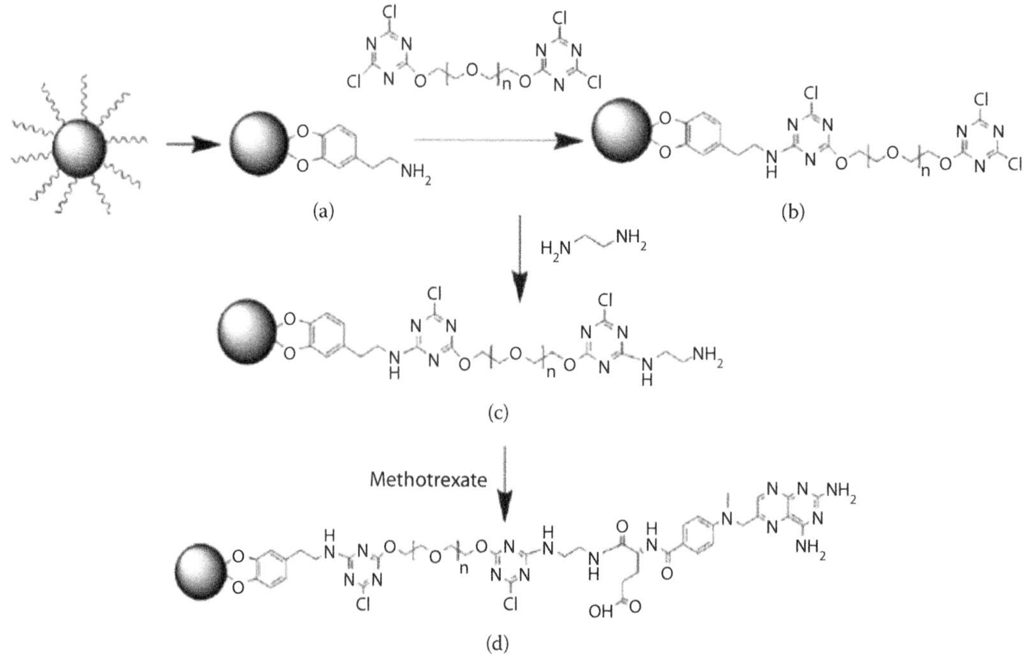

FIGURE 2.18
Chemical addition of methotrexate to a magnetite nanoparticle is depicted in four steps: (a), (b), (c), and (d). (Reproduced with permission from Young, K.L., *J. Mater. Chem.*, 19, 6400, 2009.)

human breast cancer and cervical cells expressing the folate receptor, higher levels are internalized than in negative control cells. This type of bifunctional nanoparticle is superparamagnetic and useful for imaging studies and targeted drug delivery.

Nanoparticles made of the milk protein β-casein have been used to entrap hydrophobic drugs such as mitoxantrone.[414] Protein-based nanoparticles may be useful vehicles for delivering nonpolar antimitotics such as MTX or capecitabine orally or topically with high bioavailability and possibly less toxicity.

Retinoids have been used for prevention and treatment of NMSC. Topical retinoids can be photolabile and irritating. Drawbacks of systemic retinoids include side effects such as lipid abnormalities, hematologic abnormalities, depression, teratogenicity, mucositis, and gastrointestinal disease. Retinyl acetate–loaded poly(ethylene glycol)-4-methoxy-cinnamoylphthaloylchitosan nanoparticles show persistence on murine skin, retention for up to 24 hours, and increased photostability.[415] Submicron emulsions of silica-coated lecithin all-*trans*-retinol nanoparticles showed high skin retention, localization to the epidermis, and less than 1% penetration in the dermis.[416] Deeper permeation into ex *vivo* human skin layers is seen with retinyl palmitate in a 215 nm nanoparticle with a poly(d,L-lactide) wall.[417] Polymeric nanocapsules of tretinoin show enhanced stability and antitumor activity against acute promyelocytic leukemia.[418] It may be possible to administer a systemic form for the prevention of NMSC. Retinoid release can be triggered enzymatically.[419] Phospholipid prodrugs containing retinoids in liposomal nanoparticles were synthesized with diameters of 94–118 nm. In the presence of phospholipase A2, the prodrugs were hydrolyzed, releasing active retinoids at IC50 values of 3–19 μM for colon cancer cell lines (HT-29, Colo205). No cell death was seen in the absence of phospholipase A2.

Nanoemulsified vitamin A has twofold improved oral bioavailability compared to standard oil-based formulations.[420] Topical isotretinoin delivery has also been studied. Isotretinoin-loaded SLNs made with glycerol distearate (Type I) EP, glyceryl distearate NF, and Tween80 and soybean lecithin were evaluated in a rodent model.[421] Isotretinoin in the solid lipid nanoparticle formulation accumulated in the skin but did not show systemic uptake. Topical isotretinoin that is not systemically absorbed may confer the same chemopreventive advantages of systemic retinoids for nonmelanoma skin cancer with minimal toxicity.

All-*trans*-retinoic acid is poorly soluble and has poor bioavailability. Nanodisks enriched with all-*trans*-retinoic acid were more effective than empty nanodisks and naked all-*trans*-retinoic acid in mediating cell-cycle arrest and apoptosis in mantle cell lymphoma.[422]

Topical delivery of therapeutic compounds may be enhanced by microneedle delivery. Examples include photosensitizers such as 5-ALA, vaccines, fluorescent dyes, and therapeutic antibodies.[423–428] Microneedles may be useful for the delivery of painless topical anesthesia before surgical intervention. Conversely, sampling of contents of suspicious neoplasms using a topically applied microneedle array may be helpful in the diagnosis of malignant skin lesions.[429] Fiberoptic microneedles may be useful for the transdermal delivery of light for phototherapy or PDT.[430] Nanoscale optical fiber–embedded light delivery fabrics could be used to deliver phototherapy in a precise temporal and spatial pattern.[431]

In summary, the diagnosis of skin cancer can be enhanced by nano-based imaging techniques, mapping techniques, and genome/proteome/transcriptome analysis. Therapy of skin cancer can be targeted with nano-enhanced topical delivery, systemic delivery, and PDT (Figure 2.19).

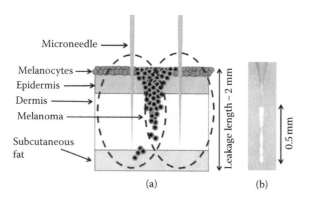

FIGURE 2.19
Schematic image of a melanoma being photothermally treated by laterally diffused laser energy from fiber optic microneedles. (a) The dashed lines denote the light diffusion profile and corresponding treatment zone. (b) Color microscope image of a fiber optic microneedle delivering red laser light ($\lambda = 633$ nm, 5 mm leakage length in air). (From Kosoglu, M.A., *J. Biomech. Eng.*, 132, 091014, 2010. With permission. Copyright 2010 by ASME.)

2.6 Viral Warts

2.6.1 Epidemiology

Cutaneous warts are found in one-fifth of school-age children.[432] The prevalence of cutaneous warts declines with age, whereas genital warts become common in adolescence and adulthood. Human papilloma virus (HPV), the cause of viral warts, is found in 20%–45% of men and women.[432] The subpopulation of men who have sex with men has one of the highest rates of HPV—73% are estimated to have anal HPV.[433] This population is also at increased risk for HIV. The combination of anal HPV and HIV puts them at risk for anal squamous intraepithelial lesions.

Most lesions spontaneously regress within 1–2 years from onset.[434] Although cutaneous warts are primarily considered a nuisance, HPV infections of the cervix are a significant factor contributing to the development of cervical cancer.

2.6.2 Pathophysiology

HPV is a double-stranded DNA virus.[435] The virus itself is nano sized, approximately 55 nm in diameter. It is spread by person-to-person contact, fomites, and autoinoculation. HPV is a hardy virus, resistant to heat and desiccation.[436] It may even be aerosolized during laser surgery. Infections often occur at sites of trauma such as the fingers, hands, elbows, or mucosal surfaces, where the basal keratinocytes are readily exposed to the virus.[434]

An HPV infection of the basal keratinocyte sets the scene for a verruca.[435] As the basal keratinocytes divide, some cells remain as basal keratinocytes, maintaining the infection. Within keratinocytes that are destined to mature, viral proteins are synthesized and the capsid is assembled. Depending on the strain, proteins that promote the degradation of tumor suppressor p53 or inhibit retinoblastoma tumor suppressor protein may be produced, leading to increased and disordered growth.[434] By the time the keratinocytes are ready to be shed, the viron is completely assembled and ready to infect the next host. Furthermore, HPV is present in the active lesion along with the perilesional tissue.

Microscopic findings vary by site and strain. When cytoplasmic vacuoles surround the nucleus, the cell is said to take on a koilocytic appearance characteristic of HPV infection. Orthokeratosis, parakeratosis, acanthosis, and elongated rete ridges are frequently seen.

2.6.3 Clinical Features

Cutaneous warts are described by their location and clinical appearance.[434] Common, palmar and plantar warts are associated with HPV-1 (plantar), 2, and 4. Common warts are hyperkeratotic, exophytic papules, or nodules most commonly found on the fingers, hands, elbows, and knees. Palmar and plantar warts are thicker, endophytic papules that are often painful when pressed or walked upon. When palmar and plantar warts coalesce, they can take on a mosaic pattern. Thrombosed capillaries are frequently seen as punctate black dots within the common palmar and plantar warts.

Flat warts are found on the hands, arms, and face. They appear as skin-colored to pink slightly elevated papules with smooth, flat tops. Flat warts are associated with HPV-3 and 10. Butcher's warts are associated with HPV-7 and are seen in meat and fish handlers, but are not a zoonosis. The lesions are verrucous papules on the dorsal, palmar, or periungual aspects of the hands.

The common mucosal warts are condyloma acuminata and oral warts. Condyloma acuminata or anogenital warts are skin colored, brown, or whitish discrete sessile exophytic papillomas. HPV-6 and 11 are most frequently detected in these warts. Although HPV-6 and 11 are considered "low-risk" subtypes, they can be associated with verrucous carcinoma, specifically a Buschke–Lowenstein tumor. Oral warts appear as small, soft, pink or white, slightly elevated papules and plaques on the buccal, gingival, or labial mucosa, tongue, or hard palate. Oral warts can be associated with HPV-6 and 11, whereas Heck's disease, oral florid papillomatosis (increased prevalence in American Indians and Inuits), is associated with HPV-13 and 32). Finally, there are several "high risk" HPV subtypes, such as 16 and 18, known to be associated with malignant transformation of condyloma to Bowen's disease (SCC in situ) and SCC (Figure 2.20).

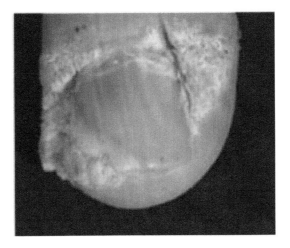

FIGURE 2.20
Verruca vulgaris. Human papiloma virus infection of the perionychial skin folds.

2.6.4 Current Treatment

Although most warts will resolve themselves, they can be painful (as is the case for palmar and plantar warts) and disfiguring to patients.[437] Podophyllotoxin (POD), imiquimod, and salicylic acid are topical therapies that can be applied by the patient at home. Topical sensitizers can be applied in the office. Local destruction of tissue by cryotherapy, electrosurgery, excision, and laser vaporization can be done in the office. Cryotherapy is preferred mainly because it does not require anesthesia and rarely causes scarring.[434] The HPV vaccine is effective in preventing many genital HPV infections; however, it does not target the HPV strains commonly found in cutaneous warts.

POD is a topically applied cytotoxic agent that prevents cell division by blocking the formation of the mitotic spindle.[19] Imiquimod functions as a ligand for toll-like receptors 7 and 8 to activate monocytes, macrophages, and antigen-presenting cells.[437] Topical sensitizers, such as diphenylcyclopropenone, are thought to increase local antigen presentation and thus accelerate clearance of HPV by the immune system. Imiquimod and topical sensitizers are thought to facilitate the natural immune response to HPV infections. As expected, all of these topical agents cause irritation.

The quadrivalent HPV vaccine contains recombinant capsid proteins from HPV-6, 11, 16, and 18.[438] As stated above, these strains rarely cause common warts. Furthermore, because the vaccine induces immunity to capsid protein, it is unlikely to clear existing infections, though there are some reports of lesion clearance associated with vaccination.[439]

Current therapy of viral warts provides only an incremental improvement over no therapy at all. A Cochrane review of 60 randomized-control trials for the treatment of common warts showed that topical therapy with salicylic acid is effective in up to 73% of cases.[440] In comparison, 48% of cases responded to placebo treatment. The same review showed that cryotherapy is no more effective than topical therapy with salicylic acid.

2.6.5 Applications in Nanotechnology

As listed above, the most common sexually transmitted infections seen by dermatologists are due to mucosal HPV infections. Once an individual contracts HPV, there is no way to eradicate the virus from the body. Multiple HPV subtypes are directly responsible for the development of neoplasia such as cervical cancer and verrucous carcinoma.[441] Nanotechnology is currently being investigated and used via three routes: diagnostics immunization, and treatment.

2.6.5.1 Diagnosis

The determination of HPV types has clinical diagnostic value and is an important factor in the assessment of risk of cancer development in patients who exhibit evidence of HPV infection. Currently, there is strong interest in the development of new bioassay techniques for a wide variety of applications, such as gene identification, gene mapping, DNA sequencing, and medical diagnostics. Presently, there are four main methods for gene identification: PCR, ELISA, microarrays, and nanoparticle agglutination-based techniques. All of these methods, though sensitive and reliable, suffer from several disadvantages. PCR technique is time consuming and expensive, requires skilled personnel, and cannot be easily automated. In addition, since it is based on target DNA amplification, it gives only semiquantitative information. Microarrays also rely on previous target DNA amplification to ensure optimal sensitivity, are expensive, and also cannot be easily automated. ELISA-like techniques reveal the presence of a specific antigen (or antibody) by detecting the reaction of antigen–antibody but exclude multiplexing, i.e., the detection of different genotypes simultaneously.

Nanotechnology is currently being used to improve on the limitations of current diagnostic modalities. One group developed a plastic modular chip suitable for one-shot HPV diagnostics.[442] The device comprised two modular and disposable plastic units. The first module was represented by a polydimethylsiloxane (PDMS) microreactor suitable for detecting the presence of the virus. The second unit was a PDMS microwell array that allowed for virus genotyping by a colorimetric assay, based on DNA hybridization technology requiring simple inspection by the naked eye. The two modules were shown to be easily matched to reusable hardware, enabling the heating/cooling processes and the real-time detection of HPV. This device may represent a low-cost tool for HPV diagnostics, thereby favoring the prediction of cancer risk in patients.

HPV DNA chip detection based on the light scattering of aggregated silica nanoparticle probes has also been investigated to enhance serotyping.[443] HPV DNA is sandwiched between the capture DNA immobilized on the chip and the probe DNA immobilized on the plain silica nanoparticle. The spot where the sandwich reaction occurs appears bright white and is readily distinguishable to the naked eye. Scanning electron microscopy images can then clearly show the aggregation of the silica nanoparticle probes and detect target DNA. The demonstrated capability to detect a disease-related target DNA with direct visualization without using a complex detection instrument provides the prerequisite for the development of portable testing kits for genotyping (Figure 2.21).

In another study, dissolvable, organic, and biofunctional nanocrystals were used for the quantitative detection of PCR products.[444] Fluorescein diacetate a precursor of fluorescein, was used to create a stable, nanosized colloid with an interface for coupling streptavidin molecules. By incorporating these particulate labels, quantitative detection of biotinylated HPV DNA was amplified in a standard PCR procedure. This approach resulted in high selectivity, short incubation times, and a sensitivity up to 147 times greater than that obtained from directly fluorescent-labeled streptavidins, providing for rapid detection of small amounts of nucleic acids and requiring fewer PCR cycles.

Last, QDs have been used to examine HPV serotypes in oral SCC via in situ hybridization (ISH).[445] The detection of high-risk HPV serotypes (16/18) using QD ISH was compared with conventional ISH. Seven cases out of 21 (33.3%) were positive for QD ISH, while 1 out of 21 (4.8%) was positive for ISH. The difference between these two methods was statistically significant, showing that QD might be an efficient method for determining HPV infection and HPV-associated oral SCC.

2.6.5.2 Vaccination

There are currently two HPV vaccines, Gardasil and Ceravix, that help prevent infection with certain species of HPV associated with the development of cervical cancer, genital warts, and some less-common cancers. Both vaccines protect against the two HPV types (HPV-16 and HPV-18) that cause 70% of cervical cancers and cause most HPV-induced genital and head and neck cancers; Gardasil also protects against the two HPV types (HPV-6 and HPV-11) that cause 90% of genital warts. In addition, Gardasil has been shown to prevent potential precursors to vulvar, vaginal, penile, and anal cancers. HPV vaccines are expected to protect against HPV-induced cancers of these areas as well as HPV-induced oral cancers.

Nanotechnology can potentially allow for improved vaccination delivery as well as improved immunogenicity. In one study,[446] a DNA vaccine for HPV was developed. A recombinant baculovirus bearing human endogenous retrovirus (HERV) envelope protein was used as a nanocarrier to deliver the HPV-16L1 DNA vaccine (AcHERV-HP16L1). Mice were injected intramuscularly with 107 particles of the constructs, with two boosts at

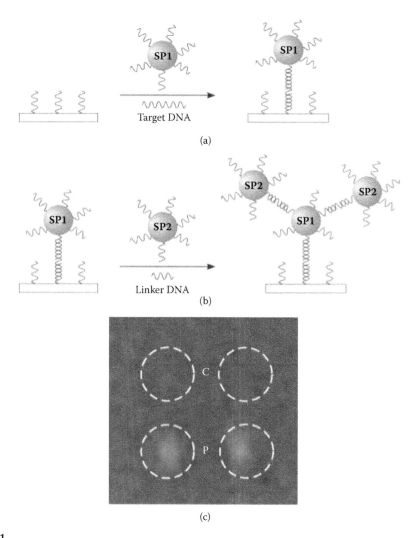

FIGURE 2.21
DNA-bearing silica nanoparticles in conjunction with a DNA chip allow for human papilloma virus test results
to be read by the human eye. (a) Target DNA allows for the binding of silica nanoparticles to a DNA chip.
(b) Additional silica nanoparticles can then be added onto the already bound base nanoparticle. (c) The end
result is a spot test that can be read with the human eye or, in this case, a desktop scanner. (From Piao, J.Y.,
Talanta, 80, 967, 2009. With permission.)

2-week intervals. Compared with Gardasil (25 µL/dose), the AcHERV-HP16L1 immunized
mice showed analogous high levels of humoral immunity in IgG/IgA and in neutraliza-
tion of HPV pseudovirions. Combined immunization (prime with AcHERV-HP16L1 and
boost with Gardasil) induced slightly higher neutralizing activity. As compared to the
group treated with Gardasil, the mice immunized with AcHERV-HP16L1 showed 450- and
490-fold increase in the IFN-γ at 5 and 20 weeks after the first priming, respectively. The
advantages of the AcHERV-HP16L1 vaccine over Gardasil included higher cellular immunogenicity, considerably lower production cost, and comparable safety, suggesting that the
AcHERV-HP16L1 may be useful as both a preventive and therapeutic vaccine.

As a means of creating improved delivery systems to enhance vaccine uptake, a dry-
coated densely packed microprojection array skin patch (Nanopatch) was developed and

compared to intramuscular injection for Gardasil in mice.[447] Neutralizing antibody titers were sustained out to 16 weeks postvaccination, and, for comparable doses of vaccine, mildly higher titers were observed with patch delivery than with intramuscular delivery via needle/syringe. This study offered that the use of the Nanopatch has the potential to overcome the limitation of needle- and syringe-delivered vaccines (Figures 2.22 and 2.23).

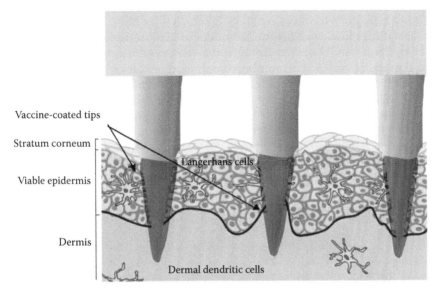

FIGURE 2.22
Microneedles penetrate the superficial layers of the skin to deliver human papilloma virus vaccine. The coated needle tips deliver the vaccine to the regions rich in antigen-presenting cells. (From Corbett, H.J., *PloS One*, 5, e13460, 2010. With permission under the creative commons attribution license.)

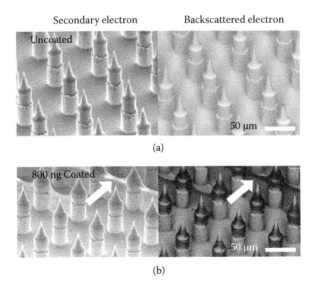

FIGURE 2.23
Scanning electron micrographs of microneedle arrays. Backscatter images of the uncoated needles (a) appear lighter than the needles coated (b) with human papilloma virus vaccine. (From Corbett, H.J., *PloS One*, 5, e13460, 2010. With permission under the creative commons attribution license.)

2.6.5.3 Therapy

As listed earlier, there are multiple treatment modalities to treat active, clinically evident lesions, ranging from destructive to immunomodulatory. Nanotechnology has been used to enhance and improve some of these treatments.

2.6.5.3.1 Podophyllotoxin

POD can inhibit the growth of epithelial cells infected by HPV in the epidermis and has been used as the first-line drug for the treatment of genital warts. However, the standard topical preparations of POD such as tincture and cream can result in a systemic absorption and severe skin irritation.[35] SLNs have been used to allow for epidermal targeting of POD, to allow for higher accumulation epidermis with transdermal permeation. The high accumulation of POD is beneficial to treat genital warts and reduce the systemic absorption and dermal irritation. In one study,[448] POD-loaded SLNs showed increased accumulative amount of POD in porcine skin—3.48 times over 0.15% tincture in *vitro*—and exhibited a strong localization of POD within the epidermis. The penetration of POD SLNs appeared to follow two pathways: along the stratum corneum and the hair follicle route, both due to the small size and permeation enhancement of the SLN soybean lecithin (Figure 2.24).

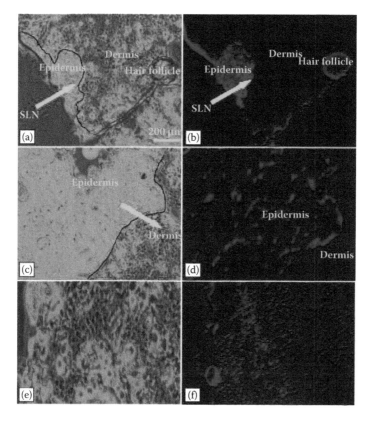

FIGURE 2.24
Podophyllotoxin-loaded solid lipid nanoparticles (P-SLNs) deliver drug throughout the epidermis. Microscopic and flourescent images of porcine skin treated with P-SLN show entry of the nanoparticle into all layers of the epidermis, including the portions of the epidermis present within the hair follicle (a and b; enlarged in c and d). In comparison (e and f), application of a tincture of podophyllotoxin results in the deposition of the toxin to nontarget tissues. (From Chen, H., *J. Control. Release*, 110, 296, 2006. With permission.)

In a randomized double-blinded study, POD SLNs gel and routine POD gel preparation was applied, respectively, for treatment of 97 volunteer patients with recurrent condyloma acuminatum.[449] Clinical clearance of condyloma patients in the first treatment course with POD SLN gel reached 97.1%, close to that with the routine preparation of 90.6%; however, the nanoparticle preparation significantly reduced the recurrence rate and adverse effect. This clinical study suggests that POD delivered via SLN gel has the potential to be used clinically to effectively clear condyloma acuminatum and reduce its recurrence rate with only mild, tolerable adverse effect.

2.6.5.3.2 Cantharidin

Cantharidin is a chemical agent derived from blister beatle, *Lytta vesicatoria*, used for decades as a blistering agent for the treatment of warts and molluscum contagiosum. Its utility as an antitumor agent has also been studied though systemic toxicity has limited its use. For this application, incorporation into nanovehicles has been pursued.[450,451] In one study,[450] cantharidin was encapsulated into PEGylated liposomes. This platform was evaluated against human breast cancer MCF-7 cells *in vitro* and its systemic toxicity *in vivo* in mice. Cantharidin liposomes labeled with octreotide were also studied as a means of lowering the dose of cantharidin without minimizing efficacy. The cytotoxic activity of PEGylated liposomal cantharidin was drastically reduced compared with free cantharidin *in vitro*. Octreotide-labeled PEGylated liposomal cantharidin induced cell death by specifically targeting somatostatin receptors in MCF-7 cells. Liposomal cantharidin had significantly less systemic toxicity than free cantharidin *in vivo* and also exhibited a high efficacy against antitumor growth in nude mice. These results suggest that liposomal platforms may be useful in mitigating toxicity of agents that are normally restricted due to their adverse effects without sacrificing therapeutic efficacy.

2.7 Benign Nevi

2.7.1 Epidemiology

Acquired nevi are common lesions that form during early childhood. Congenital nevi are present in 1% of neonates,[452] and acquired nevi increase through childhood, peaking between the ages of 20 and 29. In some individuals, they may continue to appear into adulthood.[453] In one study in the United Kingdom, nevi were found to continue to erupt in adult life. It also found that the mean number for nevi >2 mm in diameter in 754 healthy women between the ages of 18 and 46 was 57.[299,454] One study conducted on 743 children in Colorado found that the density of melanocytic nevi is greater in chronically sun-exposed areas of the body than in those that are intermediately exposed. Other factors that have been found to be positively associated with nevus prevalence are male sex, light skin, light brown hair, blue/green eyes, freckling, and the propensity to burn slightly and tan lightly compared with other reactions to sun exposure.[455,456]

2.7.2 Pathophysiology

Melanocytic nevi are thought to originate from cells called melanoblasts that migrate from the neural crest to the epidermis. Melanocytic nevi result from proliferation of altered melanocytes, or nevus cells, of monoclonal origin.[453] Nevus cells form small epidermal or dermal collections called nests. There are three broad categories of nevi, based on the location of the melanocytes. Junctional nevi are composed of nevus cells located at the dermal–epidermal junction. A compound nevus is a mole in which nevus cells are present both in the epidermis and dermis. Junctional nevi may change into a compound nevus. And third, dermal nevi are composed of nevus cells that are located solely within the dermis.[299,457]

There are two main models for nevus development. The first model suggests that a differentiated melanocyte acquires a mutation that results in a nevogenic transformation. This transformation allows the melanocyte to become more proliferative and produce cells that are less melanized and have the capacity to invade into the dermis. This theory is based on evidence that shows that nevi start out in the dermis as junctional nevi and penetrate further down into the dermis as they age. A second model suggests that a melanocyte precursor cell is the cell of origin. A mutation in this stem cell interferes with the normal replacement of differentiated melanocytes and causes an accumulation of nevomelanocytic cells. Mutations might cause deviation of nevomelanocytes from their normal pathway of dermis to epidermis, resulting in a variety of melanocytic neoplasms.

2.7.3 Clinical Features

Most nevi are tan to brown and are <6 mm with round shape and sharp borders. Junctional nevi are macular, or slightly elevated. Color can vary from uniform brown to black pigmentation, and they have a round or oval border. Compound nevi are slightly elevated or verrucous, and symmetric with a round or oval border. They are uniformly flesh colored or brown and become more elevated with age. Unlike junctional nevi, hair may be present, and a white halo may form. Dermal nevi are typically flesh or brown colored dome-shaped papules. For patients with multiple nevi, there is an increased risk for the development of cutaneous melanoma. In addition, because melanocytic nevi are less likely to occur as patient's age, there is a higher suspicion for melanoma in older patients with new nevi (Figure 2.25).[299,457,458]

2.7.4 Histopathology

The histopathologic features of acquired nevi show well-formed nests of melanocytes present in junctional and compound nevi. In the dermis of compound and intradermal nevi, sheets and cords of nevocytes are present. In the intradermal nevus, there is a well-demarcated zone of separation of the upper dermis, just below the epidermis, that is free of nevomelanocytes.[299,453,457]

2.7.5 Current Treatment

The majority of acquired nevomelanocytic nevi do not require treatment. Indications for removal of benign-appearing lesions include cosmetic concerns or irritation. Any lesion with worrisome features needs to be excised for histopathologic evaluation. Periodic examination of affected individuals for atypical or dysplastic melanocytic nevi and melanoma is recommended. Complete removal of nevi is best accomplished by excision. Incisional biopsy is used only when necessary, for lesions that cannot be excised easily but need histopathologic diagnosis. Destructive therapies—such as electrodessication, cryotherapy, dermabrasion, and laser—should be avoided, because they do not provide tissue for histopathology. Minimizing UV exposure is the most effective means of limiting the progression to malignancy.[454–456,459]

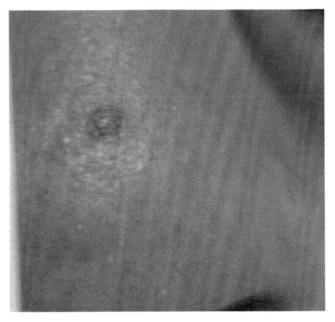

FIGURE 2.25
Junctional nevus: these lesions are completely flat or minimally elevated, and are typically symmetric.

2.7.6 Applications in Nanotechnology

2.7.6.1 Diagnosis

Traditional diagnosis of benign nevi in the skin is through history, physical examination, and occasionally through biopsy. Tracking and mapping nevi has traditionally been done using visual inspection, photography, and imaging techniques such as confocal microscopy.[313,330,460] These methods do not typically track the surface contours of nevi. Smart fabrics using nanocircuitry may be able to scan and map nevi on patients. Textiles embedded with nanoscale optical fibers have been shown in the laboratory to function as wearable cameras.[177,178,461,462] Fibers detect light levels in their local area and transmit them along the fiber length to a detector. Signals from all fibers in a woven fabric are collated and processed digitally to reconstruct a complete image. This type of device could be adapted for a type of form-fitting body-contoured fabric that can map pigmented lesions. The density of the weave would predict the theoretic resolution of the images.

Methods incorporating nanotechnology for the diagnosis of benign nevi include arrays measuring gene expression.[219,223,333,340,341,349,362,381,429,463] It has been determined that benign nevi have differing levels of DNA methylation when compared to melanoma.[333,334,361–363] Methylation status can be measured with a variety of technologies that include nanowire transistors.[464] Furthermore, biomarkers can directly be visualized using QDs or other fluorochromes (Figure 2.26).[357–360,375]

2.7.6.2 Treatment

Current methods for the management of benign nevi include observation, shave removal, or surgical excision. Surgical excision can be associated with pain, scarring, and disfigurement. In at least one study, excision, followed by dermabrasion and reintroduction of an

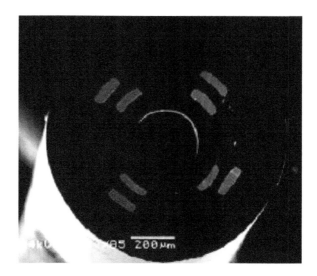

FIGURE 2.26
Scanning electron micrographs of a fiber cross section. Meters long lightweight flexible fibers integrating up to eight optical detectors with feature sizes down to 100 nm can be fabricated. (From Sorin, F., *Nano Lett.*, 9, 2630, 2009. With permission. Copyright© 2009 American Chemical Society.)

autologous cell suspension, resulted in minimal scarring in an infant with a large facial congenital melanocytic nevus.[465]

Methods to anesthetize the skin using topical delivery of local anesthetic may reduce pain associated with surgical procedures. Nanodelivery of topical anesthetic has been facilitated with liposomal delivery and by microneedle delivery.[234,426]

Laser ablation of pigmented lesions is fraught with controversy and not widely performed.[466–469] Destructive treatments of nevi have included application of a copper ion and acid solution (Solcoderm). Delivery of these types of agents could be improved with SLNs, but reports of poor cosmetic outcome mimicking pseudomelanoma have also been reported.[470–474] Selective photothermolysis using targeting-ablative agents, such as gold nanoshells or selective siRNA with homing markers specific for nevus cells, may circumvent these problems.[203,378388,398]

Methods to repair the skin with PDT and photochemical bonding have been shown to reduce scarring and speed healing.[475–480] Other methods to repair the skin include using nonchemical biocompatible and biodegradable van der Waal's based adhesives; bioadhesives from mussel proteins in inkjet-patterned arrays; and nanohooks.[375,480–494]

Surgical procedures such as excision may entail scarring. Scars and keloids may be managed with topical and intralesional therapy with imiquimod, IFN, corticosteroids, and 5-FU.[495–498] Topical penetration of these agents may be enhanced in nanodelivery vehicles. Radiation, pressure therapy, silicone gel sheets, and gene therapy have also been proposed for the treatment of keloids.[499,500] In addition, bleeding and scar formation may benefit from hemostatic agents such as oxidized cellulose nanofibers.[501–503]

The diagnosis of nevi and their distinction from other benign lesions (such as SKs) and malignant lesions (such as melanoma) may involve sophisticated imaging and gene expression analysis technologies that capitalize on nanotechnology. Treatment of nevi may involve PDT, gene therapy, or surgical excision. Methods to minimize scarring might include nonsuture bonding techniques or delivery of medication to reduce scar and keloid formation.

2.8 Psoriasis

2.8.1 Epidemiology

Psoriasis is seen in 1.5% – 3% of individuals of European descent[504] and in 1.5% of the African-American population.[505] In other populations, psoriasis is virtually nonexistent. China, for example, has a prevalence of psoriasis between 0% and 0.3%.[506] Although psoriasis can occur at any age, its peak age of onset is bimodal.[507] The first peak is at 16–22 years and the second is at 57–60 years. The younger group tends to have a more severe disease and is more likely to have a family member with psoriasis. The older group has a milder, indolent course.

Genetic factors play a significant role in determining a patient's risk of developing psoriasis along with the age of onset, distribution, and severity.[508] Environmental factors such as trauma, infection, drugs, hormonal changes, alcohol, smoking, stress, and immune status also play a role in the development of psoriasis.[509]

2.8.2 Pathophysiology

Microscopic examination of psoriatic lesions reveals hyperkeratosis, vasodilation, and a leukocytic infiltration. Although the initiating events are unknown, a variety of cytokines and inflammatory mediators released by keratinocytes play a role in establishing lesions.[510] Transforming growth factor-α promotes hyperkeratosis in an autocrine manner. IFN-γ and TNF-α recruit T-cells and neutrophils. VEGF promotes angiogenesis. Dendritic cells as well as Th17 cells, a T-helper cell subset that releases proinflammatory factors, contribute to the immune dysfunction seen in psoriasis.

Genome-wide association studies have identified nine chromosomal loci related to psoriasis. The PSORS1 locus has the strongest association of the set. It accounts for 30%–50% of heritability of psoriasis and is strongly associated with guttate psoriasis. The PSORS loci include portions of MHC proteins (PSORS1), keratinocyte desmosomes (PSORS1), the epidermodysplasia verruciformis locus (PSORS2),[512] the epidermal differentiation complex (PSORS4) and the Crohn's disease locus (PSORS8). The overlap in the PSORS8 locus with the Crohn's disease locus may explain why Crohn's disease is more prevalent in patients with psoriasis.[512]

2.8.3 Clinical Features

Chronic plaque psoriasis is the most common form of psoriasis. Guttate, pustular, and erythrodermic psoriasis are less common. The erythema, thickening, and scale correlate to the vasodilation, inflammation, and hyperkeratosis seen on histologic examination. The lesions of chronic plaque psoriasis are well-demarcated erythematous papules and plaques with a silver scale. The scalp, elbows, knees, hands, and feet are the most common sites of involvement. Nail involvement is common and presents as distal onycholysis, pitting, and "oil spot" phenomenon. When plaques clear, postinflammatory hyper- or hypopigmentation may be present. Pinpoint bleeding of the elongated dermal blood vessels following removal of a scale is called an Auspitz sign. Occasionally, a pale ring of blanched skin called Woronoff's ring surrounds a lesion.

Flexural or reverse psoriasis involves the axilla, inguinal crease, intergluteal cleft, and inframammary region. In this form of psoriasis, shiny thin red-pink plaques predominate over the thick scaling plaques seen in chronic plaque psoriasis. Guttate psoriasis is most common in children. Two-thirds of cases follow upper respiratory tract infections, and half of cases have serologic markers of streptococcal infection.[509] Erythrodermic psoriasis is characterized by generalized erythema and scaling and can be considered a medical emergency.

Pustular psoriasis consists of sterile neutrophil-filled pustules. It can be triggered by pregnancy, cessation of corticosteroid therapy, and infections.[513] There are several variants, including palmoplantar pustular psoriasis including acrodermatitis continua of Hallopeau, generalized pustular psoriasis of von Zumbusch (a medical emergency clinically characterized by lakes of pus, fever, erythroderma, and possibly hypocalcemia), and impetigo herpetiformis (pustular psoriasis of pregnancy).

Five to thirty percent of patients with psoriasis have psoriatic arthritis.[514] Psoriatic arthritis is an inflammatory and erosive arthropathy. Asymmetric oligoarthritis is the most common form. Other recognized types include symmetric arthritis, spondylitis arthritis, distal interphalangeal arthritis, and arthritis mutilans.[509] When the distal and proximal interphalangeal joints are involved, the localized inflammation can lead to "sausage digits." In contrast to rheumatoid arthritis, serologic findings are unusual.

The severity of psoriasis is determined by the extent of cutaneous involvement and the location of involvement. For example, moderate to severe psoriasis affects greater than 5%–10% of the skin or includes lesions on the face, palms, or soles (Figure 2.27).

2.8.4 Current Treatment

For mild to moderate psoriasis, topical corticosteroids and vitamin D analogs are first-line treatment.[513] In addition, calcineurin inhibitors, retinoids, tar, and anthralin can be applied topically. For more severe disease, phototherapy with narrowband UVB, MTX, or cyclosporin may be added. When these interventions fail, biological therapy can be considered. For many patients, successful treatment is not defined by complete remission but by partial remission and return to a functional status.[515]

Topical corticosteroids are the mainstay of topical treatment of psoriasis. Corticosteroids exert their immunosuppressive and anti-inflammatory effects through transcription regulation.[19] Following administration of corticosteroids, inflammatory cytokines levels, COX-2 activity, inducible nitric oxide synthase activity, prostaglandins levels, and histamine release are all decreased. Corticosteroids also affect metabolism, electrolyte balance, and growth. When administering corticosteroids topically, skin atrophy is an important side effect. The face, neck, flexures, and genitalia are particular susceptible to steroid-induced atrophy, and therefore corticosteroids should be used sparingly in these regions. High-potency, extensive, and generalized corticosteroid therapy can suppress the hypothalamic–pituitary–adrenal axis as well as cause myopathy, cataracts, and osteoporosis.

Calcipotriol/calcipotriene, calcitriol, and tacalcitol are vitamin D_3 analogs that affect keratinocyte differentiation and alter the immunologic profile of psoriasis. Long-term therapy at low doses is safe and moderately effective.[516] When betamethasone is added, better clearance and faster onset of action is observed. Rebounding or worsening of disease may follow termination of therapy. Furthermore, high doses can interfere with calcium homeostasis.

Tacrolimus and picrolimus are calcineurin inhibitors available as topical preparations. Blocking calcineurin prevents activation of T cells by inhibiting transcription of ILs 2 through

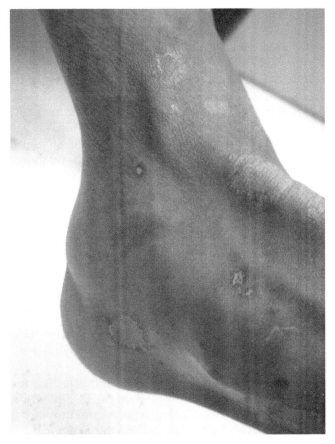

FIGURE 2.27
Psoriasis. Scaly silver plaques with an erythematous border are the hallmark of psoriasis. Both the size and multitude of plaques contribute to the morbidity of this disease.

5 and IFN-γ and TNF-α in T cells.[517] Because calcineurin inhibitors cause less atrophy than topical steroids, they are valuable for treating lesions on the face, neck, and genitalia.

Topical tazarotene, a retinoid, is effective in reducing psoriasis.[513] As in acne, topical retinoids normalize keratinocyte differentiation by activating the RAR (see Section 2.2.4). Because retinoids can be irritating, they are often used in conjunction with topical steroids.

Tar, which is derived from carbonized coal, is postulated to decrease mitotic rate and thus ameliorate the symptoms of psoriasis.[516] It is available in a myriad of formulations from crude coal tar to 1% coal tar suspended in esterified fatty acids. However, there is insufficient data to compare the many preparations. The most common side effect is folliculitis. When used in the anogenital region[509] or in combination with repeated high-dose UVB therapy,[518] tar can increase the risk of developing skin cancer.

Anthralin, a synthetic derivative of araroba tree bark, is a topical agent with anti-proliferative and anti-inflammatory properties.[509] Its use is limited by skin irritation and staining of skin and clothing. Several preparations of anthralin have been devised to reduce skin irritation, including low-dose, microcrystalline, and corticosteroid-containing preparations.

Phototherapy using UV light is effective for treating psoriasis.[519] Phototherapy regimens differ in the wavelength of UV light selected, the number of treatment sessions, and the use of a psoralen. A common regimen for psoriasis is thrice weekly narrowband UVB (311 nm) phototherapy. When narrowband UVB is unsuccessful, PUVA (psoralens pretreatment followed by UVA phototherapy) can be tried. Psoralens are a class of topically and orally administered drugs that bind to pyrimidine bases when exposed to UVA light. The end result is decreased keratinocyte proliferation, VEGF production, and ICAM-1 expression. Although it is more effective than narrowband UVB therapy, PUVA has a higher carcinogenic risk than narrowband UVB and is therefore a second-line treatment.[518]

MTX is cytotoxic to lymphoid cells and thus exerts an anti-inflammatory effect in psoriasis.[19] MTX inhibits de novo synthesis of thymidine and is therefore toxic to all cells, especially those with a rapid turnover. However, because MTX lacks specificity in its site of action, there are many side effects: hepatic cirrhosis, nephrotoxicity, interstitial pneumonitis, leukopenia, anemia, thrombocytopenia, and alopecia, to name a few.[513] Nonetheless, in severe cases of psoriasis, the benefits of MTX can outweigh the risks. Use of this drug requires close follow-up paired with frequent monitoring of liver, renal, hematologic, and immune status.

Cyclosporine acts on the same pathway as calcineurin inhibitors but further downstream.[509] The use of cyclosporine is suggested in severe psoriasis because of cyclosporine's adverse effects, most notably nephrotoxicity, hypertension, and immune suppression. A benefit of cyclosporine is its rapid onset of action. Cyclosporine is metabolized by p450 and therefore may be affected by medications that serve as p450 inducers or blockers.

Biological therapies for psoriasis can be helpful and effective for recalcitrant disease.[521] Infliximab, etanercept, and adalimumab target TNF-α, while alefacept targets T-cell protein CD2. A newer addition to this class, ustekinumab, targets the p40 of IL-12 and IL-23, part of the TH17 pathway. All of the biological agents require parenteral administration, because they are digested by gastric enzymes following oral administration.

Even with the wide variety of available treatments, complete clearance of psoriasis is considered an unrealistic expectation.[521] Topical corticosteroids are rapid acting but can be used only for so long before thinning skin and hypopigmentation appear.[515] The other topical agents have a slow onset of action. Patients with severe psoriasis must balance the beneficial and adverse effects of systemic immunosuppressive drugs. UV therapy, an effective monotherapy, still requires access to specialized equipment as well as an arduous treatment regimen of three times a week.[521] These limitations are certainly felt by patients, 40% of whom are frustrated with the ineffectiveness of their current psoriasis treatment.[522]

2.8.5 Applications in Nanotechnology

2.8.5.1 Diagnosis

Psoriasis affects up to 3% of the population. It is a chronic inflammatory disease that can affect the skin and the joints and is believed to be caused by a pathological T-cell-mediated immune response to an unidentified autoantigen.[135,523–530]

The diagnosis of psoriasis has traditionally been a clinical one with recognition of characteristic skin lesions. Occasionally, when the diagnosis is in question, skin biopsies may be helpful. Recent attention in psoriasis has focused on genetic studies and the use of biomarkers both for the diagnosis of psoriasis and for monitoring therapy. Predicting susceptibility to psoriasis lies in the realm of bioinformatics. In one study of 451,724 SNPs in 2,798 samples, it was determined that as few as 20 iteratively selected SNPs were able to predict psoriasis susceptibility with 68% accuracy.[523] OCT may also be useful in the diagnosis of psoriasis, the differentiation of nail psoriasis from onychomycosis, and for monitoring response to antipsoriatic therapy.[531]

The genes responsible for psoriasis map to a family of genes involved in chronic inflammation. These genes have also been implicated in AD, Crohn's disease, and rheumatoid arthritis. For example, genes involved in the IL-23 pathway have been implicated in psoriasis, ulcerative colitis, and Crohn's disease.[532–537] Genes in the endoplasmic reticulum aminopeptidase 1 pathway are associated with Crohn's disease and psoriasis. Other examples include IL-12 and TYK2.[538] Therapies that target IL-12 and IL-23 are showing early success in the management of plaque-type psoriasis vulgaris. Hence, some of the therapies that have shown benefit in one condition have shown crossover benefit in the others.

Currently, the diagnosis of psoriatic arthritis is challenging, based on history, examination, and rarely on imaging techniques such as radiography and magnetic resonance. In psoriatic arthritis, surrogate markers of disease status such as soluble markers of cartilage and bone metabolism can be used to monitor the effects of treatment. For example, the soluble marker matrix metalloprotease-3 (MMP-3) and melanoma inhibitory activity (MIA) (a marker for chondrocyte anabolism) have been used as a surrogate marker for efficacy of adalimumab in the treatment of psoriatic arthritis.[539] Early in adalimumab therapy, levels of MMP-3 fall, and MIA rise compared to placebo. Such screening can be helpful in the early phases of clinical trials to show efficacy. They can also be modified for an outpatient population to indirectly monitor response to therapy.

2.8.5.2 Treatment

Nanotechnology for the management of psoriasis involves enhancing drug delivery, promoting targeting of therapy, and reducing toxicity and side effects. Delivery of topical therapy can be mediated by encapsulation in SLNs. For topical steroids, this enhances deposition in the epidermis and papillary dermis and minimizes systemic absorption, thus avoiding side effects such as atrophy and telangiectasia.[146,150,151] Topical preparations of cyclosporine also persist in the epidermis and dermal–epidermal junction, where they exert their action on psoriatic patches and plaques without systemic toxicity.[540] MTX is typically given orally but suffers from side effects including hepatotoxicity. Topically administered MTX in SLNs may mitigate these concerns.[404,411,541–545]

Antibodies with biotherapeutic potential, the so-called biologics, such as adalimumab, efalizumab, infliximab, etanercept, and ustekinumab, are essentially nanoparticles (5 × 5 × 20 nm) that have been specifically engineered to target certain receptors.[370] Biologics have proven effective in treating psoriasis with greater efficacy and fewer side effects than many other systemic agents.[527,546–549]

Because one component of psoriasis is keratinocyte hyperproliferation, some therapies involving siRNAs that favor apoptosis of keratinocytes may confer benefits in the management of psoriasis.[153,339,397,532,537,550–553]

Personalized medicine can play a role in psoriasis therapy.[463,554,555] Genetic variation can alter patient response to therapy and affect outcomes. Causes of this variability may include host factors such as psoriatic disease subset, specific disease genetic basis, concomitant illnesses, concomitant medications, and the patient's ability to metabolize and respond to medication. Furthermore, some patients may not tolerate therapies because of vulnerability to medication toxicity. Identifying correct medications or combinations of medications suitable for a particular patient's psoriasis and general health profile will lead to tailored therapy, which is less costly, more effective, and has fewer side effects. For example, MTX is a standard therapy for psoriasis. Benefits of MTX therapy include generally high efficacy, low cost, and convenient weekly dosing. The drawbacks of MTX include toxicity and variable patient responsiveness.

Although the mechanism of action of MTX action is still unclear, at least five key genes are believed to be involved in its metabolic pathway.[555,556] It is transported into cells via the folate carrier SLC19A1, and is actively transported out of the cell by ABCC1 and ABCG2, and its final activity is mediated by the adenosine receptors ADORA1 and ADORA2a. In one study, Warren et al. evaluated 374 patients with chronic plaque psoriasis and evaluated their MTX transmembrane transporters and adenosine receptors. Using single-nucleotide polymorphism (SNP) date, they determined that SNPs within the efflux transporter genes ABC1 (ATP-binding cassette, subfamily C, member 1) and ABCG2 (ATP-binding cassette, subfamily G, member 2) are associated with a good response to MTX. One particular SNP in ABCC1, called rs246240, with two copies, is associated with susceptibility to MTX toxicity (Figure 2.28).

Cyclosporine A is a cyclic polypeptide immunosuppressive medication used for the management of psoriasis. Drawbacks of cyclosporine A include poor and variable oral bioavailability. Challenges include high lipophilicity, low intestinal permeability, P-glycoprotein-mediated efflux from enterocytes, and significant enterohepatic metabolism.[524,540] Oily solutions of cyclosporine A have bioavailability ranging from 1% to 89%. Self-microemulsifying concentrates of cyclosporine A have higher bioavailability; particle size averages 14.3 ± 4.5 nm. Glyceryl monooleate/poloxamer 407 cubic nanoparticles in the 180 nm size range entrap cyclosporine with high efficiency (85%) and have high and predictable bioavailability.

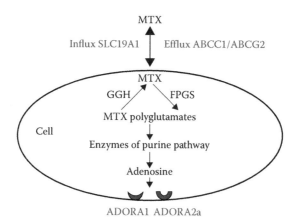

FIGURE 2.28
Methotrexate (MTX) metabolic pathway. Before exerting its effect on the cell, MTX must enter into the cell. At the same time that SLC19A1 is transporting MTX into the cell, ABCC1/ABCG2 are pumping it out. Once inside the cell, MTX is activated by polyglutamation. Activated MTX polyglutamates are then able to inhibit folate, purine, and pyrimidine pathways. Variations in the activity of these enzymes can lead to MTX toxicity at standard doses. (From Warren, R.B., *J. Invest. Dermatol.* 128, 1925, 2008. With permission.)

A combination of the vitamin D analog calcipotriol and MTX has been incorporated in a nanostructured lipid carrier for topical delivery.[543] Challenges to this combination include different polarities of the two drugs, which have differing partition coefficients of nearly four orders of magnitude. SLNs allow for controlled release and increased bioavailability. One type of SLN is the nanostructured lipid carrier. It is produced by mixing solid lipids with incompatible lipids in a special structure.[542]

In addition to inflammation, psoriasis is associated with hyperproliferation of keratinocytes. This may be mediated by fibroblast growth factor receptor 2 (FGFR2).[552] The miRNA MiR-125b is downregulated in psoriasis and is associated with keratinocyte proliferation. Enhancing expression of MiR-125b might interfere with excess keratinocyte proliferation.

One subset of T cells, Th17, may be central in psoriasis and other autoimmune diseases.[527,535–537,557–562] Th17 cells secrete a variety of proinflammatory cytokines, such as IL-17 A-F, IL-6, and TNF-α. Specific inhibition of IL-17 using a monoclonal antibody has shown neutralizing activity against IL-17A and has shown a favorable risk/benefit profile in phase I and phase II clinical studies of psoriasis.[536]

Retinoids have shown improvement in psoriasis but suffer from drawbacks of side effects such as mucositis, depression, dyslipidemias, cheilitis, xerosis, and teratogenicity. Topical delivery of retinoids such as all-*trans*-retinoic acid and isotretinoin-using nanodisks, or lipid or polymeric nanoparticles, shows enhanced cutaneous localization, decreased systemic penetration, and presumably a better side effect profile. Formulations using these agents may be suitable for consideration in trials on models of psoriasis.[421,563]

Angiogenesis is a hallmark of inflammation and is central to the pathogenesis of psoriasis. Fumarate esters, long known to have antipsoriatic activity, were recently found to act via angiogenesis inhibition. Fumarates have been compared to MTX in a multicenter trial. Fumarates may also cause immunosuppression and may be associated with increased risk of malignancy. Fumarate has been coupled to superparamagnetic nanocrystals that may be bifunctional theragnostic agents: they may be useful for the imaging (diagnosis) and treatment (therapy) of diseases responsive to fumarates, including psoriasis.[373,541,564–572] Fulvenes such as titanocene and triarylmethane dyes have shown antipsoriatic activity. The delivery of titanocene is enhanced when it is coated onto electrospun nanofibers.[573,574]

Microneedles may be useful for the delivery of drugs, including antibodies, to psoriatic lesions; they may be useful for immunotherapy of psoriasis, for sampling of psoriatic lesions for diagnosis, and for the delivery of PDT to psoriatic lesions. Microneedles may also deliver painless anesthesia to allow for less painful phototherapy or PDT.[375,424,425,429,575–581]

Psoralens can be delivered to psoriatic plaques with topical nanoparticulate psoralen, resulting in less systemic toxicity. PDT in psoriasis is hampered by variable penetration of the photosensitizer through scale in plaques. This could be improved by the use of keratolytics, nanoparticulate delivery topically or systemically, or the use of microneedles. Photosensitizers include 8-methoxypsoralen, hypericin, ALA, and methylene blue.

In summary, psoriasis is a chronic inflammatory disease. Nanotechnology may be able to map and track psoriatic skin lesions with nano-optical fibers. These same fibers may be useful for PDT of psoriasis. Targeted delivery of antipsoriatic drugs can enhance their specificity and efficacy while reducing their toxicity.

2.9 Epidermoid Cysts

2.9.1 Epidemiology

Epidermoid cysts are common and most frequently affect young and middle-aged adults. They are rare in childhood.[195] Epidermoid cysts commonly result from implantation of epidermis into dermis. They are the most common cutaneous cyst and are derived from epidermis, or the epithelium of a hair follicle. Epithelium is enclosed within the dermis, and this cystic enclosure becomes filled with keratin and lipid-rich debris. This can occur from trauma or surgery. They can also be caused by a blocked pilosebaceous pore, sometimes occurring adjacent to a body piercing.[195,582] Typically, there is no malignant potential; however, there have been cases reported of malignant tumors arising within an epidermoid cyst.[583] If a punctum is present, it is the plugged pilosebaceous unit, and cheesy debris may be expressed. Epidermoid cysts due to a developmental defect include those associated with Gardner's syndrome. These are characterized by their appearance in childhood.[584]

2.9.2 Clinical Features

Lesions are usually solitary and appear as mobile dermal or subcutaneous nodules with a central punctum. Cysts not associated with trauma are usually located on the upper chest, upper back, neck, or head. Trauma-induced cysts are more common on the palms, soles, and buttocks. They can be skin colored, yellow, or white in appearance. Cyst contents are typically cream-colored with a consistency and smell of curdled milk. The odor is quite distinct. If the epidermoid cyst ruptures into the surrounding tissue, an inflammatory reaction ensues. This may enlarge the nodule, and it is also associated with pain. Secondary polymicrobial infections are common as well.[458] Histologically, the cyst wall is formed by stratified squamous epithelium, with a granular layer present—a feature that helps delineate epidermoid cysts from pilar cysts. The content of the cyst is keratinaceous material and can appear as large, linear pink flakes.[458]

2.9.3 Current Treatments

To prevent recurrence, removal of the entire cyst wall is needed. To control small inflamed lesions, intralesional steroids can be used. If the cyst becomes inflamed, painful, or purulent, infection must be considered. Incision and drainage of the cyst, with oral antibiotics to treat *S. aureus* should be started. However, excision of the lesion should be deferred until the inflammation has decreased.[195,458]

2.9.4 Applications in Nanotechnology

2.9.4.1 Diagnosis

Epidermal inclusion cysts are typically evaluated by physical examination. They often contain a central punctum, which aids in diagnosis. Occasionally, they may be confused for other cystic lesions such as steatocystoma and ganglion cysts, benign tumors such as lipomas, vascular lesions, and rarely malignancies such as a metastasis or a localized malignancy. Imaging may be helpful in distinguishing cysts from vascular lesions and other solid tumors.[207] Fine-needle aspiration may help distinguish benign from malignant cysts.[585,586] DNA diagnosis with biomarkers may help to identify syndromes associated with cysts (i.e., Gardner's syndrome).[587–589] The use of arrays might be helpful in making this determination. Imaging studies such as ultrasound, specialized optical techniques, and radiologic techniques might combine diagnostic modalities with therapy.[130,313,330,374,376,377,590]

2.9.4.2 Treatment

Cysts that become inflamed might be treated with systemic antimicrobials. Chitosan-encapsulated antimicrobials tend to have a better systemic and topical delivery profile than their non-nanoencapsulated counterparts; nitric oxide in nanoencapsulated forms is better at penetrating abscesses and deep tissue infections.[173,591–593] Intralesional treatments that have been used include copper ion acid solutions (Solcoderm) but can create a pigmented scar that resembles a pseudomelanoma.[471,472,474,594,595] Epidermoid cysts might benefit from PDT, which has shown some benefit in the treatment of acne by targeting the sebaceous apparatus.[596–603] Liposomal methylene blue hydrogels have been used to target acne vulgaris in the photodynamic setting, and may show promise in cyst treatment. Fordyce spots and sebaceous hyperplasia have also been successfully treated with 5-ALA PDT, suggesting a preferential accumulation of the photosensitizer in sebum. Topical formulations of 8-methoxypsoralen have been made in SLNs and have shown reduced systemic toxicity and good skin delivery. The use of 8-methoxypsoralen has been used in the inhibition of *P. acnes* in the sebaceous gland.[602] Isotretinoin systemically has been used to treat severe cystic acne. Topical formulations of isotretinoin in nanoparticles have shown high retention in the skin and less systemic absorption.[421] Whether these would accumulate in superficial cysts is worthy of study.

Microneedles may also be useful for the delivery of drugs or anesthetic for the management of cysts.[424,426,575–578,604,605] For example, sclerosing agents such as 20% tricholoracetic acid and 30 mg/mL tetracycline have been shown to reduce risk of recurrence of cysts.

Surgically removed cysts may scar less with the use of topical scar therapy such as 5-FU, or with the use of minimally scarring photochemical tissue-bonding techniques.[475,477,481,487] Nanotechnology may permit surgical repair with minimal scarring, including the use of photochemical bonding, or biocompatible Van der Waal's adhesives in place of sutures.[477,479,480,487]

Nano-based methods will be useful in the rapid, minimally invasive or noninvasive diagnosis of cysts. Therapies for cysts using nanotechnology may be noninvasive and minimally scarring.

2.10 Seborrheic Keratosis

2.10.1 Epidemiology

SK is a benign skin growth that originates from keratinocytes. It is the most common benign epidermal tumor, and they are seen more often as people age.[606] They are more common in middle-aged individuals but can also arise as early as adolescence. One study of 100 Australian adults revealed that the prevalence of SKs increases from 12% in 15- to 25-year olds to 100% in those aged above 50 years.[607] Another European case-control study, which was conducted in the Netherlands, examined 966 skin cancer patients (BCC, SCC, and melanoma) for SKs.[608] In this study, 72% of patients were found to have SK, the number increasing from 69% in 50- to 59-year olds to 90% in 70- to 79-year olds. However, selection bias may have influenced the reported prevalence of SK. Most individuals with SKs have a positive family history for these lesions. Though more than half of SKs were found to be clonal in nature, they have no malignant potential.[609]

2.10.2 Pathophysiology

The pathogenesis of an SK is not completely understood. The available epidemiological studies support a possible causal role of UV light in the development of SK. SKs are found more commonly on areas of exposed skin. In a study of 303 Korean males,[610] as well as the study of Australian adults mentioned earlier, SKs were found to be more common as a function of skin surface area on the exposed areas of the body. Activating mutations in the gene encoding the tyrosine kinase receptor FGFR3 has been found in 40% of the hyperkeratotic and acanthotic subtypes, providing some insight into their development.[611] In addition, FGFR3 mutations were found in as many as 85% of adenoid SK.[612–614] SKFGFR3 belongs to a class of transmembrane tyrosine kinase receptors involved in signal transduction to regulate cell growth, differentiation, and migration. It has also been found that the mechanism of hyperpigmentation in SK involves the overexpression of endothelin (ET)-1. ET-1 is a strong keratinocyte-derived mitogen and melanogen for human melanocytes. ET secretion by keratinocytes may be augmented by local increased expression of TNF-α and endothelin-converting enzyme-1α.[615]

2.10.3 Clinical Features

SKs are usually asymptomatic but can be itchy. They are most frequent on the face and upper trunk.[195,610] The lesions usually begin as a well-circumscribed, dull, flat, tan, or brown patch. SKs can have a sharply marginated elevated edge, and because only the top layers of the epidermis are involved, they often can appear as though they are tacked on to the skin as they grow. Most SKs have fewer hairs than the skin they arise from. In contrast to melanocytic nevi, SKs do not reflect light and have plugged follicular orifices on the surface, appearing as a cerebriform pattern of fine fissures.[195] Acute eruptive SKs can sometimes signify underlying malignancy and termed the sign of Leser Tre'lat.[606]

2.10.4 Histopathology

There are several subtypes of SKs: benign squamous (solid), adenoid (reticulated), hyperkeratotic, pigmented, clonal, plane, and irritated seborrheic. Each type shows varying degrees of acanthosis, pigmentation, horn pseudocyst formation, papillomatosis, and vascular involvement when viewed histologically.[616] Typically, proliferation of monomorphous keratinocytes and melanocytes are seen, with the formation of horny cysts. The most common pathological type is the solid variant, where a mass of immature keratinocytes are seeen above the level of the surrounding epidermis.[195] The hallmark of the histopathologic findings is pseudohorn cysts, acanthosis, papillomatosis, and hyperkeratosis (Figure 2.29).

2.10.5 Current Treatment

Because these lesions are benign, they do not require treatment. However, treatment is often pursued for cosmetic purposes. The best approach is curettage after slight freezing. Light electrocautery can also be used, but this precludes histopathologic verification of

FIGURE 2.29
Seborrheic keratosis (SK). A well-circumscribed tan patch with elevated margins giving a "tacked on" appearance. Note the crebiform appearance of fine fissures within the SK.

the lesion and should be used only by an experienced diagnostician. If melanoma is suspected, an excisional biopsy is needed, because shave biopsy does not permit evaluation of the level of invasion.[195]

2.10.6 Application in Nanotechnology

2.10.6.1 Diagnosis

SK can be diagnosed clinically. In cases of ambiguity, specialized optical imaging techniques in isolation, or in combination with specific nanotags may be helpful.[131,208,209,313,314,322,324] Arrays that detect genes or gene-expression products unique to SK may be helpful in determining a diagnosis.[614,617,618] Arrays can be minimally invasive using tape stripping or microneedles for tissue analysis.[330,340,429]

2.10.6.2 Treatment

Current treatments include topical keratolytics such as ammonium lactate,[619] hydroxy acids,[243] tricholoracetic acid,[231,620] retinoids such as tazarotene,[621] and destructive methods such as cryotherapy,[621] curettage,[622] and ablative laser therapy.[623,624] Shave removal and occasionally excision are used to remove SKs. Pruritic lesions are occasionally treated with antihistamines or topical corticosteroids. Mutations seen in SK include FGFR3 and PIK3CA.[617] FGFR3 mutations appear to be somatic mutations and associated with age and with localization on the head and neck suggesting cumulative UV exposure as the cause.[614,618] SKs also have lower levels of the SRK inhibitor Srcasm in areas of proliferation, and higher levels of Srcasm in areas of differentiation.[625] Targeting these mutations with, for example, nanoencapsulated siRNA applied topically or delivered systemically may provide alternative treatments for SK.

Treatments for SK may also include phototherapy with 532 nm laser light following color enhancement with either a red marker or ferric subsulfate.[623] Phototherapy with nanoemulsion-based delivery of chromophore combined with nano-optical garments may allow for the simultaneous mapping and therapy of SK.[176–178,430,431,462]

Examples of keratolytics used in the treatment of SK include 50% urea under occlusion, ammonium lactate, etretinate, acitretin, tazarotene, trichloroacetic acid, and α-hydroxyl acid.[243,619,621,626] Vitamin D analogs such as tacalcitol, calcipotriol, and maxacalcitol have been used in the treatment of SK.[627] Protease digestion with trypsin has shown scar-free dissolution of SK of the lower extremity.[628] Iron oxide liposomes containing trypsin have been developed for diagnostic imaging and drug release but may have utility in the treatment of SK.[629] Polymeric nanoparticles containing methyl trypsin have been used as enzymatic cleaners for contact lenses but may prove useful in the management of SK.[630] Antiproliferative agents such as 5-FU have been used to treat SK.[225] Curettage and hemostasis with oxidized cellulose have also been used for improvement of the appearance of SK.[622]

A copper and acid solution (Solcoderm, a combination of 5-FU and salicylic acid) has been used for the treatment of SK.[470,471,474,631,632] Topical induction and elicitation of irritant-induced sensitization has been shown to help in the treatment of SK.[633] In one case study, a giant SK on the frontal scalp was successfully treated with 5-FU alone.[634] One patient with pityriasis rubra pilaris and eruptive SK improved after systemic retinoid therapy.[635] Mixtures of 5-FU and retinoic acid have been used to treat SK. Nanoparticulate delivery of 5-FU will need to be compared to standard formulations to determine efficacy and tolerability.[159,401–403,407,411,636,637]

In the cases where eruptive SKs are associated with malignancy, the use of specialized imaging or arrays relying on nanoengineering could be useful for the diagnosis of underlying tumors.[638–655]

Nanotechnology-based methods will be used in the future for the sensitive and specific diagnosis of SK and for the delivery of targeted topical or systemic therapy for their removal.

2.11 Dermatophytosis

2.11.1 Epidemiology

The variety of lay terms used to describe dermatophyte infections underscores the ubiquity of this class of dermatologic disease. The nontrivial cost of treatment for ringworm, athlete's foot, jock itch, thrush, and other cutaneous fungal infections, coupled with their pervasive nature, constitutes a serious dermatologic concern.[656] Medicare expends $43 million annually on the treatment of onychomycosis (infection of the nail) alone.[657] Dermatophyte infection extends across different age demographics with adolescents and adults most often affected.[660] However, tinea capitis is more common in children. Athletes who engage in contact sports are also at increased risk for infection.[661]

The prevalence of fungal pathogens varies by geographic region and fluctuates over time. *Trichophyton rubrum* is the most common cause of tinea pedis, nail infection, tinea cruris, and tinea corpus worldwide.[662] In contrast, *M. canis* is the common cause of tinea capitis in Europe, while in the United States *T. tonsurans* is the pathogen in almost all cases. Over the past century, *T. rubrum* and *T. tonsurans* have edged out *Epidermophyton floccosum* and *M. audouinii* to become the most common cutaneous fungal pathogens.[660] These changes in fungal pathogens parallel changes in living conditions, availability of treatments, and migration patterns.[661]

2.11.2 Pathophysiology

The ability of dermatophytes to thrive on the skin is due to a combination of fungal, host, and environmental factors. Unlike other fungi, dermatophytes produce keratinases that degrade keratin, mannins that locally inhibit the immune system, and adhesions that attach to carbohydrates on the skin.[662] Important host factors include lipids, sebum, and the immune system. Lipids are metabolized by the infecting fungi, sebum inhibits fungal growth and immune response prevents infections.[658] Environmental factors such as temperature and humidity also prove to be essential for fungal growth.[658]

Thirty species drawn from the genera Trichophyton, Microsporum, and Epidermophyton account for almost all dermatophytoses.[660] Transmission of these dermatophytes can occur through human or animal contact or from the soil.[658] Incubation is typically 1–3 weeks. Fungal species show trophism for different regions of the body. *T. rubrum* is common on the trunk and limbs, while *T. tonsurans* is conventionally found on the head.[660] *E. floccosum* preferentially grows in the groin and on the feet and cannot invade hair. In addition, *Candida* species are common causes of tinea pedis and onychomycosis.

2.11.3 Clinical Features

Dermatophytoses are distinguished and named by their location: tinea corporis for the body, barbae for beard, faciei for face, cruris for groin, capitis for head, pedis for feet; manuum for hand, and unguium for nail. The infection manifests as expanding annular lesions. Scale, erythema, and, in rare instances, pustules are seen at the border of the lesion. Patients may report that the involved areas are pruritic and/or painful. If lesions are treated with topical corticosteroids, they can lose their inflammatory character (erythema and scale). Diagnosis is aided by skin scrapings for microscopy and culture (Figure 2.30).[665]

Some sites of dermatophyte infection present with additional features.[658] Tinea cruris involves the inner thigh but tends to spare the scrotum. Tinea manuum of the palms is hyperkeratotic, involves the digits, but spares the nail. Tinea barbae can present with inflammation, follicular pustules, abscesses, sinus tracts, and even bacterial super-infection. Tinea capitis can cause discrete areas of alopecia. In severe cases of tinea capitis, a boggy, balding, pustular eruption called a kerion can form. This form can cause systemic signs of illness and local lymphadenopathy. Tinea unguium or onychomycosis causes thickening and yellowing of the nail as well as onycholysis. Tinea pedis can provide an entry point for bacterial infections, especially in immunocompromised patients.[663]

2.11.4 Current Treatment

Antifungals, both topical and systemic, are the cornerstone of treatment. Topical therapy has few side effects and is available in over-the-counter formulations but often requires a prolonged course of treatment.[663] The topical azoles, clotrimazole and miconazole, are the first-line treatment. Second-line treatment is the topical allylamine terbinafine. The azoles and allylamines inhibit ergosterol synthesis and thus fungal-wall formation.

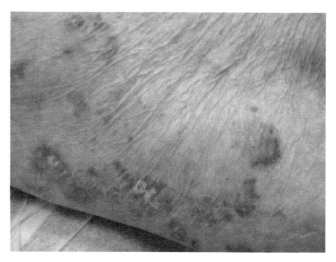

FIGURE 2.30
Dermatophytosis. Expanding erythematous annular lesions with a trailing scale are classic signs of dermatophytosis.

Systemic terbinafine, itraconazole, fluconazole, and griseofulvin are used in resistant cases and are also often necessary in cases of tinea manuum, capitis, and unguium.[658] Unlike the azoles and allylamines, griseofulvin inhibits microtubule function within the fungi. Nonpharmacologic interventions, such as keeping feet dry for tinea pedis or shaving and debridement for tinea barbae, can also be effective. Although most cases of tinea can be effectively treated with available therapy, it can take upward of 2 weeks with systemic therapy and upward of 4 weeks with topical therapy to clear the infection. This is true even in simple cases.

2.11.5 Applications in Nanotechnology

The pursuit of developing nanomaterials for the treatment of fungal-derived skin disease has diverged into two specific foci—one using materials that at the nanoscale have inherent antimicrobial properties; the second incorporating known antifungal therapeutics into nanovehicles such as liposomes or nanoparticles to enhance efficacy and improve bioavailability and pharmacokinetics. These two are not mutually exclusive, as antifungal nanomaterials are being used in conjunction with antifungal agent nanocarriers to capitalize on their synergy and differing mechanisms of action.

2.11.5.1 Antifungal Nanomaterials

Nanoparticles that consist of metals such as silver and metal oxides, or biopolymers such as chitosan may be promising agents for antifungal applications. These nanoparticles have the ability to bind to fungal cell walls causing increased membrane permeability and disruption through direct interactions, or to generate damaging free radicals. Mammalian cells are able to phagocytose nanoparticles and can subsequently degrade these particles by lysozomal fusion, reducing toxicity and free radical damage. This may allow for the selectivity of the same nanoparticle to promote tissue-forming cell functions while also inhibiting bacterial functions that lead to infection.

2.11.5.1.1 Nano-Silver

For centuries, silver (Ag) has been used for the treatment of burns and wounds to prevent infection.[664] Although the mechanism of its antimicrobial effects is not entirely known, it has been proposed that silver and silver ions (such as $AgNO_3$) penetrate bacterial cell walls and membranes via interaction with sulfur-containing proteins or thiol groups.[665] Once inside the cell, $AgNO_3$ targets and damages bacterial DNA and respiratory enzymes, leading to loss of the cell's replicating abilities and ultimately cell death.[664,666] The small size relative to the large surface area of silver nanoparticles (AgNPs) makes them better able to penetrate bacterial cell walls and membranes, and as a result the antimicrobial effect is directly dependent on nanoparticle size and shape.[664,667] Smaller nanoparticles (<10 nm) as well as triangular or truncated nanoparticles are more effective than larger particles that are round or rod shaped (Figure 2.31).[665]

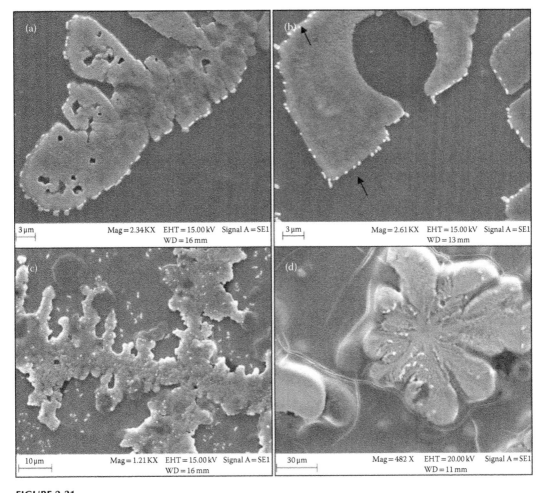

FIGURE 2.31
Scanning electron micrograph of *E. coli* treated with silver chitosan nanoparticles. (a) After 30, (b) 60, and (c) 90 minutes binding of the nanopaticle to the cell membrane and progressive disintegration of the cell over the observation window is observed. (d) In contrast, *E. coli* treated with chitosan alone does not show signs of cellular degradation after 90 minutes of treatment. (From Sanpui, P., *Int. J. Food Microbiol.*, 124, 142, 2008. With permission.)

The stronger interaction of AgNPs with microbial surfaces might also allow for lower Ag concentrations as compared to current silver agents and may limit silver's toxicity, such as argyria. However, these benefits are largely theoretical, and the adverse effects of silver nanoparticles have yet to be fully characterized.[664] In one study, the antifungal activity of AgNPs toward *Candida albicans* was shown and found to be similar to the antifungal activity in the minimal inhibitory concentration values of amphotericin.[668] The mode of action of AgNPs was shown to be due to its ability to dissipate the membrane potential of *C. albicans*. The results indicated AgNPs have a similar mechanism of action to that of amphotericin B, which forms transmembrane pores causing leakage of cell constituents and eventually cell death. In addition, investigators showed that AgNPs have either no effect or a weakly cytolytic effect on human erythrocytes, while amphotericin B incurred a higher hemolytic activity suggesting that AgNPs may be a safer treatment. These data have been further corroborated by other investigators.[669–671]

2.11.5.1.2 Chitosan

Chitosan is a natural polysaccharide biopolymer derived from chitin, which is the principal structural component of the crustacean exoskeleton. The antimicrobial properties of chitosan result from its polycationic character in a weakly acidic pH, which favors interaction with negatively charged microbial cell walls and cytoplasmic membranes, resulting in decreased osmotic stability, membrane disruption, and eventual leakage of intracellular elements.[666,672,673] In addition, chitosan is able to enter the nuclei of bacteria and fungi and inhibit mRNA and protein synthesis by binding to microbial DNA.[673,674] When nanoscaled, chitosan has a higher surface to volume ratio, translating into higher surface charge density, increased affinity to bacteria and fungi, and greater antimicrobial activity.[674]

In one study, the in *vitro* antifungal activity of low-molecular weight chitosan (LMWC) against 105 clinical *Candida* isolates was measured.[675] LMWC exhibited a significant antifungal activity, inhibiting more than 89.9% of the clinical isolates examined (68.6% of which was completely inhibited). The species included several fluconazole-resistant strains and less susceptible species such as *C. glabrata*, suggesting chitosan may be a useful tool in the setting of antifungal drug resistance. Other studies have further supported the potential utility of chitosan as an antifungal agent,[676–678] though chitosan nanoparticles have yet to be fully explored as a topical antifungal agent (Figure 2.32).

2.11.5.1.3 Titanium Dioxide

Titanium dioxide (TiO_2) forms active oxygen species when exposed to UV light, a process called photocatalysis. These oxygen species, including hydrogen peroxide and hydroxyl

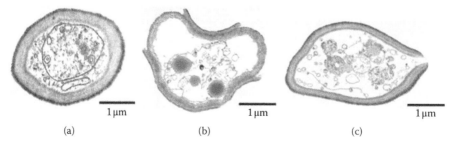

(a)   (b)   (c)

FIGURE 2.32
Gold Nanoparticles inflict structural damage to *C. albicans*. *C. albicans* incubated in (a) media alone, (b) 170, (c) 400 µg/mL of nano-Ag for 24 hours showing significant damage to the fungal envelope in nano-Ag treated fungi. (From Kim, K.J., *Biometals*, 22, 235, 2009. With permission.)

radicals, damage bacterial cell membranes resulting in cell death.[679,680] This antimicrobial property has been used in water and air purification and recently has been investigated against pathogenic and opportunistic microorganisms.[681]

TiO_2 has also been combined with silver to create TiO_2-Ag nanoparticles (TiO_2-AgNPs) that were tested against Gram-negative and Gram-positive bacteria and various fungi that are responsible for opportunistic infection and colonization of medical devices.[681] TiO_2-AgNPs showed better antifungal activity against *C. albicans* and Aspergillus *fumigatus* when compared to AgNPs alone and conventional antifungals such as fluconazole.[682]

2.11.5.1.4 Soybean Oil-Derived Nanoemulsions

Antimicrobial nanoemulsions are stable o/w emulsions composed of nanometer-sized, positively charged droplets that have broad-spectrum activity against enveloped viruses, fungi, and bacteria.[113–115,682–690] NB-002 is one such nanoemulsion that is currently being investigated as an antifungal agent. It contains the cationic quaternary ammonium compound cetylpyridinium chloride oriented at its oil–water interface, which stabilizes the nanoemulsion droplets and contributes to the anti-infective activity. The NB-002 nanoemulsion was shown to exert broad, uniform fungicidal activity against numerous dermatophyte species including *T. rubrum*, *T. mentagrophytes*, and *E. floccosum* as well as the yeast *C. albicans*, the primary causes of human fungal hair, skin, and nail disease.[684,691,692] Its ability to serve as a topical treatment has been evaluated ex *vivo* with human cadaver skin samples, showing that NB-002 can enter through a transfollicular route to enter the epidermal and dermal layers and that lateral diffusion occurs along tissue planes to sites distal from the application site.[687,693] Clinical studies of NB-002 for the treatment of onychomycosis have provided supporting results in its use as an effective topical treatment.[694]

2.11.5.2 Antifungal Encapsulated Nanomaterials

Liposomes have been used to both enhance efficacy and improve delivery of antifungal agents. The best example and application of liposomal encapsulation is with amphotericin B. This is one of the mot effective antifungal agents, causing direct fungicidal activity by binding to ergosterol in fungal cell membranes and causing pore formation and ultimately cell death. However, the clinical use of this medication is limited due to severe, dose-limiting side effects. Amphotericin B, due to its lipophilic nature, was an ideal candidate for liposomal encapsulation and was actually one of the first liposomal drugs to enter clinical trials in the late 1980s and continues to be a hot area of pursuit even today.[695–710] Several examples of products that have emerged from the initial studies include Abelcet®, Amphotec®, and Ambisome®.[703,704,711] However, the use of intravenous liposomal amphotericin B in the treatment of superficial dermatomycoses would be considered extreme, and no topical formulation is currently available. This example does set the template though for the use of liposomes to deliver topical antifungal agents, such as imidazoles (miconazole), triazoles (fluconazole and econazole), and polyenes (nystatin).

2.11.5.2.1 Miconazole Nitrate

Miconazole nitrate is a broad-spectrum antifungal agent of the imidazole group. It acts by means of a combination of two mechanisms: ergosterol biosynthesis inhibition, which causes lysis of fungal cell membranes due to the changes in both membrane integrity and fluidity, and direct membrane damage of the fungal cells. The drug is primarily used as a topical treatment for cutaneous mycoses. However, poor skin penetration capability presents a problem in the treatment of cutaneous diseases by topical application.

In one study,[712] miconazole nitrate was encapsulated in closed lamellar vesicles and parameters such as lipid composition, type of lipid, charge, microscopic appearance, size, and degree of drug entrapment were evaluated. Higher permeation of miconazole across the skin and greater retention in the skin with liposomal systems suggested that the liposomes could penetrate and form depots in upper skin layers. The liposomal phospholipids likely allowed for improved permeation and retention in the skin as they were analogous to natural constituent of skin lipids.

2.11.5.2.2 Fluconazole

Targeted and focused delivery of fluconazole by conventional therapy is a major impediment in achieving its therapeutic efficacy against skin infections, such as cutaneous candidiasis. Liposomal technology has been used to allow for topical delivery. In one study,[713] fluconazole-loaded vesicular liposomes and niosomes were generated and incorporated into carbopol gel (1%; w/w) for sustained, localized application. Animal models of cutaneous candidiasis in immunosuppressed albino rats were evaluated as a model for preclinical efficacy. The in *vivo* localization studies showed that liposomal gel produced 14.2-fold higher drug accumulation, compared with plain gel, while a 3.3-fold increase was appreciated in the case of an equivalent-dose application in the form of niosomal gel. The greatest clinical and microscopic improvement was seen with the liposomal gel, suggesting that this formulation may be an effective treatment of cutaneous candidal infections.

2.11.5.2.3 Econazole

Econazole was in fact one of the first antifungal agents to be incorporated into liposomes. In 1984, CILAG (now Janessen-CILAG) initiated the development of a liposome-based topical formulation of econazole and showed a reproducible manufacturing technique with potential for economic scale-up.[714] Clinical investigations revealed a hastened onset of effect due to increased concentration of active drug substance level in the stratum corneum, though treatment duration could be reduced. Pevaryl® Lipogel became the first approved liposome product and proved the feasibility of liposomal systems.[714]

Since then, other formulations of liposomal econazole have been pursued.[715–718] In one controlled, double-blinded trial, the efficacy of liposomal econazole was evaluated as a drug-carrier system.[715] Treatment either encompassed once-daily application for 14 days of econazole liposome gel 1%, branded econazole cream 1%, or a generic clotrimazole cream 1%. Higher cure rates based on microscopy and culture were appreciated in the econazole liposome gel treatment group. In addition, tolerability was considered slightly better in the econazole liposome gel group, compared with the econazole cream and clotrimazole cream treatment groups.

2.11.5.2.4 Nystatin

Nystatin was the first polyene antifungal agent to be identified and used. It shares a similar mechanism of action to amphotericin B, as it acts by binding to ergosterol, a component of the fungal cell membrane, resulting in abnormal membrane permeability and ultimately cell death. Nystatin has broad activity against numerous fungal species including *Candida*, Histoplasma, and *Aspergillus*. Use has been limited to topical administration as systemic side effects and toxicity are severe. Liposomal encapsulation has been pursued to enable systemic delivery in the setting of disseminated disease[719–723]; however, the use of nanoencapsulation to enhance topical administration has yet to be evaluated.

In summary, the utilization of nanomaterials as both improved vehicles for known anti-fungal agents, as well as emerging, novel therapies themselves, are already underway. Nanotechnology has a clear role in the anti-infective arena.

References

1. Elias PM. Stratum corneum defensive functions: an integrated view. *J Invest Dermatol* 2005; 125: 183–200.
2. Bos JD and Meinardi MM. The 500 Dalton rule for the skin penetration of chemical compounds and drugs. *Exp Dermatol* 2000; 9: 165–9.
3. Staples M, Daniel K, Cima MJ, and Langer R. Application of micro- and nano-electromechanical devices to drug delivery. *Pharm Res* 2006; 23: 847–63.
4. Kupper TS. Immune and inflammatory processes in cutaneous tissues. Mechanisms and speculations. *J Clin Invest* 1990; 86: 1783–9.
5. Williams IR and Kupper TS. Immunity at the surface: homeostatic mechanisms of the skin immune system. *Life Sci* 1996; 58: 1485–507.
6. Cevc G. Transfersomes, liposomes and other lipid suspensions on the skin: permeation enhancement, vesicle penetration, and transdermal drug delivery. *Crit Rev Ther Drug Carrier Syst* 1996; 13: 257–388.
7. Nasir A. Nanotechnology and dermatology: part I-potential of nanotechnology. *Clin Dermatol* 2010; 28: 458–66.
8. Cevc G and Vierl U. Nanotechnology and the transdermal route: a state of the art review and critical appraisal. *J Control Release* 2010; 141: 277–99.
9. Farokhzad OC. Nanotechnology for drug delivery: the perfect partnership. *Expert Opin Drug Deliv* 2008; 5: 927–9.
10. Zaenglein AL and Thiboutot DM. Chapter 37. Acne vulgaris. In: Bolognia J, Jorizzo JL, Rapini RP, eds. *Dermatology*. London, New York: Mosby, 2003.
11. White GM. Recent findings in the epidemiologic evidence, classification, and subtypes of acne vulgaris. *J Am Acad Dermatol* 1998; 39: S34–7.
12. Layton AM. *Disorders of the Sebaceous Glands.* In: Burns T, Breathnach S, Cox N, and Griffiths C, eds. *Rook's Textbook of Dermatology*. West Sussex, United Kingdom, Wiley-Blackwell, 2010: 1–89.
13. Jeremy AHT, Holland DB, Roberts SG, Thomson KF, and Cunliffe WJ. Inflammatory events are involved in acne lesion initiation. *J Investig Dermatol* 2003; 121: 20–7.
14. Bellew S, Thiboutot D, and Del Rosso JQ. Pathogenesis of acne vulgaris: What's new, what's interesting and what may be clinically relevant. *J Drugs Dermatol* 2011; 10: 582–5.
15. Choudhry R, Hodgins MB, Van der Kwast TH, Brinkmann AO, and Boersma WJ. Localization of androgen receptors in human skin by immunohistochemistry: implications for the hormonal regulation of hair growth, sebaceous glands and sweat glands. *J Endocrinol* 1992; 133: 467—75.
16. Deplewski D and Rosenfield RL. Role of hormones in pilosebaceous unit development. *Endocr Rev* 2000; 21: 363–92.
17. Lee L. The clinical spectrum of neonatal lupus. *Arch Dermatol Res* 2009; 301: 107–10.
18. Tsatsou F and Zouboulis CC. Acne vulgaris. In: Lebwohl M, ed. *Treatment of Skin Disease: Comprehensive Therapeutic Strategies*. Philadelphia, PA: Mosby/Elsevier, 2006: xxiv, 723 p.
19. Burkhart C, Morrell D, and Goldsmith L. Chapter 65. Dermatological pharmacology. In: Goodman LS, Brunton LL, Chabner B, and Knollmann BC, eds. *Goodman & Gilman's the Pharmacological Basis of Therapeutics*. New York: McGraw-Hill, 2011.
20. Thiboutot D, Gollnick H, Bettoli V, Dréno B, Kang S, and Leyden JJ et al. New insights into the management of acne: an update from the Global Alliance to Improve Outcomes in Acne Group. *J Am Acad Dermatol* 2009; 60: S1–50.

21. Lee LA. Neonatal lupus: clinical features and management. *Paediatr Drugs* 2004; 6: 71–8.
22. Mevorach D, Elchalal U, and Rein AJ. Prevention of complete heart block in children of mothers with anti-SSA/Ro and anti-SSB/La autoantibodies: detection and treatment of first-degree atrioventricular block. *Curr Opin Rheumatol* 2009; 21: 478–82.
23. Brucato A. Prevention of congenital heart block in children of SSA-positive mothers. *Rheumatology* 2008; 47: iii35–7.
24. Levine RM and Rasmussen JE. Intralesional corticosteroids in the treatment of nodulocystic acne. *Arch Dermatol* 1983; 119: 480–1.
25. Gumbo T. Chapter 56. Chemotherapy of tuberculosis, mycobacterium avium complex disease, and leprosy. In: Goodman LS, Brunton LL, Chabner B, Knollmann BC, eds. *Goodman & Gilman's the Pharmacological Basis of Therapeutics*. New York: McGraw-Hill, 2011.
26. Hædersdal M, Togsverd-Bo K, and Wulf HC. Evidence-based review of lasers, light sources and photodynamic therapy in the treatment of acne vulgaris. *J Eur Acad Dermatol Venereol* 2008; 22: 267–78.
27. Inglese MJ, Fleischer AB, Jr., Feldman SR, and Balkrishnan R. The pharmacoeconomics of treatment: Where are we heading? *J Dermatolog Treat* 2008; 19: 27–37.
28. Smith EV, Grindlay DJ, and Williams HC. What's new in acne? An analysis of systematic reviews published in 2009–2010. *Clin Exp Dermatol* 2011; 36: 119–22; quiz 23.
29. Trapasso E, Cosco D, Celia C, Fresta M, and Paolino D. Retinoids: new use by innovative drug-delivery systems. *Expert Opin Drug Deliv* 2009; 6: 465–83.
30. Yentzer BA, McClain RW, and Feldman SR. Do topical retinoids cause acne to "flare"? *J Drugs Dermatol* 2009; 8: 799–801.
31. Shah SK and Alexis AF. Acne in skin of color: practical approaches to treatment. *J Dermatolog Treat* 2010; 21: 206–11.
32. Merkviladze N, Gaidamashvili T, Tushurashvili P, Ekaladze E, and Jojua N. The efficacy of topical drugs in treatment of noninflammatory acne vulgaris. *Georgian Med News* 2010: 9: 46–50.
33. Eichenfield LF, Fowler JF, Jr., Fried RG, Friedlander SF, Levy ML, and Webster GF. Perspectives on therapeutic options for acne: an update. *Semin Cutan Med Surg* 2010; 29: 13–6.
34. Lott R, Taylor SL, O'Neill JL, Krowchuk DP, and Feldman SR. Medication adherence among acne patients: a review. *J Cosmet Dermatol* 2010; 9: 160–6.
35. Wiesenthal A, Hunter L, Wang SG, Wickliffe J, and Wilkerson M. Nanoparticles: small and mighty. *Int J Dermatol* 2011; 50: 247–54.
36. Bettoli V, Sarno O, Zauli S, Borghi A, Minghetti S, Ricci M et al. [What's new in acne? New therapeutic approaches]. *Ann Dermatol Venereol* 2010; 137 Suppl 2: S81–5.
37. Korting HC and Schafer-Korting M. Carriers in the topical treatment of skin disease. *Handb Exp Pharmacol* 2010: 197: 435–68.
38. James KA, Burkhart CN, and Morrell DS. Emerging drugs for acne. *Expert Opin Emerg Drugs* 2009; 14: 649–59.
39. Chourasia R and Jain SK. Drug targeting through pilosebaceous route. *Curr Drug Targets* 2009; 10: 950–67.
40. Date AA, Naik B, and Nagarsenker MS. Novel drug delivery systems: potential in improving topical delivery of antiacne agents. *Skin Pharmacol Physiol* 2006; 19: 2–16.
41. Kumar A, Agarwal SP, Ahuja A, Ali J, Choudhry R, and Baboota S. Preparation, characterization, and in vitro antimicrobial assessment of nanocarrier based formulation of nadifloxacin for acne treatment. *Pharmazie* 2011; 66: 111–4.
42. Dominguez-Delgado CL, Rodriguez-Cruz IM, Escobar-Chavez JJ, Calderon-Lojero IO, Quintanar-Guerrero D, and Ganem A. Preparation and characterization of triclosan nanoparticles intended to be used for the treatment of acne. *Eur J Pharm Biopharm* 2011; 79: 102–107.
43. Castro GA, Oliveira CA, Mahecha GA, and Ferreira LA. Comedolytic effect and reduced skin irritation of a new formulation of all-trans retinoic acid-loaded solid lipid nanoparticles for topical treatment of acne. *Arch Dermatol Res* 2011; 303: 513–20.
44. Inui S, Aoshima H, Nishiyama A, and Itami S. Improvement of acne vulgaris by topical fullerene application: unique impact on skin care. *Nanomedicine* 2011; 7: 238–41.

45. Czermak P, Steinle T, Ebrahimi M, Schmidts T, and Runkel F. Membrane-assisted production of S1P loaded SLNs for the treatment of acne vulgaris. *Desalination* 2010; 250: 1132–5.
46. Chanda N, Kattumuri V, Shukla R, Zambre A, Katti K, Upendran A et al. Bombesin functionalized gold nanoparticles show in vitro and in vivo cancer receptor specificity. *Proc Natl Acad Sci USA* 2010; 107: 8760–5.
47. de Leeuw J, de Vijlder HC, Bjerring P, and Neumann HAM. Liposomes in dermatology today. *J Eur Acad Dermatol Venereol* 2009; 23: 505–16.
48. Castro GA and Ferreira LA. Novel vesicular and particulate drug delivery systems for topical treatment of acne. *Expert Opin Drug Deliv* 2008; 5: 665–79.
49. Bernard E, Dubois JL, and Wepierre J. Importance of sebaceous glands in cutaneous penetration of an antiandrogen: target effect of liposomes. *J Pharm Sci* 1997; 86: 573–8.
50. Vogt A, Mandt N, Lademann J, Schaefer H, and Blume-Peytavi U. Follicular targeting—a promising tool in selective dermatotherapy. *J Investig Dermatol Symp Proc* 2005; 10: 252–5.
51. Azeem A, Khan ZI, Aqil M, Ahmad FJ, Khar RK, and Talegaonkar S. Microemulsions as a surrogate carrier for dermal drug delivery. *Drug Dev Ind Pharm* 2009; 35: 525–47.
52. Choi MJ and Maibach HI. Liposomes and niosomes as topical drug delivery systems. *Skin Pharmacol Physiol* 2005; 18: 209–19.
53. Redziniak G. Liposomes and skin: past, present, future. *Pathol Biol* 2003; 51: 279–81.
54. Schroeter A, Engelbrecht T, Neubert RHH, and Goebel ASB. New nanosized technologies for dermal and transdermal drug delivery. A review. *J Biomed Nanotechno* 2010; 6: 511–28.
55. Maghraby GM, Barry BW, and Williams AC. Liposomes and skin: From drug delivery to model membranes. *Eur J Pharm Sci* 2008; 34: 203–22.
56. Dutil M. Benzoyl peroxide: enhancing antibiotic efficacy in acne management. *Skin Therapy Lett* 2010; 15: 5–7.
57. Patel VB, Misra A, and Marfatia YS. Clinical assessment of the combination therapy with liposomal gels of tretinoin and benzoyl peroxide in acne. *AAPS PharmSciTech* 2001; 2: E-TN4.
58. Fluhr JW, Barsom O, Gehring W, and Gloor M. Antibacterial efficacy of benzoyl peroxide in phospholipid liposomes. A vehicle-controlled, comparative study in patients with papulopustular acne. *Dermatology* 1999; 198: 273–7.
59. Kinney MA, Yentzer BA, Fleischer AB, Jr., and Feldman SR. Trends in the treatment of acne vulgaris: are measures being taken to avoid antimicrobial resistance? *J Drugs Dermatol* 2010; 9: 519–24.
60. Ingram JR, Grindlay DJ, and Williams HC. Management of acne vulgaris: an evidence-based update. *Clin Exp Dermatol* 2010; 35: 351–4.
61. Shanmugam S, Song CK, Nagayya-Sriraman S, Baskaran R, Yong CS, Choi HG et al. Physicochemical characterization and skin permeation of liposome formulations containing clindamycin phosphate. *Arch Pharm Res* 2009; 32: 1067–75.
62. Honzak L and Sentjure M. Development of liposome encapsulated clindamycin for treatment of acne vulgaris. *Pflugers Archiv-Eur J Physiol* 2000; 440: R44–5.
63. Skalko N, Cajkovac M, and Jalsenjak I. Liposomes with clindamycin hydrochloride in the therapy of Acne-vulgaris. *Int J Pharm* 1992; 85: 97–101.
64. Hagiwara Y, Arima H, Miyamoto Y, Hirayama F, and Uekama K. Preparation and pharmaceutical evaluation of liposomes entrapping salicylic acid/gamma-cyclodextrin conjugate. *Chem Pharm Bull* 2006; 54: 26–32.
65. Motwani MR, Rhein LD, and Zatz JL. Deposition of salicylic acid into hamster sebaceous glands. *J Cosmet Sci* 2004; 55: 519–31.
66. Gollnick H. Current concepts of the pathogenesis of acne—implications for drug treatment. *Drugs* 2003; 63: 1579–96.
67. Ioele G, Cione E, Risoli A, Genchi G, and Ragno G. Accelerated photostability study of tretinoin and isotretinoin in liposome formulations. *Int J Pharm* 2005; 293: 251–60.
68. Sinico C, Manconi M, Peppi M, Lai F, Valenti D, and Fadda AM. Liposomes as carriers for denual delivery of tretinoin: in vitro evaluation of drug penneation and vesicle-skin interaction. *J Control Release* 2005; 103: 123–36.

69. Schaferkorting M, Korting HC, and Ponceposchl E. Liposomal tretinoin for uncomplicated acne-vulgaris. *Clin Investig* 1994; 72: 1086–91.

70. Masini V, Bonte F, Meybeck A, and Wepierre J. Cutaneous bioavailability in hairless rats of tretinoin in liposomes or gel. *J Pharm Sci* 1993; 82: 17–21.

71. Brisaert M, Gabriels M, Matthijs V, and Plaizier-Vercammen J. Liposomes with tretinoin: a physical and chemical evaluation. *J Pharm Biomed Anal* 2001; 26: 909–17.

72. Patel VB, Misra A, and Marfatia YS. Topical liposomal gel of tretinoin for the treatment of acne: research and clinical implications. *Pharm Dev Technol* 2000; 5: 455–64.

73. Kumar R, Singh B, Bakshi G, and Katare OP. Development of liposomal systems of Finasteride for topical applications: design, characterization, and in vitro evaluation. *Pharm Dev Technol* 2007; 12: 591–601.

74. Tabbakhian M, Tavakoli N, Jaafari MR, and Daneshamouz S. Enhancement of follicular delivery of finasteride by liposomes and niosomes. 1. In vitro permeation and in vivo deposition studies using hamster flank and ear models. *Int J Pharm* 2006; 323: 1–10.

75. Gruber DM, Sator MO, Joura EA, Kokoschka EM, Heinze G, and Huber JC. Topical cyproterone acetate treatment in women with acne—a placebo-controlled trial. *Arch Dermatol* 1998; 134: 459–63.

76. Munster U, Nakamura C, Haberland A, Jores K, Mehnert W, Rummel S et al. RU 58841-myristate—prodrug development for topical treatment of acne and androgenetic alopecia. *Pharmazie* 2005; 60: 8–12.

77. Bernard E, Dubois JL, and Wepierre J. Percutaneous absorption of a new antiandrogen included in liposomes or in solution. *Int J Pharm* 1995; 126: 235–43.

78. Battmann T, Bonfils A, Branche C, Humbert J, Goubet F, Teutsch G et al. RU-58841, a new specific topical antiandrogen—a candidate of choice for the treatment of acne, androgenetic alopecia and hirsutism. *J Steroid Biochem Mol Biol* 1994; 48: 55–60.

79. Azeem A, Anwer MK, and Talegaonkar S. Niosomes in sustained and targeted drug delivery: some recent advances. *J Drug Target* 2009; 17: 671–89.

80. Manconi M, Sinico C, Valenti D, Lai F, and Fadda AM. Niosomes as carriers for tretinoin. III. A study into the in vitro cutaneous delivery of vesicle-incorporated tretinoin. *Int J Pharm* 2006; 311: 11–9.

81. Manconi M, Valenti D, Sinico C, Lai F, Loy G, and Fadda AM. Niosomes as carriers for tretinoin. II. Influence of vesicular incorporation on tretinoin photostability. *Int J Pharm* 2003; 260: 261–72.

82. Manconi M, Sinico C, Valenti D, Loy G, and Fadda AM. Niosomes as carriers for tretinoin. I. Preparation and properties. *Int J Pharm* 2002; 234: 237–48.

83. Souto EB and Doktorovova S. Solid lipid nanoparticle formulations: pharmacokinetic and biopharmaceutical aspects in drug delivery. *Methods Enzymo; Liposomes, Pt F* 2009: 464: 105–29.

84. Kuchler S, Herrmann W, Panek-Minkin G, Blaschke T, Zoschke C, Kramer KD et al. SLN for topical application in skin diseases-characterization of drug-carrier and carrier-target interactions. *Int J Pharm* 2010; 390: 225–33.

85. Muller RH, Mader K, and Gohla S. Solid lipid nanoparticles (SLN) for controlled drug delivery—a review of the state of the art. *Eur J Pharm Biopharm* 2000; 50: 161–77.

86. Mehnert W and Mader K. Solid lipid nanoparticles: production, characterization and applications. *Adv Drug Deliv Rev* 2001; 47: 165–96.

87. Pardeike J, Hommoss A, and Muller RH. Lipid nanoparticles (SLN, NLC) in cosmetic and pharmaceutical dermal products. *Int J Pharm* 2009; 366: 170–84.

88. Lademann J, Otberg N, Jacobi U, Hoffman RM, and Blume-Peytavi U. Follicular penetration and targeting. *J Investig Dermatol Symp Proc* 2005; 10: 301–3.

89. Lademann J, Otberg N, Richter H, Weigmann HJ, Lindemann U, Schaefer H et al. Investigation of follicular penetration of topically applied substances. *Skin Pharmacol Appl Skin Physiol* 2001; 14 Suppl 1: 17–22.

90. Castro GA, Coelho A, Oliveira CA, Mahecha GAB, Orefice RL, and Ferreira LAM. Formation of ion pairing as an alternative to improve encapsulation and stability and to reduce skin irritation of retinoic acid loaded in solid lipid nanoparticles. *Int J Pharm* 2009; 381: 77–83.

91. Castro GA, Ferreira LAM, Orefice RL, and Buono VTL. Characterization of a new solid lipid nanoparticle formulation containing retinoic acid for topical treatment of acne. *Powder Diffr* 2008; 23: S30–5.
92. Carafa M, Marianecci C, Salvatorelli M, Di Marzio L, Cerreto F, Lucania G et al. Formulations of retinyl palmitate included in solid lipid nanoparticles: characterization and influence on light-induced vitamin degradation. *J Drug Deliv Sci Tec* 2008; 18: 119–24.
93. Castro GA, Orefice RL, Vilela JMC, Andrade MS, and Ferreira LAM. Development of a new solid lipid nanoparticle formulation containing retinoic acid for topical treatment of acne. *J Microencapsul* 2007; 24: 395–407.
94. Shah KA, Joshi MD, and Patravale VB. Biocompatible microemulsions for fabrication of glyceryl monostearate solid lipid nanoparticles (SLN) of tretinoin. *J Biomed Nanotechnol* 2009; 5: 396–400.
95. Stecova J, Mehnert W, Blaschke T, Kleuser B, Sivaramakrishnan R, Zouboulis CC et al. Cyproterone acetate loading to lipid nanoparticles for topical acne treatment: particle characterisation and skin uptake. *Pharm Res* 2007; 24: 991–1000.
96. Radin S, Chen T, and Ducheyne P. The controlled release of drugs from emulsified, sol gel processed silica microspheres. *Biomaterials* 2009; 30: 850–8.
97. Taglietti M, Hawkins CN, and Rao J. Novel topical drug delivery systems and their potential use in acne vulgaris. *Skin Therapy Lett* 2008; 13: 6–8.
98. Yilmaz E and Bengisu M. Drug entrapment in silica microspheres through a single step sol-gel process and in vitro release behavior. *J Biomed Mater Res B Appl Biomater* 2006; 77: 149–55.
99. Bikowski J and Del Rosso JQ. Benzoyl peroxide microsphere cream as monotherapy and combination treatment of acne. *J Drugs Dermatol* 2008; 7: 590–5.
100. Del Rosso J and Kircik L. Comparison of the tolerability of benzoyl peroxide microsphere wash versus a gentle cleanser, when used in combination with a clindamycin and tretinoin gel: A multicenter, investigator-blind, randomized study. *J Am Acad Dermatol* 2009; 60: AB17.
101. Leyden JJ, Nighland M, Rossi AB, and Ramaswamy R. Irritation potential of tretinoin gel microsphere pump versus adapalene plus benzoyl peroxide gel. *J Drugs Dermatol* 2010; 9: 998–1003.
102. Pflucker F, Wendel V, Hohenberg H, Gartner E, Will T, Pfeiffer S et al. The human stratum corneum layer: an effective barrier against dermal uptake of different forms of topically applied micronised titanium dioxide. *Skin Pharmacol Appl Skin Physiol* 2001; 14 Suppl 1: 92–7.
103. Wester RC, Patel R, Nacht S, Leyden J, Melendres J, and Maibach H. Controlled release of benzoyl peroxide from a porous microsphere polymeric system can reduce topical irritancy. *J Am Acad Dermatol* 1991; 24: 720–6.
104. Jelvehgari M, Siahi-Shadbad MR, Azarmi S, Martin GP, and Nokhodchi A. The microsponge delivery system of benzoyl peroxide: Preparation, characterization and release studies. *Int J Pharm* 2006; 308: 124–32.
105. Embil K and Nacht S. The Microsponge Delivery System (MDS): A topical delivery system with reduced irritancy incorporating multiple triggering mechanisms for the release of actives. *J Microencapsul* 1996; 13: 575–88.
106. Nyirady J, Lucas C, Yusuf M, Mignone P, and Wisniewski S. The stability of tretinoin in tretinoin gel microsphere 0.1%. *Cutis* 2002; 70: 295–8.
107. Torok HM and Pillai R. Safety and efficacy of micronized tretinoin gel (0.05%) in treating adolescent acne. *J Drugs Dermatol* 2011; 10: 647–52.
108. Lucky AW and Sugarman J. Comparison of micronized tretinoin gel 0.05% and tretinoin gel microsphere 0.1% in young adolescents with acne: a post hoc analysis of efficacy and tolerability data. *Cutis* 2011; 87: 305–10.
109. Ramos-e-Silva M, Carvalho J, Marques-Costa J, Goldsztajn K, and Carneiro S. Tretinoin microsphere 0.1% gel for acne patients. *J Am Acad Dermatol* 2011; 64: AB5.
110. Pariser D, Bucko A, Fried R, Jarratt MT, Kempers S, Kircik L et al. Tretinoin gel microsphere pump 0.04% plus 5% benzoyl peroxide wash for treatment of acne vulgaris: morning/morning regimen is as effective and safe as morning/evening regimen. *J Drugs Dermatol* 2010; 9: 805–13.
111. Eichenfield LF, Matiz C, Funk A, and Dill SW. Study of the efficacy and tolerability of 0.04% tretinoin microsphere gel for preadolescent acne. *Pediatrics* 2010; 125: E1316–23.

112. Dosik JS, Homer K, and Arsonnaud S. Cumulative irritation potential of adapalene 0.1% cream and gel compared with tretinoin microsphere 0.04% and 0.1%. *Cutis* 2005; 75: 238–43.

113. Pengon S, Limmatvapirat C, and Limmatvapirat S. Preparation and evaluation of antimicrobial nanoemulsion containing herbal extracts. *Drug Metab Rev* 2009; 41: 85.

114. Ziani K, Chang YH, McLandsborough L, and McClements DJ. Influence of surfactant charge on antimicrobial efficacy of surfactant-stabilized thyme oil nanoemulsions. *J Agric Food Chem* 2011; 59: 6247–55.

115. Hemmila MR, Mattar A, Taddonio MA, Arbabi S, Hamouda T, Ward PA et al. Topical nanoemulsion therapy reduces bacterial wound infection and inflammation after burn injury. *Surgery* 2010; 148: 499–509.

116. Wu X and Guy RH. Applications of nanoparticles in topical drug delivery and in cosmetics. *J Drug Deliv Sci Tec* 2009; 19: 371–84.

117. Pannu J, Martin A, McCarthy A, Sutcliffe J, and Ciotti S. NB-003 activity against propionibacterium acnes in a pig skin model. *J Am Acad Dermatol* 2010; 62: AB14.

118. Kang K, Polster AM, Nedorost, Susan T, Stevens SR, and Cooper KD. Chapter 13 – Atopic dermatitis. In: Bolognia J, Jorizzo JL, Rapini RP, eds. *Dermatology*. London, New York: Mosby, 2003.

119. Laughter D, Istvan JA, Tofte SJ, and Hanifin JM. The prevalence of atopic dermatitis in Oregon schoolchildren. *J Am Acad Dermatol* 2000; 43: 649–55.

120. Luoma R, Koivikko A, and Viander M. Development of asthma, allergic rhinitis and atopic dermatitis by the age of five years: A prospective study of 543 newborns. *Allergy* 1983; 38: 339–46.

121. Kemp AS. Cost of illness of atopic dermatitis in children: a societal perspective. *Pharmacoeconomics* 2003; 21: 105–13.

122. Kelsay K. Management of sleep disturbance associated with atopic dermatitis. *J Allergy Clin Immunol* 2006; 118: 198–201.

123. Williams HC. Atopic dermatitis. *N Engl J Med* 2005; 352: 2314–24.

124. Williams H and Flohr C. How epidemiology has challenged 3 prevailing concepts about atopic dermatitis. *J Allergy Clin Immunol* 2006; 118: 209–13.

125. Bieber T. Atopic dermatitis. *Ann Dermatol* 2010; 22: 125–37.

126. Cookson WOCM and Moffatt MF. The genetics of atopic dermatitis. *Curr Opin Allergy Clin Immunol* 2002; 2: 383.

127. Krakowski AC and Eichenfield LF. Atopic dermatitis. In: Lebwohl M, ed. *Treatment of Skin Disease: Comprehensive Therapeutic Strategies*. Philadelphia, PA: Mosby/Elsevier, 2006: xxiv, 723 p.

128. Meduri NB, Vandergriff T, Rasmussen H, and Jacobe H. Phototherapy in the management of atopic dermatitis: a systematic review. *Photodermatol Photoimmunol Photomed* 2007; 23: 106–12.

129. Richtig E, Ahlgrimm-Siess V, Arzberger E, and Hofmann-Wellenhof R. Noninvasive differentiation between mamillary eczema and Paget disease by in vivo reflectance confocal microscopy on the basis of two case reports. *Br J Dermatol* 2011; 165: 440–1.

130. Gonzalez S, Gonzalez E, White WM, Rajadhyaksha M, and Anderson RR. Allergic contact dermatitis: correlation of in vivo confocal imaging to routine histology. *J Am Acad Dermatol* 1999; 40: 708–13.

131. Zhang EZ, Povazay B, Laufer J, Alex A, Hofer B, Pedley B et al. Multimodal photoacoustic and optical coherence tomography scanner using an all optical detection scheme for 3D morphological skin imaging. *Biomed Opt Express* 2011; 2: 2202–15.

132. Braconi D, Bernardini G, and Santucci A. Post-genomics and skin inflammation. *Mediators Inflamm* 2010; 2010: pii: 364823.

133. Piche E, Hafner HM, Hoffmann J, and Junger M. [FOITS (fast optical in vivo topometry of human skin): new approaches to 3-D surface structures of human skin]. *Biomed Tech (Berl)* 2000; 45: 317–22.

134. Berardesca E, Barbareschi M, Veraldi S, and Pimpinelli N. Evaluation of efficacy of a skin lipid mixture in patients with irritant contact dermatitis, allergic contact dermatitis or atopic dermatitis: a multicenter study. *Contact Dermatitis* 2001; 45: 280–5.

135. Wilsmann-Theis D, Hagemann T, Jordan J, Bieber T, and Novak N. Facing psoriasis and atopic dermatitis: are there more similarities or more differences? *Eur J Dermatol* 2008; 18: 172–80.

136. Pople PV and Singh KK. Targeting tacrolimus to deeper layers of skin with improved safety for treatment of atopic dermatitis. *Int J Pharm* 2010; 398: 165–78.
137. Singh KK and Pople P. Safer than safe: lipid nanoparticulate encapsulation of tacrolimus with enhanced targeting and improved safety for atopic dermatitis. *J Biomed Nanotechnol* 2011; 7: 40–1.
138. Schmidt M, Martin SF, Freudenberg MA, and Goebeler M. Animal models for nickel allergy. *Nat Nanotechnol* 2011; 6: 533.
139. Vemula PK, Anderson RR, and Karp JM. Nanoparticles reduce nickel allergy by capturing metal ions. *Nat Nanotechnol* 2011; 6: 291–5.
140. Aschberger K, Johnston HJ, Stone V, Aitken RJ, Tran CL, Hankin SM et al. Review of fullerene toxicity and exposure—appraisal of a human health risk assessment, based on open literature. *Regul Toxicol Pharmacol* 2010; 58: 455–73.
141. Dellinger A, Zhou Z, Lenk R, MacFarland D, and Kepley CL. Fullerene nanomaterials inhibit phorbol myristate acetate-induced inflammation. *Exp Dermatol* 2009; 18: 1079–81.
142. Gao J, Wang Y, Folta KM, Krishna V, Bai W, Indeglia P et al. Polyhydroxy fullerenes (fullerols or fullerenols): beneficial effects on growth and lifespan in diverse biological models. *PloS One* 2011; 6: e19976.
143. Henry TB, Petersen EJ, and Compton RN. Aqueous fullerene aggregates (nC(60)) generate minimal reactive oxygen species and are of low toxicity in fish: a revision of previous reports. *Curr Opin Biotechnol* 2011; 22: 533–7.
144. Kolosnjaj J, Szwarc H, and Moussa F. Toxicity studies of fullerenes and derivatives. *Adv Exp Med Biol* 2007; 620: 168–80.
145. Uo M, Akasaka T, Watari F, Sato Y, and Tohji K. Toxicity evaluations of various carbon nanomaterials. *Dent Mater J* 2011; 30: 245–63.
146. Kalariya M, Padhi BK, Chougule M, and Misra A. Clobetasol propionate solid lipid nanoparticles cream for effective treatment of eczema: formulation and clinical implications. *Indian J Exp Biol* 2005; 43: 233–40.
147. Baspinar Y, Keck CM, and Borchert HH. Development of a positively charged prednicarbate nanoemulsion. *Int J Pharm* 2010; 383: 201–8.
148. Buder K, Knuschke P, and Wozel G. Evaluation of methylprednisolone aceponate, tacrolimus and combination thereof in the psoriasis plaque test using sum score, 20-MHz-ultrasonography and optical coherence tomography. *Int J Clin Pharmacol Ther* 2010; 48: 814–20.
149. Horie RT, Sakamoto T, Nakagawa T, Ishihara T, Higaki M, and Ito J. Stealth-nanoparticle strategy for enhancing the efficacy of steroids in mice with noise-induced hearing loss. *Nanomedicine (Lond)* 2010; 5: 1331–40.
150. Joseph AM, Biggs T, Garr M, Singh J, and Lederle FA. Stealth steroids. *N Engl J Med* 1991; 324: 62.
151. Zhang J and Smith E. Percutaneous permeation of betamethasone 17-valerate incorporated in lipid nanoparticles. *J Pharm Sci* 2011; 100: 896–903.
152. Hofkens W, van den Hoven JM, Pesman GJ, Nabbe KC, Sweep FC Storm G et al. Safety of glucocorticoids can be improved by lower yet still effective dosages of liposomal steroid formulations in murine antigen-induced arthritis: comparison of prednisolone with budesonide. *Int J Pharm* 2011; 416: 493–8.
153. Momekova D, Rangelov S, Yanev S, Nikolova E, Konstantinov S, Romberg B et al. Long-circulating, pH-sensitive liposomes sterically stabilized by copolymers bearing short blocks of lipid-mimetic units. *Eur J Pharm Sci* 2007; 32: 308–17.
154. Sun BK and Tsao H. Small RNAs in development and disease. *J Am Acad Dermatol* 2008; 59: 725–37; quiz 38–40.
155. Uchida T, Kanazawa T, Kawai M, Takashima Y, and Okada H. Therapeutic effects on atopic dermatitis by anti-RelA short interfering RNA combined with functional peptides Tat and AT1002. *J Pharmacol Exp Ther* 2011; 338: 443–50.
156. Azuma M, Ritprajak P, and Hashiguchi M. Topical application of siRNA targeting cutaneous dendritic cells in allergic skin disease. *Methods Mol Biol* 2010; 623: 373–81.
157. Dernedde J, Rausch A, Weinhart M, Enders S, Tauber R, Licha K et al. Dendritic polyglycerol sulfates as multivalent inhibitors of inflammation. *Proc Natl Acad Sci USA* 2010; 107: 19679–84.

158. Toyama T, Matsuda H, Ishida I, Tani M, Kitaba S, Sano S et al. A case of toxic epidermal necrolysis-like dermatitis evolving from contact dermatitis of the hands associated with exposure to dendrimers. *Contact Dermatitis* 2008; 59: 122–3.

159. Tripathi PK, Khopade AJ, Nagaich S, Shrivastava S, Jain S, and Jain NK. Dendrimer grafts for delivery of 5-fluorouracil. *Pharmazie* 2002; 57: 261–4.

160. Ward BB, Dunham T, Majoros IJ, and Baker JR, Jr. Targeted dendrimer chemotherapy in an animal model for head and neck squamous cell carcinoma. *J Oral Maxillofac Surg* 2011; 69: 2452–9.

161. Buckley DA. Staphylococcus aureus endocarditis as a complication of acupuncture for eczema. *Br J Dermatol* 2011; 164: 1405–6.

162. Cheng TO, Lee RJ, and McIlwain JC. Subacute bacterial endocarditis following ear acupuncture. *Int J Cardiol* 1985; 8: 97.

163. Evans P. Acupuncture and endocarditis. *J Am Board Fam Pract* 2002; 15: 432–3; author reply 433.

164. Jefferys DB, Smith S, Brennand-Roper DA, and Curry PV. Acupuncture needles as a cause of bacterial endocarditis. *Br Med J (Clin Res Ed)* 1983; 287: 326–7.

165. Lee RJ and McIlwain JC. Subacute bacterial endocarditis following ear acupuncture. *Int J Cardiol* 1985; 7: 62–3.

166. Nambiar P and Ratnatunga C. Prosthetic valve endocarditis in a patient with Marfan's syndrome following acupuncture. *J Heart Valve Dis* 2001; 10: 689–90.

167. Scheel O, Sundsfjord A, Lunde P, and Andersen BM. Endocarditis after acupuncture and injection— treatment by a natural healer. *JAMA* 1992; 267: 56.

168. Bath-Hextall FJ, Birnie AJ, Ravenscroft JC, and Williams HC. Interventions to reduce Staphylococcus aureus in the management of atopic eczema: an updated Cochrane review. *Br J Dermatol* 2011; 164: 228.

169. Fluhr JW, Breternitz M, Kowatzki D, Bauer A, Bossert J, Elsner P et al. Silver-loaded seaweed-based cellulosic fiber improves epidermal skin physiology in atopic dermatitis: safety assessment, mode of action and controlled, randomized single-blinded exploratory in vivo study. *Exp Dermatol* 2010; 19: e9–15.

170. Gauger A. Silver-coated textiles in the therapy of atopic eczema. *Curr Probl Dermatol* 2006; 33: 152–64.

171. Gauger A, Mempel M, Schekatz A, Schafer T, Ring J, and Abeck D. Silver-coated textiles reduce Staphylococcus aureus colonization in patients with atopic eczema. *Dermatology* 2003; 207: 15–21.

172. Haug S, Roll A, Schmid-Grendelmeier P, Johansen P, Wuthrich B, Kundig TM et al. Coated textiles in the treatment of atopic dermatitis. *Curr Probl Dermatol* 2006; 33: 144–51.

173. Juenger M, Ladwig A, Staecker S, Arnold A, Kramer A, Daeschlein G et al. Efficacy and safety of silver textile in the treatment of atopic dermatitis (AD). *Curr Med Res Opin* 2006; 22: 739–50.

174. Kramer A, Guggenbichler P, Heldt P, Junger M, Ladwig A, Thierbach H et al. Hygienic relevance and risk assessment of antimicrobial-impregnated textiles. *Curr Probl Dermatol* 2006; 33: 78–109.

175. Ricci G, Patrizi A, Bellini F, and Medri M. Use of textiles in atopic dermatitis: care of atopic dermatitis. *Curr Probl Dermatol* 2006; 33: 127–43.

176. Fujimura T, Takagi Y, Sugano I, Sano Y, Yamaguchi N, Kitahara T et al. Real-life use of underwear treated with fabric softeners improves skin dryness by decreasing the friction of fabrics against the skin. *Int J Cosmet Sci* 2011; 33: 566–71.

177. Ruff Z, Shemuly D, Peng X, Shapira O, Wang Z, and Fink Y. Polymer-composite fibers for transmitting high peak power pulses at 1.55 microns. *Opt Express* 2010; 18: 15697–703.

178. Sorin F, Lestoquoy G, Danto S, Joannopoulos JD, and Fink Y. Resolving optical illumination distributions along an axially symmetric photodetecting fiber. *Opt Express* 2010; 18: 24264–75.

179. Sorin F, Shapira O, Abouraddy AF, Spencer M, Orf ND, Joannopoulos JD et al. Exploiting collective effects of multiple optoelectronic devices integrated in a single fiber. *Nano Lett* 2009; 9: 2630–5.

180. Welzel J, Bruhns M, and Wolff HH. Optical coherence tomography in contact dermatitis and psoriasis. *Arch Dermatol Res* 2003; 295: 50–5.

181. Gupta AK, Cooper EA, Feldman SR, and Fleischer AB. A survey of office visits for actinic keratosis as reported by NAMCS, 1990–1999. *Cutis* 2002; 70: 8–13.
182. Lebwohl M. Actinic keratosis: epidemiology and progression to squamous cell carcinoma. *Br J Dermatol* 2003; 149: 31–3.
183. Salasche SJ. Epidemiology of actinic keratoses and squamous cell carcinoma. *J Am Acad Dermatol* 2000; 42: S4–7.
184. Ackerman AB and Mones JM. Solar (actinic) keratosis is squamous cell carcinoma. *Br J Dermatol* 2006; 155: 9–22.
185. Park HR, Min SK, Cho HD, Kim KH, Shin HS, and Park YE. Expression profiles of p63, p53, survivin, and hTERT in skin tumors. *J Cutan Pathol* 2004; 31: 544–9.
186. Soufir N, Moles JP, Vilmer C, Moch C, Verola O, Rivet J et al. P16 UV mutations in human skin epithelial tumors. *Oncogene* 1999; 18: 5477–81.
187. Mortier L, Marchetti P, Delaporte E, Martin de Lassalle E, Thomas P, Piette F et al. Progression of actinic keratosis to squamous cell carcinoma of the skin correlates with deletion of the 9p21 region encoding the p16(INK4a) tumor suppressor. *Cancer Lett* 2002; 176: 205–14.
188. Leffell DJ. The scientific basis of skin cancer. *J Am Acad Dermatol* 2000; 42: S18–22.
189. Anwar J, Wrone DA, Kimyai-Asadi A, and Alam M. The development of actinic keratosis into invasive squamous cell carcinoma: evidence and evolving classification schemes. *Clin Dermatol* 2004; 22: 189–96.
190. Marks R, Rennie G, and Selwood TS. Malignant transformation of solar keratoses to squamous cell carcinoma. *Lancet* 1988; 1: 795–7.
191. Dodson JM, DeSpain J, Hewett JE, and Clark DP. Malignant potential of actinic keratoses and the controversy over treatment. A patient-oriented perspective. *Arch Dermato* 1991; 127: 1029–31.
192. Green A and Battistutta D. Incidence and determinants of skin-cancer in a high-risk Australian population. *Int J Cancer* 1990; 46: 356–61.
193. Marks R, Foley P, Goodman G, Hage BH, and Selwood TS. Spontaneous remission of solar keratoses—the case for conservative management. *Br J Dermatol* 1986; 115: 649–55.
194. Holmes C, Foley P, Freeman M, and Chong AH. Solar keratosis: epidemiology, pathogenesis, presentation and treatment. *Australas J Dermatol* 2007; 48: 67–76.
195. Quinn AG and Perkins W. Non-melanoma skin cancer and other epidermal skin tumours. In: Burns T, Breathnach S, Cox N, Griffiths C eds. *Rook's Textbook of Dermatology*. West Sussex, United Kingdom, Wiley-Blackwell, 2010: 1–48.
196. Drake LA, Ceilley RI, Cornelison RL, Dobes WL, Dorner W, Goltz RW et al. Guidelines of care for actinic keratoses. Committee on Guidelines of Care. *J Am Acad Dermatol* 1995; 32: 95–8.
197. Dinehart SM. The treatment of actinic keratoses. *J Am Acad Dermatol* 2000; 42: 25–8.
198. Detlef Klaus G. Topical chemotherapy with 5-fluorouracil: A review. *J Am Acad Dermatol* 1981; 4: 633–49.
199. Stanley MA. Imiquimod and the imidazoquinolones: mechanism of action and therapeutic potential. *Clin Exp Dermatol* 2002; 27: 571–7.
200. Fritsch C, Goerz G, and Ruzicka T. Photodynamic therapy in dermatology. *Arch Dermatol* 1998; 134: 207–14.
201. Touma D, Yaar M, Whitehead S, Konnikov N, and Gilchrest BA. A trial of short incubation, broad-area photodynamic therapy for facial actinic keratoses and diffuse photodamage. *Arch Dermatol* 2004; 140: 33–40.
202. Szeimies RM, Karrer S, Radakovic-Fijan S, Tanew A, Calzavara-Pinton PG, Zane C et al. Photodynamic therapy using topical methyl 5-aminolevulinate compared with cryotherapy for actinic keratosis: a prospective, randomized study. *J Am Acad Dermatol* 2002; 47: 258–62.
203. Akay BN, Kocyigit P, Heper AO, and Erdem C. Dermatoscopy of flat pigmented facial lesions: diagnostic challenge between pigmented actinic keratosis and lentigo maligna. *Br J Dermatol* 2010; 163: 1212–7.
204. Bae Y, Son T, Stuart Nelson J, Kim JH, Choi EH, and Jung B. Dermatological feasibility of multimodal facial color imaging modality for cross-evaluation of facial actinic keratosis. *Skin Res Technol* 2011; 17: 4–10.

205. Ciudad C, Avilés JA, Suárez R, and Lázaro P. Diagnostic utility of dermoscopy in pigmented actinic keratosis. *Actas* Dermosifiliogr 2011; 102(8): 623–6. [in Spanish]

206. Cuellar F, Vilalta A, Puig S, Palou J, Salerni G, and Malvehy J. New dermoscopic pattern in actinic keratosis and related conditions. *Arch Dermatol* 2009; 145: 732.

207. Forsea AM, Carstea EM, Ghervase L, Giurcaneanu C, and Pavelescu G. Clinical application of optical coherence tomography for the imaging of non-melanocytic cutaneous tumors: a pilot multi-modal study. *J Med Life* 2010; 3: 381–9.

208. Jorgensen TM, Tycho A, Mogensen M, Bjerring P, and Jemec GB. Machine-learning classification of non-melanoma skin cancers from image features obtained by optical coherence tomography. *Skin Res Technol* 2008; 14: 364–9.

209. Konig K, Speicher M, Buckle R, Reckfort J, McKenzie G, Welzel J et al. Clinical optical coherence tomography combined with multiphoton tomography of patients with skin diseases. *J Biophotonics* 2009; 2: 389–97.

210. Mogensen M and Jemec GB. Diagnosis of nonmelanoma skin cancer/keratinocyte carcinoma: a review of diagnostic accuracy of nonmelanoma skin cancer diagnostic tests and technologies. *Dermatol Surg* 2007; 33: 1158–74.

211. Mogensen M, Thrane L, Joergensen TM, Andersen PE, and Jemec GB. Optical coherence tomography for imaging of skin and skin diseases. *Semin Cutan Med Surg* 2009; 28: 196–202.

212. Peris K, Micantonio T, Piccolo D, and Fargnoli MC. Dermoscopic features of actinic keratosis. *J Ger Soc Dermatol* 2007; 5: 970–6.

213. Pock L, Drlik L, and Hercogova J. Dermatoscopy of pigmented actinic keratosis—a striking similarity to lentigo maligna. *Int J Dermatol* 2007; 46: 414–6.

214. von Felbert V, Neis M, Megahed M, and Spoler F. Imaging of actinic porokeratosis by optical coherence tomography (OCT). *Hautarzt* 2008; 59: 877–9. [in German]

215. Zalaudek I, Ferrara G, Leinweber B, Mercogliano A, D'Ambrosio A, and Argenziano G. Pitfalls in the clinical and dermoscopic diagnosis of pigmented actinic keratosis. *J Am Acad Dermatol* 2005; 53: 1071–4.

216. Zalaudek I, Giacomel J, Argenziano G, Hofmann-Wellenhof R, Micantonio T, Di Stefani A et al. Dermoscopy of facial nonpigmented actinic keratosis. *Br J Dermatol* 2006; 155: 951–6.

217. Haas P, Then P, Wild A, Grange W, Zorman S, Hegner M et al. Fast quantitative single-molecule detection at ultralow concentrations. *Anal Chem* 2010; 82: 6299–302.

218. Husale S, Grange W, Karle M, Burgi S, and Hegner M. Interaction of cationic surfactants with DNA: a single-molecule study. *Nucleic Acids Res* 2008; 36: 1443–9.

219. McKendry R, Zhang J, Arntz Y, Strunz T, Hegner M, Lang HP et al. Multiple label-free biodetection and quantitative DNA-binding assays on a nanomechanical cantilever array. *Proc Natl Acad Sci USA* 2002; 99: 9783–8.

220. Schumakovitch I, Grange W, Strunz T, Bertoncini P, Guntherodt HJ, and Hegner M. Temperature dependence of unbinding forces between complementary DNA strands. *Biophys J* 2002; 82: 517–21.

221. Ra SH, Li X, and Binder S. Molecular discrimination of cutaneous squamous cell carcinoma from actinic keratosis and normal skin. *Mod Pathol* 2011; 24(7):963–73.

222. Torres A, Storey L, Anders M, Bulbulian BJ, Jin J, Raghavan S et al. Microanalysis of aberrant gene expression in actinic keratosis: effect of the toll-like receptor-7 agonist imiquimod. *Br J Dermatol* 2007; 157(6):1132–47.

223. Nindl I, Dang C, Forschner T, Kuban RJ, Meyer T, Sterry W et al. Identification of differentially expressed genes in cutaneous squamous cell carcinoma by microarray expression profiling. *Mol Cancer* 2006; 5: 30.

224. Babilas P, Knobler R, Hummel S, Gottschaller C, Maisch T, Koller M et al. Variable pulsed light is less painful than light-emitting diodes for topical photodynamic therapy of actinic keratosis: a prospective randomized controlled trial. *Br J Dermatol* 2007; 157: 111–7.

225. Benoldi D, Pezzarossa E, Alinovi A, Labrini G, Marcheselli W, de Panfilis G et al. [Retinoic acid and 5-FU mixture in the topical treatment of several skin diseases (author's transl)]. *Ateneo Parmense Acta Biomed* 1980; 51: 187–92. [in Italian]

226. Elmets CA, Viner JL, Pentland AP, Cantrell W, Lin HY, Bailey H et al. Chemoprevention of non-melanoma skin cancer with celecoxib: a randomized, double-blind, placebo-controlled trial. *J Natl Cancer Inst* 2010; 102: 1835–44.

227. Coleman WP, 3rd, Yarborough JM, and Mandy SH. Dermabrasion for prophylaxis and treatment of actinic keratoses. *Dermatol Surg* 1996; 22: 17–21.

228. Foote JA, Ranger-Moore JR, Einspahr JG, Saboda K, Kenyon J, Warneke J et al. Chemoprevention of human actinic keratoses by topical DL-alpha-tocopherol. *Cancer Prev Res (Phila)* 2009; 2: 394–400.

229. Gordon LG, Scuffham PA, van der Pols JC, McBride P, Williams GM, and Green AC. Regular sunscreen use is a cost-effective approach to skin cancer prevention in subtropical settings. *J Invest Dermatol* 2009; 129: 2766–71.

230. Halldin CB, Paoli J, Sandberg C, Ericson MB, and Wennberg AM. Transcutaneous electrical nerve stimulation for pain relief during photodynamic therapy of actinic keratoses. *Acta Dermato-Venereologica* 2008; 88: 311–3.

231. Hantash BM, Stewart DB, Cooper ZA, Rehmus WE, Koch RJ, and Swetter SM. Facial resurfacing for nonmelanoma skin cancer prophylaxis. *Arch Dermatol* 2006; 142: 976–82.

232. Hermanns JF, Pierard GE, and Quatresooz P. Erlotinib-responsive actinic keratoses. *Oncol Rep* 2007; 18: 581–4.

233. Hess LM, Saboda K, Malone DC, Salasche S, Warneke J, and Alberts DS. Adherence assessment using medication weight in a phase IIb clinical trial of difluoromethylornithine for the chemoprevention of skin cancer. *Cancer Epidemiol Biomarkers Prev* 2005; 14: 2579–83.

234. Holmes MV, Dawe RS, Ferguson J, and Ibbotson SH. A randomized, double-blind, placebo-controlled study of the efficacy of tetracaine gel (Ametop) for pain relief during topical photodynamic therapy. *Br J Dermatol* 2004; 150: 337–40.

235. Iyer S, Friedli A, Bowes L, Kricorian G, and Fitzpatrick RE. Full face laser resurfacing: therapy and prophylaxis for actinic keratoses and non-melanoma skin cancer. *Lasers Surg Med* 2004; 34: 114–9.

236. Moore AR and Willoughby DA. Hyaluronan as a drug delivery system for diclofenac: a hypothesis for mode of action. *Int J Tissue React* 1995; 17: 153–6.

237. Naylor MF, Boyd A, Smith DW, Cameron GS, Hubbard D, and Neldner KH. High sun protection factor sunscreens in the suppression of actinic neoplasia. *Arch Dermatol* 1995; 131: 170–5.

238. Paoli J, Halldin C, Ericson MB, and Wennberg AM. Nerve blocks provide effective pain relief during topical photodynamic therapy for extensive facial actinic keratoses. *Clin Exp Dermatol* 2008; 33: 559–64.

239. Peng YM, Peng YS, Lin Y, Moon T, and Baier M. Micronutrient concentrations in paired skin and plasma of patients with actinic keratoses: effect of prolonged retinol supplementation. *Cancer Epidemiol Biomarkers Prev* 1993; 2: 145–50.

240. Smit JV, de Sevaux RG, Blokx WA, van de Kerkhof PC, Hoitsma AJ, and de Jong EM. Acitretin treatment in (pre)malignant skin disorders of renal transplant recipients: histologic and immunohistochemical effects. *J Am Acad Dermatol* 2004; 50: 189–96.

241. Ulrich C, Jurgensen JS, Degen A, Hackethal M, Ulrich M, Patel MJ et al. Prevention of nonmelanoma skin cancer in organ transplant patients by regular use of a sunscreen: a 24 months, prospective, case-control study. *Br J Dermatol* 2009; 161 Suppl 3: 78–84.

242. Szeimies RM, Bichel J, Ortonne JP, Stockfleth E, Lee J, and Meng TC. A phase II dose-ranging study of topical resiquimod to treat actinic keratosis. *Br J Dermatol* 2008; 159: 205–10.

243. Van Scott EJ and Yu RJ. Alpha hydroxy acids: procedures for use in clinical practice. *Cutis* 1989; 43: 222–8.

244. Watson AB. Preventative effect of etretinate therapy on multiple actinic keratoses. *Cancer Detect Prev* 1986; 9: 161–5.

245. Wiegell SR, Haedersdal M, and Wulf HC. Cold water and pauses in illumination reduces pain during photodynamic therapy: a randomized clinical study. *Acta Dermato-Venereologica* 2009; 89: 145–9.

246. Wulf HC, Pavel S, Stender I, and Bakker-Wensveen CA. Topical photodynamic therapy for prevention of new skin lesions in renal transplant recipients. *Acta Dermato-Venereologica* 2006; 86: 25–8.

247. Zhan H and Zheng H. The role of topical cyclo-oxygenase-2 inhibitors in skin cancer: treatment and prevention. *Am J Clin Dermatol* 2007; 8: 195–200.
248. Zhao B and He YY. Recent advances in the prevention and treatment of skin cancer using photodynamic therapy. *Expert Rev Anticancer Ther* 2010; 10: 1797–809.
249. Alvi IA, Madan J, Kaushik D, Sardana S, Pandey RS, and Ali A. Comparative study of transferosomes, liposomes, and niosomes for topical delivery of 5-fluorouracil to skin cancer cells: preparation, characterization, in-vitro release, and cytotoxicity analysis. *Anticancer Drugs* 2011; 22: 774–82.
250. Meyskens FL, Jr. and McLaren CE. Chemoprevention, risk reduction, therapeutic prevention, or preventive therapy? *J Natl Cancer Inst* 2010; 102: 1815–7.
251. Jirakulaporn T, Endrizzi B, Lindgren B, Mathew J, Lee PK, and Dudek AZ. Capecitabine for skin cancer prevention in solid organ transplant recipients. *Clin Transplant* 2011; 25: 541–8.
252. Bartels P, Yozwiak M, Einspahr J, Saboda K, Liu Y, Brooks C et al. Chemopreventive efficacy of topical difluoromethylornithine and/or triamcinolone in the treatment of actinic keratoses analyzed by karyometry. *Anal Quant Cytol Histol* 2009; 31: 355–66.
253. Bath-Hextall F, Leonardi-Bee J, Somchand N, Webster A, Delitt J, and Perkins W. Interventions for preventing non-melanoma skin cancers in high-risk groups. *Cochrane Database Syst Rev* 2007: 17: CD005414.
254. Hughes MC, Williams GM, Fourtanier A, and Green AC. Food intake, dietary patterns, and actinic keratoses of the skin: a longitudinal study. *Am J Clin Nutr* 2009; 89: 1246–55.
255. Ali H, Shirode AB, Sylvester PW, and Nazzal S. Preparation, characterization, and anticancer effects of simvastatin-tocotrienol lipid nanoparticles. *Int J Pharm* 2010; 389: 223–31.
256. Dorgan JF, Boakye NA, Fears TR, Schleicher RL, Helsel W, Anderson C et al. Serum carotenoids and alpha-tocopherol and risk of nonmelanoma skin cancer. *Cancer Epidemiol Biomarkers Prev* 2004; 13: 1276–82.
257. Hatanaka J, Chikamori H, Sato H, Uchida S, Debari K, Onoue S et al. Physicochemical and pharmacological characterization of alpha-tocopherol-loaded nano-emulsion system. *Int J Pharm* 2010; 396: 188–93.
258. Luo Y, Zhang B, Whent M, Yu LL, and Wang Q. Preparation and characterization of zein/chitosan complex for encapsulation of alpha-tocopherol, and its in vitro controlled release study. *Colloids Surf B Biointerfaces* 2011; 85: 145–52.
259. Moddaresi M, Brown MB, Tamburic S, and Jones SA. Tocopheryl acetate disposition in porcine and human skin when administered using lipid nanocarriers. *J Pharm Pharmacol* 2010; 62: 762–9.
260. Moddaresi M, Brown MB, Zhao Y, Tamburic S, and Jones SA. The role of vehicle-nanoparticle interactions in topical drug delivery. *Int J Pharm* 2010; 400: 176–82.
261. Moddaresi M, Tamburic S, Williams S, Jones SA, Zhao Y, and Brown MB. Effects of lipid nanocarriers on the performance of topical vehicles in vivo. *J Cosmet Dermatol* 2009; 8: 136–43.
262. Murugeshu A, Astete C, Leonardi C, Morgan T, and Sabliov CM. Chitosan/PLGA particles for controlled release of alpha-tocopherol in the GI tract via oral administration. *Nanomedicine (Lond)* 2011; 6: 1513–28.
263. Shukat R and Relkin P. Lipid nanoparticles as vitamin matrix carriers in liquid food systems: on the role of high-pressure homogenisation, droplet size and adsorbed materials. *Colloids Surf B Biointerfaces* 2011; 86: 119–24.
264. Song C and Liu S. A new healthy sunscreen system for human: solid lipid nanoparticles as carrier for 3,4,5-trimethoxybenzoylchitin and the improvement by adding Vitamin E. *Int J Biol Macromol* 2005; 36: 116–9.
265. Sylvester PW, Kaddoumi A, Nazzal S, and El Sayed KA. The value of tocotrienols in the prevention and treatment of cancer. *J Am Coll Nutr* 2010; 29: 324S–33S.
266. Taepaiboon P, Rungsardthong U, and Supaphol P. Vitamin-loaded electrospun cellulose acetate nanofiber mats as transdermal and dermal therapeutic agents of vitamin A acid and vitamin E. *Eur J Pharm Biopharm* 2007; 67: 387–97.
267. Trombino S, Cassano R, Muzzalupo R, Pingitore A, Cione E, and Picci N. Stearyl ferulate-based solid lipid nanoparticles for the encapsulation and stabilization of beta-carotene and alpha-tocopherol. *Colloids Surf B Biointerfaces* 2009; 72: 181–7.

268. Wissing SA and Muller RH. A novel sunscreen system based on tocopherol acetate incorporated into solid lipid nanoparticles. *Int J Cosmet Sci* 2001; 23: 233–43.

269. Zigoneanu IG, Astete CE, and Sabliov CM. Nanoparticles with entrapped alpha-tocopherol: synthesis, characterization, and controlled release. *Nanotechnology* 2008; 19: 105606.

270. Andersen SM, Rosada C, Dagnaes-Hansen F, Laugesen IG, de Darko E, Dam TN et al. Topical application of valrubicin has a beneficial effect on developing skin tumors. *Carcinogenesis* 2010; 31: 1483–90.

271. Batra R, Sundararajan S, and Sandramouli S. Topical diclofenac gel for the management of periocular actinic keratosis. *Ophthal Plast Reconstr Surg* 2011; 28: 1–3.

272. Martinez JC and Otley CC. The management of melanoma and nonmelanoma skin cancer: a review for the primary care physician. *Mayo Clin Proc* 2001; 76: 1253–65.

273. ACS. Cancer facts and figures 2011. Atlanta, GA: American Cancer Society, 2011.

274. Diepgen TL and Mahler V. The epidemiology of skin cancer. *Br J Dermatol* 2002; 146: 1–6.

275. Albert MR and Weinstock MA. Keratinocyte carcinoma. *CA Cancer J Clin* 2003; 53: 292–302.

276. Brash DE, Rudolph JA, Simon JA, Lin A, McKenna GJ, Baden HP et al. A role for sunlight in skin cancer: UV-induced p53 mutations in squamous cell carcinoma. *Proc Natl Acad Sci USA* 1991; 88: 10124–8.

277. Ridky TW. Nonmelanoma skin cancer. *J Am Acad Dermatol* 2007; 57: 484–501.

278. Ichihashi M, Ueda M, Budiyanto A, Bito T, Oka M, Fukunaga M et al. UV-induced skin damage. *Toxicology* 2003; 189: 21–39.

279. Bendesky A, Michel A, Sordo M, Calderón-Aranda ES, Acosta-Saavedra LC, Salazar AM et al. DNA damage, oxidative mutagen sensitivity, and repair of oxidative DNA damage in nonmelanoma skin cancer patients. *Environ Mol Mutagen* 2006; 47: 509–17.

280. Daya-Grosjean L and Sarasin A. The role of UV induced lesions in skin carcinogenesis: an overview of oncogene and tumor suppressor gene modifications in xeroderma pigmentosum skin tumors. *Mutat Res/Fund Mol Mech Mutagenesis* 2005; 571: 43–56.

281. Xie J, Murone M, Luoh SM, Ryan A, Gu Q, Zhang C et al. Activating smoothened mutations in sporadic basal-cell carcinoma. *Nature* 1998; 391: 90–2.

282. Couve-Privat S, Bouadjar B, Avril MF, Sarasin A, and Daya-Grosjean L. Significantly high levels of ultraviolet-specific mutations in the smoothened gene in basal cell carcinomas from DNA repair-deficient xeroderma pigmentosum patients. *Cancer Res* 2002; 62: 7186–9.

283. Rubin AI, Chen EH, and Ratner D. Basal-cell carcinoma. *N Engl J Med* 2005; 353: 2262–9.

284. Morgan MB. Basal cell carcinoma: variants and challenges. In: Morgan MB, Hamill JR, Spencer JM, eds. *Atlas of Mohs and Frozen Section Cutaneous Pathology*. New York: Springer, 2010: 79–104.

285. Hendi A, Martinez JC, and Martinez J-C. *Nonmelanoma Skin Cancer Atlas of Skin Cancers*. Berlin, Heidelberg: Springer, 2011: 23–76.

286. Gross K, Kircik L, and Kricorian G. 5% 5-Fluorouracil cream for the treatment of small superficial Basal cell carcinoma: efficacy, tolerability, cosmetic outcome, and patient satisfaction. *Dermatol Surg* 2007; 33: 433–40.

287. Kaprealian T, Rembert J, Margolis LW, and Yom SS. Skin cancer. In: Hansen EK, Roach M, eds. *Handbook of Evidence-Based Radiation Oncology*. New York: Springer, 2010: 3–25.

288. Berwick M. Melanoma epidemiology. In: Bosserhoff A, ed. Melanoma Development. Vienna: Springer, 2011: 35–55.

289. Whiteman DC, Pavan WJ, and Bastian BC. The melanomas: a synthesis of epidemiological, clinical, histopathological, genetic, and biological aspects, supporting distinct subtypes, causal pathways, and cells of origin. *Pigment Cell Melanoma Res* 2011; 24(5): 879–97.

290. Howlader N, Noone A, Krapcho M, Neyman N, Aminou R, Waldron W et al. (eds). *SEER Cancer Statistics Review, 1975–2008*. Bethesda, MD, National Cancer Institute, 2011.

291. Geller AC, Miller DR, Annas GD, Demierre MF, Gilchrest BA, and Koh HK. Melanoma incidence and mortality among US whites, 1969–1999. *JAMA* 2002; 288: 1719–20.

292. Lachiewicz AM, Berwick M, Wiggins CL, and Thomas NE. Epidemiologic support for melanoma heterogeneity using the surveillance, epidemiology, and end results program. *J Invest Dermatol* 2008; 128: 1340–2.

293. Whiteman DC, Watt P, Purdie DM, Hughes MC, Hayward NK, and Green AC. Melanocytic nevi, solar keratoses, and divergent pathways to cutaneous melanoma. *J Natl Cancer Inst* 2003; 95: 806–12.

294. Carli P and Palli D. Re: Melanocytic nevi, solar keratoses, and divergent pathways to cutaneous melanoma. *J Natl Cancer Inst* 2003; 95: 1801.

295. Whiteman DC, Parsons PG, and Green AC. p53 expression and risk factors for cutaneous melanoma: a case-control study. *Int J Cancer* 1998; 77: 843–8.

296. Green A. A theory of site distribution of melanomas: Queensland, Australia. *Cancer Causes Control* 1992; 3: 513–6.

297. Green A, MacLennan R, and Siskind V. Common acquired nevi and the risk of malignant melanoma. *Int J Cancer* 1985; 35: 297–300.

298. Carli P, Massi D, Santucci M, Biggeri A, and Giannotti B. Cutaneous melanoma histologically associated with a nevus and melanoma de novo have a different profile of risk: results from a case-control study. *J Am Acad Dermatol* 1999; 40: 549–57.

299. Newton Bishop JA. Lentigos, melanocytic naevi and melanoma. In: Burns T, Breathnach S, Cox N, and Griffiths C, eds. *Rook's Textbook of Dermatology*. West Sussex, United Kingdom: Wiley-Blackwell, 2010: 1–57.

300. Bastian BC, Olshen AB, LeBoit PE, and Pinkel D. Classifying melanocytic tumors based on DNA copy number changes. *Am J Pathol* 2003; 163: 1765–70.

301. Goodson AG and Grossman D. Strategies for early melanoma detection: approaches to the patient with nevi. *J Am Acad Dermatol* 2009; 60: 719–35.

302. Puccio FB and Chian C. Acral junctional nevus versus acral lentiginous melanoma in situ: a differential diagnosis that should be based on clinicopathologic correlation. *Arch Pathol Lab Med* 2011; 135: 847–52.

303. Ivan D and Prieto VG. An update on reporting histopathologic prognostic factors in melanoma. *Arch Pathol Lab Med* 2011; 135: 825–9.

304. Gershenwald JE, Soong SJ, Balch CM, and American Joint Committee on Cancer (AJCC) Melanoma Staging Committee. 2010 TNM staging system for cutaneous melanoma...and beyond. *Ann Surg Oncol* 2010; 17: 1475–7.

305. Balch CM, Gershenwald JE, Soong SJ, Thompson JF, Atkins MB, Byrd DR et al. Final version of 2009 AJCC melanoma staging and classification. *J Clin Oncol* 2009; 27: 6199–206.

306. Tawbi H and Nimmagadda N. Targeted therapy in melanoma. *Biologics* 2009; 3: 475–84.

307. Markovic SN. Malignant melanoma in the 21st century, part 2: staging, prognosis, and treatment. *Mayo Clin Proc* 2007; 82: 490–513.

308. Gasent Blesa JM, Grande Pulido E, Alberola Candel V, and Provencio Pulla M. Melanoma: from darkness to promise. *Am J Clin Oncol* 2011; 34: 179–87.

309. Gogas H, Ioannovich J, Dafni U, Stavropoulou-Giokas C, Frangia K, Tsoutsos D et al. Prognostic significance of autoimmunity during treatment of melanoma with interferon. *N Engl J Med* 2006; 354: 709–18.

310. Hauschild A, Agarwala SS, Trefzer U, Hogg D, Robert C, Hersey P et al. Results of a phase III, randomized, placebo-controlled study of sorafenib in combination with carboplatin and paclitaxel as second-line treatment in patients with unresectable stage III or stage IV melanoma. *J Clin Oncol* 2009; 27: 2823–30.

311. Vultur A, Villanueva J, and Herlyn M. Targeting BRAF in advanced melanoma: a first step toward manageable disease. *Clin Cancer Res* 2011; 17: 1658–63.

312. Berry E, Handley JW, Fitzgerald AJ, Merchant WJ, Boyle RD, Zinov'ev NN et al. Multispectral classification techniques for terahertz pulsed imaging: an example in histopathology. *Med Eng Phys* 2004; 26: 423–30.

313. Hong H, Sun J, and Cai W. Anatomical and molecular imaging of skin cancer. *Clin Cosmet Investig Dermatol* 2008; 1: 1–17.

314. Joseph CS, Yaroslavsky AN, Neel VA, Goyette TM, and Giles RH. Continuous wave terahertz transmission imaging of nonmelanoma skin cancers. *Lasers Surg Med* 2011; 43: 457–62.

315. Mitobe K, Manabe M, Yoshimura N, and Kurabayashi T. Imaging of epithelial cancer in sub-terahertz electromagnetic wave. *Conf Proc IEEE Eng Med Biol Soc* 2005; 1: 199–200.

316. Ney M and Abdulhalim I. Modeling of reflectometric and ellipsometric spectra from the skin in the terahertz and submillimeter waves region. *J Biomed Opt* 2011; 16: 067006.

317. Pickwell E, Cole BE, Fitzgerald AJ, Pepper M, and Wallace VP. In vivo study of human skin using pulsed terahertz radiation. *Phys Med Biol* 2004; 49: 1595–607.

318. Tewari P, Taylor ZD, Bennett D, Singh RS, Culjat MO, Kealey CP et al. Terahertz imaging of biological tissues. *Stud Health Technol Inform* 2011; 163: 653–7.

319. Wallace VP, Fitzgerald AJ, Pickwell E, Pye RJ, Taday PF, Flanagan N et al. Terahertz pulsed spectroscopy of human Basal cell carcinoma. *Appl Spectrosc* 2006; 60: 1127–33.

320. Wallace VP, Fitzgerald AJ, Shankar S, Flanagan N, Pye R, Cluff J et al. Terahertz pulsed imaging of basal cell carcinoma ex vivo and in vivo. *Br J Dermatol* 2004; 151: 424–32.

321. Woodward RM, Wallace VP, Pye RJ, Cole BE, Arnone DD, Linfield EH et al. Terahertz pulse imaging of ex vivo basal cell carcinoma. *J Invest Dermatol* 2003; 120: 72–8.

322. Gambichler T, Regeniter P, Bechara FG, Orlikov A, Vasa R, Moussa G et al. Characterization of benign and malignant melanocytic skin lesions using optical coherence tomography in vivo. *J Am Acad Dermatol* 2007; 57: 629–37.

323. Hamdoon Z, Jerjes W, Upile T, and Hopper C. Optical coherence tomography-guided photody-namic therapy for skin cancer: case study. *Photodiagnosis Photodyn Ther* 2011; 8: 49–52.

324. Mogensen M, Joergensen TM, Nurnberg BM, Morsy HA, Thomsen JB, Thrane L et al. Assessment of optical coherence tomography imaging in the diagnosis of non-melanoma skin cancer and benign lesions versus normal skin: observer-blinded evaluation by dermatologists and pathologists. *Dermatol Surg* 2009; 35: 965–72.

325. Mogensen M, Jorgensen TM, Thrane L, Nurnberg BM, and Jemec GB. Improved quality of optical coherence tomography imaging of basal cell carcinomas using speckle reduction. *Exp Dermatol* 2010; 19: e293–5.

326. Mogensen M, Nurnberg BM, Forman JL, Thomsen JB, Thrane L, and Jemec GB. In vivo thick-ness measurement of basal cell carcinoma and actinic keratosis with optical coherence tomog-raphy and 20-MHz ultrasound. *Br J Dermatol* 2009; 160: 1026–33.

327. Mogensen M, Nurnberg BM, Thrane L, Jorgensen TM, Andersen PE, and Jemec GB. How his-tological features of basal cell carcinomas influence image quality in optical coherence tomog-raphy. *J Biophotonics* 2011; 4: 544–51.

328. Mogensen M, Thrane L, Jorgensen TM, Andersen PE, and Jemec GB. OCT imaging of skin can-cer and other dermatological diseases. *J Biophotonics* 2009; 2: 442–51.

329. Morsy H, Kamp S, Thrane L, Behrendt N, Saunder B, Zayan H et al. Optical coherence tomog-raphy imaging of psoriasis vulgaris: correlation with histology and disease severity. *Arch Dermatol Res* 2010; 302: 105–11.

330. Patel JK, Konda S, Perez OA, Amini S, Elgart G, and Berman B. Newer technologies/techniques and tools in the diagnosis of melanoma. *Eur J Dermatol* 2008; 18: 617–31.

331. Strasswimmer J, Pierce MC, Park BH, Neel V, and de Boer JF. Polarization-sensitive optical coherence tomography of invasive basal cell carcinoma. *J Biomed Opt* 2004; 9: 292–8.

332. Welzel J, Lankenau E, Birngruber R, and Engelhardt R. Optical coherence tomography of the human skin. *J Am Acad Dermatol* 1997; 37: 958–63.

333. Conway K, Edmiston SN, Khondker ZS, Groben PA, Zhou X, Chu H et al. DNA-methylation profiling distinguishes malignant melanomas from benign nevi. *Pigment Cell Melanoma Res* 2011; 24: 352–60.

334. Millington GW. Epigenetics and dermatological disease. *Pharmacogenomics* 2008; 9: 1835–50.

335. Backmann N, Zahnd C, Huber F, Bietsch A, Pluckthun A, Lang HP et al. A label-free immuno-sensor array using single-chain antibody fragments. *Proc Natl Acad Sci USA* 2005; 102: 14587–92.

336. Zhou W, Ma Y, Yang H, Ding Y, and Luo X. A label-free biosensor based on silver nanopar-ticles array for clinical detection of serum p53 in head and neck squamous cell carcinoma. *Int J Nanomedicine* 2011; 6: 381–6.

337. Ma Z, Swede H, Cassarino D, Fleming E, Fire A, and Dadras SS. Up-regulated Dicer expression in patients with cutaneous melanoma. *PloS One* 2011; 6: e20494.
338. Sand M, Gambichler T, Sand D, Altmeyer P, Stuecker M, and Bechara FG. Immunohistochemical expression patterns of the miRNA-processing enzyme Dicer in cutaneous malignant melanomas, benign melanocytic nevi and dysplastic melanocytic nevi. *Eur J Dermato* 2011; 21: 18–21.
339. Watt SA, Pourreyron C, Purdie K, Hogan C, Cole CL, Foster N et al. Integrative mRNA profiling comparing cultured primary cells with clinical samples reveals PLK1 and C20orf20 as therapeutic targets in cutaneous squamous cell carcinoma. *Oncogene* 2011; 30: 4666–77.
340. Wachsman W, Morhenn V, Palmer T, Walls L, Hata T, Zalla J et al. Noninvasive genomic detection of melanoma. *Br J Dermatol* 2011; 164: 797–806.
341. Lang JM, Shennan M, Njauw JC, Luo S, Bishop JN, Harland M et al. A flexible multiplex bead-based assay for detecting germline CDKN2A and CDK4 variants in melanoma-prone kindreds. *J Invest Dermatol* 2011; 131: 480–6.
342. Nan H, Qureshi AA, Prescott J, De Vivo I, and Han J. Genetic variants in telomere-maintaining genes and skin cancer risk. *Hum Genet* 2011; 129: 247–53.
343. Ibarrola-Villava M, Pena-Chilet M, Fernandez LP, Aviles JA, Mayor M, Martin-Gonzalez M et al. Genetic polymorphisms in DNA repair and oxidative stress pathways associated with malignant melanoma susceptibility. *Eur J Cancer* 2011; 47: 2618–25.
344. Fernandez AA, Garcia R, Paniker L, Trono D, and Mitchell DL. An experimental population study of nucleotide excision repair as a risk factor for UVB-induced melanoma. *Photochem Photobiol* 2011; 87: 335–41.
345. Aalborg J, Morelli JG, Byers TE, Mokrohisky ST, and Crane LA. Effect of hair color and sun sensitivity on nevus counts in white children in Colorado. *J Am Acad Dermatol* 2010; 63: 430–9.
346. Sharma S, Neale MH, Di Nicolantonio F, Knight LA, Whitehouse PA, Mercer SJ et al. Outcome of ATP-based tumor chemosensitivity assay directed chemotherapy in heavily pre-treated recurrent ovarian carcinoma. *BMC Cancer* 2003; 3: 19.
347. Cree IA. Chemosensitivity testing as an aid to anti-cancer drug and regimen development. *Recent Results Cancer Res* 2003; 161: 119–25.
348. Kurbacher CM and Cree IA. Chemosensitivity testing using microplate adenosine triphosphate-based luminescence measurements. *Methods Mol Med* 2005; 110: 101–20.
349. Parker KA, Glaysher S, Polak M, Gabriel FG, Johnson P, Knight LA et al. The molecular basis of the chemosensitivity of metastatic cutaneous melanoma to chemotherapy. *J Clin Pathol* 2010; 63: 1012–20.
350. Ugurel S, Tilgen W, and Reinhold U. Chemosensitivity testing in malignant melanoma. *Recent Results Cancer Res* 2003; 161: 81–92.
351. Xie J, Qureshi AA, Li Y, and Han J. ABO blood group and incidence of skin cancer. *PloS One* 2010; 5: e11972.
352. Gajewski TF. Molecular profiling of melanoma and the evolution of patient-specific therapy. *Semin Oncol* 2011; 38: 236–42.
353. Sznol M. Molecular markers of response to treatment for melanoma. *Cancer J* 2011; 17: 127–33.
354. Bichakjian CK, Halpern AC, Johnson TM, Hood AF, Grichnik JM, Swetter SM et al. Guidelines of care for the management of primary cutaneous melanoma. *J Am Acad Dermatol* 2011; 65: 1032–47.
355. Veenstra HJ, Vermeeren L, Valdes Olmos RA, and Nieweg OE. The additional value of lymphatic mapping with routine SPECT/CT in unselected patients with clinically localized melanoma. *Ann Surg Oncol* 2011; 19: 1018–23.
356. Ballou B, Ernst LA, Andreko S, Harper T, Fitzpatrick JA, Waggoner AS et al. Sentinel lymph node imaging using quantum dots in mouse tumor models. *Bioconjug Chem* 2007; 18: 389–96.
357. Tanaka E, Choi HS, Fujii H, Bawendi MG, and Frangioni JV. Image-guided oncologic surgery using invisible light: completed pre-clinical development for sentinel lymph node mapping. *Ann Surg Oncol* 2006; 13: 1671–81.
358. Erogbogbo F, Tien CA, Chang CW, Yong KT, Law WC, Ding H et al. Bioconjugation of luminescent silicon quantum dots for selective uptake by cancer cells. *Bioconjug Chem* 2011; 22: 1081–8.

359. Erogbogbo F, Yong KT, Roy I, Hu R, Law WC, Zhao W et al. In vivo targeted cancer imaging, sentinel lymph node mapping and multi-channel imaging with biocompatible silicon nanocrystals. *ACS Nano* 2011; 5: 413–23.

360. Pons T, Pic E, Lequeux N, Cassette E, Bezdetnaya L, Guillemin F et al. Cadmium-free CuInS2/ZnS quantum dots for sentinel lymph node imaging with reduced toxicity. *ACS Nano* 2010; 4: 2531–8.

361. Mazar J, DeBlasio D, Govindarajan SS, Zhang S, and Perera RJ. Epigenetic regulation of microRNA-375 and its role in melanoma development in humans. *FEBS Lett* 2011; 585: 2467–76.

362. Shinojima Y, Terui T, Hara H, Kimura M, Igarashi J, Wang X et al. Identification and analysis of an early diagnostic marker for malignant melanoma: ZAR1 intra-genic differential methylation. *J Dermatol Sci* 2010; 59: 98–106.

363. Sigalotti L, Fratta E, Bidoli E, Covre A, Parisi G, Colizzi F et al. Methylation levels of the "long interspersed nucleotide element-1" repetitive sequences predict survival of melanoma patients. *J Transl Med* 2011; 9: 78.

364. Barnes EA, Heidtman KJ, and Donoghue DJ. Constitutive activation of the shh-ptc1 pathway by a patched1 mutation identified in BCC. *Oncogene* 2005; 24: 902–15.

365. LoRusso PM, Rudin CM, Reddy JC, Tibes R, Weiss GJ, Borad MJ et al. Phase I trial of hedgehog pathway inhibitor vismodegib (GDC-0449) in patients with refractory, locally advanced or metastatic solid tumors. *Clin Cancer Res* 2011; 17: 2502–11.

366. Stanton BZ and Peng LF. Small-molecule modulators of the Sonic Hedgehog signaling pathway. *Mol Biosyst* 2010; 6: 44–54.

367. De Smaele E, Ferretti E, and Gulino A. Vismodegib, a small-molecule inhibitor of the hedgehog pathway for the treatment of advanced cancers. *Curr Opin Investig Drugs* 2010; 11: 707–18.

368. Palumbo A, Hauler F, Dziunycz P, Schwager K, Soltermann A, Pretto F et al. A chemically modified antibody mediates complete eradication of tumours by selective disruption of tumour blood vessels. *Br J Cancer* 2011; 104: 1106–15.

369. Khan ZA, Chan BM, Uniyal S, Barbin YP, Farhangkhoee H, Chen S et al. EDB fibronectin and angiogenesis—a novel mechanistic pathway. *Angiogenesis* 2005; 8: 183–96.

370. Luo J, Wu SJ, Lacy ER, Orlovsky Y, Baker A, Teplyakov A et al. Structural basis for the dual recognition of IL-12 and IL-23 by ustekinumab. *J Mol Biol* 2010; 402: 797–812.

371. Hodi FS, O'Day SJ, McDermott DF, Weber RW, Sosman JA, Haanen JB et al. Improved survival with ipilimumab in patients with metastatic melanoma. *N Engl J Med* 2010; 363: 711–23.

372. Chapman PB, Hauschild A, Robert C, Haanen JB, Ascierto P, Larkin J et al. Improved survival with vemurafenib in melanoma with BRAF V600E mutation. *N Engl J Med* 2011; 364: 2507–16.

373. Amiri H, Mahmoudi M, and Lascialfari A. Superparamagnetic colloidal nanocrystal clusters coated with polyethylene glycol fumarate: a possible novel theranostic agent. *Nanoscale* 2011; 3: 1022–30.

374. Caldorera-Moore ME, Liechty WB, and Peppas NA. Responsive theranostic systems: integration of diagnostic imaging agents and responsive controlled release drug delivery carriers. *Acc Chem Res* 2011; 44: 1061–70.

375. Gittard SD, Miller PR, Boehm RD, Ovsianikov A, Chichkov BN, Heiser J et al. Multiphoton microscopy of transdermal quantum dot delivery using two photon polymerization-fabricated polymer microneedles. *Faraday Discuss* 2011; 149: 171–85; discussion 227–45.

376. Puri A, Blumenthal R. Polymeric lipid assemblies as novel theranostic tools. *Acc Chem Res* 2011; 44: 1071–9.

377. Yoo D, Lee JH, Shin TH, and Cheon J. Theranostic magnetic nanoparticles. *Acc Chem Res* 2011; 44: 863–74.

378. Lu W, Xiong C, Zhang G, Huang Q, Zhang R, Zhang JZ et al. Targeted photothermal ablation of murine melanomas with melanocyte-stimulating hormone analog-conjugated hollow gold nanospheres. *Clin Cancer Res* 2009; 15: 876–86.

379. Sanmartin O and Guillen C. Images in clinical medicine. Fluorescence diagnosis of subclinical actinic keratoses. *N Engl J Med* 2008; 358: e21.

380. Longo C, Gambara G, Espina V, Luchini A, Bishop B, Patanarut AS et al. A novel biomarker harvesting nanotechnology identifies Bak as a candidate melanoma biomarker in serum. *Exp Dermatol* 2011; 20: 29–34.

381. Solassol J, Guillot B, and Maudelonde T. [Circulating prognosis markers in melanoma: proteomic profiling and clinical studies]. *Ann Biol Clin (Paris)* 2011; 69: 151–7.
382. Sotiriou E, Apalla Z, and Ioannides D. Complete resolution of a squamous cell carcinoma of the skin using intralesional 5-aminolevulinic acid photodynamic therapy intralesional PDT for SCC. *Photodermatol Photoimmunol Photomed* 2010; 26: 269–71.
383. Seki T, Fang J, and Maeda H. Enhanced delivery of macromolecular antitumor drugs to tumors by nitroglycerin application. *Cancer Sci* 2009; 100: 2426–30.
384. Bao W, Chen M, Zhao X, Kumar R, Spinnler C, Thullberg M et al. PRIMA-1Met/APR-246 induces wild-type p53-dependent suppression of malignant melanoma tumor growth in 3D culture and in vivo. *Cell Cycle* 2011; 10: 301–7.
385. Malki A, Pulipaka AB, Evans SC, and Bergmeier SC. Structure–activity studies of quinuclidinone analogs as anti-proliferative agents in lung cancer cell lines. *Bioorg Med Chem Lett* 2006; 16: 1156–9.
386. Kigasawa K, Kajimoto K, Nakamura T, Hama S, Kanamura K, Harashima H et al. Noninvasive and efficient transdermal delivery of CpG-oligodeoxynucleotide for cancer immunotherapy. *J Control Release* 2011; 150: 256–65.
387. Cerkovnik P, Novakovic BJ, Stegel V, and Novakovic S. Tumor vaccine composed of C-class CpG oligodeoxynucleotides and irradiated tumor cells induces long-term antitumor immunity. *BMC Immunol* 2010; 11: 45.
388. Chen Y, Bathula SR, Yang Q, and Huang L. Targeted nanoparticles deliver siRNA to melanoma. *J Invest Dermatol* 2010; 130: 2790–8.
389. Antonini D, Russo MT, De Rosa L, Gorrese M, Del Vecchio L, and Missero C. Transcriptional repression of miR-34 family contributes to p63-mediated cell cycle progression in epidermal cells. *J Invest Dermatol* 2010; 130: 1249–57.
390. Chen Y, Zhu X, Zhang X, Liu B, and Huang L. Nanoparticles modified with tumor-targeting scFv deliver siRNA and miRNA for cancer therapy. *Mol Ther* 2010; 18: 1650–6.
391. Deiters A. Small molecule modifiers of the microRNA and RNA interference pathway. *AAPS J* 2010; 12: 51–60.
392. Howell PM, Jr., Li X, Riker AI, and Xi Y. MicroRNA in melanoma. *Ochsner J* 2010; 10: 83–92.
393. Montgomery RL, van Rooij E. Therapeutic advances in MicroRNA targeting. *J Cardiovasc Pharmacol* 2011; 57: 1–7.
394. Sun Q, Cong R, Yan H, Gu H, Zeng Y, Liu N et al. Genistein inhibits growth of human uveal melanoma cells and affects microRNA-27a and target gene expression. *Oncol Rep* 2009; 22: 563–7.
395. Strillacci A, Griffoni C, Valerii MC, Lazzarini G, Tomasi V, and Spisni E. RNAi-based strategies for cyclooxygenase-2 inhibition in cancer. *J Biomed Biotechnol* 2010; 2010: 828045.
396. Chen L, Zhu Y, Li H, Wang GL, Wu YY, Lu YX et al. Knockdown of TSPAN1 by RNA silencing and antisense technique inhibits proliferation and infiltration of human skin squamous carcinoma cells. *Tumori* 2010; 96: 289–95.
397. Davis ME, Zuckerman JE, Choi CH, Seligson D, Tolcher A, Alabi CA et al. Evidence of RNAi in humans from systemically administered siRNA via targeted nanoparticles. *Nature* 2010; 464: 1067–70.
398. Yen MC, Lin CC, Chen YL, Huang SS, Yang HJ, Chang CP et al. A novel cancer therapy by skin delivery of indoleamine 2,3-dioxygenase siRNA. *Clin Cancer Res* 2009; 15: 641–9.
399. Alshamsan A, Haddadi A, Hamdy S, Samuel J, El-Kadi AO, Uludag H et al. STAT3 Silencing in dendritic cells by siRNA polyplexes encapsulated in PLGA nanoparticles for the modulation of anticancer immune response. *Mol Pharm* 2010.
400. Alshamsan A, Hamdy S, Haddadi A, Samuel J, El-Kadi AO, Uludag H et al. STAT3 knockdown in B16 melanoma by siRNA lipopolyplexes induces bystander immune response in vitro and in vivo. *Transl Oncol* 2011; 4: 178–88.
401. Nagarwal RC, Singh PN, Kant S, Maiti P, and Pandit JK. Chitosan nanoparticles of 5-fluorouracil for ophthalmic delivery: characterization, in-vitro and in-vivo study. *Chem Pharm Bull (Tokyo)* 2011; 59: 272–8.

402. Yan C, Gu J, Guo Y, and Chen D. In vivo biodistribution for tumor targeting of 5-fluoro-uracil (5-FU) loaded N-succinyl-chitosan (Suc-Chi) nanoparticles. *Yakugaku Zasshi* 2010; 130:801–4.

403. Zhu L, Ma J, Jia N, Zhao Y, and Shen H. Chitosan-coated magnetic nanoparticles as carriers of 5-fluorouracil: preparation, characterization and cytotoxicity studies. *Colloids Surf B Biointerfaces* 2009; 68:1–6.

404. Majoros IJ, Williams CR, Becker A, and Baker JR Jr. Methotrexate delivery via folate targeted dendrimer-based nanotherapeutic platform. *Wiley Interdiscip Rev Nanomed Nanobiotechnol* 2009; 1:502–10.

405. Ross JF, Chaudhuri PK, and Ratnam M. Differential regulation of folate receptor isoforms in normal and malignant tissues in vivo and in established cell lines. Physiologic and clinical implications. *Cancer* 1994; 73:2432–43.

406. Saba NF, Wang X, Muller S, Tighiouart M, Cho K, Nie S et al. Examining expression of folate receptor in squamous cell carcinoma of the head and neck as a target for a novel nanotherapeutic drug. *Head Neck* 2009; 31:475–81.

407. Zhang Y, Li J, Lang M, Tang X, Li L, and Shen X. Folate-functionalized nanoparticles for controlled 5-Fluorouracil delivery. *J Colloid Interface Sci* 2011; 354:202–9.

408. Chen H, Gu Y, Hub Y, and Qian Z. Characterization of pH- and temperature-sensitive hydrogel nanoparticles for controlled drug release. *PDA J Pharm Sci Technol* 2007; 61:303–13.

409. Hadjikirova M, Troyanova P, and Simeonova M. Nanoparticles as drug carrier system of 5-fluorouracil in local treatment of patients with superficial basal cell carcinoma. *J BUON* 2005; 10:517–21.

410. Soto-Castro D, Cruz-Morales JA, Ramirez Apan MT, and Guadarrama P. Synthesis of non-cytotoxic poly(ester-amine) dendrimers as potential solubility enhancers for drugs: methotrexate as a case study. *Molecules* 2010; 15:8082–97.

411. Singka GS, Samah NA, Zulfakar MH, Yurdasiper A, and Heard CM. Enhanced topical delivery and anti-inflammatory activity of methotrexate from an activated nanogel. *Eur J Pharm Biopharm* 2010; 76: 275–81.

412. Young KL, Xu C, Xie J, and Sun S. Conjugating Methotrexate to magnetite (Fe_3O_4) nanoparticles via trichloro-s-triazine. *J Mater Chem* 2009; 19: 6400–6.

413. Kohler N, Sun C, Wang J, and Zhang M. Methotrexate-modified superparamagnetic nanoparticles and their intracellular uptake into human cancer cells. *Langmuir* 2005; 21: 8858–64.

414. Shapira A, Assaraf YG, and Livney YD. Beta-casein nanovehicles for oral delivery of chemotherapeutic drugs. *Nanomedicine* 2010; 6: 119–26.

415. Arayachukeat S, Wanichwecharungruang SP, and Tree-Udom T. Retinyl acetate-loaded nanoparticles: dermal penetration and release of the retinyl acetate. *Int J Pharm* 2011; 404: 281–8.

416. Ghouchi Eskandar N, Simovic S, and Prestidge CA. Nanoparticle coated submicron emulsions: sustained in-vitro release and improved dermal delivery of all-trans-retinol. *Pharm Res* 2009; 26: 1764–75.

417. Teixeira Z, Zanchetta B, Melo BA, Oliveira LL, Santana MH, Paredes-Gamero EJ et al. Retinyl palmitate flexible polymeric nanocapsules: characterization and permeation studies. *Colloids Surf B Biointerfaces* 2010; 81: 374–80.

418. Ourique AF, Azoubel S, Ferreira CV, Silva CB, Marchiori MC, Pohlmann AR et al. Lipid-core nanocapsules as a nanomedicine for parenteral administration of tretinoin: development and in vitro antitumor activity on human myeloid leukaemia cells. *J Biomed Nanotechnol* 2010; 6: 214–23.

419. Pedersen PJ, Adolph SK, Subramanian AK, Arouri A, Andresen TL, Mouritsen OG et al. Liposomal formulation of retinoids designed for enzyme triggered release. *J Med Chem* 2010; 53: 3782–92.

420. Taha E, Ghorab D, and Zaghloul AA. Bioavailability assessment of vitamin A self-nanoemulsified drug delivery systems in rats: a comparative study. *Med Princ Pract* 2007; 16: 355–9.

421. Liu J, Hu W, Chen H, Ni Q, Xu H, and Yang X. Isotretinoin-loaded solid lipid nanoparticles with skin targeting for topical delivery. *Int J Pharm* 2007; 328: 191–5.

422. Singh AT, Evens AM, Anderson RJ, Beckstead JA, Sankar N, Sassano A et al. All trans retinoic acid nanodisks enhance retinoic acid receptor mediated apoptosis and cell cycle arrest in mantle cell lymphoma. *Br J Haematol* 2010; 150: 158–69.
423. Bal SM, Kruithof AC, Zwier R, Dietz E, Bouwstra JA, Lademann J et al. Influence of microneedle shape on the transport of a fluorescent dye into human skin in vivo. *J Control Release* 2010; 147: 218–24.
424. Donnelly RF, Morrow DI, Fay F, Scott CJ, Abdelghany S, Singh RR et al. Microneedle-mediated intradermal nanoparticle delivery: potential for enhanced local administration of hydrophobic pre-formed photosensitisers. *Photodiagnosis Photodyn Ther* 2010; 7: 222–31.
425. Donnelly RF, Morrow DI, McCarron PA, David Woolfson A, Morrissey A, Juzenas P et al. Microneedle arrays permit enhanced intradermal delivery of a preformed photosensitizer. *Photochem Photobiol* 2009; 85: 195–204.
426. Li X, Zhao R, Qin Z, Zhang J, Zhai S, Qiu Y et al. Microneedle pretreatment improves efficacy of cutaneous topical anesthesia. *Am J Emerg Med* 2010; 28: 130–4.
427. Mikolajewska P, Donnelly RF, Garland MJ, Morrow DI, Singh TR, Iani V et al. Microneedle pre-treatment of human skin improves 5-aminolevulininc acid (ALA)- and 5-aminolevulinic acid methyl ester (MAL)-induced PpIX production for topical photodynamic therapy without increase in pain or erythema. *Pharm Res* 2010; 27: 2213–20.
428. Song JM, Kim YC, Barlow PG, Hossain MJ, Park KM, Donis RO et al. Improved protection against avian influenza H5N1 virus by a single vaccination with virus-like particles in skin using microneedles. *Antiviral Res* 2010; 88: 244–7.
429. Miller PR, Gittard SD, Edwards TL, Lopez DM, Xiao X, Wheeler DR et al. Integrated carbon fiber electrodes within hollow polymer microneedles for transdermal electrochemical sensing. *Biomicrofluidics* 2011; 5: 13415.
430. Kosoglu MA, Hood RL, Chen Y, Xu Y, Rylander MN, and Rylander CG. Fiber optic microneedles for transdermal light delivery: ex vivo porcine skin penetration experiments. *J Biomech Eng* 2010; 132: 091014.
431. Shapira O, Kuriki K, Orf ND, Abouraddy AF, Benoit G, Viens JF et al. Surface-emitting fiber lasers. *Opt Express* 2006; 14: 3929–35.
432. Plasmeijer EI, Struijk L, Bouwes Bavinck JN, and Feltkamp MCW. Epidemiology of cutaneous human papillomavirus infections. In: Stockfleth E, Ulrich C, eds. *Skin Cancer after Organ Transplantation*. US: Springer, 2009: 143–57.
433. Chin-Hong PV and Palefsky JM. Human papillomavirus anogenital disease in HIV-infected individuals. *Dermatol Ther* 2005; 18: 67–76.
434. Kirnbauer R, Lenz P, and Okun MM. Chapter 78 - Human papillomavirus. In: Bolognia J, Jorizzo JL, Rapini RP, eds. *Dermatology*. London, New York: Mosby, 2003.
435. Steben M and Duarte-Franco E. Human papillomavirus infection: Epidemiology and pathophysiology. *Gynecol Oncol* 2007; 107: S2–5.
436. Roden RBS, Lowy DR, and Schiller JT. Papillomavirus is resistant to desiccation. *J Infect Dis* 1997; 176: 1076–9.
437. Ahmed I. Viral warts. In: Lebwohl M, ed. *Treatment of Skin Disease: Comprehensive Therapeutic Strategies*. Philadelphia, PA: Mosby/Elsevier, 2006.
438. Brown DR, Kjaer SK, Sigurdsson K, Iversen O-E, Hernandez-Avila M, Wheeler CM et al. The impact of quadrivalent human papillomavirus (HPV; types 6, 11, 16, and 18) L1 virus-like particle vaccine on infection and disease due to oncogenic nonvaccine HPV types in generally HPV-naive women aged 16–26 years. *J Infect Dis* 2009; 199: 926–35.
439. Lee HJ, Kim JK, Kim DH, and Yoon MS. Condyloma accuminatum treated with recombinant quadrivalent human papillomavirus vaccine (types 6, 11, 16, 18). *J Am Acad Dermatol* 2011; 64: e130–2.
440. Gibbs S and Harvey I. Topical treatments for cutaneous warts. *Cochrane Database Syst Rev* 2006; 9: CD001781.
441. Hsueh PR. Human papillomavirus, genital warts, and vaccines. *J Microbiol Immunol Infect* 2009; 42: 101–6.

442. Vecchio G, Sabella S, Tagliaferro L, Menegazzi P, Di Bello MP, Brunetti V et al. Modular plastic chip for one-shot human papillomavirus diagnostic analysis. *Anal Biochem* 2010; 397: 53–9.

443. Piao JY, Park EH, Choi K, Quan B, Kang DH, Park PY et al. Direct visual detection of DNA based on the light scattering of silica nanoparticles on a human papillomavirus DNA chip. *Talanta* 2009; 80: 967–73.

444. Chan CP, Tzang LC, Sin KK, Ji SL, Cheung KY, Tam TK et al. Biofunctional organic nanocrystals for quantitative detection of pathogen deoxyribonucleic acid. *Anal Chim Acta* 2007; 584: 7–11.

445. Xue J, Chen H, Fan M, Zhu F, Diao L, Chen X et al. Use of quantum dots to detect human papillomavirus in oral squamous cell carcinoma. *J Oral Pathol Med* 2009; 38: 668–71.

446. Lee HJ, Park N, Cho HJ, Yoon JK, Van ND, Oh YK et al. Development of a novel viral DNA vaccine against human papillomavirus: AcHERV-HP16L1. *Vaccine* 2010; 28: 1613–9.

447. Corbett HJ, Fernando GJ, Chen X, Frazer IH, and Kendall MA. Skin vaccination against cervical cancer associated human papillomavirus with a novel micro-projection array in a mouse model. *PloS One* 2010; 5: e13460.

448. Chen H, Chang X, Du D, Liu W, Liu J, Weng T et al. Podophyllotoxin-loaded solid lipid nanoparticles for epidermal targeting. *J Control Release* 2006; 110: 296–306.

449. Xie FM, Zeng K, Chen ZL, Li GF, Lin ZF, Zhu XL et al. [Treatment of recurrent condyloma acuminatum with solid lipid nanoparticle gel containing podophyllotoxin: a randomized double-blinded, controlled clinical trial]. *Nan Fang Yi Ke Da Xue Xue Bao* 2007; 27: 657–9.

450. Chang CC, Liu DZ, Lin SY, Liang HJ, Hou WC, Huang WJ et al. Liposome encapsulation reduces cantharidin toxicity. *Food Chem Toxicol* 2008; 46: 3116–21.

451. Li H, Fang Q, Zhang H, Zang C, Zhang B, Nie Q et al. [Preparation and characterization of nonionic surfactant vesicle of cantharidin]. *Zhongguo Zhong Yao Za Zhi* 2010; 35: 2546–50.

452. Tannous ZS, Mihm MC, Jr., Sober AJ, and Duncan LM. Congenital melanocytic nevi: clinical and histopathologic features, risk of melanoma, and clinical management. *J Am Acad Dermatol* 2005; 52: 197–203.

453. Bolognia J, Jorizzo JL, and Rapini RP. *Dermatology*. St. Louis, MO, London: Mosby Elsevier, 2008.

454. Silva Idos S, Higgins CD, Abramsky T, Swanwick MA, Frazer J, Whitaker LM et al. Overseas sun exposure, nevus counts, and premature skin aging in young English women: a population-based survey. *J Invest Dermatol* 2009; 129: 50–9.

455. Dodd AT, Morelli J, Mokrohisky ST, Asdigian N, Byers TE, and Crane LA. Melanocytic nevi and sun exposure in a cohort of Colorado children: anatomic distribution and site-specific sunburn. *Cancer Epidemiol Biomarkers Prev* 2007; 16: 2136–43.

456. Bauer J, Büttner P, Wiecker TS, Luther H, and Garbe C. Risk factors of incident melanocytic nevi: a longitudinal study in a cohort of 1,232 young German children. *Int J Cancer* 2005; 115: 121–6.

457. Smith MA. Dysplastic nevus. In: Usatine R, ed. *The Color Atlas of Family Medicine*. New York: McGraw-Hill Medical, 2009: xviii, 1095 p.

458. Thomas Valencia D SNA and Lee Ken K. Chapter 118. Benign epithelial tumors, hamartomas, and hyperplasias. In: Goldsmith LA, Fitzpatrick TB, eds. *Fitzpatrick's Dermatology in General Medicine*. New York: McGraw-Hill Professional, 2012:p. 1054–67.

459. Kelly JW, Rivers JK, MacLennan R, Harrison S, Lewis AE, and Tate BJ. Sunlight: a major factor associated with the development of melanocytic nevi in Australian schoolchildren. *J Am Acad Dermatol* 1994; 30: 40–8.

460. Longo C, Rito C, Beretti F, Cesinaro AM, Pineiro-Maceira J, Seidenari S et al. De novo melanoma and melanoma arising from pre-existing nevus: in vivo morphologic differences as evaluated by confocal microscopy. *J Am Acad Dermatol* 2011; 65: 604–14.

461. Abouraddy AF, Shapira O, Bayindir M, Arnold J, Sorin F, Hinczewski DS et al. Large-scale optical-field measurements with geometric fibre constructs. *Nat Mater* 2006; 5: 532–6.

462. Benoit G, Kuriki K, Viens JF, Joannopoulos JD, and Fink Y. Dynamic all-optical tuning of transverse resonant cavity modes in photonic bandgap fibers. *Opt Lett* 2005; 30: 1620–2.

463. Galasso M, Elena Sana M, and Volinia S. Non-coding RNAs: a key to future personalized molecular therapy? *Genome Med* 2010; 2: 12.

464. Maki WC, Mishra NN, Cameron EG, Filanoski B, Rastogi SK, and Maki GK. Nanowire-transistor based ultra-sensitive DNA methylation detection. *Biosens Bioelectron* 2008; 23: 780–7.
465. O'Neill TB, Rawlins J, Rea S, and Wood F. Treatment of a large congenital melanocytic nevus with dermabrasion and autologous cell suspension (ReCELL®): a case report. *J Plast Reconstr Aesthet Surg* 2011; 64(12): 1672–6.
466. Gottschaller C, Hohenleutner U, and Landthaler M. Metastasis of a malignant melanoma 2 years after carbon dioxide laser treatment of a pigmented lesion: case report and review of the literature. *Acta Dermato-Venereologica* 2006; 86: 44–7.
467. Helsing P, Mork G, and Sveen B. Ruby laser treatment of congenital melanocytic naevi—a pessimistic view. *Acta Dermato-Venereologica* 2006; 86: 235–7.
468. Jones CE and Nouri K. Laser treatment for pigmented lesions: a review. *J Cosmet Dermatol* 2006; 5: 9–13.
469. Lee HW, Ahn SJ, Lee MW, Choi JH, Moon KC, and Koh JK. Pseudomelanoma following laser therapy. *J Eur Acad Dermatol Venereol* 2006; 20: 342–4.
470. Burri P. Treatment of naevi and warts by topical chemotherapy with Solcoderm. *Dermatologica* 1984; 168 Suppl 1: 52–7.
471. Engelberg IS, Ronnen M, Suster S, Schewach-Millet M, Stempler D, and Schibi-Brilliant G. Effects of Solcoderm. *Int J Dermatol* 1986; 25: 606–7.
472. Grunwald MH, Gat A, and Amichai B. Pseudomelanoma after Solcoderm treatment. *Melanoma Res* 2006; 16: 459–60.
473. Haim S and Cohen A. Solcoderm treatment of epidermal growths including intradermal nevi. *Dermatologica* 1984; 168 Suppl 1: 46–8.
474. Labhardt WC. An overview of clinical experience with Solcoderm. *Dermatologica* 1984; 168 Suppl 1: 31–2.
475. Franco RA, Dowdall JR, Bujold K, Amann C, Faquin W, Redmond RW et al. Photochemical repair of vocal fold microflap defects. *Laryngoscope* 2011; 121: 1244–51.
476. Ibusuki S, Halbesma GJ, Randolph MA, Redmond RW, Kochevar IE, and Gill TJ. Photochemically cross-linked collagen gels as three-dimensional scaffolds for tissue engineering. *Tissue Eng* 2007; 13: 1995–2001.
477. Kamegaya Y, Farinelli WA, Vila Echague AV, Akita H, Gallagher J, Flotte TJ et al. Evaluation of photochemical tissue bonding for closure of skin incisions and excisions. *Lasers Surg Med* 2005; 37: 264–70.
478. Wang Y, Kochevar IE, Redmond RW, and Yao M. A light-activated method for repair of corneal surface defects. *Lasers Surg Med* 2011; 43: 481–9.
479. Yao M, Yaroslavsky A, Henry FP, Redmond RW, and Kochevar IE. Phototoxicity is not associated with photochemical tissue bonding of skin. *Lasers Surg Med* 2010; 42: 123–31.
480. Boehm RD, Chen B, Gittard SD, Chichkov BC, Monteiro-Riveire NA, Nasir A et al. Two photon polymerization-micromolding of microscale barbs for medical applications. *J Mater Sci* 2011; http://www.tandfonline.com/doi/ref/10.1080/01694243.2012.693828. Accessed July 11, 2013.
481. Boesel LF, Greiner C, Arzt E, and del Campo A. Gecko-inspired surfaces: a path to strong and reversible dry adhesives. *Adv Mater* 2010; 22: 2125–37.
482. Geim AK, Dubonos SV, Grigorieva IV, Novoselov KS, Zhukov AA, and Shapoval SY. Microfabricated adhesive mimicking gecko foot-hair. *Nat Mater* 2003; 2: 461–3.
483. Gravish N, Wilkinson M, Sponberg S, Parness A, Esparza N, Soto D et al. Rate-dependent frictional adhesion in natural and synthetic gecko setae. *J R Soc Interface* 2010; 7: 259–69.
484. Kim S, Kustandi TS, and Yi DK. Synthesis of artificial polymeric nanopillars for clean and reusable adhesives. *J Nanosci Nanotechnol* 2008; 8: 4779–82.
485. Lee H, Lee BP, and Messersmith PB. A reversible wet/dry adhesive inspired by mussels and geckos. *Nature* 2007; 448: 338–41.
486. Lee J and Fearing RS. Contact self-cleaning of synthetic gecko adhesive from polymer microfibers. *Langmuir* 2008; 24: 10587–91.
487. Mahdavi A, Ferreira L, Sundback C, Nichol JW, Chan EP, Carter DJ et al. A biodegradable and biocompatible gecko-inspired tissue adhesive. *Proc Natl Acad Sci USA* 2008; 105: 2307–12.

488. Parness A, Soto D, Esparza N, Gravish N, Wilkinson M, Autumn K et al. A microfabricated wedge-shaped adhesive array displaying gecko-like dynamic adhesion, directionality and long lifetime. *J R Soc Interface* 2009; 6: 1223–32.

489. Russell AP and Higham TE. A new angle on clinging in geckos: incline, not substrate, triggers the deployment of the adhesive system. *Proc Biol Sci* 2009; 276: 3705–9.

490. Schargott M. A mechanical model of biomimetic adhesive pads with tilted and hierarchical structures. *Bioinspir Biomim* 2009; 4: 026002.

491. Yamaguchi T, Gravish N, Autumn K, and Creton C. Microscopic modeling of the dynamics of frictional adhesion in the gecko attachment system. *J Phys Chem B* 2009; 113: 3622–8.

492. Yanik MF. Towards gecko-feet-inspired bandages. *Trends Biotechnol* 2009; 27: 1–2.

493. Zeng H, Pesika N, Tian Y, Zhao B, Chen Y, Tirrell M et al. Frictional adhesion of patterned surfaces and implications for gecko and biomimetic systems. *Langmuir* 2009; 25: 7486–95.

494. Waite JH, Hansen DC, and Little KT. The glue protein of ribbed mussels (Geukensia demissa): a natural adhesive with some features of collagen. *J Comp Physiol B* 1989; 159: 517–25.

495. Park TH, Seo SW, Kim JK, and Chang CH. Clinical characteristics of facial keloids treated with surgical excision followed by intra- and postoperative intralesional steroid injections. *Aesthetic Plast Surg* 2011; 36: 169–73.

496. Que SK and Bergstrom KG. What's new in treatment of keloids? New applications for common therapies, new treatments to come. *J Drugs Dermatol* 2011; 10: 548–51.

497. Sidle DM and Kim H. Keloids: prevention and management. *Facial Plast Surg Clin North Am* 2011; 19: 505–15.

498. Viera MH, Caperton CV, and Berman B. Advances in the treatment of keloids. *J Drugs Dermatol* 2011; 10: 468–80.

499. Lee WJ, Kim YO, Choi IK, Rah DK, and Yun CO. Adenovirus-relaxin gene therapy for keloids: implication for reversing pathological fibrosis. *Br J Dermatol* 2011; 165: 673–7.

500. Sakuraba M, Takahashi N, Akahoshi T, Miyasaka Y, and Suzuki K. Use of silicone gel sheets for prevention of keloid scars after median sternotomy. *Surg Today* 2011; 41: 496–9.

501. Cullen B. The role of oxidized regenerated cellulose/collagen in chronic wound repair. Part 2. *Ostomy Wound Manage* 2002; 48: 8–13.

502. Isogai A, Saito T, and Fukuzumi H. TEMPO-oxidized cellulose nanofibers. *Nanoscale* 2011; 3: 71–85.

503. Sundaram CP and Keenan AC. Evolution of hemostatic agents in surgical practice. *Indian J Urol* 2010; 26: 374–8.

504. Gelfand JM, Weinstein R, Porter SB, Neimann AL, Berlin JA, and Margolis DJ. Prevalence and treatment of psoriasis in the United Kingdom: a population-based study. *Arch Dermatol* 2005; 141: 1537–41.

505. Gelfand JM, Stern RS, Nijsten T, Feldman SR, Thomas J, Kist J et al. The prevalence of psoriasis in African Americans: results from a population-based study. *J Am Acad Dermatol* 2005; 52: 23–6.

506. Yip SY. The prevalence of psoriasis in the Mongoloid race. *J Am Acad Dermatol* 1984; 10: 965–8.

507. Ferrándiz C, Pujol RM, García-Patos V, Bordas X, and Smandía JA. Psoriasis of early and late onset: a clinical and epidemiologic study from Spain. *J Am Acad Dermatol* 2002; 46: 867–73.

508. Duffy DL, Spelman LS, and Martin NG. Psoriasis in Australian twins. *J Am Acad Dermatol* 1993; 29: 428–34.

509. Griffiths CEM and Barker JNWN. Psoriasis. In: Burns T, Breathnach S, Cox N, Griffiths C, eds. *Rook's Textbook of Dermatology*. West Sussex, United Kingdom: Wiley-Blackwell, 2010: 1–89.

510. Nestle FO, Kaplan DH, and Barker J. Psoriasis. *N Engl J Med* 2009; 361: 496–509.

511. Ramoz N, Rueda L-A, Bouadjar B, Favre M, and Orth G. A Susceptibility locus for epidermodysplasia verruciformis, an abnormal predisposition to infection with the oncogenic human papillomavirus type 5, maps to chromosome 17qter in a region containing a psoriasis locus. 1999; 112: 259–63.

512. Christophers E. Psoriasis—epidemiology and clinical spectrum. *Clin Exp Dermatol* 2001; 26: 314–20.

513. van de Kerkhof PC and Schalkwijk J. Chapter 9. Psoriasis. In: Bolognia J, Jorizzo JL, Rapini RP, eds. *Dermatology*. London, New York: Mosby, 2003.

514. Cantini F, Niccoli L, Nannini C, Kaloudi O, Bertoni M, and Cassarà E. Psoriatic arthritis: a systematic review. *Int J Rheum Dis* 2010; 13: 300–17.
515. Winterfield L, Menter A, Gordon K, and Gottlieb A. Psoriasis treatment: current and emerging directed therapies. *Ann Rheum Dis* 2005; 64: ii87.
516. Lebwohl MG, van de Kerkhof P. and Psoriasis. In: Lebwohl M, ed. *Treatment of Skin Disease: Comprehensive Therapeutic Strategies*. Philadelphia, PA: Mosby/Elsevier, 2006: xxiv, 723 p.
517. Assmann T and Ruzicka T. New immunosuppressive drugs in dermatology (mycophenolate mofetil, tacrolimus): unapproved uses, dosages, or indications. *Clin Dermatol* 2002; 20: 505–14.
518. Stern R, Zierler S, and Parrish J. Skin carcinoma in patients with psoriasis treated with topical tar and artificial ultraviolet radiation. *Lancet* 1980; 315: 732–5.
519. Stern RS. Psoralen and ultraviolet a light therapy for psoriasis. *N Engl J Med* 2007; 357: 682–90.
520. Brimhall AK, King LN, Licciardone JC, Jacobe H, and Menter A. Safety and efficacy of alefacept, efalizumab, etanercept and infliximab in treating moderate to severe plaque psoriasis: a meta-analysis of randomized controlled trials. *Br J Dermatol* 2008; 159: 274–85.
521. Al-Suwaidan SN and Feldman SR. Clearance is not a realistic expectation of psoriasis treatment. *J Am Acad Dermatol* 2000; 42: 796–802.
522. Krueger G, Koo J, Lebwohl M, Menter A, Stern RS, and Rolstad T. The impact of psoriasis on quality of life: results of a 1998 National Psoriasis Foundation patient-membership survey. *Arch Dermatol* 2001; 137: 280–4.
523. Al Robaee AA. Molecular genetics of Psoriasis (Principles, technology, gene location, genetic polymorphism and gene expression). *Int J Health Sci (Qassim)* 2010; 4: 103–27.
524. O'Rielly DD and Rahman P. Pharmacogenetics of psoriasis. *Pharmacogenomics* 2011; 12: 87–101.
525. Roberson ED and Bowcock AM. Psoriasis genetics: breaking the barrier. *Trends Genet* 2010; 26: 415–23.
526. Ryan C, Menter A, and Warren RB. The latest advances in pharmacogenetics and pharmacogenomics in the treatment of psoriasis. *Mol Diagn Ther* 2010; 14: 81–93.
527. Sobell JM, Kalb RE, and Weinberg JM. Management of moderate to severe plaque psoriasis (part 2): clinical update on T-cell modulators and investigational agents. *J Drugs Dermatol* 2009; 8: 230–8.
528. Wagner EF, Schonthaler HB, Guinea-Viniegra J, and Tschachler E. Psoriasis: what we have learned from mouse models. *Nat Rev Rheumatol* 2010; 6: 704–14.
529. Weger W. An update on the diagnosis and management of psoriatic arthritis. *G Ital Dermatol Venereol* 2011; 146: 1–8.
530. Zippin JH. The genetics of psoriasis. *J Drugs Dermatol* 2009; 8: 414–7.
531. Aydin SZ, Ash Z, Del Galdo F, Marzo-Ortega H, Wakefield RJ, Emery P et al. Optical coherence tomography: a new tool to assess nail disease in psoriasis? *Dermatology* 2011; 222: 311–3.
532. D'Elios MM, Del Prete G, and Amedei A. Targeting IL-23 in human diseases. *Expert Opin Ther Targets* 2010; 14: 759–74.
533. Duffin KC and Krueger GG. Genetic variations in cytokines and cytokine receptors associated with psoriasis found by genome-wide association. *J Invest Dermatol* 2009; 129: 827–33.
534. Elder JT. Genome-wide association scan yields new insights into the immunopathogenesis of psoriasis. *Genes Immun* 2009; 10: 201–9.
535. Guttman-Yassky E, Lowes MA, Fuentes-Duculan J, Zaba LC, Cardinale I, Nograles KE et al. Low expression of the IL-23/Th17 pathway in atopic dermatitis compared to psoriasis. *J Immunol* 2008; 181: 7420–7.
536. Kagami S, Rizzo HL, Lee JJ, Koguchi Y, and Blauvelt A. Circulating Th17, Th22, and Th1 cells are increased in psoriasis. *J Invest Dermatol* 2010; 130: 1373–83.
537. Tokura Y, Mori T, and Hino R. Psoriasis and other Th17-mediated skin diseases. *J UOEH* 2010; 32: 317–28.
538. Strange A, Capon F, Spencer CC, Knight J, Weale ME, Allen MH et al. A genome-wide association study identifies new psoriasis susceptibility loci and an interaction between HLA-C and ERAP1. *Nat Genet* 2010; 42: 985–90.

539. van Kuijk AW, DeGroot J, Koeman RC, Sakkee N, Baeten DL, Gerlag DM et al. Soluble biomarkers of cartilage and bone metabolism in early proof of concept trials in psoriatic arthritis: effects of adalimumab versus placebo. *PloS One* 2010; 5: pii: e12556.

540. Kim ST, Jang DJ, Kim JH, Park JY, Lim JS, Lee SY et al. Topical administration of cyclosporin A in a solid lipid nanoparticle formulation. *Pharmazie* 2009; 64: 510–4.

541. Barker JN. Methotrexate or fumarates: which is the best oral treatment for psoriasis? *Br J Dermatol* 2011; 164: 695.

542. Concannon C, Hennelly DA, Noott S, and Sarker DK. Nanoemulsion encapsulation and in vitro SLN models of delivery for cytotoxic methotrexate. *Curr Drug Discov Technol* 2010; 7: 123–36.

543. Lin YK, Huang ZR, Zhuo RZ, and Fang JY. Combination of calcipotriol and methotrexate in nanostructured lipid carriers for topical delivery. *Int J Nanomedicine* 2010; 5: 117–28.

544. Mukesh U, Kulkarni V, Tushar R, and Murthy RS. Methotrexate loaded self stabilized calcium phosphate nanoparticles: a novel inorganic carrier for intracellular drug delivery. *J Biomed Nanotechnol* 2009; 5: 99–105.

545. Paliwal R, Rai S, and Vyas SP. Lipid drug conjugate (LDC) nanoparticles as autolymphotrophs for oral delivery of methotrexate. *J Biomed Nanotechnol* 2011; 7: 130–1.

546. Benson JM, Sachs CW, Treacy G, Zhou H, Pendley CE, Brodmerkel CM et al. Therapeutic targeting of the IL-12/23 pathways: generation and characterization of ustekinumab. *Nat Biotechnol* 2011; 29: 615–24.

547. Ayroldi E, Bastianelli A, Cannarile L, Petrillo MG, Delfino DV, and Fierabracci A. A pathogenetic approach to autoimmune skin disease therapy: psoriasis and biological drugs, unresolved issues, and future directions. *Curr Pharm Des* 2011; 17(29):3176–90.

548. Raval K, Lofland JH, Waters H, and Piech CT. Disease and treatment burden of psoriasis: examining the impact of biologics. *J Drugs Dermatol* 2011; 10: 189–96.

549. Reich AK, Burden AD, Eaton JN, and Hawkins NS. Efficacy of biologics in the treatment of moderate to severe psoriasis: a network meta-analysis of randomised controlled trials. *Br J Dermatol* 2011; 166(1):179–88.

550. Bak RO and Mikkelsen JG. Regulation of cytokines by small RNAs during skin inflammation. *J Biomed Sci* 2010; 17: 53.

551. Leng RX, Pan HF, Tao JH, and Ye DQ. IL-19, IL-20 and IL-24: potential therapeutic targets for autoimmune diseases. *Expert Opin Ther Targets* 2011; 15: 119–26.

552. Xu N, Brodin P, Wei T, Meisgen F, Eidsmo L, Nagy N et al. MiR-125b, a microRNA downregulated in psoriasis, modulates keratinocyte proliferation by targeting FGFR2. *J Invest Dermatol* 2011; 131: 1521–9.

553. Alam MR, Ming X, Fisher M, Lackey JG, Rajeev KG, Manoharan M et al. Multivalent cyclic RGD conjugates for targeted delivery of small interfering RNA. *Bioconjug Chem* 2011; 22: 1673–81.

554. Al-Hoqail IA. Personalized medicine in psoriasis: concept and applications. *Curr Vasc Pharmacol* 2010; 8: 432–6.

555. Woolf RT and Smith CH. How genetic variation affects patient response and outcome to therapy for psoriasis. *Expert Rev Clin Immunol* 2010; 6: 957–66.

556. Warren RB, Smith RL, Campalani E, Eyre S, Smith CH, Barker JN et al. Genetic variation in efflux transporters influences outcome to methotrexate therapy in patients with psoriasis. *J Invest Dermatol* 2008; 128: 1925–9.

557. Armstrong AW, Voyles SV, Armstrong EJ, Fuller EN, and Rutledge JC. A tale of two plaques: convergent mechanisms of T-cell-mediated inflammation in psoriasis and atherosclerosis. *Exp Dermatol* 2011; 20: 544–9.

558. Egesi A, Sun G, Khachemoune A, and Rashid RM. Statins in skin: research and rediscovery, from psoriasis to sclerosis. *J Drugs Dermatol* 2010; 9: 921–7.

559. Eyerich S, Onken AT, Weidinger S, Franke A, Nasorri F, Pennino D et al. Mutual antagonism of T cells causing psoriasis and atopic eczema. *N Engl J Med* 2011; 365: 231–8.

560. Nakajima H, Nakajima K, Tarutani M, Morishige R, and Sano S. Kinetics of circulating Th17 cytokines and adipokines in psoriasis patients. *Arch Dermatol Res* 2011; 303: 451–5.

561. Shibata S, Tada Y, Kanda N, Nashiro K, Kamata M, Karakawa M et al. Possible roles of IL-27 in the pathogenesis of psoriasis. *J Invest Dermatol* 2010; 130: 1034–9.

562. Ward NL, Loyd CM, Wolfram JA, Diaconu D, Michaels CM, and McCormick TS. Depletion of antigen-presenting cells by clodronate liposomes reverses the psoriatic skin phenotype in KC-Tie2 mice. *Br J Dermatol* 2011; 164: 750–8.

563. Akyol M and Ozcelik S. Non-acne dermatologic indications for systemic isotretinoin. *Am J Clin Dermatol* 2005; 6: 175–84.

564. Arbiser JL. Fumarate esters as angiogenesis inhibitors: key to action in psoriasis? *J Invest Dermatol* 2011; 131: 1189–91.

565. Bovenschen HJ, Langewouters AM, and van de Kerkhof PC. Dimethylfumarate for psoriasis: pronounced effects on lesional T-cell subsets, epidermal proliferation and differentiation, but not on natural killer T cells in immunohistochemical study. *Am J Clin Dermato* 2010; 11: 343–50.

566. Chen AF and Kirsner RS. Mechanisms of drug action: the potential of dimethylfumarate for the treatment of neoplasms. *J Invest Dermatol* 2011; 131: 1181.

567. Fallah Arani S, Neumann H, Hop WC, and Thio HB. Fumarates vs. methotrexate in moderate to severe chronic plaque psoriasis: a multicentre prospective randomized controlled clinical trial. *Br J Dermatol* 2011; 164: 855–61.

568. Jennings L and Murphy G. Squamous cell carcinoma as a complication of fumaric acid ester immunosuppression. *J Eur Acad Dermatol Venereol* 2009; 23: 1451.

569. Meili-Butz S, Niermann T, Fasler-Kan E, Barbosa V, Butz N, John D et al. Dimethyl fumarate, a small molecule drug for psoriasis, inhibits Nuclear Factor-kappaB and reduces myocardial infarct size in rats. *Eur J Pharmacol* 2008; 586: 251–8.

570. Meissner M, Doll M, Hrgovic I, Reichenbach G, Konig V, Hailemariam-Jahn T et al. Suppression of VEGFR2 expression in human endothelial cells by dimethylfumarate treatment: evidence for anti-angiogenic action. *J Invest Dermatol* 2011; 131: 1356–64.

571. Mrowietz U, Rostami-Yazdi M, Neureither M, and Reich K. [15 years of fumaderm: fumaric acid esters for the systemic treatment of moderately severe and severe psoriasis vulgaris]. *J Ger Soc Dermatol* 2009; 7 Suppl 2: S3–16.

572. Wain EM, Darling MI, Pleass RD, Barker JN, and Smith CH. Treatment of severe, recalcitrant, chronic plaque psoriasis with fumaric acid esters: a prospective study. *Br J Dermatol* 2010; 162: 427–34.

573. Chen P, Wu QS, Ding YP, Chu M, Huang ZM, and Hu W. A controlled release system of titanocene dichloride by electrospun fiber and its antitumor activity in vitro. *Eur J Pharm Biopharm* 2010; 76: 413–20.

574. Berrios RL and Arbiser JL. Novel antiangiogenic agents in dermatology. *Arch Biochem Biophys* 2011; 508: 222–6.

575. Al-Qallaf B and Das DB. Optimizing microneedle arrays to increase skin permeability for transdermal drug delivery. *Ann N Y Acad Sci* 2009; 1161: 83–94.

576. Badran MM, Kuntsche J, and Fahr A. Skin penetration enhancement by a microneedle device (Dermaroller) in vitro: dependency on needle size and applied formulation. *Eur J Pharm Sci* 2009; 36: 511–23.

577. Chen H, Zhu H, Zheng J, Mou D, Wan J, Zhang J et al. Iontophoresis-driven penetration of nanovesicles through microneedle-induced skin microchannels for enhancing transdermal delivery of insulin. *J Control Release* 2009; 139: 63–72.

578. Li G, Badkar A, Nema S, Kolli CS, and Banga AK. In vitro transdermal delivery of therapeutic antibodies using maltose microneedles. *Int J Pharm* 2009; 368: 109–15.

579. Quan FS, Kim YC, Compans RW, Prausnitz MR, and Kang SM. Dose sparing enabled by skin immunization with influenza virus-like particle vaccine using microneedles. *J Control Release* 2010; 147: 326–32.

580. Sivamani RK, Stoeber B, Wu GC, Zhai H, Liepmann D, and Maibach H. Clinical microneedle injection of methyl nicotinate: stratum corneum penetration. *Skin Res Technol* 2005; 11: 152–6.

581. Zhou CP, Liu YL, Wang HL, Zhang PX, and Zhang JL. Transdermal delivery of insulin using microneedle rollers in vivo. *Int J Pharm* 2010; 392: 127–33.

582. Wolff K, Johnson RA, and Fitzpatrick TB. *Fitzpatrick's Color Atlas and Synopsis of Clinical Dermatology*. New York, Toronto: McGraw-Hill Medical, 2009.

583. Bhatt V, Evans M, and Malins TJ. Squamous cell carcinoma arising in the lining of an epidermoid cyst within the sublingual gland: a case report. *Br J Oral Maxillofac Surg* 2008; 46: 683–5.

584. Leppard B and Bussey HJR. Epidermoid cysts, polyposis coli and Gardner's syndrome. *Br J Surg* 1975; 62: 387–93.

585. Garcia-Rojo B, Garcia-Solano J, Sanchez-Sanchez C, Montalban-Romero S, Martinez-Parra D, and Perez-Guillermo M. On the utility of fine-needle aspiration in the diagnosis of primary scalp lesions. *Diagn Cytopathol* 2001; 24: 104–11.

586. Jehle KS, Shakir AJ, and Sayegh ME. Squamous cell carcinoma arising in an epidermoid cyst. *Br J Hosp Med (Lond)* 2007; 68: 446.

587. Juhn E and Khachemoune A. Gardner syndrome: skin manifestations, differential diagnosis and management. *Am J Clin Dermatol* 2010; 11: 117–22.

588. Egbert BM, Price NM, and Segal RJ. Steatocystoma multiplex. Report of a florid case and a review. *Arch Dermatol* 1979; 115: 334–5.

589. Feinstein A, Friedman J, and Schewach-Millet M. Pachyonychia congenita. *J Am Acad Dermatol* 1988; 19: 705–11.

590. Kah JC, Olivo M, Chow TH, Song KS, Koh KZ, Mhaisalkar S et al. Control of optical contrast using gold nanoshells for optical coherence tomography imaging of mouse xenograft tumor model in vivo. *J Biomed Opt* 2009; 14: 054015.

591. Manca ML, Loy G, Zaru M, Fadda AM, and Antimisiaris SG. Release of rifampicin from chitosan, PLGA and chitosan-coated PLGA microparticles. *Colloids Surf B Biointerfaces* 2008; 67: 166–70.

592. Friedman A, Blecher K, Sanchez D, Tuckman-Vernon C, Gialanella P, Friedman JM et al. Susceptibility of Gram-positive and -negative bacteria to novel nitric oxide-releasing nanoparticle technology. *Virulence* 2011; 2: 217–21.

593. Han G, Martinez LR, Mihu MR, Friedman AJ, Friedman JM, and Nosanchuk JD. Nitric oxide releasing nanoparticles are therapeutic for Staphylococcus aureus abscesses in a murine model of infection. *PloS One* 2009; 4: e7804.

594. Brenner S. Treatment of pilar cyst with solcoderm solution. *J Am Acad Dermatol* 1986; 14: 145.

595. Ronnen M, Suster S, and Klin B. Treatment of epidermal cysts with Solcoderm (a copper ion and acid solution). *Clin Exp Dermatol* 1993; 18: 500–3.

596. Fadel M, Salah M, Samy N, and Mona S. Liposomal methylene blue hydrogel for selective photodynamic therapy of acne vulgaris. *J Drugs Dermatol* 2009; 8: 983–90.

597. Horfelt C, Stenquist B, Halldin CB, Ericson MB, and Wennberg AM. Single low-dose red light is as efficacious as methyl-aminolevulinate—photodynamic therapy for treatment of acne: clinical assessment and fluorescence monitoring. *Acta Derm Venereol* 2009; 89: 372–8.

598. Kim SK, Do JE, Kang HY, Lee ES, and Kim YC. Combination of topical 5-aminolevulinic acid-photodynamic therapy with carbon dioxide laser for sebaceous hyperplasia. *J Am Acad Dermatol* 2007; 56: 523–4.

599. Kim YJ, Kang HY, Lee ES, and Kim YC. Treatment of Fordyce spots with 5-aminolaevulinic acid-photodynamic therapy. *Br J Dermatol* 2007; 156: 399–400.

600. Kosaka S, Miyoshi N, Akilov OE, Hasan T, and Kawana S. Targeting of sebaceous glands by delta-aminolevulinic acid-based photodynamic therapy: an in vivo study. *Lasers Surg Med* 2011; 43: 376–81.

601. Na JI, Kim SY, Kim JH, Youn SW, Huh CH, and Park KC. Indole-3-acetic acid: a potential new photosensitizer for photodynamic therapy of acne vulgaris. *Lasers Surg Med* 2011; 43: 200–5.

602. Osswald F, Gloor M, Steinbacher M, and Franke M. [Inhibition of propionibacteria in the sebaceous gland duct using photochemotherapy and 8-methoxypsoralen]. *Zeitschrift fur Hautkrankheiten* 1980; 55: 30–4.

603. Perrett CM, McGregor J, Barlow RJ, Karran P, Proby C, and Harwood CA. Topical photodynamic therapy with methyl aminolevulinate to treat sebaceous hyperplasia in an organ transplant recipient. *Arch Dermatol* 2006; 142: 781–2.

604. Donnelly RF. Re: Microneedle-mediated intradermal delivery of 5-aminolevulinic acid. *J Control Release* 2008; 129: 153.

605. Lee K, Lee CY, and Jung H. Dissolving microneedles for transdermal drug administration prepared by stepwise controlled drawing of maltose. *Biomaterials* 2011; 32: 3134–40.

606. Hafner C and Vogt T. Seborrheic keratosis. *J Dtsch Dermatol Ges* 2008; 6: 664–77.

607. Yeatman JM, Kilkenny M, and Marks R. The prevalence of seborrhoeic keratoses in an Australian population: does exposure to sunlight play a part in their frequency? *Br J Dermatol* 1997; 137: 411–4.

608. Kennedy C, Bajdik CD, Willemze R, de Gruijl FR, and Bouwes Bavinck JN. The influence of painful sunburns and lifetime sun exposure on the risk of actinic keratoses, seborrheic warts, melanocytic nevi, atypical nevi, and skin cancer. *J Investig Dermatol* 2003; 120: 1087–93.

609. Kitamura Y, Nakamura H, Hirota S, Adachi S, Ozaki K, and Asada H. Clonal nature of seborrheic keratosis demonstrated by using the polymorphism of the human androgen receptor locus as a marker. *J Invest Dermatol* 2001; 116: 506–10.

610. Kwon OS, Hwang EJ, Bae JH, Park HE, Lee JC, Youn JI et al. Seborrheic keratosis in the Korean males: causative role of sunlight. *Photodermatol Photoimmunol Photomed* 2003; 19: 73–80.

611. Logie A, Dunois-Larde C, Rosty C, Levrel O, Blanche M, Ribeiro A et al. Activating mutations of the tyrosine kinase receptor FGFR3 are associated with benign skin tumors in mice and humans. *Hum Mol Genet* 2005; 14: 1153–60.

612. Hafner C, van Oers JMM, Hartmann A, Landthaler M, Stoehr R, Blaszyk H et al. High frequency of FGFR3 mutations in adenoid seborrheic keratoses. *J Invest Dermatol* 2006; 126: 2404–7.

613. Hafner C, Hartmann A, Real FX, Hofstaedter F, Landthaler M, and Vogt T. Spectrum of FGFR3 mutations in multiple intraindividual seborrheic keratoses. *J Invest Dermatol* 2007; 127: 1883–5.

614. Hafner C, Hartmann A, van Oers JM, Stoehr R, Zwarthoff EC, Hofstaedter F et al. FGFR3 mutations in seborrheic keratoses are already present in flat lesions and associated with age and localization. *Mod Pathol* 2007; 20: 895–903.

615. Imokawa G, Manaka I, Kadono S, Kawashima M, and Kobayashi T. The mechanism of hyperpigmentation in seborrhoeic keratosis involves the high expression of endothelin-converting enzyme-1 alpha and TNF-alpha, which stimulate secretion of endothelin 1. *Br J Dermatol* 2001; 145: 895–903.

616. Elgart GW. Seborrheic keratoses, solar lentigines, and lichenoid keratoses. Dermatoscopic features and correlation to histology and clinical signs. *Dermatol Clin* 2001; 19: 347–57.

617. Hafner C, Landthaler M, Mentzel T, and Vogt T. FGFR3 and PIK3CA mutations in stucco keratosis and dermatosis papulosa nigra. *Br J Dermatol* 2010; 162: 508–12.

618. Hafner C, Lopez-Knowles E, Luis NM, Toll A, Baselga E, Fernandez-Casado A et al. Oncogenic PIK3CA mutations occur in epidermal nevi and seborrheic keratoses with a characteristic mutation pattern. *Proc Natl Acad Sci USA* 2007; 104: 13450–4.

619. Klaus MV, Wehr RF, Rogers RS, 3rd, Russell TJ, and Krochmal L. Evaluation of ammonium lactate in the treatment of seborrheic keratoses. *J Am Acad Dermatol* 1990; 22: 199–203.

620. Chun EY, Lee JB, and Lee KH. Focal trichloroacetic acid peel method for benign pigmented lesions in dark-skinned patients. *Dermatol Surg* 2004; 30: 512–6; discussion 6.

621. Herron MD, Bowen AR, and Krueger GG. Seborrheic keratoses: a study comparing the standard cryosurgery with topical calcipotriene, topical tazarotene, and topical imiquimod. *Int J Dermatol* 2004; 43: 300–2.

622. Mohs FE. Seborrheic keratoses. Scarless removal by curettage and oxidized cellulose. *JAMA* 1970; 212: 1956–8.

623. Culbertson GR. 532-nm diode laser treatment of seborrheic keratoses with color enhancement. *Dermatol Surg* 2008; 34: 525–8; discussion 8.

624. Polder KD, Landau JM, Vergilis-Kalner IJ, Goldberg LH, Friedman PM, and Bruce S. Laser eradication of pigmented lesions: a review. *Dermatol Surg* 2011; 37: 572–95.

625. Meulener MC, Ayli EE, Elenitsas R, and Seykora JT. Decreased Srcasm expression in hyperproliferative cutaneous lesions. *J Cutan Pathol* 2009; 36: 291–5.

626. Burkhart CG and Burkhart CN. Use of a keratolytic agent with occlusion for topical treatment of hyperkeratotic seborrheic keratoses. *Skinmed* 2008; 7: 15–8.

627. Mitsuhashi Y, Kawaguchi M, Hozumi Y, and Kondo S. Topical vitamin D3 is effective in treating senile warts possibly by inducing apoptosis. *J Dermatol* 2005; 32: 420–3.
628. Fein H, Maytin EV, Mutasim DF, and Bailin PL. Topical protease therapy as a novel method of epidermal ablation: preliminary report. *Dermatol Surg* 2005; 31: 139–47; discussion 47–8.
629. Jeng J, Lin MF, Cheng FY, Yeh CS, and Shiea J. Using high-concentration trypsin-immobilized magnetic nanoparticles for rapid in situ protein digestion at elevated temperature. *Rapid Commun Mass Spectrom* 2007; 21: 3060–8.
630. Jimenez N, Galan J, Vallet A, Egea MA, and Garcia ML. Methyl trypsin loaded poly(D,L-lactide-coglycolide) nanoparticles for contact lens care. *J Pharm Sci* 2010; 99: 1414–26.
631. Feuerman EJ, Katzenelson V, and Halevy S. Solcoderm in the treatment of solar and seborrheic keratoses. *Dermatologica* 1984; 168 Suppl 1: 33–5.
632. Hettich R. Solcoderm as a tool for the plastic surgeon. The treatment of verrucae. *Dermatologica* 1984; 168 Suppl 1: 36–42.
633. Burkhart CG. Irritant-induced enhancement of induction and elicitation of sensitization for the treatment of seborrheic keratoses. *Int J Dermatol* 2006; 45: 1240–2.
634. Tsuji T and Morita A. Giant seborrheic keratosis on the frontal scalp treated with topical fluorouracil. *J Dermatol* 1995; 22: 74–5.
635. Schwengle LE and Rampen FH. Eruptive seborrheic keratoses associated with erythrodermic pityriasis rubra pilaris. Possible role of retinoid therapy. *Acta Dermato-Venereologica* 1988; 68: 443–5.
636. Du B, Yan Y, Li Y, Wang S, and Zhang Z. Preparation and passive target of 5-fluorouracil solid lipid nanoparticles. *Pharm Dev Technol* 2010; 15: 346–53.
637. Yassin AE, Anwer MK, Mowafy HA, El-Bagory IM, Bayomi MA, and Alsarra IA. Optimization of 5-flurouracil solid-lipid nanoparticles: a preliminary study to treat colon cancer. *Int J Med Sci* 2010; 7: 398–408.
638. Bolke E, Gerber PA, Peiper M, Knoefel WT, Cohnen M, Matuschek C et al. Leser-Trelat sign presenting in a patient with ovarian cancer: a case report. *J Med Case Rep* 2009; 3: 8583.
639. Constantinou C, Dancea H, and Meade P. The sign of Leser-Trelat in colorectal adenocarcinoma. *Am Surg* 2010; 76: 340–1.
640. da Costa Franca AF, Siqueira NS, Carvalheira JB, Saad MJ, and Souza EM. Acanthosis nigricans, tripe palms and the sign of Leser-Trelat in a patient with a benign hepatic neoplasia. *J Eur Acad Dermatol Venereol* 2007; 21: 846–8.
641. da Rosa AC, Pinto GM, Bortoluzzi JS, Duquia RP, and de Almeida HL, Jr. Three simultaneous paraneoplastic manifestations (ichthyosis acquisita, Bazex syndrome, and Leser-Trelat sign) with prostate adenocarcinoma. *J Am Acad Dermatol* 2009; 61: 538–40.
642. Dasanu CA and Alexandrescu DT. Bilateral Leser-Trelat sign mirroring lung adenocarcinoma with early metastases to the contralateral lung. *South Med J* 2009; 102: 216–8.
643. Fetil E, Ozkan S, Gurler N, Kusku E, Arda F, and Gunes AT. Recurrent Leser-Trelat sign associated with two malignancies. *Dermatology* 2002; 204: 254–5.
644. Fink AM, Filz D, Krajnik G, Jurecka W, Ludwig H, and Steiner A. Seborrhoeic keratoses in patients with internal malignancies: a case-control study with prospective accrual of patients. *J Eur Acad Dermatol Venereol* 2009; 23: 1316–9.
645. Kameya S, Noda A, Isobe E, and Watanabe T. The sign of Leser-Trelat associated with carcinoma of the stomach. *Am J Gastroenterol* 1988; 83: 664–6.
646. Kilickap S and Yalcin B. Images in clinical medicine. The sign of Leser-Trelat. *N Engl J Med* 2007; 356: 2184.
647. Kluger N and Guillot B. Sign of Leser-Trelat with an adenocarcinoma of the prostate: a case report. *Cases J* 2009; 2: 8868.
648. Li M, Yang LJ, Zhu XH, Zhang YS, Sun H, Jiang PD et al. The Leser-Trelat sign is associated with nasopharyngeal carcinoma: case report and review of cases reported in China. *Clin Exp Dermatol* 2009; 34: 52–4.
649. Moore RL and Devere TS. Epidermal manifestations of internal malignancy. *Dermatol Clin* 2008; 26: 17–29, vii.

650. Ponti G, Luppi G, Losi L, Giannetti A, and Seidenari S. Leser-Trelat syndrome in patients affected by six multiple metachronous primitive cancers. *J Hematol Oncol* 2010; 3: 2.

651. Rubegni P, Mandato F, Mourmouras V, Danielli R, and Fimiani M. False Leser-Trelat sign. *Int J Dermatol* 2009; 48: 912–3.

652. Safa G and Darrieux L. Leser-Trelat sign without internal malignancy. *Case Rep Oncol* 2011; 4: 175–7.

653. Safai B, Grant JM, and Good RA. Cutaneous manifestation of internal malignancies (II): the sign of Leser-Trelat. *Int J Dermatol* 1978; 17: 494–5.

654. Siedek V, Schuh T, and Wollenberg A. Leser-Trelat sign in metastasized malignant melanoma. *Eur Arch Otorhinolaryngol* 2009; 266: 297–9.

655. Yavasoglu I, Kadikoylu G, and Bolaman Z. The Leser-Trelat sign is a associated with acute myeloid leukemia. *Ann Hematol* 2011; 90: 363.

656. Chren M-M. Costs of therapy for dermatophyte infections. *J Am Acad Dermatol* 1994; 31: S103–6.

657. Arenas-Guzman R, Tosti A, Hay R, and Haneke E. Pharmacoeconomics: an aid to better decision-making. *J Eur Acad Dermatol Venereol* 2005; 19: 34–9.

658. Sobera JO and Elewski BE. Chapter 76. Fungal diseases. In: Bolognia J, Jorizzo JL, Rapini RP, eds. *Dermatology*. London, New York: Mosby, 2003.

659. Adams BB. *Sports Dermatology*. New York: Springer, 2006.

660. Seebacher C, Bouchara JP, and Mignon B. Updates on the epidemiology of dermatophyte infections. *Mycopathologia* 2008; 166: 335–52.

661. Svejgaard EL. Epidemiology of dermatophytes in Europe. *Int J Dermatol* 1995; 34: 525–8.

662. Vermout S, Tabart J, Baldo A, Mathy A, Losson B, and Mignon B. Pathogenesis of dermatophytosis. *Mycopathologia* 2008; 166: 267–75.

663. Fuller LC. Tinea pedis and skin dermatophytosis. In: Lebwohl M, ed. *Treatment of Skin Disease: Comprehensive Therapeutic Strategies*. Philadelphia, PA: Mosby/Elsevier, 2006: xxiv, 723 p.

664. Mahendra R, Yadav A, and Gade A. Silver nanoparticles as a new generation of antimicrobials. *Biotechnol Adv* 2009; 27: 76–83.

665. Pal S, Tak YK, and Song JM. Does the antibacterial activity of silver nanoparticles depend on the shape of the nanoparticles? A study of the Gram-negative bacterium *Escherechia coli*. *Appl Environ Microbiol* 2007; 27: 1712–20.

666. Sanpui P, Murugadoss A, Prasad PV, Ghosh SS, and Chattopadhyay A. The antibacterial properties of a novel chitosan-Ag-nanoparticle composite. *Int J Food Microbiol* 2008; 124: 142–6.

667. Ruparelia JP, Chatterjee AK, Duttagupta SP, and Mukherji S. Strain specificity in antimicrobial activity of silver and copper nanoparticles. *Acta Biomater* 2008; 4: 707–16.

668. Kim KJ, Sung WS, Suh BK, Moon SK, Choi JS, Kim J et al. Antifungal activity and mode of action of silver nano-particles on Candida albicans. *Biometals* 2009; 22: 235–42.

669. Paulo CS, Vidal M, and Ferreira LS. Antifungal nanoparticles and surfaces. *Biomacromolecules* 2010; 11: 2810–7.

670. Panacek A, Kolar M, Vecerova R, Prucek R, Soukupova J, Krystof V et al. Antifungal activity of silver nanoparticles against Candida spp. *Biomaterials* 2009; 30: 6333–40.

671. Esteban-Tejeda L, Malpartida F, Esteban-Cubillo A, Pecharroman C, and Moya JS. The antibacterial and antifungal activity of a soda-lime glass containing silver nanoparticles. *Nanotechnology* 2009; 20: 085103.

672. Banergee M, Mallick S, Paul A, Chattopadhyay A, and Ghosh S. Heightened reactive oxygen species generation in the antimicrobial activity of three component iodinated chitosan-silver nanoparticle composite. *Langmuir* 2010; 26: 5901–8.

673. Ma Y, Zhou T, and Zhao C. Preparation of chitosan-nylon-6 blended membranes containing silver ions as antibacterial materials. *Carbohydr Res* 2008; 343: 230–7.

674. Qi L, Xu Z, Jiang X, Hu C, and Zou X. Preparation and antibacterial activity of chitosan nanoparticles. *Carbohydr Res* 2005; 339: 2693–700.

675. Alburquenque C, Bucarey SA, Neira-Carrillo A, Urzua B, Hermosilla G, and Tapia CV. Antifungal activity of low molecular weight chitosan against clinical isolates of Candida spp. *Med Mycol* 2010; 48: 1018–23.

676. Albasarah YY, Somavarapu S, Stapleton P, and Taylor KMG. Chitosan-coated antifungal formulations for nebulisation. *J Pharm Pharmacol* 2010; 62: 821–8.
677. Li RC, Guo ZY, and Jiang PA. Synthesis, characterization, and antifungal activity of novel quaternary chitosan derivatives. *Carbohydr Res* 2010; 345: 1896–900.
678. Kulikov SN, Tiurin IuA, Fassakhov RS, and Varlamov VP. [Antibacterial and antimycotic activity of chitosan: mechanisms of action and role of the structure]. *Zh Mikrobiol Epidemiol Immunobiol* 2009: 91–7.
679. Kwak S, Kim SH, and Kim SS. Hybrid organic/inorganic reverse osmosis (RO) membrane for bactericidal anti-fouling. 1. Preparation and characterization of TiO nanoparticle self-assembled aromatic polyamide thin-film-composite (TFC) membrane. *Environ Sci Technol* 2001; 35: 2388–94.
680. Kim SH, Kwak S, Sohn B, and Park TH. Design of TiO_2 nanoparticle self-assembled aromatic polyamide thin-film-composite (TFC) membrane as an approach to solve biofouling problem. *J Membrane Sci* 2003; 211: 157–65.
681. Martinez-Gutierrez F, Olive PL, Banuelos A, Orrantia E, Nino N, Sanchez EM et al. Synthesis, characterization, and evaluation of antimicrobial and cytotoxic effect of silver and titanium nanoparticles. *Nanomedicine* 2010; 6: 681–8.
682. Karthikeyan R, Amaechi BT, Rawls HR, and Lee VA. Antimicrobial activity of nanoemulsion on cariogenic Streptococcus mutans. *Arch Oral Biol* 2011; 56:437–45.
683. Ciotti S, Eisma R, Ma L, and Baker JR. In-vitro skin penetration of novel antimicrobial nanoemulsion formulations containing antifungal agents. *J Invest Dermatol* 2009; 129: S78.
684. Fothergill AW, McCarthy DI, Sutcliffe JA, and Rinaldi MG. Antifungal activity of NB-002 a topical nanoemulsion, against rare fungal pathogens of onychomycosis. *J Am Acad Dermatol* 2009; 60: AB117.
685. Jones T, Ijzerman M, and Flack M. A randomized, double-blind, vehicle-controlled trial of a novel topical antifungal nanoemulsion (NB-002) in subjects with distal subungual onychomycosis. *J Am Acad Dermatol* 2009; 60: AB102.
686. McCarthy A, Pannu J, Ciotti S, Hamouda T, Sutcliffe J, and Baker JR. Antimicrobial activity of nanoemulsion against clinical isolates from cystic fibrosis patients. Pediatr Pulm 2009: S32: 339.
687. Hamouda T, Flack M, and Baker J. Development of a novel topically applied antifungal agent (NB-002) based on nanoemulsion technology. *J Am Acad Dermatol* 2008; 58: AB90.
688. Teixeira PC, Leite GM, Domingues RJ, Silva J, Gibbs PA, and Ferreira JP. Antimicrobial effects of a microemulsion and a nanoemulsion on enteric and other pathogens and biofilms. *Int J Food Microbiol* 2007; 118: 15–9.
689. Hamouda T, Myc A, Donovan B, Shih AY, Reuter JD, and Baker JR. A novel surfactant nanoemulsion with a unique non-irritant topical antimicrobial activity against bacteria, enveloped viruses and fungi. *Microbiol Res* 2001; 156: 1–7.
690. Hamouda T, Hayes MM, Cao ZY, Tonda R, Johnson K, Wright DC et al. A novel surfactant nanoemulsion with broad-spectrum sporicidal activity against Bacillus species. *J Infect Dis* 1999; 180: 1939–49.
691. Pannu J, McCarthy A, Martin A, Hamouda T, Ciotti S, Fothergill A et al. NB-002, a novel nanoemulsion with broad antifungal activity against dermatophytes, other filamentous fungi, and Candida albicans. *Antimicrob Agents Chemother* 2009; 53: 3273–9.
692. Pannu J, Sutcliffe J, Ma LF, and Ciotti S. Antifungal activity and mechanism of action of NB-002, a novel topical antifungal, against the major pathogens of onychomycosis. *J Am Acad Dermatol* 2009; 60: AB114.
693. Jones T, Flack M, Ijzerman M, and Baker J. Safety, tolerance, and pharmacokinetics of topical nanoemulsion (NB-002) for the treatment of onychomycosis. *J Am Acad Dermatol* 2008; 58: AB83.
694. Ijzerman M, Baker J, Flack M, and Robinson P. Efficacy of topical nanoemulsion (NB-002) for the treatment of distal subungual onychomycosis: a randomized, double-blind, vehicle-controlled trial. *J Am Acad Dermatol* 2010; 62: AB76.

695. Shim YH, Kim YC, Lee HJ, Bougard F, Dubois P, Choi KC et al. Amphotericin B aggregation inhibition with novel nanoparticles prepared with poly(epsilon-caprolactone)/poly(n,n-dimethylamino-2-ethyl methacrylate) diblock copolymer. *J Microbiol Biotechnol* 2011; 21: 28–36.
696. Sheikh S, Ali SM, Ahmad MU, Ahmad A, Mushtaq M, Paithankar M et al. Nanosomal Amphotericin B is an efficacious alternative to Ambisome for fungal therapy. *Int J Pharm* 2010; 397: 103–8.
697. Burgess BL, Cavigiolio G, Fannucchi MV, Illek B, Forte TM, and Oda MN. A phospholipid-apolipoprotein A-I nanoparticle containing amphotericin B as a drug delivery platform with cell membrane protective properties. *Int J Pharm* 2010; 399: 148–55.
698. Shao K, Huang RQ, Li JF, Han LA, Ye LY, Lou JN et al. Angiopep-2 modified PE-PEG based polymeric micelles for amphotericin B delivery targeted to the brain. *J Control Release* 2010; 147: 118–26.
699. Jung SH, Lim DH, Lee JE, Jeong KS, Seong H, and Shin BC. Amphotericin B-entrapping lipid nanoparticles and their in vitro and in vivo characteristics. *Eur J Pharm Sci* 2009; 37: 313–20.
700. Amaral AC, Bocca AL, Ribeiro AM, Nunes J, Peixoto DL, Simioni AR et al. Amphotericin B in poly(lactic-co-glycolic acid) (PLGA) and dimercaptosuccinic acid (DMSA) nanoparticles against paracoccidioidomycosis. *J Antimicrob Chemother* 2009; 63: 526–33.
701. Fukui H, Koike T, Saheki A, Sonoke S, Tomii Y, and Seki J. Evaluation of the efficacy and toxicity of amphotericin B incorporated in lipid nano-sphere (LNS (R)). *Int J Pharm* 2003; 263: 51–60.
702. Ritter J. Amphotericin B and its lipid formulations. *Mycoses* 2002; 45: 34–8.
703. Bekersky I, Boswell GW, Hiles R, Fielding RM, Buell D, and Walsh TJ. Safety, toxicokinetics and tissue distribution of long-term intravenous liposomal amphotericin B (AmBisome): a 91-day study in rats. *Pharm Res* 2000; 17: 1494–502.
704. Bekersky I, Boswell GW, Hiles R, Fielding RM, Buell D, and Walsh TJ. Safety and toxicokinetics of intravenous liposomal amphotericin B (AmBisome) in beagle dogs. *Pharm Res* 1999; 16: 1694–701.
705. Johnson EM, Ojwang JO, Szekely A, Wallace TL, and Warnock DW. Comparison of in vitro antifungal activities of free and liposome-encapsulated nystatin with those of four amphotericin B formulations. *Antimicrob Agents Chemother* 1998; 42: 1412–6.
706. Gulati M, Bajad S, Singh S, Ferdous AJ, and Singh M. Development of liposomal amphotericin B formulation. *J Microencapsul* 1998; 15: 137–51.
707. Hiemenz JW and Walsh TJ. Lipid formulations of amphotericin B. *J Lipos Res* 1998; 8: 443–67.
708. Hillery AM. Supramolecular lipidic drug delivery systems: From laboratory to clinic—a review of the recently introduced commercial liposomal and lipid-based formulations of Amphotericin B. *Adv Drug Deliver Rev* 1997; 24: 345–63.
709. Anstey NM, Stewart LM, Packard M, Graney WF, and Bartlett JA. Open-label titration study of the safety of RMP-7 in patients with the acquired immune deficiency syndrome. *Int J Antimicrob Agents* 1996; 6: 183–7.
710. Joly V, Farinotti R, Saintjulien L, Cheron M, Carbon C, and Yeni P. In-vitro renal toxicity and in-vivo therapeutic efficacy in experimental murine cryptococcosis of amphotericin-B (Fungizone) associated with Intralipid. *Antimicrob Agents Chemother* 1994; 38: 177–83.
711. Bekersky I, Fielding RM, Buell D, and Lawrence I. Lipid-based amphotericin B formulations: from animals to man. *Pharm Sci Technolo Today* 1999; 2: 230–6.
712. Bhalekar MR, Pokharkar V, Madgulkar A, and Patil N. Preparation and evaluation of miconazole nitrate-loaded solid lipid nanoparticles for topical delivery. *AAPS PharmSciTech* 2009; 10: 289–96.
713. Gupta M, Goyal AK, Paliwal SR, Paliwal R, Mishra N, Vaidya B et al. Development and characterization of effective topical liposomal system for localized treatment of cutaneous candidiasis. *J Liposome Res* 2010; 20: 341–50.
714. Naeff R. Feasibility of topical liposome drugs produced on an industrial scale. *Adv Drug Deliver Rev* 1996; 18: 343–7.
715. Korting HC, Klovekorn W, Klovekorn G, Orfanos C, Heilgemeir E, Hornstein M et al. Comparative efficacy and tolerability of econazole liposomal gel 1%, branded econazole conventional cream 1% and generic clotrimazole cream 1% in tinea pedis. *Clin Drug Invest* 1997; 14: 286–93.

716. Korting HC, Patzak U, Schaller M, and Maibach HI. A model of human cutaneous candidosis based on reconstructed human epidermis for the light and electron microscopic study of pathogenesis and treatment. *J Infection* 1998; 36: 259–67.

717. Qi XR, Liu MH, Liu HY, Maitani Y, and Nagai T. Topical econazole delivery using liposomal gel. *Stp Pharma Sci* 2003; 13: 241–5.

718. Wasan EK, Cogswell S, Berger S, Waterhouse D, and Bally MB. A parenteral econazole formulation using a novel micelle-to-liposome transfer method: in vitro characterization and tumor growth delay in a breast cancer xenograft model. *Pharm Res* 2006; 23: 2575–85.

719. Adis International Limited. Nystatin—liposomal. AR 121, Nyotran. *Drugs R D* 1999; 1: 181–3.

720. Groll AH, Petraitis V, Petraitiene R, Field-Ridley A, Calendario M, Bacher J et al. Safety and efficacy of multilamellar liposomal nystatin against disseminated candidiasis in persistently neutropenic rabbits. *Antimicrob Agents Chemother* 1999; 43: 2463–7.

721. Moribe K and Maruyama K. Pharmaceutical design of the liposomal antimicrobial agents for infectious disease. *Curr Pharm Des* 2002; 8: 441–54.

722. Wallace TL, Paetznick V, Cossum PA, LopezBerestein G, Rex JH, and Anaissie E. Activity of liposomal nystatin against disseminated Aspergillus fumigatus infection in neutropenic mice. *Antimicrob Agents Chemother* 1997; 41: 2238–43.

723. 4. Wasan KM, Ramaswamy M, Cassidy SM, Kazemi M, Strobel FW, and Thies RL. Physical characteristics and lipoprotein distribution of liposomal nystatin in human plasma. *Antimicrob Agents Chemother* 1997; 41: 1871–5.

3

Nanotechnology Applications in Ophthalmology

Eman Elhawy, MD and John Danias, MD, PhD

CONTENTS

3.1 Introduction

The human eye is a complex, highly specialized body organ. It is the sensory organ for vision that contains the light-sensitive retina and the refractive system that focuses light onto the retina. It is connected to the central nervous system through the optic nerve that carries the visual information to higher visual centers that allows us to see. Although a thorough discussion of the eye anatomy and physiology is beyond the scope of this review, appreciating some of the basic anatomy and physiology is important in understanding the applications of nanotechnology in the field of ophthalmology. The eye anatomy can be simplified by describing the eye as a sphere with a diameter of approximately 23 mm, with its wall consisting of three layers (Figure 3.1). The outer layer consists of the transparent cornea anteriorly and the white sclera posteriorly. The middle layer (called uvea) is highly vascular and consists of the iris and the ciliary body anteriorly and choroid posteriorly. The innermost layer, the retina, contains the light-sensitive photoreceptors (rods and cones), intermediate neurons (bipolar and amacrine cells), and the retinal ganglion cells (RGCs) whose axons form the optic nerve. The inside of the eye contains the lens and is filled by the gel-like vitreous (posteriorly) and the aqueous humor (anteriorly).

The cornea is a transparent avascular tissue. Its main function is refractive, as it provides 60%–90% (depending on age) of the refractive power (ability to focus light) of the eye. It consists of three distinct layers: the epithelium (on the outside), the stroma (in the middle),

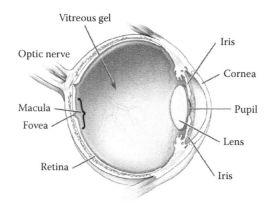

FIGURE 3.1
Anatomy of the eye. (Eye diagram showing the macula and fovea [black and white]. Ref#: NEA09.) (Courtesy of National Eye Institute, National Institutes of Health [NEI/NIH].)

and the endothelium (toward the inside of the eye). The epithelium and endothelium attach to thickened basement membranes (Bowman's and Decement's membrane, respectively), while the stroma is made up of neatly arranged collagen fibers in bundles with interspersed keratocytes (modified fibroblasts). Transparency is dependent on collagen organization, hydration, and avascularity. An optimal water level in the corneal stroma is actively maintained by the endothelium. Although both epithelial cells and keratocytes can regenerate following injury, the endothelium cannot. Since it controls corneal hydration, the cornea becomes opaque when endothelial cell density falls below critical levels. Deep injuries can also lead to scarring of the corneal stroma, thus compromising corneal transparency by disturbing the arrangement of collagen fibers. The three corneal layers also constitute a barrier to the diffusion of medications from outside toward the inside of the eye.

The tear film covers the anterior surface of the eye and serves to protect and lubricate the cornea. The tear film consists of three parts: the innermost part is a mucoid layer formed by the mucous-secreting goblet cells (present in the conjunctiva). The middle layer consists of the aqueous part formed by accessory lacrimal glands in the conjunctiva and contains dissolved nutrients, oxygen, and anti-infective molecules. The outermost layer of the tear film (the lipid layer) is formed by the meibomian glands in the eyelid. Its main function is stabilization of the tears and evaporation prevention by decreasing tear film surface tension. Medications applied to the surface of the eye are rapidly diluted in tears and readily eliminated by constant tear flow through the nasolacrimal passages.

The posterior continuation of the cornea is the avascular, white (opaque) sclera. The corneoscleral junction, also known as the limbus, gives on the outside attachment to the conjunctiva (a thin transparent vascular membrane) that covers the sclera and reflects to line the inner aspect of the eyelids. Subcojuctival injections are often used as depot for local treatment. The limbus is important, because it contains the vascular circle that supplies nutrients to the periphery of the avascular cornea as well as the stem cells that continuously repopulate the corneal epithelium.

The iris is the disc-shaped tissue that is responsible for the eye color when the eye is viewed en face (see also Figures 3.1 and 3.2). The central opening is the pupil. Contraction of the opposing autonomically innervated iris dilator and sphincter muscles regulates pupil diameter. The peripheral base of the iris together with the peripheral cornea define

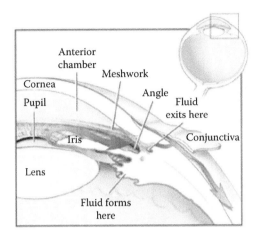

FIGURE 3.2
Angle of the eye. (A clear fluid flows continuously in and out of the anterior chamber and nourishes nearby tissues. The fluid leaves the chamber at the open angle where the cornea and iris meet. When the fluid reaches the angle, it flows through a spongy meshwork and leaves the eye. Ref#: NEA11.) (Courtesy of National Eye Institute, National Institutes of Health [NEI/NIH].)

the anterior chamber angle (Figure 3.2) that contains the trabecular meshwork (a sponge-like structure that regulates outflow of aqueous humor [see Figure 3.2]).

The ciliary body (Figure 3.1) is the middle portion of the uveal tract. It is responsible for secretion of the aqueous humor, which carries nutrients to the avascular anterior segment structures (cornea and lens) and maintains intraocular pressure. The ciliary body also gives attachment to the zonules that suspend the lens in place.

The posterior continuation of the ciliary body is the choroid, which is highly vascular. It provides oxygen and nutrition to the outer half of the retina.

The human crystalline lens is an avascular clear biconvex tissue with refractive power and is responsible for accommodation (ability of the eye to change its focal plane). It is suspended by the suspensory ligaments that attach to the ciliary body and help regulate accommodation. It consists of the lens capsule (outside), which is lined anteriorly with lens epithelium. Lens fibers, formed by maturing lens epithelial cells, make the bulk of the lens. The oldest fibers lie toward the center, forming the nucleus, while the cortex is formed by the relatively recently developed fibers. Over a lifetime lens fibers become continuously compacted. This eventually results in loss of transparency, which is clinically called a cataract.

The retina can be physiologically divided into the outer retinal pigment epithelium (RPE) and the inner neural retina (separation of these two layers is termed retinal detachment). The neural retina is organized so that different neural cell populations form distinct layers (Figure 3.3), with the photoreceptors occupying the outermost location (closest to RPE and the sclera). Thus the light has to transverse the entire retina to stimulate the photoreceptors. The density of photoreceptors decreases from a maximum at the macula (close to the posterior pole of the eye) to the retinal periphery (close to the ciliary body). The maximal density is found in an area in the center of the macula, with a diameter of 1.5 mm, called the fovea. The orderly retinal cell arrangement is altered there, with only the photoreceptors remaining (all other cells have been pushed to the side).

The optic nerve, which is formed by the axons of the RGCs, emanates from the posterior pole of the eye. Axons are covered with myelin, encased within protective glial tissue and

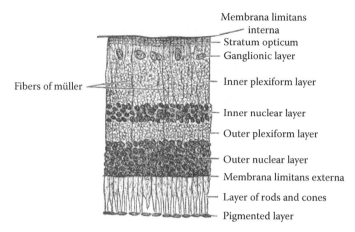

FIGURE 3.3
Retina anatomy. (From Lewis, W.H., ed., *Grey's Anatomy*, 20th ed., Lea and Febiger, Philadelphia, PA, 1918.)

surrounded by meninges. The two optic nerves from both eyes fuse at the chiasm (intra-cranial) where approximately 50% of the axons decussate (those coming from the nasal side of retina) into the opposite optic tract, to synapse at the lateral geniculate nucleus of the thalamus. The visual signal is then conveyed to the primary visual cortex (V1) in the occipital lobe of the brain and from there to higher visual centers.

3.2 Pathology of the Eye

Because of its complexity in organization and functional specialization, the human eye is prone to different pathological processes, which can be congenital or acquired. Congenital diseases usually arise as a result of failure to properly execute specific morphogenetic events that are critical for eye and visual system development.

Acquired eye disease pathology includes infectious, inflammatory, degenerative, and neoplastic conditions. A detailed description of the various pathologies of specific tissues within the eye is clearly beyond the scope of this review. References to specific conditions will be made in the following sections as they pertain to possible applications of nanotechnology.

3.3 Nanomedicine Applications in Ophthalmology

Treatment of ocular disorders involves the use of systemic or more often topically applied medications in the form of eyedrops, peri- or intraocular injections, and surgery. Drawbacks of such "conventional" therapy include systemic toxicity, [1–5] rapid elimination of topically applied medications through the lacrimal passages,[6] poor penetration through the ocular surface,[7,8] and the risk of infection associated with surgery and intra-ocular injections.[9]

The emerging use of nanotechnology in ophthalmology attempts to address the issues associated with the use of conventional therapy and provides new methods for early diagnosis and monitoring of disease status. Recent advances of nano-ophthalmology extend in the fields of drug delivery (including gene delivery), the use of biofeedback systems and sensors, and in the field of regenerative medicine, where it promises to revolutionize our approach and potentially cure previously blinding diseases.[9,10]

This chapter discusses nanotechnology applications used to improve diagnosis in ophthalmology, drug delivery systems, and tissue engineering use. Finally, it briefly covers new experimental approaches aimed at promoting tissue healing using nanotechnology.

3.3.1 Uses of Nanotechnology to Improve Diagnosis

3.3.1.1 Quantum Dots Use in Imaging in the Eye

Imaging is critical in the diagnosis of many ocular diseases. Because of its transparency, the eye allows for imaging that helps identify and quantify the extent of disease and get a better understanding of ongoing disease processes. Technologic advancements have greatly improved our ability for multimodal imaging.

Intravenous contrast media such as fluorescein injected in fluorescein angiography are particularly useful to allow imaging of the vascular lumen and detect leakage, which is the hallmark of many eye diseases. Nano-sized contrast materials have been explored as adjuncts to such *in vivo* imaging. Quantum dots (QDs) are 2–20 nm crystals composed of a semiconductor core of binary alloys—such as cadmium selenide, cadmium sulfide, indium arsenide, and indium phosphide—and are usually coated with a semiconductor shell of a material with certain optical properties such as zinc sulfide.[11] Coating usually aims to decrease QD toxicity. QDs exhibit strong light absorption, photostability, and high fluorescence with emission that can be tuned by modifying the QD size.[12]

Researchers have experimented using QDs to identify changes in the transparent vitreous. Preliminary results have shown that QDs performed better than conventional dyes such as triamcinolone, which is generally used to stain the vitreous during surgery.[13] Thus, QDs could be developed as a way to visualize the transparent (and thus invisible) vitreous during vitrectomy. Their performance for such visualization is superior to the use of triamcinolone when tested on enucleated porcine eyes.[13] Apart from potential clinical applications, QDs are extensively used to understand the physiology of eye tissues. A recent interesting finding suggests that the eye has a lymphatic drainage system, and this finding was based on tracing of QDs to submandibular lymph nodes after intraocular injection.[14]

Other types of nanoparticles coupled to metals such as gold nanorods have also been experimentally used *in vivo* as contrast agents in ocular coherence tomography (OCT) with promising results, because they produce strong OCT signals when intravenously injected. The successful application as contrast medium to visualize the retinal details in living mice[15,16] suggests that such an approach may be used clinically in the future.

However, there are some concerns about QD safety and toxicity when clinically used. Because QDs are complexes of heavy metals, they may be potentially toxic to human cells by affecting normal enzyme function through interfering with their cofactor activity.[17] QD toxicity varies widely depending on the nature of the material used in manufacturing, the size of the dots, the concentration used, and coating bioactivity.[18] Cadmium and selenium are the most toxic components used in the QD core complex. Industry now is investigating the use of cadmium-free QDs to eliminate potential toxicity.[18] Elimination of the QDs after their use is also an issue, as QDs are "trapped" in tissues, potentially causing long-term toxicity.

3.3.1.2 Use of Nanobiosensors as a Diagnostic and Therapeutic Device

Sensors have been used for a long time in industry and in medicine to monitor changes in specific environments. Nanotechnology has enabled new methods for building nanometer-sized sensors. Such sensors are particularly relevant when trying to detect changes at the cellular and even molecular level without disturbing normal human body function.

Depending on the nature of the "sensor" component, we can categorize nanosensors into many subtypes: there are *synthetic* sensors made of specific particles such as those based on carbon nanotubes; *chemical* sensors that have the ability to detect changes in wavelength, light, and speed of different molecules; and *biological* sensors that use a biologically sensitive element such as an enzyme, antibody, nucleic acid, or even whole cells to detect the change in question. Sensors usually produce a specific signal, which is transmitted to an amplifier that amplifies the signal before further analysis. Such analysis can occur locally or be performed at another device or computer after transmitting the signal wirelessly.

In the field of ophthalmology, nanobiosensors equipped with a feedback mechanism can offer not only diagnostic but also therapeutic applications. Such sensors can be used to monitor a specific function or parameter and to start or stop treatment based on received data.[19]

The use of nanobiosensors thus offers the potential of disease prevention when a disease can be successfully treated, before the development of clinical symptoms and signs.[20]

Retinopathy of prematurity (ROP) is a potentially blinding neonatal condition. It develops when premature infants are exposed to increased amount of oxygen. In the immature retina, high oxygen elicits the development of abnormal blood vessels that can result in retinal hemorrhages, retinal detachment, and retinal fibrosis. When diagnosed late, ROP is irreversible and leads to blindness.[21]

Although no nanobiosensors have been developed yet specifically for ROP, it is envisioned that ROP will be one of the first conditions whose outcome can be modified using this approach. A preterm infant at risk for ROP could, for example, be administered a therapeutic agent to prevent further reactive oxygen species production or to initiate DNA repair. The therapy would be coupled with an oxidant biosensor to detect reactive oxygen species.[22] If no reactive oxygen species are detected, then the therapy will be retained within the nanobiosensor complex. If, however, reactive oxygen species were detected in the retina, the therapeutic agent would be delivered in a specific dosing pattern until the biosensor determined that such administration is no longer needed. Similarly, patients who have glaucoma may one day be able to receive medications that will be released from nanoparticles only if biosensors detect elevated intraocular pressure. In this manner, nanobiosensor technology can offer a unique combination of both diagnostic and therapeutic tools for various ocular diseases.

3.3.2 Uses of Nanotechnology in Treating Ocular Conditions

3.3.2.1 Drug Delivery Systems

The eye presents a unique challenge for drug delivery. Unlike other areas of the body, most medications are traditionally administered by topical application. A number of factors diminish their bioavailability. Drug elimination through lacrimal passages and local drug metabolism obviously affect local drug concentration. In addition, the corneal and sclera barrier properties further diminish intraocular penetration. In addition, because tears drain to the back of the pharynx, topically applied medications can easily get into the systemic circulation through the respiratory mucosa bypassing hepatic metabolism. Thus, systemic side effects have to be considered during drug delivery to the eye.[23]

Medications and carriers should also avoid impeding vision and causing discomfort to assure patient compliance. Therefore, the ideal ocular drug delivery system should be able to maximize ocular penetration and minimize medication elimination in addition to being nontoxic, biodegradable, reproducible, and cost-effective in delivering drugs in a timely manner to site-specific targets.

The most extensively studied nanoscience application in ophthalmology is drug delivery. Such drug delivery systems improve the pharmacokinetics of common eye medications by prolonging drug contact time, offering controlled release, increasing the ability to penetrate ocular tissues, and specifically targeting diseased cells while eliminating systemic side effects.[24]

To simplify and better understand different nanotechnology-based drug delivery systems, we can group them into the following four broad categories that share some characteristics (see also Table 3.1):[10]

1. Drug delivery systems that use colloidal carriers[25] or polymer-based *sustained release* to extend the release of the medicine.[24]

2. Drug delivery systems that use *dendrimers* (which have a 3D structure composed of a central core unit and branches) to serve as a repository for the pharmacological agent.[24]

3. Drug delivery systems that use *carriers to encapsulate the agent* delivered.[9] These include liposomes, niosomes, and nanobeads delivering encapsulated cells.

4. Drug delivery systems that use *nanoparticles* (either composite polymeric or solid nanoparticles) with the deliverable medication bound to the surface.

3.3.2.1.1 Sustained Release Systems

3.3.2.1.1.1 Microemulsions Microemulsions are mixtures of oil and water stabilized with a surfactant (sometimes a cosurfactant may also be required). They are easily prepared and thermodynamically stable. They can be used to deliver poorly soluble medications to the eye. They also increase bioavailability, because they enhance penetration capacity.

TABLE 3.1

Different Nanoscale Drug Delivery Systems with Potential or Current Use in Ophthalmology

Group	Drug Delivery System	Excipient Composition
1	Microemulsions	Water, oil, surfactant, and cosurfactant
	Nanosuspensions	Colloidal hydrophobic particles stabilized with surfactant and may contain some polymers to enhance suspension's functions
2	Dendrimers	Multivalent core with multiple radiating branches that can be tagged with specific markers to target-specific tissues
3	Liposomes	Phospholipid bilayer
	Niosomes	Surfactant vesicles
4	Nanoparticles	Polymers, lipid, or metal

Moreover, microemulsions given as eyedrops have increased contact time due to their ability to bind to the corneal surface by adsorption. They thus form a reservoir for the medication in the precorneal tear film that allows delivery over extended periods and avoids washout by tears and blinking. Surfactants are also easily sterilized and stored due to their thermodynamically stable nature.[8,24]

The choice of surfactant is, however, challenging. Most ionic surfactants are toxic to ocular tissues. Thus, only nonionic surfactants can practically be used for ocular drug delivery.[8]

Microemulsions have been investigated *in vivo* on healthy volunteers and also for delivery of multiple medications over the past 15 years. They improve ocular penetration for poorly soluble medications such as everolimus used as a local immunosuppressive to prevent corneal graft rejection.[26] Timolol microemulsions using lecithin as a surfactant increase bioavailability 3.5-fold compared to traditionally formulated timolol eyedrops.[27] Pilocarpine microemulsion dosed twice daily obtained the same biological effect as traditionally formulated pilocarpine given four times daily.[28] Other microemulsions that have been investigated for ocular use include dexamethasone[29] and chloramphenicol.[30]

Tacrolimus microemulsion used to suppress immunity and prevent graft rejection is now approved by U.S. Food and Drug Administration (FDA). Cyclosporine microemulsion eyedrops used to treat a variety of eye conditions such as dry eye are also commercially available.

3.3.2.1.1.2 Nanosuspensions   Nanosuspensions are submicron-sized pure colloidal dispersions of hydrophobically active nanomolecules stabilized by a surfactant. They have been used to facilitate the ocular penetration of hydrophobic medications such as steroids and nonsteroidal anti-inflammatory drugs. The addition of polymers (such as Eudragit) can further enhance the characteristics of the nanosuspensions by inducing controlled and prolonged medication release.[24,31]

Both *in vivo* and *in vitro* studies have been performed on various nanosuspensions for ophthalmic use; these include studies with prednisolone,[32] dexamethasone,[32] hydrocortisone,[32] cloricromene (a coumarin derivative experimentally used to treat uveitis),[33] flurbiprofen,[34] ibuprofen,[35] and diclophenac.[36] These studies showed superior intraocular bioavailability compared to commercially available formulations.

3.3.2.1.2 Dendrimers

Dendrimers (from the Greek *dendron* or tree) represent a separate group of drug carrier systems. They are composed of a multivalent core with multiple radiating branches with high-density terminals. Dendrimers work as a repository carrying the agent to be delivered either between branches or in the central core.[10] The nature of the molecule surface can be modified to target specific tissues by the addition of specific bioactive groups. Dendrimers have been extensively investigated as controlled release delivery agents[37–39] and are considered to be one of the most promising nanomedicine applications.

In ophthalmology, dendrimers have been used as drug delivery agents for topically applied medications, photosensitizers in photodynamic therapy (PDT), and as agents to enhance transfection for gene therapy.

For example, polyamiodamine dendrimers have been used for ocular delivery of pilocarpine and tropicamide, providing improved drug bioavailability with no ocular toxicity.[40]

Choroidal neovascularization (CNV) results from abnormal vascular growth from the choroid into the retina. It is the principal reason for vision loss in age-related macular degeneration (ARMD).

PDT can be used to occlude these abnormal vessels. However, standard photosensitizers such as verteporfin did not achieve dramatic results because of a relative lack of photosensitization efficacy due to aggregation. This diminished the effect of photoirradiation.

Dendrimers have been used as photosensitizers in PDT for the treatment of ARMD using a porphyrin core and polybenzyl ether branches (dendrons) using micelles as vehicles for delivery. Use of these dendrimers circumvented the problem of aggregation of classic photosensitizes achieving a 280-fold increase in photoirradiation efficacy without long-term phototoxicity.[41,42] The same technology has been adopted for use in PDT for treatment of corneal neovascularization caused by other diseases.[42]

Although PDT was one of the first effective treatments for CNV, it has now mostly been replaced by anti-VEGF treatment. Currently, the gold standard for treatment of neovascular ARMD involves monthly intraocular injection of anti-VEGF antibodies. Sense vascular endothelial growth factor (VEGF) oligonucleotides in lipid lysine dendrimers, which possess anti-VEGF activity, have also been successfully tested in animal models.[43] The ability to inhibit mRNA transcription correlated with the number of positive charges on the dendrimer surface. Significant reductions in VEGF activity were observed for approximately 4–6 months without any observable toxicity.[43,44]

3.3.2.1.3 Encapsulating Agents for Drug Delivery

3.3.2.1.3.1 Liposomes Liposomes are lipid vesicles that consist of a phospholipid bilayer and can carry the deliverable medication or genetic material in the inner core (aqueous). Lipophilic medications can also be carried within the phospholipid bilayer itself.[10] They have been intensively investigated for use in various medical applications including cancer therapy. Liposome surface can be tagged to specifically target tissues that reduce systemic side effects. However, liposomes also have certain drawbacks such as instability, lack of reproducibility, and limited drug carrying capacity.[24]

Ocular applications include the use of liposome-encapsulated acyclovir to treat viral keratitis. These topically applied liposomes offer the advantages of prolonged residual time and decreased elimination, but they have the disadvantage of corneal toxicity that is induced by endocytosis of liposome vesicles.[45]

Liposomes have been investigated (in animal models) for use as a carrier for angiogenic vessel homing peptide (APRPG liposomal SU5416) in the treatment of CNV in neovascular ARMD. They hold the potential for allowing the formulation of treatments that will require only a single intravitreal injection to treat CNV, as opposed to the current practice of monthly intravitreal injections of antibodies to VEGF.[46] Their large advantage for intraocular delivery is their stability in the vitreous, which allows them to deliver medications with high efficacy for prolonged period of time.[47] Liposomes also have unique characteristics that make them potentially useful not only in drug delivery but also in imaging.

Neovascular disorders (where abnormal blood vessels develop inside the eye due to retinal ischemia—such as in diabetes—in retinal vein occlusion, and in some retinal degeneration subtypes) represent a group of challenging ocular diseases. Methods for monitoring the disease status to provide precise and timely management are needed. Fluorescence-labeled cationic liposomes were successfully used to monitor choroidal neovascular membranes in mice. This application carries a great potential for allowing follow-up patients with similar disorders, since it appears to be both safe and effective.[48]

3.3.2.1.3.2 Niosomes   Niosomes are nonionic surfactant vesicles that encapsulate both hydrophilic and hydrophobic drugs to deliver them to the target site.[49,50] They can be coated with mucoadhesive material such as carbopol or chitosan to enhance their bioavailability.[51] Niosomes are generally chemically stable; cholesterol can also be added to enhance their stability against plasma and plasma proteins.

In vitro and *in vivo* experiments have tested their efficacy and ocular toxicity for delivering timolol,[28] acetazolamide,[50,52,53] and gentamycine.[52] No ocular toxicity was noted, and prolonged activity with biphasic release of the medicine was reported for niosomes when administered topically.[52]

Current (fluconazole) antifungal eyedrops used in treating fungual corneal ulcers have a very short half-life resulting in poor patient compliance. Niosomes were recently investigated *in vitro* for delivering the medication to the infected cells. Such application proved to be safe and effective.[48] If further developed, this application may help in eradicating one of the causes of blindness in developing countries.

3.3.2.1.3.3 Using Nanobeads to Deliver Intraocular Drugs in Encapsulated Cells   The concept of intraocular encapsulated cell-based drug delivery is to introduce genetically engineered cells that produce the protein of interest inside the eye. As a result, no repeat injections will be required for treatment. The cells are introduced inside beads or nanotubes, so that the immune surveillance will not detect them. Seven weeks of action was achieved in an experiment introducing cells producing ciliary neurotrophic factor encapsulated inside polymer membrane capsules and implanted by intravitreal injection. This approach may be of great use in the future for conditions requiring repeat administration of the therapeutic agent.[54]

3.3.2.1.4 Nanoparticles

The use of nanoparticles in drug delivery systems represents an expanding frontier in ophthalmic nanomedicine. Nanoparticles can act as a drug delivery device with precise drug release profiles sensitive to environmental or chemical signals and can be naturally eliminated by the body when the pharmacologic effect has subsided.[55]

Nanoparticles are ideal for ocular drug delivery. Discovered in the 1970s, nanoparticles can be described as molecules with a size ranging from 10 to 1000 nm. They are manufactured with a therapeutic product (small molecule, peptide, or nucleic acid) entrapped, encapsulated, attached, or dissolved in the nanoparticle carrier. Their small size allows them to overcome anatomic barriers[56] and efficiently be transported intracellularly compared with their predecessors, the microparticles.[57] Research is currently focused on improving specificity of cellular targeting, understanding intacellular trafficking, and achieving precise and controlled therapeutic release for extended periods.

The two major categories of nanoparticles are solid lipid and polymeric nanoparticles. Others are gold nanoparticles that are generally used as probes and recently investigated to suppress neovascularization.[58] During their evolution, major changes took place in the manufacture of nanoparticles by coating them using substances (opsonins) that increase their cellular uptake either by opsonization[59] or by coating them with a polymer such as polyethylene glycol (PEG) that prevents or decreases their opsonization.

3.3.2.1.4.1 Solid Lipid Nanoparticles   These are colloidal carriers constructed of a physiologic solid lipid matrix such as triglycerides, fatty acids, or steroids.[60] Their composition affords biocompatibility, controlled drug release, and protection for many types of therapeutic agents in their hydrophobic core. They have the advantage of lower ocular toxicity than that associated with the polymeric forms and thus more patient compliance.

3.3.2.1.4.2 Polymeric Nanoparticles These are composed of the active compound combined with a polymeric carrier.[61] They possess the added advantage of a higher degree of encapsulation, which minimizes drug toxicity and enhances bioavailability, as the therapeutic agent is more contained in the core of the carrier.[60,62] The main disadvantages to pharmaceutical use are the low drug-loading capacities for certain forms of nanoparticles and the lack of standardized large-scale production techniques. However, polymeric nanoparticles can avoid opsonization by stealth effect, thus increasing their bioavailability and decreasing their elimination.[63]

Polymeric nanoparticles have been used to treat eye diseases. For example, poly(alkyl cyanoacrylate) nanoparticles have been shown to increase corneal penetration of both lipophilic and hydrophilic drugs. This application was limited because of the corneal epithelial toxicity associated with the use of this carrier.[64] Poly-ε-caprolactone nanocapsules and nanoparticles have been used to deliver indomethacin, metipranolol, and betaxolol with less ocular toxicity and increased drug bioavailability. The success of nanoparticles for ophthalmic drug delivery largely depends on optimizing the lipophylic/hydrophilic properties of the polymer carrier system.[65]

Chitosan nanoparticles were investigated on rabbits for delivering cyclosporine A, showing prolonged intraocular bioavailability with minimal systemic absorption and ocular side effects. Polylactic acid (PLA) nanospheres have been used for intravitreal delivery of acyclovir either through injection or just by local administration as eyedrops. Coating the spheres with PEG improves the bioavailability of the drug in the eye.[66–68]

It is worth mentioning that chitosan and PLA nanospheres are used to develop ocular implants for prolonged control of intraocular viral infections such as CMV retinitis. Since they are biodegradable, they overcome the need for surgical removal of the traditionally used implants with nonbiodegradable polymers.[67] Nanoparticles used for drug delivery to the posterior segment of the eye also have the potential to prolong the duration of action of the injected drug thus decreasing the number of injections needed to sustain drug release and avoid potential complications associated with repeated intraocular injections. Promising results were achieved *in vitro* using ganciclovir-loaded albumin nanoparticles for intravitreal injections to treat cytomegalovirus retinitis.[67,69] The same nanoparticles were used to deliver fomivirsen for the same condition.[70]

RPE has showed enhanced capacity for nanoparticles uptake.[71] This ability suggests the potential for nanoparticles to be used for the treatment of retinal degenerations and explains the success of nanoparticles for gene delivery to the posterior segment of the eye.[71]

3.3.2.1.4.3 Nanoparticle Use in Gene Therapy One of the potentially most important applications of nanoparticles is in the field of gene therapy, since it overcomes the disadvantages associated with the use of viral vectors. The foremost area of concern with the use of viral vectors is systemic safety. In addition, despite the immune-privileged status of the eye, repeated introduction of viral vectors can be immunogenic leading to potentially devastating inflammation and sight-threatening consequences. Certain viral vectors can also theoretically undergo transformation to the wild type through recombination with other active viruses in the body leading to wild-type infectivity. Furthermore, failure to adequately limit tropism can result in ectopic expression in tissues other than the target one, which may have unintended cosequences.[72]

Albumin nanoparticles carrying DNA were FDA approved as adjuvant in advanced breast cancer. Nanoparticles carry a great potential for use in the treatment of hereditary retinal degeneration and Leber's hereditary optic neuropathy. Recent studies using viral carriers showed that subretinal injection for gene therapy using recombinant adeno-associated virus 2 carrying the *RPE65* gene was safe when injected outside the foveal area. It also proved effective in improving extrafoveal rod and cone function.[10,73]

3.3.2.2 Tissue Engineering (Regenerative Medicine) in Ophthalmology

Regenerative medicine represents the ability to modify the tissue environment to optimize its function or regain a lost function. By using tissue engineering on a nanoscale, we can potentially introduce cells to replace dead or dying cells, suppress the growth of aberrant tissue that otherwise will cause undesired effects, or simply modify the environment to help cells regain their normal function.[74]

Applications of tissue engineering in ophthalmology include corneal endothelial cell transplantation, RGC repair, and nanofiber scaffoldings to help optic nerve fiber regeneration.[74] Nanostructured cell sheet engineering can be used to grow corneal endothelial cells. A transfer medium is then used to collect the cells for implantation on Decement's membrane.[75] Platinum nanowires can assess the function of ganglion cells without interfering with blood flow in capillaries.[76] They also have the potential to be used as stents to help reperfuse retinal blood capillaries in the future.[74]

The use of scaffolds to introduce cells in the eye to help regenerate or guide ganglion cell axonal growth is one of the very promising applications of regenerative nanomedicine in the eye.[74] Scaffolds are developed from synthetic, natural, or biological materials. The scaffold has to dissolve after carrying the implanted cells or their axons to the desired location. Although synthetic scaffolds offer the advantage of being custom designed with desired properties that meet the requirements of the implanted cells, the issue with the use of synthetic material as PLAs, poly(lactic-*co*-glycolic) acid, and methyl methacrylate is the immune reaction developed inside the eye that might lead to the unfavorable effect of scarring. Synthetic biological scaffolds do not lead to adverse immune reactions and are designed using self-assembling biological materials such as peptides, DNA, and RNA, which form a nanofiber meshwork to guide the growth of cells.[77–79] At the same time, some of the peptides that are not recognized by the immune system as foreign materials can still be cleared by tissue enzymes, in a controlled way without causing adverse tissue reactions.[77–79]

Scaffolds have also been investigated for use in arresting retinal detachment, restoring the cornea clarity, and even in experimental eye transplantation, where they are used as a bridge to guide regenerating implanted axons to remnants of the recipient optic nerve.[74,80]

3.3.3 Nanomedicine Applications in Larger Devices and Processes

3.3.3.1 Visual Prosthesis

Visual prostheses represent a group of devices that are implanted inside the eye or other parts of the visual system in an attempt to recover visual ability lost due to retinal, optic nerve, or even central nervous disease. The available devices are based on integrated circuit devices and currently face many problems, including implant toxicity, degradation, and tissue response to the implanted device that can cause inflammation, neuronal irritation, or even thermal damage.[10,81,82]

Nanotechnology is trying to eliminate problems associated with conventional devices by developing nanoelectrodes that act as sensors, processors, transducers, and also provide the energy needed for the prosthesis function. Carbon nanotubes are currently being investigated for the use in nanoelectrode devices, due to their high tensile strength, ultra light weight, and chemical and thermal stability, together with their ability to conduct electrical impulses.[10] The development of such ultra small high-tech implants will not only decrease tissue reaction and enhance thermal dissipation but also improve the prosthesis functionality by providing the potential of higher-resolution images and nanoscale fuel cells.[83]

3.3.3.2 Nanogold Particle Use in Connective Tissue Welding

Laser-activated gold nanoparticles were recently investigated *in vitro* for connective tissue welding. Tissue welding is a technique used to help rapid connective tissue healing without the use of exogenous proteins or polymers in a process called soldering.[84]

These experiments have succeeded in healing porcine lens capsules using laser-activated gold nanoparticles that result in strictly localized denaturation of lens capsule proteins and efficient healing of the treated area. The photothermal effects were localized and caused fusion of the lens capsule patch to the recipient lens capsule.[84]

These results may be used, together with the development of polymer lens refilling, as a treatment of presbyopia.[84] Polymer lens refilling is an investigational procedure developed to restore functional accommodation. Accommodation (the ability to change the eye focal power at will) is lost with age due to lens nuclear sclerosis.[17] Lens refilling can be performed by creating a small opening in the lens capsule through which the lens matter can be evacuated and replaced by a transparent, pliable polymer. Although this approach is feasible from a technical standpoint, one of the difficulties remaining is sealing the capsular opening after insertion of the polymer. Using the new welding technology, we can overcome this limitation. This approach can thus potentially become a reality and help overcome presbyopia.[17]

3.4 Summary

Given the unique nature of the eye and the challenging features of most eye diseases, nanotechnology holds great potential for both diagnosis and treatment.

Nanotechnology potential solutions to eye disease diagnosis and management include development of monitoring devices, contrast agents, drug delivery systems, tissue engineering, scaffold, components of electronic implants, and methods for tissue healing.

Among these potential applications, drug delivery is the most advanced to date, with some of these systems already in clinical practice. However, most of the other potential applications of nanotechnology are not far behind in the development process. It is thus safe to say that nanotechnology will soon have dramatic effect in how we detect and treat eye disease.

References

1. De Laey JJ. Systemic toxicity of sympathomimetic and parasympatholytic eye drops. *Bull Soc Belge Ophtalmol* 1979;186:21–5.
2. Netter P, Sirbat D, and Trechot P. [Systemic effects of ophthalmic solutions]. *Bull Soc Ophtalmol Fr* 1985;Spec No:255–75.

3. Milkowski S. [General complications after local pharmacological ophthalmological treatment]. *Wiad Lek* 1968;21(23):2127–32.

4. Bodd E and Lunde PK. [Systemic side effects of topical ophthalmic drugs]. *Tidsskr Nor Laegeforen* 1990;110(11):1375.

5. Lama PJ. Systemic reactions associated with ophthalmic medications. *Ophthalmol Clin North Am* 2005;18(4):569–84.

6. Byrro RM, de Oliveira Fulgencio G, da Silva Cunha A, Jr., Cesar IC, Chellini PR, and Pianetti GA. Determination of ofloxacin in tear by HPLC-ESI-MS/MS method: comparison of ophthalmic drug release between a new mucoadhesive chitosan films and a conventional eye drop formulation in rabbit model. *J Pharm Biomed Anal* 2012;70:544–8.

7. Sieg JW and Robinson JR. Mechanistic studies on transcorneal permeation of pilocarpine. *J Pharm Sci* 1976;65(12):1816–22.

8. Vandamme TF. Microemulsions as ocular drug delivery systems: recent developments and future challenges. *Prog Retin Eye Res* 2002;21(1):15–34.

9. Thomson H and Lotery A. The promise of nanomedicine for ocular disease. *Nanomedicine (London)* 2009;4(6):599–604.

10. Nguyen P, Meyyappan M, and Yiu SC. Applications of nanobiotechnology in ophthalmology—Part I. *Ophthalmic Res* 2010;44(1):1–16.

11. Bailey RE and Nie S. Alloyed semiconductor quantum dots: tuning the optical properties without changing the particle size. *J Am Chem Soc* 2003;125(23):7100–6.

12. Drummen GP. Quantum dots—from synthesis to applications in biomedicine and life sciences. *Int J Mol Sci* 2010;11(1):154–63.

13. Yamamoto S, Manabe N, Fujioka K, Hoshino A, and Yamamoto K. Visualizing vitreous using quantum dots as imaging agents. *IEEE Trans Nanobiosci* 2007;6(1):94–8.

14. Tam AL, Gupta N, Zhang Z, and Yucel YH. Quantum dots trace lymphatic drainage from the mouse eye. *Nanotechnology* 2011;22(42):425101.

15. Awdeh AdlZ RM, Perez VL, Rugerri M, and Gambhir S. Optical coherence molecular imaging using gold nanorods in living mice eyes. *ARVO Meeting Abstracts* 2011:455/D1102.

16. Tong L, Wei Q, Wei A, and Cheng JX. Gold nanorods as contrast agents for biological imaging: optical properties, surface conjugation and photothermal effects. *Photochem Photobiol* 2009;85(1):21–32.

17. True LD and Gao X. Quantum dots for molecular pathology: their time has arrived. *J Mol Diagn* 2007;9(1):7–11.

18. Hardman R. A toxicologic review of quantum dots: toxicity depends on physicochemical and environmental factors. *Environ Health Perspect* 2006;114(2):165–72.

19. Prow TW, Rose WA, Wang N, Reece LM, Lvov Y, and Leary JF. Biosensor-controlled gene therapy/drug delivery with nanoparticles for nanomedicine. *Proc SPIE* 2005;5692:199–208.

20. Leary JF. Nanotechnology: what is it and why is small so big? *Can J Ophthalmol* 2010;45(5):449–56.

21. Chen ML, Guo L, Smith LE, Dammann CE, and Dammann O. High or low oxygen saturation and severe retinopathy of prematurity: a meta-analysis. *Pediatrics* 2010;125(6):e1483–92.

22. Prow T, Grebe R, Merges C et al. Nanoparticle tethered antioxidant response element as a biosensor for oxygen induced toxicity in retinal endothelial cells. *Mol Vis* 2006;12:616–25.

23. Eugen Barbu LV, Nevell TG, and Tsibouklis J. Polymeric materials for ophthalmic drug delivery: trends and perspectives. *J Mater Chem* 2006;16(34):234–58.

24. Wadhwa S and Paliwal R, Paliwal SR, and Vyas SP. Nanocarriers in ocular drug delivery: an update review. *Curr Pharm Des* 2009;15(23):2724–50.

25. Gupta S and Vyas SP. Carbopol/chitosan based pH triggered in situ gelling system for ocular delivery of timolol maleate. *Sci Pharm* 2010;78(4):959–76.

26. Baspinar Y, Bertelmann E, Pleyer U, Buech G, Siebenbrodt I, and Borchert HH. Corneal permeation studies of everolimus microemulsion. *J Ocul Pharmacol Ther* 2008;24(4):399–402.

27. Gasco MR, Gallarate M, Trotta M, Bauchiero L, Gremmo E, and Chiappero O. Microemulsions as topical delivery vehicles: ocular administration of timolol. *J Pharm Biomed Anal* 1989;7(4):433–9.

28. Garty N and Lusky M. Pilocarpine in submicron emulsion formulation for treatment of ocular hypertension: a phase II clinical trial. *Invest Ophthalmol Vis Sci* 1994;35(4):2175.
29. Fialho SL and da Silva-Cunha A. New vehicle based on a microemulsion for topical ocular administration of dexamethasone. *Clin Experiment Ophthalmol* 2004;32(6):626–32.
30. Lv FF, Li N, Zheng LQ, and Tung CH. Studies on the stability of the chloramphenicol in the microemulsion free of alcohols. *Eur J Pharm Biopharm* 2006;62(3):288–94.
31. Patravale VB, Date AA, and Kulkarni RM. Nanosuspensions: a promising drug delivery strategy. *J Pharm Pharmacol* 2004;56(7):827–40.
32. Kassem MA, Abdel Rahman AA, Ghorab MM, Ahmed MB, and Khalil RM. Nanosuspension as an ophthalmic delivery system for certain glucocorticoid drugs. *Int J Pharm* 2007;340(1–2):126–33.
33. Pignatello R, Ricupero N, Bucolo C, Maugeri F, Maltese A, and Puglisi G. Preparation and characterization of Eudragit retard nanosuspensions for the ocular delivery of cloricromene. *AAPS Pharm Sci Tech* 2006;7(1):E27.
34. Pignatello R, Bucolo C, Spedalieri G, Maltese A, and Puglisi G. Flurbiprofen-loaded acrylate polymer nanosuspensions for ophthalmic application. *Biomaterials* 2002;23(15):3247–55.
35. Pignatello R, Bucolo C, Ferrara P, Maltese A, Puleo A, and Puglisi G. Eudragit RS100 nanosuspensions for the ophthalmic controlled delivery of ibuprofen. *Eur J Pharm Sci* 2002;16(1–2):53–61.
36. Agnihotri SM and Vavia PR. Diclofenac-loaded biopolymeric nanosuspensions for ophthalmic application. *Nanomedicine* 2009;5(1):90–5.
37. Schatzlein AG, Zinselmeyer BH, Elouzi A et al. Preferential liver gene expression with polypropylenimine dendrimers. *J Control Release* 2005;101(1–3):247–58.
38. Yang H and Lopina ST. Penicillin V-conjugated PEG-PAMAM star polymers. *J Biomater Sci Polym Ed* 2003;14(10):1043–56.
39. Esfand R and Tomalia DA. Poly(amidoamine) (PAMAM) dendrimers: from biomimicry to drug delivery and biomedical applications. *Drug Discov Today* 2001;6(8):427–36.
40. Vandamme TF and Brobeck L. Poly(amidoamine) dendrimers as ophthalmic vehicles for ocular delivery of pilocarpine nitrate and tropicamide. *J Control Release* 2005;102(1):23–38.
41. Ideta R, Tasaka F, Jang WD et al. Nanotechnology-based photodynamic therapy for neovascular disease using a supramolecular nanocarrier loaded with a dendritic photosensitizer. *Nano Lett* 2005;5(12):2426–31.
42. Sugisaki K, Usui T, Nishiyama N et al. Photodynamic therapy for corneal neovascularization using polymeric micelles encapsulating dendrimer porphyrins. *Invest Ophthalmol Vis Sci* 2008;49(3):894–9.
43. Marano RJ, Wimmer N, Kearns PS et al. Inhibition of *in vitro* VEGF expression and choroidal neovascularization by synthetic dendrimer peptide mediated delivery of a sense oligonucleotide. *Exp Eye Res* 2004;79(4):525–35.
44. Marano RJ, Toth I, Wimmer N, Brankov M, and Rakoczy PE. Dendrimer delivery of an anti-VEGF oligonucleotide into the eye: a long-term study into inhibition of laser-induced CNV, distribution, uptake and toxicity. *Gene Ther* 2005;12(21):1544–50.
45. Law SL, Huang KJ, and Chiang CH. Acyclovir-containing liposomes for potential ocular delivery. Corneal penetration and absorption. *J Control Release* 2000;63(1–2):135–40.
46. Honda M, Asai T, Umemoto T, Araki Y, Oku N, and Tanaka M. Suppression of choroidal neovascularization by intravitreal injection of liposomal SU5416. *Arch Ophthalmol* 2011;129(3):317–21.
47. Maeda N, Takeuchi Y, Takada M, Sadzuka Y, Namba Y, and Oku N. Anti-neovascular therapy by use of tumor neovasculature-targeted long-circulating liposome. *J Control Release* 2004;100(1):41–52.
48. Kaur IP, Rana C, Singh M, Bhushan S, Singh H, and Kakkar S. Development and evaluation of novel surfactant-based elastic vesicular system for ocular delivery of fluconazole. *J Ocul Pharmacol Ther* 2012;28(5): 484–96.
49. Cortesi R, Esposito E, Corradini F et al. Non-phospholipid vesicles as carriers for peptides and proteins: production, characterization and stability studies. *Int J Pharm* 2007;339(1–2):52–60.
50. Vyas SP, Mysore N, Jaitely V, and Venkatesan N. Discoidal niosome based controlled ocular delivery of timolol maleate. *Pharmazie* 1998;53(7):466–9.

51. Aggarwal D and Kaur IP. Improved pharmacodynamics of timolol maleate from a mucoadhesive niosomal ophthalmic drug delivery system. *Int J Pharm* 2005;290(1–2):155–9.

52. Abdelbary G and El-Gendy N. Niosome-encapsulated gentamicin for ophthalmic controlled delivery. *AAPS PharmSciTech* 2008;9(3):740–7.

53. Guinedi AS, Mortada ND, Mansour S, and Hathout RM. Preparation and evaluation of reverse-phase evaporation and multilamellar niosomes as ophthalmic carriers of acetazolamide. *Int J Pharm* 2005;306(1–2):71–82.

54. Tao W. Application of encapsulated cell technology for retinal degenerative diseases. *Expert Opin Biol Ther* 2006;6(7):717–26.

55. Schmid G. *Nanoparticles: From Theory to Application*. 2nd ed. Wiley-VCH, Weinheim, 2010.

56. Arayne MS, Sultana N, and Noor Us-S. Fabrication of solid nanoparticles for drug delivery. *Pak J Pharm Sci* 2007;20(3):251–9.

57. Alonso MJ and Sanchez A. Biodegradable nanoparticles as new transmucosal drug carriers. *ACS Symposium Series Washington DC* 2004:283–95.

58. Kim JH, Kim MH, Jo DH, Yu YS, and Lee TG. The inhibition of retinal neovascularization by gold nanoparticles via suppression of VEGFR-2 activation. *Biomaterials* 2011;32(7):1865–71.

59. Aggarwal P, Hall JB, McLeland CB, Dobrovolskaia MA, and McNeil SE. Nanoparticle interaction with plasma proteins as it relates to particle biodistribution, biocompatibility and therapeutic efficacy. *Adv Drug Deliv Rev* 2009;61(6):428–37.

60. Pedraz JL, Orive G, Igartua M et al. Nanotechnologies for the Life Sciences. In: *Solid Lipid and Polymeric Nanoparticles for Drug Delivery*. Wiley-VCH, Weinheim, 2007.

61. Lamprecht A. *Nanotherapeutics: Drug Delivery Concepts in Nanoscience*. Pan Stanford Publishing, Singapore, 2009:241–6.

62. Chan JM, Valencia PM, Zhang L, Langer R, and Farokhzad OC. Polymeric nanoparticles for drug delivery. *Methods Mol Biol* 2010;624:163–75.

63. Gref R, Minamitake Y, Peracchia MT, Trubetskoy V, Torchilin V, and Langer R. Biodegradable long-circulating polymeric nanospheres. *Science* 1994;263(5153):1600–3.

64. Zimmer A and Kreuter J. Microspheres and nanoparticles used in ocular delivery systems. *Adv Drug Deliv Rev* 1995;1:61–73.

65. Calvo P, Vila-Jato JL, and Alonso MJ. Comparative *in vitro* evaluation of several colloidal systems, nanoparticles, nanocapsules, and nanoemulsions, as ocular drug carriers. *J Pharm Sci* 1996;85(5):530–6.

66. De Campos AM, Sanchez A, and Alonso MJ. Chitosan nanoparticles: a new vehicle for the improvement of the delivery of drugs to the ocular surface. Application to cyclosporin A. *Int J Pharm* 2001;224(1–2):159–68.

67. Merodio M, Espuelas MS, Mirshahi M, Arnedo A, and Irache JM. Efficacy of ganciclovir-loaded nanoparticles in human cytomegalovirus (HCMV)-infected cells. *J Drug Target* 2002;10(3):231–8.

68. Giannavola C, Bucolo C, Maltese A et al. Influence of preparation conditions on acyclovir-loaded poly-d,l-lactic acid nanospheres and effect of PEG coating on ocular drug bioavailability. *Pharm Res* 2003;20(4):584–90.

69. Merodio M, Arnedo A, Renedo MJ, and Irache JM. Ganciclovir-loaded albumin nanoparticles: characterization and *in vitro* release properties. *Eur J Pharm Sci* 2001;12(3):251–9.

70. Irache JM, Merodio M, Arnedo A, Camapanero MA, Mirshahi M, and Espuelas S. Albumin nanoparticles for the intravitreal delivery of anticytomegaloviral drugs. *Mini Rev Med Chem* 2005;5(3):293–305.

71. Bejjani RA, BenEzra D, Cohen H et al. Nanoparticles for gene delivery to retinal pigment epithelial cells. *Mol Vis* 2005;11:124–32.

72. BioSide Lines. Viral vectors and biological safety. University of Wisconsin-Madison. The Office of Biological Safety. 2011;36:1-4. http://www.ehs.wisc.edu/documents/bio-Biosidelines_June_2011.pdf. Accessed July 10, 2013.

73. Jacobson SG, Cideciyan AV, Ratnakaram R et al. Gene therapy for Leber congenital amaurosis caused by RPE65 mutations: safety and efficacy in 15 children and adults followed up to 3 years. *Arch Ophthalmol* 2012;130(1):9–24.

74. Ellis-Behnke R and Jonas JB. Redefining tissue engineering for nanomedicine in ophthalmology. *Acta Ophthalmol* 2011;89(2):e108–14.

75. Yang J, Yamato M, Nishida K et al. Cell delivery in regenerative medicine: the cell sheet engineering approach. *J Control Release* 2006;116(2):193–203.

76. Llinás RR, Walton KD, Nakao M, Hunter I, and Anquetil PA. Neurovascular central nervous recording/stimulating system: using nanotechnology probes. *J Nanopart Res* 2005;7:111–27.

77. De Laporte L and Shea LD. Matrices and scaffolds for DNA delivery in tissue engineering. *Adv Drug Deliv Rev* 2007;59(4–5):292–307.

78. Ellis-Behnke RG, Liang YX, You SW et al. Nano neuro knitting: peptide nanofiber scaffold for brain repair and axon regeneration with functional return of vision. *Proc Natl Acad Sci USA* 2006;103(13):5054–9.

79. Guo J, Su H, Zeng Y et al. Reknitting the injured spinal cord by self-assembling peptide nanofiber scaffold. *Nanomedicine* 2007;3(4):311–21.

80. Ellenberg D, Shi J, Jain S et al. Impediments to eye transplantation: ocular viability following optic-nerve transection or enucleation. *Br J Ophthalmol* 2009;93(9):1134–40.

81. Weiland JD, Liu W, and Humayun MS. Retinal prosthesis. *Annu Rev Biomed Eng* 2005;7:361–401.

82. Colodetti L, Weiland JD, Colodetti S et al. Pathology of damaging electrical stimulation in the retina. *Exp Eye Res* 2007;85(1):23–33.

83. Nguyen-Vu TD, Chen H, Cassell AM, Andrews RJ, Meyyappan M, and Li J. Vertically aligned carbon nanofiber architecture as a multifunctional 3-D neural electrical interface. *IEEE Trans Biomed Eng* 2007;54(6 Pt 1):1121–8.

84. Ratto F, Matteini P, Rossi F et al. Photothermal effects in connective tissues mediated by laser-activated gold nanorods. *Nanomedicine* 2009;5(2):143–51.

4

Nanotechnology Applications in Urology

Himanshu Aggarwal, MD and Barry A. Kogan, MD

CONTENTS

4.1 Introduction

The National Nanotechnology Initiative defines nanotechnology as the science of materials and phenomena in the range of 1–100 nm (10^{-9} m). Nanomedicine is the application of nanotechnology to medicine and has the potential to impact numerous aspects of healthcare.[1] Much of nanotechnology in medicine is based on the use of nanoscale particles. Because of their small size, these particles can be transferred through blood vessels enabling them to easily come in contact with tissues and cellular molecules. These particles can be conjugated with functional molecules, including disease-specific ligands, antibodies, anticancer drugs, and imaging probes. This can be done with a single molecule or several molecules/ligands simultaneously. This allows them to be targeted to specific cells or molecules either on the surface or interior of the targeted cells. In addition, nanocarriers have been developed to deliver drugs or imaging molecules to targeted cells.[2]

The potential applications of nanotechnology in urology are wide ranging and include prevention, early detection, improvement in diagnosis, and treatment and follow-up of various urological diseases.[3,4] In this chapter, we discuss the broad categories of nano-medicine applications in urological oncology and other urological disorders.

4.2 Urological Oncology

4.2.1 Imaging

Many types of nanoparticle-based technologies are in development for improved imaging of cancers. Some of the best examples in urology are in the field of prostate cancer. Prostate cancer is the most common cancer affecting American men with an incidence of 125 cases per 100,000 per year and accounting for approximately 12% of all newly diagnosed cancer cases.[5] Three-quarters of all cases are diagnosed in men aged 65 or older. Almost all the patients undergo prostate biopsy for diagnosis, generally due to a high serum prostate-specific antigen (PSA) level and/or an abnormal digital rectal examination. Prostate cancer patients with lymph node metastases have a poorer prognosis,[6,7] and the presence of lymph node metastases significantly alters treatment recommendations.[8,9] Current magnetic resonance imaging (MRI) technology provides images with great anatomical detail and soft-tissue contrast but is not very sensitive for the detection of lymph node metastases.[10,11] However, the sensitivity of MRI may be enhanced significantly by the use of ultrasmall superparamagnetic iron oxide particles. These nanoparticles have a monocrystalline, superparamagnetic iron oxide core, containing a dense packing of dextrans to prolong their time in circulation, and these are avidly taken up by lymph nodes in animals and humans. The mechanism of action appears to be that these nanoparticles are slowly extravasated from vascular to the interstitial space, from which they are transported to lymph nodes by way of lymphatic vessels. Disturbances in lymph flow or in nodal architecture caused by metastases lead to abnormal patterns of accumulation of lymphotropic superparamagnetic nanoparticles in lymph nodes. In the lymph nodes, these nanoparticles are internalized by macrophages, which cause changes in magnetic properties detectable by MRI.[12,13]

In one clinical study comparing MRI using lymphotropic superparamagnetic nanopar-ticles and conventional MRI in patients with localized prostate cancer, there was a marked increase in detection of lymph nodes of all sizes using the lymphotropic superparamag-netic nanoparticles. When comparing lymph nodes that were between 5 and 10 mm, the MRI using lymphotropic superparamagnetic nanoparticles had a sensitivity of 96% ver-sus 29%, specificity of 99% versus 87%, positive predictive value of 96% versus 29%, and negative predictive value of 99% versus 87% in comparison with conventional MRI. In lymph nodes <5 mm, sensitivity and positive predictive values were 41% and 78%, respec-tively, with lymphotropic superparamagnetic nanoparticles MRI versus 0% for conven-tional MRI.[12] Ferumoxtran-10 (Combidex), a dextran-coated superparamagnetic iron oxide nanoparticle, is currently in a clinical phase III trial for prostate cancer imaging.[14] A study in patients with renal cancer found that lymphotropic nanoparticle-enhanced MRI dem-onstrated 100% sensitivity and 95.7% specificity in diagnosing lymph node metastases.[15] In patients with urothelial carcinoma of urinary bladder, the sensitivity and negative predic-tive value to detect lymph nodes improved significantly with use of nanoparticle-based contrast than with precontrast imaging; that is, from 76% to 96% ($p < .001$) and from 91%

to 98% ($p < .01$), respectively. At postcontrast imaging, metastases (4–9 mm) were prospectively found in 10 of 12 normal-sized nodes (<10 mm); these metastases were not detected on precontrast images. Postcontrast images also showed lymph nodes that were missed at pelvic node dissection in two patients, implying that ferumoxtran-10-enhanced MRI may improve nodal staging in patients with bladder cancer by depicting metastases even in normal-sized lymph nodes.[16] In a study of patients with stage I testicular cancer, lymphotropic nanoparticle-enhanced MRI was found to be 88% sensitive, 92% specific, and 90% accurate for malignant lymph node detection.[17] This technology may also play a role in penile cancer, in which sentinel lymph node biopsy helps in staging and therapy. If positive lymph nodes could be highlighted visually, this technology could be of great value to surgeons attempting to locate the true "sentinel node." It is important to recognize that these techniques are very good at detecting lymph node metastases; they are not very helpful in detecting cancer outside of lymph nodes.[14] Although these studies are preliminary and require considerable expertise in interpretation, the techniques seem to have great promise.[18]

Semiconductor quantum dots (QDs) are nanometer-scale, light-emitting particles with unique optical and electronic properties such as size-tunable light emission and the ability to simultaneously excite multiple fluorescent colors. These QDs can be targeted by binding them with antibodies to tumor markers, thereby suggesting that the technique has great potential to improve cancer detection with imaging.[2] The possibilities for ultrasensitive and multiplexed imaging of molecular targets are considerable. For example, QDs conjugated to prostate-specific membrane antigen (PSMA) antibody have been successfully used to visualize micrometastasis of prostate cancers in mouse bone models. Similarly, Gao et al.[19] reported that it is feasible to simultaneously target and image prostate tumors in living animal models using bioconjugated PSMA-targeted QDs. With this technology, as few as 500,000 prostate cancer cells (0.5 mg of tumor mass) can be detected in a mouse tibia prostate cancer bone metastasis.[20] An alternative use of this technology allows for rapid quantitative detection of epithelial-to-mesenchymal transition markers, one of the first signs of androgen independence in prostate cell models and in clinical prostate cancer specimens.[20]

4.2.2 Biomarkers

Nanotechnology has also been applied to measurement of biomarkers. PSA is a serine protease that enters the serum from prostate cells. PSA levels are usually elevated in patients with prostate cancer. Currently, measurements of PSA are based on either an enzyme-linked immunosorbent assay (ELISA) or a chemiluminescent immunoassay and are able to detect PSA levels of up to 0.01 ng/mL.[21] Using a nanotech approach, Zheng et al. demonstrated that multiplexed electrical detection of PSA is possible using arrays of silicon nanowire field-effect devices. This also allows highly sensitive and selective method of detection of PSA in undiluted serum samples to a level as low as 0.9 pg/mL (0.0009 ng/mL).[22] Perhaps the most promising technique is based on bio-barcode technology. Using a bio-barcode assay, Nam et al.[23,24] reported a different nanoparticle-based ultrasensitive technique for detecting free PSA, up to 3 aM (0.00003 ng/mL) concentrations. This bio-barcode system relies on magnetic microparticle probes with antibodies that specifically bind to PSA, nanoparticle probes that are encoded with DNA that is unique to PSA, and antibodies that can sandwich the target captured by the microparticle probes. This technique is highly sensitive and has several advantages over conventional technology to detect PSA in serum. The use of the nanoparticle bio-barcodes provides a high ratio of PCR-amplifiable DNA to labeling antibody, and this substantially increases assay sensitivity. In addition, this type of assay obviates the need for complicated conjugation chemistry for attaching DNA to the labeling antibody.[24] In another similar study using on-chip barcode assay, PSA detection was accomplished with four orders of magnitude higher sensitivity compared to commercially available ELISA-based PSA. This corresponds to only 300 copies of protein analytes using 1 μL total sample volume.[25] The clinical utility of this extremely sensitive technique is still being worked out. However, in a pilot clinical study using a comparable nanotechnique, a bio-barcode assay was used to detect PSA in 18 men who had undergone radical prostatectomy for clinically localized prostate cancer. All the men had undetectable PSA by conventional methods. However, using the bio-barcode assay, PSA was detectable in all men after radical prostatectomy. In nine men, there was no rise in the level of this extremely sensitive PSA and they were found to have no evidence of recurrence over a median follow-up, of 6.3 years (range 4.8–9.1 years). The other nine had a rise in ultrasensitive PSA level on follow-up, and all developed a biochemical recurrence of their cancer, in each case, detected well before it was diagnosed by conventional PSA measurements.[26] Further and more long-term studies are needed to confirm these findings and to determine its clinical significance.

There are significant clinical opportunities related to the detection of biomarkers on cancer cells. There has been a steady move in oncology to "personalized" therapy for cancer. With this approach, therapy is targeted to the individual molecular "signature" of the cancer. Nanotechnology has great potential value in this area as well. In particular, nanoparticle probes can be used to quantify a panel of molecular markers on both intact cancer cells and tissue specimens, allowing a correlation of traditional histopathology and molecular signatures for the same material.[27] Similarly, QD-labeled oligonucleotides can be used as a new fluorescent in situ hybridization (FISH) probe and offers potential advantages over standard FISH, particularly in the identification of genes expressed at low levels. As an example, QD-conjugated oligonucleotides have been used to detect sonic hedgehog mRNA in prostate cancer cells.[20] Similarly, QD-based protein micro- and nanoarrays have been developed for detection of prostate cancer biomarkers. These enable patterning of high-density features in an extremely small area of prostate tissue along with rapid screening of a large number of molecular targets in a single experiment.[28] It is too early to determine the clinical applicability of these techniques, but they have great promise.

4.2.3 Treatment (Drug Delivery)

Most anticancer chemotherapeutic drugs are water insoluble and need an organic solvent to be administered as an injectable solution. These organic solvents are often toxic and have their own side effects. By placing chemotherapeutic agents inside soluble nanoparticles, the drugs can be delivered in a nontoxic soluble medium. This eliminates the need for toxic organic solvents.[1]

More important is targeting the drug delivery. Most common forms of chemotherapy inhibit cell division. Although cancer cells divide rapidly and are good targets, many healthy cells also divide and are affected as a side effect. Most chemotherapeutic agents do not differentiate well between cancerous and rapidly dividing normal cells, and this is the cause of systemic toxicity and adverse effects of chemotherapy (such as bone marrow suppression, cardiomyopathy, and neurotoxicity). This accounts for the narrow therapeutic index of these drugs and limits the maximal allowable dose of the drugs. In addition, rapid elimination and widespread distribution of the drugs into nontargeted tissues increase the frequency of administration that is needed. Nanotechnology offers a more targeted approach and can both increase the therapeutic index of these drugs and allow for a more sustained release of these drugs in body. Indeed, the use of nanoparticles for drug delivery and targeting is one of the most exciting and clinically important applications of cancer nanotechnology. The two main types that we will discuss are passive and active targeting.

4.2.3.1 Passive Targeting

Rapid neovascularization in growing cancerous tissues causes leaky, defective vessels and impaired lymphatic drainage. These characteristics can be used to therapeutic benefit and have been termed enhanced permeability and retention effect (EPR).[29–32] Because of this EPR, intravenous administration of nanoparticles allows accumulation of nanoparticles at tumor sites. Furthermore, these particles can be engineered to be hydrophilic by adding polymer coatings, such as polyethylene glycol (PEG), poloxamines, poloxamers, polysaccharides, and liposomes, or by using branched or block amphiphilic copolymers.[33–35] The hydrophilic coating and the small size of nanoparticles (<100 nm) help prevent their clearance by macrophages and the reticuloendothelial system.[36] This allows them to accumulate and remain where they are needed. In particular, liposomes have been studied for this purpose. An increased therapeutic effect has been found and is thought to be due to passive targeting of liposomes, leading to prolonged accumulation in tumors and the slow release of the drug from liposomes. In an experimental study in normal dog bladder, it was shown that liposomal doxorubicin injected submucosally into the bladder wall distributed very well, both in the entire bladder wall and in regional lymph nodes. Also, the doxorubicin remained at 15–100 times higher concentration in these lymph nodes for at least a week after submucosal injection, as compared to intravenous injection of free doxorubicin. This implies that submucosal injection of liposomal chemotherapy may be advantageous for accumulating the drug in bladder cancer tissue with potential use of a new modality of chemotherapy application.[37] Similarly, in dogs, an intravesical dose of paclitaxel-loaded gelatin nanoparticles achieved 2.6 times drug concentration in the urothelium and lamina propria tissue layers as compared to dogs treated with the commercially available paclitaxel formulation.[38] In a phase II clinical trial in patients with advanced unresectable or metastatic urothelial carcinoma who had not received prior chemotherapy for metastatic disease, those patients treated with intravenous pegylated liposomal doxorubicin showed clinically significant activity along with low toxicity. Hence, this formulation of the drug is of interest for further study in advanced urothelial cancers and can be combined with other active agents.[39]

Using another form of passive targeting, carboplatin was loaded onto carbon nanotubes and carbon nanofibers. These carbon nanotubes functioned as slow release drug depots that gradually released 68% of the drug over 2 weeks. This system was tested in urological tumor cell lines and was found to be more effective in the impairment of proliferation and clonogenic survival of tumor cells as compared to free carboplatin. These nanotubes are promising delivery agents for anticancer drugs that would benefit from a slow continuous release of drug.[40]

Nanotechniques may also be of benefit in delivering "natural" products. There are several natural agents that have shown promising results in cancer prevention and treatment in *in vitro* and *in vivo* studies. However, delivering effective doses and/or increasing their bioavailability have limited their usage. Nanoparticle-mediated delivery systems can enhance bioavailability and limit toxicity of these natural agents. For example, there is considerable evidence that catechins, the active ingredient present in green tea, have anticancer activity. Despite promising results in preclinical settings, the applicability of these agents in human cancers has met with only limited success, largely due to inefficient systemic delivery and bioavailability of promising agents. A recent experimental study demonstrated that encapsulation of epigallocatechin-3-gallate (EGCG), an extract of green tea, in biodegradable polylactic acid–PEG nanoparticles enhances the antitumor properties of EGCG *in vitro* and *in vivo* while using 1/10 the dosage as compared to EGCG alone.[41]

Another example is turmeric (*Curcuma longa* Linn), a crystalline compound that has been traditionally used in medicine and cuisine in India. Curcumin (diferuloylmethane) is the major active component of turmeric and has been shown to be a cancer chemopreventive in several different cancers including prostate cancer both *in vitro* and *in vivo*. In spite of its promising therapeutic index, the biological activity of curcumin is severely limited because of its poor bioavailability. In an experimental *in vitro* study, curcumin loaded in a liposomal nanocarrier delivery system showed similar efficacy as free curcumin against prostate cancer cells, with 1/10 of the dose.

Tetraiodothyroacetic acid (tetrac) is a deaminated analogue of L-thyroxine (T_4) that blocks the proangiogenesis actions of T_4 and 3,5,3′-triiodo-L-thyronine as well as other growth factors at the cell surface receptor integrin αvβ3. Since this integrin is expressed on renal cancer cells and endothelial and vascular smooth cells of tumor, the possibility exists that tetrac may act on both cell types to block the proliferative effects of thyroid hormone on tumor growth and tumor-related angiogenesis. However, intracellular uptake of tetrac reduces its activity. A new nanoparticle form of tetrac (PLGA-tetrac) prevents intracellular absorption, thereby limiting its activity to the cell surface only. It has been shown to have superior anticancer activity in *in vitro* and *in vivo* experiments in renal cell cancer as compared to free tetrac.[42]

In one last example, 7-ethyl-10-hydroxycamptothecin (SN-38), a biologically active metabolite of irinotecan hydrochloride (CPT-11), has potent antitumor activity, especially in hypervascular tumors with high levels of vascular endothelial growth factor, but has not been used clinically because it is water insoluble. Recently, a new nanodevice, NK012, has been developed. The device slowly releases SN-38 and has been shown to markedly enhance the antitumor activity of SN-38 in hypervascular metastatic renal cell cancers particularly using EPR.[43]

4.2.3.2 Active Targeting

Active targeting causes preferential accumulation of nanoparticles in a specific area. This is usually achieved by placing the therapeutic agent inside a nanoparticle and conjugating the nanoparticle surface with a targeting component of either antibody or aptamer. This active targeting can be directed to the tumor organ, the tumor itself, individual cancer cells, or even intracellular organelles inside cancer cells. In most cases, the targeting moiety is directed toward specific receptors or antigens expressed on the plasma membrane or elsewhere at the tumor site, but specific interactions may take advantage of lectin–carbohydrate, ligand–receptor, or antibody–antigen relationships.[44]

Many cancer cells overexpress certain cell membrane antigens, and this lends itself to efficient drug uptake via antibody binding to these antigens and receptor-mediated endocytosis. For example, nearly 100% of androgen-dependent prostate cancer cells express PSMA on their surface, yet there are relatively low levels of PSMA in normal prostate, kidney, brain, and small intestine tissue. Hence, PSMA is an excellent target for ligand or antibody-specific targeting.[14] This has been tried with an aptamer that binds specifically to PSMA combined with cisplatin. These cisplatin-aptamer nanoparticles showed comparable *in vitro* cytotoxicity in PSMA-positive LNCaP prostate cancer cells at 1/100 of dose as compared to free cisplatin.[45] In another example, docetaxel-encapsulated nanoparticles formulated with biocompatible and biodegradable poly(D,L-lactic-*co*-glycolic acid)-*block*-poly(ethylene glycol) (PLGA-*b*-PEG) copolymer were functionalized with the A10 2′-fluoropyrimidine RNA aptamers that recognize PSMA. These targeted nanoparticles release docetaxel directly inside the cancer cells, resulting in enhanced efficacy and decreased systemic toxicity.[46] These particles have shown enhanced cytotoxicity *in vitro* as well as enhanced antitumor effect and decreased systemic toxicity in prostate cancer xenografts.[46–49] Also using PSMA as a target, both thermally cross-linked superparamagnetic and chlorotoxin-conjugated iron oxide nanoparticles have been synthesized. These particles can deliver targeted chemotherapeutic agents directly to the prostate cancer cells *in vitro* and lead to improved cytotoxicity in prostate cancer cells.[14,50]

Other novel forms of targeting allow for dual functions. For example, a QD-aptamer-doxorubicin conjugated multifunctional system has been developed that can deliver doxorubicin to targeted prostate cancer cells, and at the same time it can sense and report the delivery of doxorubicin. This is accomplished by activating the fluorescence of QD, which enables concurrent images of the cancer cells.[51] These multifunctional nanoparticles have greater advantage as compared to conventional magnetic nanoparticles, as these can be used for both drug delivery and imaging applications.[52] In another example of dual functionality, particles with an iron oxide core were first coated with oleic acid, and then these particles were stabilized with Pluronic F-127. The oleic acid makes them dispersible in an aqueous vehicle, and Pluronic F-127 prevents protein-binding particle aggregation. The latter property reduces their clearance by the reticuloendothelial system, keeping nanoparticles in systemic circulation and thereby allowing their extravasation into tumor tissue. This has been tested with doxorubicin and paclitaxel, and it was found that using this system, the drugs were released in sustained manner over 2 weeks.[52]

Dual action can also be accomplished by combining drugs allowing for synergistic cytotoxic activity. Dual functionality also allows for two entirely different forms of therapy—for example, combining chemotherapy with immunotherapy. Unmethylated CpG oligonucleotides act as immune stimulants and have shown promise in providing improved outcomes in cancer treatment in both preclinical and clinical trials. A nanoplatform has been developed combining doxorubicin and unmethylated CpG oligonucleotides. This treatment has been shown to be effective against prostate cancer cells *in vitro* and *in vivo*. This delivery system is based on a dendrimer and a single-strand DNA-A9 PSMA RNA aptamer hybrid, and the success of this experiment shows the potential for this nanostructure system to be as a new combination approach for improving cancer treatments.[53]

4.2.3.2.1 Photodynamic Therapy

Photodynamic therapy is a minimally invasive experimental cancer therapy that depends on the buildup of a photosensitizing drug within targeted tissue. The photosensitizer is subsequently activated by light of a specific wavelength, resulting in destruction of the targeted tissue by free radicals or singlet oxygen. Successful treatment requires delivery of critical amounts of the photosensitizer into the cancerous tissue. In conventional therapy, this requires very high doses of the drug in the circulatory system and leads to side effects due to accumulation of photosensitizer in normal tissue. In a recent study, LNCaP prostate cancer cells were targeted, using a nanoparticle carrier, containing the photosensitizer hematoporphyrin, conjugated with an antibody against PSMA. This form of nanoprotein-based active targeting allowed the photosensitizer to be selectively delivered to the tumor cells, leading to considerably less photosensitizer in the circulation and hence fewer side effects.[54]

Superficial bladder cancer is ideally suited to treatment with photodynamic therapy because of the easy access to the lumen of the bladder by cystoscopy and the fact that bladder tissue is more translucent than most other human tissues. Yet, photodynamic therapy of recurrent superficial bladder cancer has never gained wide acceptance in the urological community.[55] This is likely due to the lack of selectivity of photosensitizers, toxicity, and complexity of the technical devices, all leading to inconsistent therapeutic results. Again, though, nanotechnology may enable much more selectivity of the photosensitizer. Bladder transitional cell carcinoma cells overexpress the transferrin receptor on their surface and this is a potential target. In an experimental study, it was shown that the bladder cancer cells incubated with transferrin-liposomal-aluminum phthalocyanine tetrasulfonate (AlPcS4) photosensitizer had much higher intracellular AlPcS4 levels than those incubated with liposomal-AlPcS4 alone. Similarly, in orthotopic rat bladder cancer, intravesical instillation of transferrin-liposomal-AlPcS4 photosensitizer resulted in significantly higher accumulation of AlPcS4 in tumor tissue as compared to normal urothelium and submucosa/muscle. In contrast, instillation of free AlPcS4 resulted in nonselective accumulation throughout the whole bladder. This suggests that transferrin-mediated liposomal targeting of photosensitizing drugs has potential as a tool for photodynamic therapy of superficial bladder tumors.[56]

4.2.3.3 Hyperthermia and Thermal Ablative Therapies

Hyperthermia generally refers to temperatures between 40°C and 45°C. Even this mild elevation in temperature has antitumor effects. These are achieved via influencing the tumor microenvironment, induction of apoptosis, activation of immunological processes, and induction of gene and protein synthesis.

Temperatures above 45°C are considered directly thermoablative, as they cause cell death directly by denaturing proteins. Thermal therapies have been evaluated extensively for both hyperplastic and malignant conditions of the prostate during the last few decades.[57] The therapeutic application of nanoparticles to achieve tumor heating is emerging as a novel form of "nanothermal therapy" of tumors. Although several potential hyperthermic particles—such as silver, lanthanum, and zinc nanoparticles—are available, the thermal activation properties of gold nanoparticles, magnetic nanoparticles, and carbon nanotubes have been studied most extensively and have potential for clinical applications.[58]

4.2.3.3.1 Gold Nanoparticles

Several experimental studies have been reported on the use of gold nanoparticles such as gold nanorods and nanocages for the thermal therapy of various cancers. These gold nanoparticles have been designed to be activated by near-infrared rays to produce photothermal ablation.[58] For prostate tumors, gold nanoshells activated by laser light of 810 nm wavelength have been designed. These nanoparticles have been shown to have significant hyperthermic ablative effect (temperature elevation to around 65°C) on prostate cancer cells *in vitro*[59] and produced nearly 93% tumor necrosis in mouse xenograft prostate cancer tissues.[60]

Both active and passive targeting of gold nanoparticles to achieve the photothermal ablation therapy has been reported. Several different nanoparticles (gold nanospheres and gold nanorods) have demonstrated size-dependent uptake and negligible toxicity in prostate cancer cells.[61] Targeted thermal therapy of PC3 prostate cancer cell lines using prostate-specific EphrinA1-conjugated gold nanoshells demonstrated localized thermal damage to cells that were bound to it.[62]

Gold nanoparticles also have the ability to generate high temperatures at a desired site with externally tunable control; that is, the degree of hyperthermia can be controlled externally. Even so, there is a need for monitoring of tumor response during the photothermal process. To address this issue, a multifunctional nanoparticle has been developed. This particle uses a gold nanopopcorn-based surface-enhanced Raman scattering approach for targeted sensing, nanotherapy treatment, and in situ monitoring of photothermal response during the therapy. The very high sensitivity afforded by this technique, along with the highly informative spectra characteristic of Raman spectroscopy, makes nanopopcorn-based surface-enhanced Raman scattering unique for ultrasensitive biological analysis. As an example, gold nanopopcorns were conjugated with multiple PSMA-specific targets: anti-PSMA antibody and Raman dye (Rh6G) attached to A9 RNA anti-PSMA aptamers. Rh6G-modified RNA aptamers covalently attached to the cell surface and serve a dual function as targeting molecules and Raman dye-carrying vehicles. When added to LNCaP prostate cancer cells (PSMA-positive cells), these multifunctional popcorn-shaped gold nanoparticles formed several hot spots and provided a significant enhancement of the Raman signal intensity from Rh6G-modified aptamers by several orders of magnitude. This bioassay was highly sensitive, with a detection ability of 50 cancer cells, and was rapid, taking about 30 minutes from cancer cell binding to detection and destruction of the cell due to the localized heating during near-infrared irradiation. In situ time-dependent experimental results demonstrated that, as the nanotherapy progresses, the surface-enhanced Raman scattering intensity decreases, and as a result, by monitoring the surface-enhanced Raman scattering intensity change, one can monitor the response to photothermal therapy over time.[63] Although highly complex, this nanotechnology-driven assay could have enormous potential for application in rapid, on-site targeting of cancer cells, nanotherapy treatment, and simultaneous monitoring of the nanotherapy process, which is critical to providing effective, less toxic treatment of cancer.

4.2.3.3.2 *Carbon Nanotubes*

Carbon nanotubes are another class of nanomaterials that holds great potential for extrinsically activated hyperthermia. These are cylindrical grapheme structures with diameters ranging from a few to hundreds of nanometers. These nanotubes have an extraordinary photon-to-thermal energy conversion efficiency with high absorption in the near-infrared region of the electromagnetic spectrum. Their potential has stimulated several investigations to determine whether it is possible to exploit their characteristics for anticancer therapy.[58] Although single-walled carbon nanotubes were used in early studies, recent data suggest the superiority of multiwalled carbon nanotubes. In a study on mouse xenografts, renal cancer cells were transplanted into the flanks of female athymic mice. After tumor formation, intratumor injection of multiwalled carbon nanotube solution in different concentrations was given. This was followed by an external laser treatment with 1064-nm continuous-wave laser beam at 3 W/cm^2 (spot size, 5 mm) for 30 seconds. Thermoablation was induced by using only 100 μg of multiwalled carbon nanotubes and resulted in complete ablation of tumors along with a >3.5-month durable remission in 80% of mice.[64] In another study, treatment of prostate cancer xenografts in nude mice with multiwalled carbon nanotubes demonstrated that DNA encasement (which enhances aqueous solubility) enhanced the heat emission from the nanotubes following near-infrared irradiation. A single intratumoral injection of multiwalled carbon nanotubes (100 μL of a 500 μg/mL solution) followed by an external laser irradiation at 1064 nm, 2.5 W/cm^2, resulted in complete eradication of PC3 xenograft tumors.[65]

4.2.3.3.3 Magnetic Nanoparticles

Thermal therapy using magnetic nanoparticles involves the coupling of an external magnetic field to a tumor laden with magnetic particles. This locally induced magnetic field generates high-energy photons in the vicinity of the nanoparticles. In a study using prostate cancer tumors in rat xenografts, intracellular hyperthermia was produced by nanoparticle magnetite cationic liposomes that generated heat after exposure to an alternating magnetic field. Hyperthermia also induces an immune response by activating heat shock protein 70, which induces CD3-, CD4-, and CD8-positive lymphocytes. Treatment resulted in a significant reduction in tumor size as compared to control tissues.[66,67] In another study, hyperthermia produced by these nanoparticles suppressed tumor proliferation in the bone microenvironment. However, when used in cranial bone, the method exhibited side effects in the central nerve system due to whole brain hyperthermia.[68]

A prospective phase I trial of nanoparticle-mediated magnetic thermal ablation in patients with recurrent prostate cancer did show safety and promising short-term results. The nanoparticles were injected via a transperineal route into the prostate under general anesthesia with transrectal ultrasound and fluoroscopic guidance. Alternating magnetic fields of strength 4–5 kA/m were applied to generate hyperthermia by activating these nanoparticles. Thermometry was performed, sometimes directly in the prostate and at other times in the urethra and rectum. Maximum temperatures of up to 55°C were achieved in the prostates. No systemic toxicity was observed and low magnetic fields were tolerated well by all patients. However, higher magnetic field strengths caused discomfort in the groin/perineal region. In some patients, temperatures of up to 44°C were observed at the skin level, typically in folds of the scrotal and anal region, but these were able to be managed by cooling and ventilation. Although the intensity of pain in these anatomical regions correlated with increasing magnetic field strength, there was no direct correlation of these side effects with the temperatures achieved in the prostates. No late treatment–related morbidity was observed in this study at a median follow-up of 17.5 months. A decrease of serum PSA was observed in 8 of 10 patients following treatment. However, mean duration of PSA control was only 5 months. All patients ultimately progressed during follow-up with local progression in seven and distant disease in three patients.[69–71] Although the long-term results were suboptimal in this trial designed primarily to assess safety, they do demonstrate proof of principle and modifications or repeat therapy may allow for better outcomes.

Therapeutic synergism of hyperthermia with radiation to treat prostate cancer has also been pursued for a long time.[72] By enhancing the biological effects of radiation with hyperthermia, lower radiation doses may yield equal efficacy while decreasing radiation-induced toxicity. In a phase I clinical trial, the feasibility, tolerance, and achieved temperatures of combined interstitial thermoradiotherapy were evaluated. Twenty-two patients with locally recurrent prostate cancer after definitive radiotherapy received magnetic nanoparticle thermotherapy combined with low dose of brachytherapy (90–100 Gy) as a salvage approach. Transperineal injection of a nanoparticle suspension was carried out in the same setting as permanent implantation of 125-iodine seeds under general anesthesia, guided by transrectal ultrasound and fluoroscopy. The investigators were able to place the seeds and the magnetic nanoparticles in all cases. Exposure to a low-strength alternating magnetic field (4–5 kA/m) resulted in measured and calculated maximum intraprostatic temperatures of 42.4°C (40.6°C–45.5°C) and 41.5°C (40.9°C–45.4°C), respectively, with a good correlation between intraluminal urethral temperature measurements and noninvasive calculations. Only local toxicity was observed; however, it was quite serious, with one patient developing a rectourethral fistula that required temporary colostomy.[73] Based on these studies, it can be concluded that magnetic nanoparticle thermotherapy is feasible, and hyperthermic to thermoablative temperatures can be achieved in the prostates at relatively low magnetic field strengths. However, further improvement of the temperature distribution is required by refining the implantation techniques or by adjusting the amount of nanofluid or the magnetic field strength.[74]

4.2.3.4 Gene Transfer

Genetic treatments of disease are highly desirable, but the technology of delivering the gene therapy still has significant limitations. Generally, live viruses are used to transfer genetic materials to the targets. This has its own risks and there are ethical concerns that may interfere with this. As an alternative, nucleic acid–delivering systems are being developed that incorporate viruslike functions in a single nanoparticle. Although their development is still in its infancy, it is expected that such artificial viruses will have a great impact on the advancements of gene therapeutics.[75] Indeed, in a few experimental studies, nanoparticle-mediated gene transfection has been shown to be more efficient and successful in prostate cancer cells and bladder cancer cells *in vitro* than are conventional gene transfer vectors.[76,77] In other studies using the J591 antibody that recognizes the PSMA protein, J591-targeted nanoparticles have been shown to yield an eight- to tenfold increment of gene transfection levels as compared with untargeted nanoparticles in prostate cancer cells.[78,79]

4.3 Other Applications of Nanotechnology in Urology

4.3.1 Nanoparticle-Coated Urinary Catheters

Urinary tract infections account for over 40% of all nosocomial infections, and almost all these infections are associated with indwelling catheters.[80,81] Catheter-associated infections result from an ascending bacterial colonization within the glycocalyx-enclosed biofilm on the inner and outer surface of the catheter. Free silver ions exhibit a strong antimicrobial activity against a variety of organisms irrespective of their resistance to antibiotics,[81] and silver-impregnated catheters have been shown to retard the development of this biofilm.[82] The data regarding the clinical benefits of this are controversial, with some studies showing a decrease in infection rate[83] with use of silver-impregnated catheters and others not showing any difference.[84] One problem with the currently available silver-coated catheters is that the duration of the release of bactericidal concentrations of silver ions is short, and hence the antimicrobial activity is limited to a few days only.[81,84] A nanotechnology-based technique has been shown, in the laboratory, to release a bactericidal concentration of silver ions for over a month.[84] This technology involves an even distribution of billions of nanoparticles (3–8 nm) of silver in the catheter matrix (polyurethane, silicone) using a barium sulfate carrier. Polyurethane is hygroscopic and rapidly attracts water; the interaction of electrolyte solutions with the extremely finely distributed silver throughout the polyurethane releases silver ions over a period of years to the surface of the material. The electronegatively charged surface of bacteria attracts the positively charged silver ions. The concentrations released from the polyurethane are far below the toxic concentrations for humans.[85]

Using a similar technology, a novel antibacterial hydroxyapatite nanoparticle-coated indwelling urinary catheter was shown to prevent biofilm formation and catheter-associated urinary tract infection in rabbits.[86] Of course, clinical studies need to be done to show whether these innovative catheters result in any clinically significant reduction of infections in patients with indwelling catheters.

4.3.2 Infections

Recently, biosensors have been used for detection of infection in the body. A biosensor is a device in which a biological sensing element is either intimately connected to or integrated within a transducer. These biosensors can be adapted for the detection of the infection based on the ELISA system. Nano-biosensors have been used for bacterial detection in urine of patients with urinary tract infection by using glycoconjugate-specific antibody-bound gold nanowire arrays (GNWAs) that are very sensitive and specific.[87] In this technology, anti-*Escherichia coli* antibody-bound GNWAs prepared on an anodized porous alumina template are used as the first step followed by binding of the bacteria in the urine specimen. An alkaline phosphatase-conjugated second antibody is then added to the system, and the resultant degree of binding is determined by both electrochemical and optical measurements.[87]

Catechin, an extract of green tea, is known to have anti-inflammatory and antimicrobial effects against various bacteria. However, catechin is broken down during digestion, and after oral ingestion, it has very limited bioavailability. This degradation can be reduced by coating catechin with hydroxypropyl methyl cellulose nanoparticles, resulting in enhanced absorption of catechin. In a rat preparation, the plasma concentrations of nano-catechin, as well as its anti-inflammatory and antimicrobial effect, on chronic bacterial prostatitis were determined. Plasma concentrations of active catechin metabolites were significantly higher in the nano-catechin group than those in the control group (traditional catechin), suggesting that coated nano formulation of catechin is an effective method of drug delivery.[88] Further, the nano-catechin group showed a statistically significant decrease in bacterial growth and improvement in prostatic inflammation as compared to the free catechin group.

4.3.3 Gene Probes

Rapid and accurate detection of genetic mutations can be helpful in genetic counseling. In some diseases, the current techniques for determining these mutations are cumbersome. For example, in patients with autosomal dominant polycystic kidney disease (ADPKD), current methods of genetic testing are suboptimal due to the large size of the mutated gene. A nanotechnology-based DNA assay has been developed for detection of single-nucleotide polymorphisms in a feline ADPKD model. The nanotechnology-based DNA assay uses simple methodology, is rapidly available, and can reduce the cost of the determining the mutations involved in human ADPKD and many other genetic diseases. The sensitivity of this method is very high, and for blood genomic DNA, a blood sample of as little as 0.05 mL is needed.[89] This technique can readily be adapted to the diagnosis of human ADPKD.

4.3.4 Urine Proteomic Profiling

A proteome consists of all the proteins and peptides within a particular body compartment. Proteomes are cell and tissue specific and change over time in response to different situations. Proteomics is the assessment of these proteomes. These proteomes may be directly related to the disease (e.g., IgA immune deposits in IgA nephropathy) or may result from secondary events (e.g., generation of specific cleavage products caused by metalloproteases that are upregulated during inflammatory processes in the kidney).

Urine is an ideal biosample, as it is easy to obtain and is a rich source of biomarkers for diagnostic information.[90] Numerous efforts are being made to discover, identify, and validate biological markers for the diagnosis of kidney and urinary tract diseases. For this reason, a more thorough investigation of the total protein composition of human urine is therefore required. Such a global analysis is important and can enhance our understanding of urogenital tract diseases and pathogenesis. The urinary proteome is of great interest in this respect. Current technology for proteome analysis includes two-dimensional polyacrylamide gel electrophoresis and mass spectrometry analysis. However, these techniques lack sensitivity in characterizing the urine proteome profile. Nanotechnology-based assays have been developed, which are highly sensitive and specific for proteomic profiling of human urine.[91,92]

Using this technique may help to stratify patients with hydronephrosis, so that those with obstruction sufficient to be causing renal injury can be separated from those with benign dilation. In a clinical study using nanospray liquid chromatography coupled with mass spectrometry, the urinary proteome profile of patients with unilateral high-grade hydronephrosis was compared to normal healthy controls, and there were significant differences in the number and quantity of urinary proteins in these two groups. These data are preliminary, and further studies are needed to validate these results in different diseases.[93]

In a somewhat similar study, nanotechnology-based proteome profiling showed three urinary proteins to be significantly altered in patients with bladder cancer due to chronic arsenic exposure as compared to proteins in the urine of urological patients without cancer.[94]

4.3.5 Regenerative Urology

There is no effective purely synthetic polymer-based technology available to replace or augment the urinary bladder in patients undergoing cystectomy or needing augmentation of their bladder capacity. The problems of using synthetic material for this purpose have been infection, toxicity, and issues of biocompatibility. Among the reasons for this may be the large size of the surface features of these polymeric formulations. Although seemingly small, the micrometer-sized surface features of these biomaterials do not sufficiently mimic the neighboring tissue and incite host responses, as well as limit the regenerative abilities of the surrounding natural tissue. Some recent studies have shown that these shortcomings can be overcome by using nanosized synthetic polymers. These studies demonstrated that nanostructured polymeric scaffolds (specifically, PLGA and polyether urethane) were capable of significantly enhancing the human bladder smooth muscle cell adhesion and proliferation and enhanced the production of extracellular matrix proteins in *in vitro* and *in vivo* experiments.[95–99]

4.3.6 Hemostatic Device (ENSEAL®)

With recent advances in laparoscopic surgery, efficient and reliable energy-based vascular sealing instruments have rapidly become an integral tool for most urologists, facilitating complex laparoscopic procedures. Traditional mono- and bipolar cautery devices result in inconsistent vessel sealing and generate significant heat that results in substantial thermal spread and charring. Use of nanotechnology-based methods may help improve this. A new tissue sealing and hemostasis system, ENSEAL®, significantly reduces thermal injury and inflammatory response. The ENSEAL system uses millions of nanometer-sized particles embedded in a bipolar temperature coefficient matrix (Smart Electrode Technology). The cutting mechanism is in the shape of an "I" beam, utilizing equal and high tissue compression to enhance the seal as the blade is advanced along the length of the jaw. This allows sealing and transection to occur in a single step and simultaneously regulates the current, thereby minimizing collateral thermal spread and tissue damage.[100] Electrical current flow is active only when the device jaws are closed and is modulated at the tissue–electrode interface by the nanoparticles, because they locally interrupt the electrical current when temperatures exceed 100°C. A study comparing ENSEAL Gyrus Bipolar Cutting Forceps (Gyrus ACMI, Maple Grove, MN), Ligasure 5 (Covidien, CO), and SonoSurg Ultrasonic Scissors (Olympus Surgical & Industrial America Inc., Orangeburg, NY) in a porcine model reported that the ENSEAL produces lower mean peak temperatures (86.9°C vs. 96.9°C to 180°C) and reduced mean thermal spread to surrounding tissues (1.10 mm vs. 2.78–3.23 mm).[101] ENSEAL has been shown to produce only 1 mm of lateral thermal tissue damage, the least amount of lateral thermal tissue damage of any bipolar-advanced energy modality.[102] In a case report using this technology for laparoscopic radical nephrectomy, it proved effective in performing both blunt dissection and sealing/transection of perirenal tissues (fibroadipose, peritoneal, and lymphatic). The device proved adequate for sealing and transecting small peripheral vessels less than 3 mm, and subjectively, plume production was minimal compared with previous experience with the Harmonic ACE (Ethicon Endo-Surgery Inc., Somerville, NJ) and monopolar cautery devices.[100] In another recent study, the ENSEAL Trio system was used along with cold saline irrigation for control of the vascular pedicles during bilateral non-nerve-sparing robotically assisted laparoscopic radical prostatectomy. Evaluating the specimens removed showed that the mean distance of thermal injury from the inked margin was only 0.31 mm (range 0.15–0.40 mm) demonstrating that this device works effectively under cold irrigation and suggesting that it may be able to be used to ligate the vascular pedicle of the prostate and still preserve the periprostatic nerves. This technique should improve the sexual function outcomes if used for nerve sparing radical prostatectomy.[103]

4.3.7 Erectile Dysfunction

Erectile dysfunction (ED) is a medical condition that affects more than 52% of men above age 40 years[104] and 16%–82% of prostate cancer patients treated by radical prostatectomy. Current treatments are ineffective in 50%–60% of these patients. The protein Sonic hedgehog is a critical regulator of penile smooth muscle and is decreased in models of diabetic ED and cavernous nerve injury after prostatectomy. This protein plays a significant role in peripheral nerve regeneration and has potential to be used as a regenerative therapy for the cavernosal nerves in prostatectomy patients and in other patients with neuropathy of peripheral nerves. However, efforts to regenerate the cavernous nerve have not been successful, in part due to inability to effectively deliver Sonic hedgehog protein to injured cavernosal nerves.[105] Biodegradable peptide amphiphile nanofibers are a noninvasive and effective method to deliver proteins *in vivo*, provide directional guidance to regenerating axons, and could deliver proteins over extended periods. The amphiphilic nature of the molecules in these nanofibers encourages aggregation in aqueous environments, while the β-sheet-forming segment specifically drives self-assembly into high aspect ratio nanofibers.

At appropriate concentrations, the bundling and entanglement of these nanofibers leads to the formation of a hydrogel that mimics the architecture of extracellular matrices, with tunable mechanical properties. The advantage of this methodology for protein delivery *in vivo* is that nanofibers are noninvasive, biodegradable, elicit no immune response, and form structures that can serve as a scaffold for regenerating axons. In one study, Sonic hedgehog protein was delivered to crushed cavernosal nerves in adult rats using linear biodegradable peptide amphiphile nanofibers for 6 weeks. After 6 weeks of Sonic hedgehog peptide amphiphile treatment, erectile function was improved by 58% as measured by intracavernosal pressure. Electron microscopy 6 weeks after crush injury showed abundant demyelination and axonal degeneration in controls but significantly improved cavernosal nerve morphology in the treated nerve group. Similar improvement was also seen by immunohistochemistry for glial fibrillary acidic protein. Control nerves had a 22% increase in this protein versus the treated group, implying that the Sonic hedgehog peptide amphiphile-treated cavernosal nerves had undergone improved regeneration.[105] In a similar study, it was shown that using Sonic hedgehog peptide amphiphile, there was a 9% and 19% increase in Sonic hedgehog protein at 4 and 7 days after cavernosal nerve injury and a 25% and 16% reduction in apoptosis at 4 and 7 days after injury.[106] These results suggest substantial translational potential of this methodology.

Intraurethral prostaglandin-E1 has been used for improving erectile function in patients with ED. Its use is limited by the pain associated with high doses. Topical liposomal preparations have been demonstrated to deliver higher drug concentration in the epidermis and dermis as compared to regular topical preparation in a study on different drugs. Electron microscopic observations suggest a mechanism of action, by confirming that intact liposomes penetrate the skin and deposit in the dermis acting as a slow-release depot system.[107] The same may be true in the urethra. In a small clinical study, the intraurethral application of liposomal prostaglandin-E1 produced rigidity in patients with psychogenic ED and may thus provide a therapeutic alternative in these selected patients.[108] None of these patients reported any pain or any other discomfort during administration of the drug. It may also be of benefit when given transdermally. In one study, five patients were randomized to receive either transdermal liposome encapsulated prostaglandin-E1 (0.05%) on the penis or placebo with a crossover at least 1 week later. The peak systolic flow velocities as measured by color Doppler in the deep cavernosal arteries of patients increased significantly compared with preapplication values in treated patients. The highest mean peak systolic flow velocity was achieved at 45 minutes after application of the formulation and resulted in a sevenfold increase in mean flow velocity compared with baseline values. This could serve as a new modality of drug delivery for patients with ED.[109]

4.3.8 Renal Transplant

In a study using a rat renal transplant preparation, rats treated with intravenous methylprednisolone had survivals of about 20 days, but with significant side effects, including increased protein excretion and retention of creatinine and urea. In contrast, a low dose of methylprednisolone encapsulated in bilayer liposomes, when used intravenously once weekly starting at time of transplant, prevented acute rejection and resulted in similar survival to 20 ± 7 days with minimal side effects. This study shows that the development of liposomal drug formulation has the potential to facilitate a reduction of steroidal immune-suppression in transplant recipients. Because side effects of steroids are a significant problem, this technology has great potential importance. The exact mechanism of the reduction in toxicity is unknown; however, possible mechanisms include inhibition of macrophage activation as well as a modification of antigen presentation and an activation of T-helper cells.[110]

4.3.9 Nanorobotics

Urology is a specialty that has been open to use of new technological advancements in treatment of various urological disorders. This is very evident from quick adaptation of surgical robotic technology in the treatment of various urological diseases. With advancements in robotic technology and nanotechnology simultaneously, there is the potential to develop nanorobots that could be used to deliver chemotherapy or targeted gene therapy into end organs or cells in remote parts of the body. The world's smallest untethered, controllable robot (the size of a human hair and half the length of a period) was recently developed, measuring 250×100 μm. The key to miniaturizing nanorobots is the downscaling of integrated circuit processors. It is projected that these circuits will have reduced the size of today's processors by a factor of 25 within 10 years, leading to the development of nanorobots measuring just 50 μm, small enough to get into the smallest vessels in the body.[111] The potential uses of these tiny robots remain to be determined, but their potential is amazing.

References

1. Bharali DJ and Mousa SA. Emerging nanomedicines for early cancer detection and improved treatment: current perspective and future promise. *Pharmacol Ther.* 2010;128(2):324–335.
2. Wang X, Yang L, Chen ZG, and Shin DM. Application of nanotechnology in cancer therapy and imaging. *CA Cancer J Clin.* 2008;58(2):97–110.
3. Jin S and Labhasetwar V. Nanotechnology in urology. *Urol Clin North Am.* 2009;36(2):179–188, viii.
4. Shergill IS, Rao A, Arya M, Patel H, and Gill IS. Nanotechnology: potential applications in urology. *BJU Int.* 2006;97(2):219–220.
5. Ilic D, O'Connor D, Green S, and Wilt TJ. Screening for prostate cancer: an updated Cochrane systematic review. *BJU Int.* 2011;107(6):882–891.
6. Kothari PS, Scardino PT, Ohori M, Kattan MW, and Wheeler TM. Incidence, location, and significance of periprostatic and periseminal vesicle lymph nodes in prostate cancer. *Am J Surg Pathol.* 2001;25(11):1429–1432.
7. Cheng L, Bergstralh EJ, Cheville JC et al. Cancer volume of lymph node metastasis predicts progression in prostate cancer. *Am J Surg Pathol.* 1998;22(12):1491–1500.
8. Kumar S, Shelley M, Harrison C, Coles B, Wilt TJ, and Mason MD. Neo-adjuvant and adjuvant hormone therapy for localised and locally advanced prostate cancer. *Cochrane Database Syst Rev.* 2006;(4):CD006019.
9. Messing EM, Manola J, Sarosdy M, Wilding G, Crawford ED, and Trump D. Immediate hormonal therapy compared with observation after radical prostatectomy and pelvic lymphadenectomy in men with node-positive prostate cancer. *N Engl J Med.* 9 1999;341(24):1781–1788.
10. Misselwitz B. MR contrast agents in lymph node imaging. *Eur J Radiol.* 2006;58(3):375–382.
11. Tempany CM and McNeil BJ. Advances in biomedical imaging. *JAMA.* 2001;285(5):562–567.
12. Harisinghani MG, Barentsz J, and Hahn PF et al. Noninvasive detection of clinically occult lymph-node metastases in prostate cancer. *N Engl J Med.* 2003;348(25):2491–2499.
13. Sun C, Lee JS and Zhang M. Magnetic nanoparticles in MR imaging and drug delivery. *Adv Drug Deliv Rev.* 2008;60(11):1252–1265.

14. Wang AZ, Bagalkot V, Vasilliou CC et al. Superparamagnetic iron oxide nanoparticle-aptamer bioconjugates for combined prostate cancer imaging and therapy. *Chem Med Chem.* 2008;3(9):1311–1315.
15. Guimaraes AR, Tabatabei S, Dahl D, McDougal WS, Weissleder R, and Harisinghani MG. Pilot study evaluating use of lymphotrophic nanoparticle-enhanced magnetic resonance imaging for assessing lymph nodes in renal cell cancer. *Urology.* 2008;71(4):708–712.
16. Deserno WM, Harisinghani MG, Taupitz M et al. Urinary bladder cancer: preoperative nodal staging with ferumoxtran-10-enhanced MR imaging. *Radiology.* 2004;233(2):449–456.
17. Harisinghani MG, Saksena M, Ross RW et al. A pilot study of lymphotrophic nanoparticle-enhanced magnetic resonance imaging technique in early stage testicular cancer: a new method for noninvasive lymph node evaluation. *Urology.* 2005;66(5):1066–1071.
18. Harisinghani MG, Saksena MA, Hahn PF et al. Ferumoxtran-10-enhanced MR lymphangiography: does contrast-enhanced imaging alone suffice for accurate lymph node characterization? *AJR Am J Roentgenol.* 2006;186(1):144–148.
19. Gao X, Cui Y, Levenson RM, Chung LW, and Nie S. *In vivo* cancer targeting and imaging with semiconductor quantum dots. *Nat Biotechnol.* 2004;22(8):969–976.
20. Shi C, Zhu Y, Cerwinka WH et al. Quantum dots: emerging applications in urologic oncology. *Urol Oncol.* 2008;26(1):86–92.
21. Datta P and Dasgupta A. Evaluation of an automated chemiluminescent immunoassay for complexed PSA on the Bayer ACS:180 system. *J Clin Lab Anal.* 2003;17(5):174–178.
22. Zheng G, Patolsky F, Cui Y, Wang WU, and Lieber CM. Multiplexed electrical detection of cancer markers with nanowire sensor arrays. *Nat Biotechnol.* 2005;23(10):1294–1301.
23. Nam JM, Park SJ, and Mirkin CA. Bio-barcodes based on oligonucleotide-modified nanoparticles. *J Am Chem Soc.* 2002;124(15):3820–3821.
24. Nam JM, Thaxton CS, and Mirkin CA. Nanoparticle-based bio-bar codes for the ultrasensitive detection of proteins. *Science.* 2003;301(5641):1884–1886.
25. Goluch ED, Nam JM, Georganopoulou DG et al. A bio-barcode assay for on-chip attomolar-sensitivity protein detection. *Lab Chip.* 2006;6(10):1293–1299.
26. Thaxton CS, Elghanian R, Thomas AD et al. Nanoparticle-based bio-barcode assay redefines "undetectable" PSA and biochemical recurrence after radical prostatectomy. *Proc Natl Acad Sci USA.* 2009;106(44):18437–18442.
27. Harma H, Soukka T, and Lovgren T. Europium nanoparticles and time-resolved fluorescence for ultrasensitive detection of prostate-specific antigen. *Clin Chem.* 2001;47(3):561–568.
28. Gokarna A, Jin LH, Hwang JS et al. Quantum dot-based protein micro- and nanoarrays for detection of prostate cancer biomarkers. *Proteomics.* 2008;8(9):1809–1818.
29. Matsumura Y and Maeda H. A new concept for macromolecular therapeutics in cancer chemotherapy: mechanism of tumoritropic accumulation of proteins and the antitumor agent smancs. *Cancer Res.* 1986;46(12 Pt 1):6387–6392.
30. Duncan R. The dawning era of polymer therapeutics. *Nat Rev Drug Discov.* 2003;2(5):347–360.
31. Jain RK. Delivery of molecular medicine to solid tumors: lessons from *in vivo* imaging of gene expression and function. *J Control Release.* 2001;74(1–3):7–25.
32. Jain RK. Understanding barriers to drug delivery: high resolution *in vivo* imaging is key. *Clin Cancer Res.* 1999;5(7):1605–1606.
33. Davis FF. The origin of pegnology. *Adv Drug Deliv Rev.* 2002;54(4):457–458.
34. Moghimi SM and Hunter AC. Poloxamers and poloxamines in nanoparticle engineering and experimental medicine. *Trends Biotechnol.* 2000;18(10):412–420.
35. Park EK, Lee SB, and Lee YM. Preparation and characterization of methoxy poly(ethylene glycol)/poly(epsilon-caprolactone) amphiphilic block copolymeric nanospheres for tumor-specific folate-mediated targeting of anticancer drugs. *Biomaterials.* 2005;26(9):1053–1061.
36. Gref R, Minamitake Y, Peracchia MT, Trubetskoy V, Torchilin V, and Langer R. Biodegradable long-circulating polymeric nanospheres. *Science.* 1994;263(5153):1600–1603.

37. Kiyokawa H, Igawa Y, Muraishi O, Katsuyama Y, Iizuka K, and Nishizawa O. Distribution of doxorubicin in the bladder wall and regional lymph nodes after bladder submucosal injection of liposomal doxorubicin in the dog. *J Urol.* 1999;161(2):665–667.

38. Lu Z, Yeh TK, Tsai M, Au JL, and Wientjes MG. Paclitaxel-loaded gelatin nanoparticles for intravesical bladder cancer therapy. *Clin Cancer Res.* 2004;10(22):7677–7684.

39. Winquist E, Ernst DS, Jonker D et al. Phase II trial of pegylated-liposomal doxorubicin in the treatment of locally advanced unresectable or metastatic transitional cell carcinoma of the urothelial tract. *Eur J Cancer.* 2003;39(13):1866–1871.

40. Arlt M, Haase D, Hampel S et al. Delivery of carboplatin by carbon-based nanocontainers mediates increased cancer cell death. *Nanotechnology.* 2010;21(33):335101.

41. Siddiqui IA, Adhami VM, Bharali DJ et al. Introducing nanochemoprevention as a novel approach for cancer control: proof of principle with green tea polyphenol epigallocatechin-3-gallate. *Cancer Res.* 2009;69(5):1712–1716.

42. Yalcin M, Bharali DJ, Lansing L et al. Tetraidothyroacetic acid (tetrac) and tetrac nanoparticles inhibit growth of human renal cell carcinoma xenografts. *Anticancer Res.* 2009;29(10):3825–3831.

43. Sumitomo M, Koizumi F, Asano T et al. Novel SN-38-incorporated polymeric micelle, NK012, strongly suppresses renal cancer progression. *Cancer Res.* 2008;68(6):1631–1635.

44. Allen TM. Ligand-targeted therapeutics in anticancer therapy. *Nat Rev Cancer.* 2002;2(10):750–763.

45. Dhar S, Gu FX, Langer R, Farokhzad OC, and Lippard SJ. Targeted delivery of cisplatin to prostate cancer cells by aptamer functionalized Pt(IV) prodrug-PLGA-PEG nanoparticles. *Proc Natl Acad Sci USA.* 2008;105(45):17356–17361.

46. Farokhzad OC, Cheng J, Teply BA et al. Targeted nanoparticle-aptamer bioconjugates for cancer chemotherapy *in vivo. Proc Natl Acad Sci U S A.* 2006;103(16):6315–6320.

47. Farokhzad OC, Jon S, Khademhosseini A, Tran TN, Lavan DA, and Langer R. Nanoparticle-aptamer bioconjugates: a new approach for targeting prostate cancer cells. *Cancer Res.* 2004;64(21):7668–7672.

48. Farokhzad OC, Karp JM, and Langer R. Nanoparticle-aptamer bioconjugates for cancer targeting. *Expert Opin Drug Deliv.* 2006;3(3):311–324.

49. Cheng J, Teply BA, Sherifi I et al. Formulation of functionalized PLGA-PEG nanoparticles for *in vivo* targeted drug delivery. *Biomaterials.* 2007;28(5):869–876.

50. Sun C, Fang C, Stephen Z et al. Tumor-targeted drug delivery and MRI contrast enhancement by chlorotoxin-conjugated iron oxide nanoparticles. *Nanomedicine (Lond).* 2008;3(4):495–505.

51. Bagalkot V, Zhang L, Levy-Nissenbaum E et al. Quantum dot-aptamer conjugates for synchronous cancer imaging, therapy, and sensing of drug delivery based on bi-fluorescence resonance energy transfer. *Nano Lett.* 2007;7(10):3065–3070.

52. Jain TK, Richey J, Strand M, Leslie-Pelecky DL, Flask CA, and Labhasetwar V. Magnetic nanoparticles with dual functional properties: drug delivery and magnetic resonance imaging. *Biomaterials.* 2008;29(29):4012–4021.

53. Lee IH, An S, Yu MK, Kwon HK, Im SH, and Jon S. Targeted chemoimmunotherapy using drug-loaded aptamer-dendrimer bioconjugates. *J Control Release.* 2011;155(3):435–441.

54. Jankun J. Protein-based nanotechnology: antibody conjugated with photosensitizer in targeted anticancer photoimmunotherapy. *Int J Oncol.* 2011;39(4):949–953.

55. Jichlinski P and Leisinger HJ. Photodynamic therapy in superficial bladder cancer: past, present and future. *Urol Res.* 2001;29(6):396–405.

56. Derycke AS, Kamuhabwa A, Gijsens A et al. Transferrin-conjugated liposome targeting of photosensitizer AlPcS4 to rat bladder carcinoma cells. *J Natl Cancer Inst.* 2004;96(21):1620–1630.

57. Stauffer PR. Evolving technology for thermal therapy of cancer. *Int J Hyperthermia.* 2005;21(8):731–744.

58. Krishnan S, Diagaradjane P, and Cho SH. Nanoparticle-mediated thermal therapy: evolving strategies for prostate cancer therapy. *Int J Hyperthermia.* 2010;26(8):775–789.

59. Stern JM, Stanfield J, Lotan Y, Park S, Hsieh JT, and Cadeddu JA. Efficacy of laser-activated gold nanoshells in ablating prostate cancer cells *in vitro. J Endourol.* 2007;21(8):939–943.

60. Stern JM, Stanfield J, Kabbani W, Hsieh JT, and Cadeddu JA. Selective prostate cancer thermal ablation with laser activated gold nanoshells. *J Urol.* 2008;179(2):748–753.

61. Malugin A and Ghandehari H. Cellular uptake and toxicity of gold nanoparticles in prostate cancer cells: a comparative study of rods and spheres. *J Appl Toxicol.* 2010;30(3):212–217.

62. Gobin AM, Moon JJ, and West JL. EphrinA I-targeted nanoshells for photothermal ablation of prostate cancer cells. *Int J Nanomed.* 2008;3(3):351–358.

63. Lu W, Singh AK, Khan SA, Senapati D, Yu H, and Ray PC. Gold nano-popcorn-based targeted diagnosis, nanotherapy treatment, and in situ monitoring of photothermal therapy response of prostate cancer cells using surface-enhanced Raman spectroscopy. *J Am Chem Soc.* 2010;132(51):18103–18114.

64. Burke A, Ding X, Singh R et al. Long-term survival following a single treatment of kidney tumors with multiwalled carbon nanotubes and near-infrared radiation. *Proc Natl Acad Sci USA.* 2009;106(31):12897–12902.

65. Ghosh S, Dutta S, Gomes E et al. Increased heating efficiency and selective thermal ablation of malignant tissue with DNA-encased multiwalled carbon nanotubes. *ACS Nano.* 2009;3(9):2667–2673.

66. Kawai N, Ito A, Nakahara Y et al. Anticancer effect of hyperthermia on prostate cancer mediated by magnetite cationic liposomes and immune-response induction in transplanted syngeneic rats. *Prostate.* 2005;64(4):373–381.

67. Kawai N, Ito A, Nakahara Y et al. Complete regression of experimental prostate cancer in nude mice by repeated hyperthermia using magnetite cationic liposomes and a newly developed solenoid containing a ferrite core. *Prostate.* 2006;66(7):718–727.

68. Kawai N, Futakuchi M, Yoshida T et al. Effect of heat therapy using magnetic nanoparticles conjugated with cationic liposomes on prostate tumor in bone. *Prostate.* 2008;68(7):784–792.

69. Johannsen M, Gneveckow U, Eckelt L et al. Clinical hyperthermia of prostate cancer using magnetic nanoparticles: presentation of a new interstitial technique. *Int J Hyperthermia.* 2005;21(7):637–647.

70. Johannsen M, Gneveckow U, Thiesen B et al. Thermotherapy of prostate cancer using magnetic nanoparticles: feasibility, imaging, and three-dimensional temperature distribution. *Eur Urol.* 2007;52(6):1653–1661.

71. Johannsen M, Gneveckow U, Taymoorian K et al. Morbidity and quality of life during thermotherapy using magnetic nanoparticles in locally recurrent prostate cancer: results of a prospective phase I trial. *Int J Hyperthermia.* 2007;23(3):315–323.

72. Van Vulpen M, De Leeuw AA, Raaymakers BW et al. Radiotherapy and hyperthermia in the treatment of patients with locally advanced prostate cancer: preliminary results. *BJU Int.* 2004;93(1):36–41.

73. Wust P, Gneveckow U, Johannsen M et al. Magnetic nanoparticles for interstitial thermotherapy—feasibility, tolerance and achieved temperatures. *Int J Hyperthermia.* 2006;22(8):673–685.

74. Johannsen M, Thiesen B, Wust P, and Jordan A. Magnetic nanoparticle hyperthermia for prostate cancer. *Int J Hyperthermia.* 2010;26(8):790–795.

75. Mastrobattista E, van der Aa MA, Hennink WE, and Crommelin DJ. Artificial viruses: a nanotechnological approach to gene delivery. *Nat Rev Drug Discov.* 2006;5(2):115–121.

76. Prabha S and Labhasetwar V. Critical determinants in PLGA/PLA nanoparticle-mediated gene expression. *Pharm Res.* 2004;21(2):354–364.

77. Cao ZG, Zhou SW, Sun K, Lu XB, and Luo G, and Liu JH. [Preparation and feasibility of superparamagnetic dextran iron oxide nanoparticles as gene carrier]. *Ai Zheng.* 2004;23(10):1105–1109.

78. Moffatt S, Papasakelariou C, Wiehle S, and Cristiano R. Successful *in vivo* tumor targeting of prostate-specific membrane antigen with a highly efficient J591/PEI/DNA molecular conjugate. *Gene Ther.* 2006;13(9):761–772.

79. Moffatt S and Cristiano RJ. PEGylated J591 mAb loaded in PLGA-PEG-PLGA tri-block copolymer for targeted delivery: *in vitro* evaluation in human prostate cancer cells. *Int J Pharm.* 2006;317(1):10–13.

80. Liedberg H. Catheter induced urethral inflammatory reaction and urinary tract infection. An experimental and clinical study. *Scand J Urol Nephrol Suppl.* 1989;124:1–43.
81. Samuel U and Guggenbichler JP. Prevention of catheter-related infections: the potential of a new nano-silver impregnated catheter. *Int J Antimicrob Agents.* 2004;23(Suppl 1):S75–S78.
82. Muzzi-Bjornson L and Macera L. Preventing infection in elders with long-term indwelling urinary catheters. *J Am Acad Nurse Pract.* 2011;23(3):127–134.
83. Karchmer TB, Giannetta ET, Muto CA, Strain BA, and Farr BM. A randomized crossover study of silver-coated urinary catheters in hospitalized patients. *Arch Intern Med.* 2000;160(21):3294–3298.
84. Srinivasan A, Karchmer T, Richards A, Song X, and Perl TM. A prospective trial of a novel, silicone-based, silver-coated foley catheter for the prevention of nosocomial urinary tract infections. *Infect Control Hosp Epidemiol.* 2006;27(1):38–43.
85. Guggenbichler JP, Boswald M, Lugauer S, and Krall T. A new technology of microdispersed silver in polyurethane induces antimicrobial activity in central venous catheters. *Infection.* 1999;27(Suppl 1):S16–S23.
86. Evliyaoglu Y, Kobaner M, Celebi H, Yelsel K, and Dogan A. The efficacy of a novel antibacterial hydroxyapatite nanoparticle-coated indwelling urinary catheter in preventing biofilm formation and catheter-associated urinary tract infection in rabbits. *Urol Res.* 2011;39(6):443–449.
87. Basu M, Seggerson S, Henshaw J et al. Nano-biosensor development for bacterial detection during human kidney infection: use of glycoconjugate-specific antibody-bound gold NanoWire arrays (GNWA). *Glycoconj J.* 2004;21(8–9):487–496.
88. Yoon BI, Ha US, Sohn DW et al. Anti-inflammatory and antimicrobial effects of nanocatechin in a chronic bacterial prostatitis rat model. *J Infect Chemother.* 2011;17(2):189–194.
89. Son A, Dhirapong A, Dosev DK, Kennedy IM, Weiss RH, and Hristova KR. Rapid and quantitative DNA analysis of genetic mutations for polycystic kidney disease (PKD) using magnetic/luminescent nanoparticles. *Anal Bioanal Chem.* 2008;390(7):1829–1835.
90. Fliser D, Novak J, Thongboonkerd V et al. Advances in urinary proteome analysis and biomarker discovery. *J Am Soc Nephrol.* 2007;18(4):1057–1071.
91. Lee RS, Monigatti F, Lutchman M et al. Temporal variations of the postnatal rat urinary proteome as a reflection of systemic maturation. *Proteomics.* 2008;8(5):1097–1112.
92. Tyan YC, Guo HR, Liu CY, and Liao PC. Proteomic profiling of human urinary proteome using nano-high performance liquid chromatography/electrospray ionization tandem mass spectrometry. *Anal Chim Acta.* 2006;579(2):158–176.
93. Mesrobian HG, Mitchell ME, See WA et al. Candidate urinary biomarker discovery in ureteropelvic junction obstruction: a proteomic approach. *J Urol.* 2010;184(2):709–714.
94. Tan LB, Chen KT, Tyan YC, Liao PC, and Guo HR. Proteomic analysis for human urinary proteins associated with arsenic intoxication. *Proteomics Clin Appl.* 2008;2(7–8):1087–1098.
95. Pattison M, Webster TJ, Leslie J, Kaefer M, and Haberstroh KM. Evaluating the *in vitro* and *in vivo* efficacy of nano-structured polymers for bladder tissue replacement applications. *Macromol Biosci.* 2007;7(5):690–700.
96. Pattison MA, Wurster S, Webster TJ, and Haberstroh KM. Three-dimensional, nano-structured PLGA scaffolds for bladder tissue replacement applications. *Biomaterials.* 2005;26(15):2491–2500.
97. Thapa A, Miller DC, Webster TJ, and Haberstroh KM. Nano-structured polymers enhance bladder smooth muscle cell function. *Biomaterials.* 2003;24(17):2915–2926.
98. Roth CC. Urologic tissue engineering in pediatrics: from nanostructures to bladders. *Pediatr Res.* 2010;67(5):509–513.
99. Roth CC and Kropp BP. Recent advances in urologic tissue engineering. *Curr Urol Rep.* 2009;10(2):119–125.
100. Smaldone MC, Gibbons EP, and Jackman SV. Laparoscopic nephrectomy using the EnSeal Tissue Sealing and Hemostasis System: successful therapeutic application of nanotechnology. *JSLS.* 2008;12(2):213–216.
101. Landman J, Kerbl K, Rehman J et al. Evaluation of a vessel sealing system, bipolar electrosurgery, harmonic scalpel, titanium clips, endoscopic gastrointestinal anastomosis vascular staples and sutures for arterial and venous ligation in a porcine model. *J Urol.* 2003;169(2):697–700.

102. Person B, Vivas DA, Ruiz D, Talcott M, Coad JE, and Wexner SD. Comparison of four energy-based vascular sealing and cutting instruments: a porcine model. *Surg Endosc.* 2008;22(2):534–538.
103. Zorn KC, Bhojani N, Gautam G et al. Application of ice cold irrigation during vascular pedicle control of robot-assisted radical prostatectomy: EnSeal instrument cooling to reduce collateral thermal tissue damage. *J Endourol.* 2010;24(12):1991–1996.
104. Feldman HA, Goldstein I, Hatzichristou DG, Krane RJ, and McKinlay JB. Impotence and its medical and psychosocial correlates: results of the Massachusetts Male Aging Study. *J Urol.* 1994;151(1):54–61.
105. Angeloni NL, Bond CW, Tang Y et al. Regeneration of the cavernous nerve by Sonic hedgehog using aligned peptide amphiphile nanofibers. *Biomaterials.* 2011;32(4):1091–1101.
106. Bond CW, Angeloni NL, Harrington DA, Stupp SI, McKenna KE, and Podlasek CA. Peptide amphiphile nanofiber delivery of sonic hedgehog protein to reduce smooth muscle apoptosis in the penis after cavernous nerve resection. *J Sex Med.* 2011;8(1):78–89.
107. Foldvari M, Gesztes A, and Mezei M. Dermal drug delivery by liposome encapsulation: clinical and electron microscopic studies. *J Microencapsul.* 1990;7(4):479–489.
108. Engelhardt PF, Plas E, Hubner WA, and Pfluger H. Comparison of intraurethral liposomal and intracavernosal prostaglandin-E1 in the management of erectile dysfunction. *Br J Urol.* 1998;81(3):441–444.
109. Foldvari M, Oguejiofor C, Afridi S, Kudel T, and Wilson T. Liposome encapsulated prostaglandin E1 in erectile dysfunction: correlation between *in vitro* delivery through foreskin and efficacy in patients. *Urology.* 1998;52(5):838–843.
110. Binder J, Braeutigam R, Oertl A et al. Methylprednisolone in bilayer liposomes prolongs cardiac and renal allograft survival, inhibits macrophage activation, and selectively modifies antigen presentation and T-helper cell function in rat recipients. *Transplant Proc.* 1998;30(4):1051.
111. Murphy D, Challacombe B, Khan MS, and Dasgupta P. Robotic technology in urology. *Postgrad Med J.* 2006;82(973):743–747.

5

Nanotechnology Applications in Preventive Medicine and Public Health

Julielynn Wong, MD, MPH and Sara Brenner, MD, MPH

CONTENTS

5.1 Introduction

Preventive medicine is a medical specialty that applies a population-based or systemic approach to improve the health of individuals (American College of Preventive Medicine 2012). Among the many roles of preventive medicine physicians, most work to identify high-risk subpopulations through the application of statistical methods, prevent and limit disease and injury, facilitate early diagnosis through screening and education outreach, promote health and wellness in individuals and populations, and improve the quality of health-care delivery (American College of Preventive Medicine 2011). Public health is a field that focuses on evaluating health status, preventing disease, and promoting the health of populations (Aschengrau and Seage 2003). According to the Association of

Schools of Public Health, the five core disciplines of public health are (1) epidemiology, (2) biostatistics, (3) health policy management, (4) social and community behavior, and (5) environmental health sciences (Association of Schools of Public Health 2006). Major public health issues addressed in the past century include addiction, automotive safety, cancer, cardiovascular disease, environmental health, food safety, infectious disease control, maternal and child health, oral health, public health infrastructure, and vaccines (DeBuono 2006).

Preventive medicine physicians and public health practitioners deal with diverse patient populations, especially the underserved and vulnerable (American College of Preventive Medicine 2012). The groups they serve range in scope from individuals, families, schools, work sites, military units, and local communities to regional, state, national, and international populations. Preventive medicine specialists work in the arenas of public health, managed care, occupational medicine, aerospace medicine, informatics, policy development, global health, and research, and in academic and clinical medicine across all specialties and disciplines.

Prevention strategies are typically classified into three levels: primary, secondary, and tertiary (Aschengrau and Seage 2003). Primary prevention is defined as activities that prevent the onset and reduce the incidence of disease. Examples of primary prevention include utilization of sunscreens; vaccinations; and ensuring a clean supply of air, food, and water. Secondary prevention is defined as strategies that prolong the onset and severity of symptomatic clinical disease. Examples of secondary prevention include screening practices such as detection of coronary artery disease or certain types of cancer. Tertiary prevention is defined as interventions that slow down disease progression and reduce disease morbidity and mortality. Examples of tertiary prevention are strict glycemic control in Type 1 diabetes and HIV antiretroviral therapy. Tertiary prevention strategies include clinical interventions across medical and surgical specialties. Therefore, this chapter, which focuses on public health and preventive medicine, mainly focuses on nanomedicine applications of primary and secondary prevention strategies.

5.2 Epidemiology

In 2008, the World Health Organization (WHO) compiled the top 10 causes of death worldwide, as listed in Table 5.1. Tables 5.2 and 5.3 further describe the top 10 causes of death worldwide when categorized by low- and high-income countries, respectively

5.3 Economic Impact of Disease

Although accurate estimates of the global costs of the principal causes of death are nearly impossible to ascertain, an online analysis estimated that the top 10 leading causes of death in the United States cost the American economy a total of $1.1 trillion (Allen 2012). The enormous economic and untold social costs underscore the dramatic need for more effective primary, secondary, and tertiary prevention measures.

TABLE 5.1

Top 10 Causes of Death Worldwide

Causes of Death	Deaths (millions)	Deaths (%)
Ischemic heart disease	7.25	12.8
Stroke and other cerebrovascular diseases	6.15	10.8
Lower respiratory infections	3.46	6.1
Chronic obstructive pulmonary disease	3.28	5.8
Diarrheal diseases	2.46	4.3
HIV/AIDS	1.78	3.1
Trachea, bronchus, and lung cancers	1.39	2.4
TB	1.34	2.4
Diabetes mellitus	1.26	2.2
Road traffic accidents	1.21	2.1

Source: World Health Organization. 2011. The Top 10 Causes of Death: Fact Sheet. http://www.who.int/mediacentre/factsheets/fs310/en/index.html.

TABLE 5.2

Top 10 Causes of Death in Low-Income Countries

Causes of Death	Deaths (millions)	Deaths (%)
Lower respiratory infections	1.05	11.3
Diarrheal diseases	0.76	8.2
HIV/AIDS	0.72	7.8
Ischemic heart disease	0.57	6.1
Malaria	0.48	5.2
Stroke and other cerebrovascular diseases	0.45	4.9
TB	0.40	4.3
Prematurity and low birth weight	0.30	3.2
Birth asphyxia and birth trauma	0.27	2.9
Neonatal infections	0.24	2.6

Source: World Health Organization. 2011. The Top 10 Causes of Death: Fact Sheet. http://www.who.int/mediacentre/factsheets/fs310/en/index.html.

5.4 Opportunities for Advancement through Technology

The value and the way any technology is judged is according to the range and depth of the problems that it solves.

Dillon (2010)

Few problems are as critical to our society as the leading and preventable causes of death and illness. Innovative nanomedicine technologies can and should focus on leading causes

TABLE 5.3

Top 10 Causes of Death in High-Income Countries

Causes of Death	Deaths (millions)	Deaths (%)
Ischemic heart disease	1.42	15.6
Stroke and other cerebrovascular diseases	0.79	8.7
Trachea, bronchus, and lung cancers	0.54	5.9
Alzheimer's disease and other dementias	0.37	4.1
Lower respiratory infections	0.35	3.8
Chronic obstructive pulmonary disease	0.32	3.5
Colon and rectum cancers	0.30	3.3
Diabetes mellitus	0.24	2.6
Hypertensive heart disease	0.21	2.3
Breast cancer	0.17	1.9

Source: World Health Organization. 2011. The Top 10 Causes of Death: Fact Sheet.http://www.who.int/mediacentre/factsheets/fs310/en/index.html.

of death, pain, suffering, and health-care expenditures. A preventive medicine perspective allows one to take an unbiased systems-level approach to identify where the greatest potential impact of nanomedicine applications lies both in the United States and globally. Although great public health advances have been made in the past century, more work needs to be done in the areas of child mortality, immunizations, safe water/sanitation, malaria, HIV/AIDS, tuberculosis (TB), tropical diseases, tobacco control, road safety, and global health threats (Table 5.4).

5.5 Current Nanomedicine Applications

5.5.1 Nanovaccines

Experimental nanovaccines can be delivered via nanoemulsion adjuvants, encapsulating nanoparticles, nanobeads, or micro-nanoprojections (Nandedkar 2009). Nanovaccines can be administered subcutaneously or noninvasively via oral, nasal, or cutaneous routes. Vaccines are currently developed for type 1 diabetes, hepatitis B, *Chlamydia trachomatis*, *Escherichia coli*, TB, and leishmaniasis. Additional vaccine development and applications can be found in Chapter 1.

5.5.2 Nanoemulsion Adjuvants

Needle-free nasal hepatitis B vaccines with a nanoemulsion adjuvant are reported as being safe, well tolerated, and effective in multiple animal models (Makidon et al. 2008). The nanoemulsion adjuvant contains emulsified detergent droplets, 40 nm in size, along with soybean oil, alcohol, and water. This nanoemulsion does not exhibit the irritating side

TABLE 5.4

Top 10 Global Public Health Achievements from 2001 to 2010 and Current Standard of Care

Public Health Achievement (Not Ranked in Any Order)	Current Standard of Care/Best Practices
Reductions in deaths in children aged <5 years, which are primarily due to infectious diseases (diarrhea, pneumonia, malaria, and AIDS); undernutrition; prematurity; birth asphyxia; and neonatal sepsis	Scaled-up immunizations, micronutrient supplements, access to safe water, insecticide-treated bed nets, oral rehydration therapy, antimicrobials, antimalarials and antiretrovirals, increased financial resources, strong partnerships, intensified country support, and innovations in service delivery
Vaccine-preventable diseases (i.e., measles, polio, diphtheria, tetanus, pertussis, hepatitis B, and *Haemophilus influenza* type B)	Expanded vaccination coverage globally
Access to safe water and sanitation	Water, sanitation, and hygiene initiatives to increase water and sanitation coverage, promote hygienic practices, and continuation of existing services
Malaria prevention and control	Insecticide-treated bed nets, indoor residual spraying, speedy diagnosis, timely treatment with artemisinin combination therapy, and preventive treatment during pregnancy
Prevention and control of HIV/AIDS	Scaled-up programs on provider-initiated HIV testing and counseling, prevention of mother–child HIV transmission, expanded use and availability of condoms and sterile injection gear, better blood safety, and antiretroviral therapy, particularly in low- and middle-income countries
TB control	Directly observed therapy, short-course strategy, initiation of antiretroviral therapy in patients coinfected with HIV, guided use and quality control of second-line drugs for multidrug-resistant TB
Control of neglected tropical diseases (dracunculiasis, onchocerciasis, and lymphatic filariasis)	Filters, larvicidal agents, safe water sources, public health education to prevent water contamination and encourage filtration for guinea worm disease; widespread drug treatment to prevent river blindness; antifilarial drugs
Tobacco control	MPOWER strategy package (monitor use, protect from tobacco smoke, offer to help quit, warn about risks, enforce advertising bans, and raise tobacco taxes)
Increased awareness and response for improving global road safety	Road safety advances including better road and vehicle design; speed control; compliance with using seat belts and helmets; improved public transit systems; reducing drunk driving; and more effective care of road traffic injuries at local, regional, national, and global levels
Improved preparedness and response to global health threats (i.e., AIDS, severe acute respiratory syndrome, and influenza)	Use of digital media for public health, international influenza response networks, early detection and response systems to unusual disease clusters, improved laboratory and epidemiological capabilities and training, expanded geographic availability of new diagnostic tests, regional training to reduce transmission at animal–human interfaces

Source: Centers for Disease Control and Prevention. 2011. Ten Great Public Health Achievements—Worldwide, 2001–2010. http://www.cdc.gov/mmwr/preview/mmwrhtml/mm6024a4.htm.

effects of conventional adjuvants, such as alum. The nasal vaccine is nontoxic and pain free and avoids the risks associated with intramuscular injections. This nanoemulsion vaccine is stable for 6 months and does not require refrigeration, a huge advantage for distribution in the developing world. This experimental hepatitis B vaccine series can be completed with two shots instead of the usual three shots, which is likely to improve compliance with the dosing schedule (Pautler and Brenner 2010). This nanoemulsion strategy shows potential promise for vaccines against smallpox, influenza, anthrax, and HIV (Nandedkar 2009). Sinyakov et al. (2006) used monodispersed polyacrolein nanoparticles, about 200 nm in size, as an adjuvant in mice and found that antibody response was comparable to that of the control adjuvant, bovine serum albumin alum. Oral nanoemulsion vaccines containing MAGE1-HSP70 and SEA complex proteins using a magnetic ultrasound method elicited a strong immune response and were shown to delay tumor growth and prevent tumor occurrence in mice (Ge et al. 2009). Another nanoemulsion vaccine, comprising MG7 mimotope peptide and CpG ODN 1645 adjuvant, was shown to reduce the size and rate of gastric tumor occurrence in mice (Shi et al. 2005).

5.5.3 Nanoparticle Encapsulation

The barriers to global coverage of conventional vaccines are cost per dose, poor compliance with dosing schedules, and the necessity for cold storage (Nandedkar 2009). Two billion people globally are infected with hepatitis B, which leads to approximately 600,000 deaths every year (WHO 2010). Although a highly effective vaccine exists to prevent hepatitis B, it requires two to four doses to achieve immunity, depending on the vaccine type (CDC 2012). Noncompliance with the dosing schedule can lead to reduced or no immunity to this carcinogenic agent. Experimental nanovaccines can help overcome these challenges to worldwide vaccination coverage by decreasing the dosing schedule to a single immunization and provide a vaccine that can be transported without refrigeration (Pautler and Brenner 2010). Future challenges in implementing nanovaccines include ensuring standardized nano-sized formulations during the production process and demonstrating nanovaccine efficacy in clinical trials (Nandedkar 2009).

Biocompatible and biodegradable polymer-based encapsulating nanoparticles, such as polylactide-*co*-glycolide acid (PLGA), have been developed for use as potential immunostimulatory adjuvants for polymer-based vaccines, which tend to be weakly antigenic (Nandedkar 2009). The advantages of PLGA nanoparticle encapsulation of antigens are the extensively documented safety profile, demonstrated biocompatibility, ease of delivery, protection of antigen from degradation, reduced dosing schedule due to prolonged release of antigen, and avoiding the use of inflammatory adjuvants (Kersten and Hirschberg 2004; Nandedkar 2009). An oral DNA vaccine for hepatitis B using encapsulated PLGA nanoparticles was shown to enhance immunity in mice models (He et al. 2005). *In vitro* and *in vivo* studies on mice showed that a single dose of PGLA nanoparticle encapsulating the hepatitis B surface antigen vaccine could generate antibody levels comparable to three doses of the standard hepatitis B vaccine (Feng et al. 2006).

A PGLA85/15 encapsulated nanoparticle (~200 nm, 98% encapsulation efficiency) containing a recombinant peptide, *C. trachomatis* major outer membrane protein, has been developed and tested in animals (Taha et al. 2012). This nanovaccine against the world's most common sexually transmitted disease has been shown to generate an immune response in mice (Taha et al. 2012; WHO 2012a).

An encapsulated vaccine against enterotoxigenic *E. coli*, a principal cause of diarrhea in the developing world, has been tested in mice (Deng et al. 2012). This vaccine consists of 80–200 nm nanospheres containing a fusion protein comprising a purified heat-stable enterotoxin and subunit B of a heat-labile enterotoxin. This gelatin nanovaccine induced an increased and longer immune response in mice compared to a control vaccine in a standard aluminum hydrate salt.

Studies on an Rv3619c nanovaccine against TB has been shown to elicit an effective cell-mediated response and decrease the mycobacterial burden in infected mice (Ansari et al. 2011). This nanovaccine encapsulates RD gene products, which is a T-cell antigen.

A subcutaneous single-dose nanovaccine for leishmaniasis, consisting of chitosan nanoparticles encapsulating a recombinant *Leishmania* superoxide dismutase, can generate cell-mediated immunity in mice at levels effective for disease eradication (Danesh-Bahreini et al. 2011). This vaccine in development will be a huge boon to the estimated 350 million people in 88 countries at risk for this protozoan disease.

Encapsulating vaccine antigens in bioadhesive nanoparticles can potentially improve the efficacy of nasal vaccines (Slutter et al. 2008). Particulate adjuvants can also help to shield the antigen from degradation during intramuscular delivery and can also stimulate cell-mediated immunity for viral infections (Nandedkar 2009).

5.5.4 Nanobeads

Nanobeads are solid inert beads whose surfaces are coated with antigen (Nandedkar 2009). The advantages of nanobead vaccines are that they are stable during storage, efficiently taken up by antigen-presenting cells, and require a lower dose of antigen (Gengoux and Leclerc 1995). Nanobeads exhibit size-dependent immunogenicity (Fifis et al. 2004; Scheerlinck et al. 2006). Nanobeads, 40 nm in size, have been shown to promote higher T-cell responses compared to other antigens mixed with nanobeads or antigens alone in mice (Fifis et al. 2004). Nanobeads, 50 nm in size, are known to induce both humoral and cell-mediated responses in animal models. Nanobeads have been shown to be effective preventive and therapeutic immunogens for mouse tumors (Nandedkar 2009). Systemic delivery of nanobead vaccines coated with type 1 diabetes–relevant peptide major histocompatibility complex molecules has been shown to initiate the expansion of disease-suppressing T cells (Clemente-Casares et al. 2011) in mice. This dampened the chronic autoimmune response against pancreatic β cells and reestablished glucose homeostasis.

5.5.5 Micro-Nanoprojection Patches

Needle-free skin delivery technologies are painless and can achieve more effective targeted delivery of antigen to the highly immunogenic Langerhans cells in the viable epidermis (Nandedkar 2009). In 2008, Chen et al. developed a skin patch with microneedles with nanoscale tips coated with a DNA vaccine. They applied this coated skin patch to the ears of mice and showed that it was able to successfully deliver the vaccine products to the epidermis. Using ovalbumin and Fluvax 2008, a commercial trivalent influenza vaccine, these nanopatches induced robust systemic immune responses in mice (Raphael et al. 2010). These innovative experiments suggest that needle-free skin patches with coated nanoprojections could provide pain-free, safe, and efficient vaccines.

5.5.6 Nicotine Nanovaccine

In 2008, a WHO report estimated that 5.4 million deaths worldwide were attributed to tobacco use. Tobacco cessation can significantly reduce the risk of premature death and the development of numerous tobacco-related diseases (CDC 2011a). Nicotine is a highly addictive drug, and nicotine dependence is the most prevalent form of chemical addiction in the United States (CDC 2011b). Smokers often require multiple attempts before they successfully quit. Among current American adult smokers, nearly 70% reported that they wanted to quit, and millions have attempted to quit smoking. An efficacious nicotine nanovaccine could be a huge boon to the millions of tobacco users who seek to effectively quit and would lead to billions of dollars in health-care cost savings (DeBuono 2006).

In 2011, the world's first fully integrated, synthetic nanovaccine, SEL-068, began to be tested in phase I randomized, double-blinded, and placebo-controlled clinical trials in healthy smoking and nonsmoking adults (Selecta Biosciences 2011; ClinicalTrials.gov 2012). SEL-068 is a self-assembling subcutaneous two-dose nanovaccine against nicotine for preventing relapse during tobacco cessation (Kishimoto 2012). It is theorized that the high antibody levels generated by the vaccine can bind to inhaled nicotine and prevent it from reaching the brain and setting off an addiction response (Selecta Biosciences 2011). The nanovaccine contains a biodegradable and biocompatible polymer matrix that encapsulates a synthetic toll-like receptor (TLR) agonist plus a peptide antigen, which stimulates cellular immunity (Kishimoto 2012). SEL-068 also has nicotine, a B-cell antigen, covalently bound to the nanoparticle surface. Therefore, SEL-068 incorporates both encapsulating nanoparticle and nanobead designs to generate both B- and T-cell-mediated immunities. The nanoparticles are also designed to target lymph node tissue. The controlled release of the TLR agonist minimizes inflammation, thereby raising the safety profile and facilitating the use of new adjuvants. The self-assembling nature of SEL-068 facilitates cost-effective nanovaccine production. No dose-limiting systematic toxicities were seen in initial safety and efficacy studies in monkeys. The primary outcome results of the phase I clinical trial are scheduled to be completed in the spring of 2013 (ClinicalTrials.gov 2012).

5.6 Broader Applications

5.6.1 Road Safety

Nearly 1.3 million people die on the world's roads every year, and this figure is expected to double by 2030 (CDC 2011c). Improved vehicle design is one method to reduce the forecasted global rise in road traffic deaths. Nanoscale compounds added to polymer composites are used in car parts (e.g., fenders) to make them light, rigid, durable, and flexible (Nano.gov 2012). Because of their higher mechanical strength at less weight per unit volume, carbon nanotubes are used in the manufacture of car parts that are lightweight and strong (National Nanotechnology Initiative 2008). For example, FIBRIL multiwalled carbon/plastic composite nanotubes are used to provide electrical conductivity to plastic parts of fenders and fuel lines while retaining base resin toughness (Project on Emerging Nanotechnologies 2012a). These incredibly strong carbon nanotubes help to prevent brittle failures of critical car parts during accidents. Several nano-based windscreen sealant compounds (e.g., Nano Windschutzscheiben-Versiegelung Profi-Spray-Set, Nanosafeguard Auto Glass Treatment, NanoConcept Pane Fit Sealing, TCnano Clear View, and Easy2Clean Windscreen) are commercially available. These nanosealants facilitate the clearing of rain and snow from windscreens and thereby enhance driver visibility and safety during adverse weather conditions. Glare-resistant coatings are being applied to windows and car mirrors (National Nanotechnology Initiative 2008). Nanotechnology has been used to create a highly flexible silica-based compound for tires that maximizes road grip in wet and dry conditions (Project on Emerging Nanotechnologies 2012a).

Better road design and expanded public transportation infrastructure can also save millions of lives (CDC 2011c). Nanoengineering of steel and cement materials holds great future promise for decreasing costs while increasing the performance, durability, and life span of highway and transportation infrastructure (Nano.gov 2012). Nanosensors can offer affordable and continuous monitoring of the condition and structural performance of bridges, tunnels, railroads, and roadways. Also, nano-based sensors can support a dynamic transportation monitoring and communication system to help drivers avoid collisions.

5.6.2 Food Safety

Reductions in microbial contamination have led to dramatic improvements in food safety over the past century (CDC 1999). However, even today everyone remains at risk for food-borne illnesses (WHO 2007). An estimated 76 million cases of food-borne illnesses occur each year in the United States alone, leading to over 300,000 hospitalizations and about 5,000 deaths annually (DeBuono 2006). It is estimated that one in three people will develop a food-borne illness every year. Although the total global burden remains unknown (DeBuono 2006), it is recognized that food-borne illnesses pose a significant and rising public health burden in both developing and developed countries (WHO 2007). Therefore, improvements in food safety systems are still required (DeBuono 2006).

Potential nanoscale applications could be used in safe food production, processing, and manufacturing (United States Department of Agriculture 2012). Nanotechnology applications in agricultural and veterinary sectors may enhance crop yields and livestock productivity, which may impact nutrition worldwide (Weber 2011). Nanosensors are under development to detect microbial and chemical contaminants (i.e., pathogens, pesticides, allergens, heavy metals, and particulates) and environmental characteristics (i.e., wet/dry, temperature, and color) of food prior to packaging and shipping (Nano.gov 2012; United States Department of Agriculture 2012). Nanoengineered materials are used in food containers to decrease oxygen inflow or bacterial growth to prolong shelf life. N-Coat is a commercially available transparent film with outstanding gas barrier properties for snack food packaging (Project on Emerging Nanotechnologies 2012a). Nano Plastic Wrap by SongSing Nano Technology, Co., Taiwan, uses nano-sized zinc oxide to help keep wrapped food fresh. At Texas A&M University, scientists are developing a food-compatible nanoclay film, which is impermeable to air and thereby helps to keep food fresh longer (Shaw 2011). The clay acts like bricks in a wall, and a chitin polymer functions as the mortar. This biodegradable film is much more environmentally sustainable compared to conventional plastic wrappings for food. Paper-based wraps, known as "killer paper," can be imbued with nanosilver to prevent bacterial growth. Nanodevices and data trackers can help monitor significant biological and chemical events that impact food quality during the life cycle of food products (United States Department of Agriculture 2012). These applications may prove highly valuable in the epidemiological investigation of food-borne disease outbreaks.

Food safety also starts with consumers in their homes (DeBuono 2006). Nanosensors integrated with food packaging can alert consumers to spoilage (Nano.gov 2012). Nanosilver-coated antibacterial tableware (cutlery) and kitchenware (e.g., cutting boards and salad bowls) are commercially available for individual consumers and restaurants (Project on Emerging Nanotechnologies 2012a). Airtight food storage containers and bags with nanosilver (e.g., BlueMoonGoods Fresh Box Silver Nanoparticle Food Containers, FresherLonger Miracle Food Storage Containers and Bags, and Quan Zhou Hu Zheng NanoTechnology, Co., and Nano-silver Storage Box) help food to stay fresh longer due to its inherent antimicrobial properties. Nanosilver compounds have been used in refrigerators (e.g., by Daewoo, LG, and Samsung) to minimize bacterial growth and odor. Baby milk bottles and sipping cups with antimicrobial nanosilver are also commercially available.

Food producers have some nano-sized ingredients, claiming that they bolster their effects (Nano.gov 2012; Project on Emerging Nanotechnologies 2012a). Commercially available liposomal nanospheres containing vitamin C (Lypo-Spheric vitamin C) and other supplements (B complex and minerals and glutathione) claim more efficient absorption. However, because supplements are not regulated by the Food and Drug Administration, these claims are unlikely to be verified in the peer-reviewed scientific literature.

5.6.3 Sunscreens

The incidence of skin cancers, both melanoma and nonmelanoma types, has been increasing (WHO 2012b). Approximately 2–3 million nonmelanoma skin cancers and 132,000 melanoma skin cancers occur worldwide every year. While one in every three cancers diagnosed is a skin cancer (WHO 2012b), the majority of skin cancers can be prevented with proper sun protection measures (DeBuono 2006). Nanoscale titanium dioxide or zinc oxide is utilized in over 30 sunscreen products (Project on Emerging Nanotechnologies 2012a). The advantage of these nanoparticles is that they are transparent on application while effectively reflecting ultraviolet waves for sun protection (National Nanotechnology Initiative 2008). The greater transparency is a crucial aesthetic feature and, therefore, increases user compliance and creates a commercial advantage (Monteiro-Riviere et al. 2007). Early studies indicate that there is no significant penetration of titanium dioxide nanoparticles through intact skin (Monteiro-Riviere et al. 2007; Sadrieh et al. 2010).

5.6.4 Environmental Remediation

Nanotechnology is used in sensors that can detect harmful chemical agents and in interventions designed to address chemical spills (Project on Emerging Nanotechnologies 2012b). A nanoscale dry powder has been developed to neutralize toxic vapor and liquid chemical spills. Nanoscale materials have been used in the remediation of contaminated sites of oil fields, manufacturing sites, military installations, private properties, and residences across North America and Europe. In 2004, the Environmental Protection Agency estimated that it would take approximately three decades and cost the United States approximately $250 billion to clean up its hazardous waste sites. However, the potential risks of nanoremediation processes are not well understood. Studies evaluating the potential adverse effects of engineered nanoparticles on human health are required. Further, full-scale ecosystem-wide studies are necessary to prevent any adverse environmental effects of nanoremediation processes.

5.6.5 Clean Drinking Water

Nanotechnology is enabling the production of low-cost filters that provide clean drinking water (National Nanotechnology Initiative 2008). These filters may utilize carbon nanotubes, nanoporous ceramics, magnetic nanoparticles, or other engineered nanomaterials (Hillie and Hlophe 2007). A range of nano-enabled water treatment devices are commercially available or are in advanced stages of development. The commercially available Lifesaver™ is the world's first portable water nanofiltration device. This device has a nanopore size of 15 nm and filters out protozoans, bacteria, viruses, fungi, and heavy metals and produces sterile drinking water at a cost of 0.013 c. (American) per liter (London School of Tropical Medicine and Hygiene 2007; Lifesaver 2012a). Hundreds of Lifesaver products have been distributed for humanitarian assistance in Haiti, Nigeria, and Pakistan (Lifesaver 2012b,c). In addition to providing clean drinking water, this device has been utilized to irrigate wounds and prepare infant formulas (Fox News 2008; Lifesaver 2012b,d). Another commercial product is the award-winning TATA Swach® Nanotech Water Purifier™, which uses silver nanoparticles to destroy bacteria and viruses (TATA Swach 2012). Research is ongoing on the application of nanoscale sensors to detect water contaminants, the use of titanium dioxide to neutralize *E. coli* in water, and the use of nanoscale rust to remove arsenic from drinking water (National Nanotechnology Initiative 2008). Nanoscale chitin or chitosan could be used for removing arsenic from groundwater (Da Sacco and Masotti 2010). Large-scale adoption of cost-effective nano-enabled technologies for water purification would provide substantial health and socioeconomic benefits for millions of children and adults who lack access to safe drinking water.

5.7 Quality Improvement in Health Care

Watson, the IBM supercomputer, contained nanocircuitry that facilitated its ability to integrate and analyze patient data to generate accurate diagnoses. Someday, Watson's nano-electronics might be teamed with physicians to reduce medical errors and enhance the quality of health-care delivery (Weber 2011). Nanosensors in conjunction with mobile telecommunication networks can extend the reach of medical specialists to remote, resource-constrained regions (Tibbals 2011).

Nanosensors exhibit the following advanced features: smaller size, reduced power requirements, higher sensitivity, greater selectivity, wireless operating capabilities, integration with microfluidics/electronics, and biocompatibility (Tibbals 2011). Personal wireless medical monitoring devices could detect medical conditions in vulnerable populations before they reach critical stages and thereby reduce health-care costs. Wearable nano-based biosensors embedded into fabrics are currently being prototyped, which would allow innovative visual displays of physiological parameters.

5.8 Future Directions

Public health and preventive medicine have made incredible strides in the past century, resulting in significant increases in life span and reduction of preventable deaths (CDC 2011c). The collective contributions of these interventions have led to a demographic shift from infectious diseases to chronic, noncommunicable diseases as leading causes of death worldwide. Nano-based technologies can play an important role both in communicable disease management and in addressing the rising rates of chronic conditions, such as cardiovascular disease, diabetes, cancer, and injuries.

References

Allen, B.B. 2012. $1.1 Trillion: what the 10 leading causes of death cost the U.S. economy. *24/7 Wall Street*. http://247wallst.com/2012/01/18/1-1-trillion-what-the-10-leading-causes-of-death-cost-the-u-s-economy/#ixzz2FegD3et7 (accessed December 20, 2012).

American College of Preventive Medicine. 2011. Member Profile: Fact sheet, Washington, DC: American College of Preventive Medicine.

American College of Preventive Medicine. 2012. Who We Are… Leaders in Science, Practice, and Policy of Preventive Medicine. http://www.acpm.org/resource/resmgr/Membership-Files/WhoWeAre.pdf (accessed August 1, 2012).

Ansari, M.A., Zubair, S., Mahmood, A. et al. 2011. RD antigen based nanovaccine imparts long term protection by inducing memory response against experimental murine tuberculosis. *PLoS One* 6:e22889.

Aschengrau, A. and Seage, G.R. 2003. *Essentials of Epidemiology in Public Health*. Sudbury, MA: Jones & Bartlett.

Association of Schools of Public Health. 2006. MPH Core Competency Model. http://www.asph.org/publication/MPH_Core_Competency_Model/index.html (accessed June 18, 2010).

Centers for Disease Control and Prevention (CDC). 1999. Ten great public health achievements—United States, 1900–1999. www.cdc.gov/mmwr/preview/mmwrhtml/00056796.htm (accessed August 13, 2012).

Centers for Disease Control and Prevention (CDC). 2011a. Smoking Cessation. http://www.cdc.gov/tobacco/data_statistics/fact_sheets/cessation/quitting/index.htm (accessed December 20, 2012).

Centers for Disease Control and Prevention (CDC). 2011b. Smoking Cessation. http://www.cdc.gov/tobacco/quit_smoking/how_to_quit/you_can_quit/nicotine/ (accessed December 20, 2012).

Centers for Disease Control and Prevention (CDC). 2011c. Ten Great Public Health Achievements—Worldwide 2001–2010. http://www.cdc.gov/mmwr/preview/mmwrhtml/mm6024a4.htm (accessed August 13, 2012).

Centers for Disease Control and Prevention (CDC). 2012. Hepatitis B Vaccination. http://www.cdc.gov/vaccines/vpd-vac/hepb/default.htm#vacc (accessed December 20, 2012).

Chen, X., Prow, T., Chrichton, M.L. et al. 2008. *Novel Coating of Micro-Nanoprojection Patches for Targeted Vaccine Delivery to Skin*. International Conference on Nanoscience and Nanotechnology (ICONN) held at Melbourne, Victoria, Australia, February 25–29 (abstract).

Clemente-Casares, X., Tsai, S., Yang, Y., and Santamaria P. 2011. Peptide-MHC-based nanovaccines for the treatment of autoimmunity: a "one size fits all" approach? *J Mol Med (Berl)* 89:733–42.

ClinicalTrials.gov. 2012. Safety and Pharmacodynamics of SEL-068 Vaccine in Smokers and Non-smokers. ClinicalTrials.gov—A Service of the U.S. National Institutes of Health. http://www.clinicaltrials.gov/ct2/show/NCT01478893?term=SEL-068&rank=1 (accessed December 9, 2012).

Da Sacco, L. and Masotti, A. 2010. Chitin and chitosan as multipurpose natural polymers for groundwater arsenic removal and As_2O_3 delivery in tumour therapy. *Mar Drugs* 28:1518–25.

Danesh-Bahreini, M.A., Shokri, J., Samiei, A. et al. 2011. Nanovaccine for leishmaniasis: preparation of chitosan nanoparticles containing *Leishmania* superoxide dismutase and evaluation of its immunogenicity in BALB/c mice. *Int J Nanomedicine* 6:835–42.

DeBuono, B.A. 2006. *Milestones in Public Health: Accomplishments in Public Health Over the Last 100 Years*. New York: Pfizer Global Pharmaceuticals.

Deng, G., Zeng, J., Jian M. et al. 2012. Nanoparticulated heat-stable (STa) and heat-labile B subunit (LTB) recombinant toxin improves vaccine protection against enterotoxigenic *Escherichia coli* challenge in mouse. *J Biosci Bioeng* 115(2):147–53. doi: 10.1016/ j.jbiosc.2012.09.009.

Dillon, H.F. 2010. Resolving Food and Oil at Scale. TEDxAtlanta. http://www.youtube.com/watch?v=Fj8ZkoL-_wA (accessed December 20, 2012).

Feng, L., Qi, X., Zhou, X. et al. 2006. Pharmaceutical and immunological evaluation of a single-dose hepatitis B vaccination using PLGA microspheres. *J Control Release* 112:35–42.

Fifis, T., Gamvrellis, A., Crimeen-Irwin, B. et al. 2004. Size-dependent immunogenicity: therapeutic and protective properties of nanovaccines against tumours. *J Immunol* 173:3148–54.

Fox News. 2008. 'Lifesaver' Bottle Purifies Water in Seconds. http://www.foxnews.com/story/ 0,2933,354735,00.html (accessed August 8, 2012).

Ge, W., Li, Y., Li, Z.S. et al. 2009. The antitumour immune responses induced by nanoemulsion-encapsulated MAGE1-HSP70/SEA complex protein vaccine following perioral administration route. *Cancer Immunol Immunother* 58:201–8.

Gengoux, C. and Leclerc C. 1995. *In vivo* induction of CD4+ T cell responses by antigen covalently linked to synthetic microspheres does not require adjuvant. *Int Immunol* 7:45–53.

He, X., Wang, F., Jiang, L. et al. 2005. Induction of mucosal and systemic immune response by single-dose oral immunization with biodegradable microparticles containing DNA encoding HBsAg. *J Gen Virol* 86:601–10.

Hillie, T. and Hlophe, M. 2007. Nanotechnology and the challenge of clean water. *Nat Nanotechnol* 2:663–4.

Kersten, G. and Hirschberg, H. 2004. Antigen delivery systems. *Expert Rev Vaccines* 3:453–62.

Kishimoto, T. 2012. *Rational Design of a Fully Synthetic Nanoparticle-Based Vaccine.* Paper presented at the second annual conference on vaccines and vaccination, August 20–22, Chicago/Northbrook.

Lifesaver. 2012a. Lifesaver: Working to End Water Poverty. http://www.lifesaversystems.com/ (accessed August 8, 2012).

Lifesaver. 2012b. 1.1 Billion People Are Trapped in Water Poverty. Why? http://www.lifesaver systems.com/documents/LIFESAVER%20Humanitarian%202011.pdf (accessed August 8, 2012).

Lifesaver. 2012c. What We Do: Humanitarian. http://www.lifesaversystems.com/what-we-do/ help-us-help-them (accessed August 8, 2012).

Lifesaver. 2012d. Haiti Appeal—Lifesaver Clean Water Project. http://www.lifesaversystems.com/ become-a-lifesaver/haiti-appeal (accessed August 8, 2012).

London School of Tropical Medicine and Hygiene. 2007. Report on Microbiological Tests Carried Out for Lifesaver Systems. http://www.lifesaversystems.com/documents/LSHTM.pdf (accessed August 8, 2012).

Makidon, P.E., Bielinska, A.V., Nigarekar, S.S. et al. 2008. Pre-clinical evaluation of a novel nanoemulsion-based hepatitis B mucosal vaccine. *PLoS One* 3:e2954. doi: 10.1371/journal.pone.0002954.

Monteiro-Riviere, N.A., Inman, A.O. and Ryman-Rasmussen, J.P. 2007. Dermal effects of nanoparticles. In *Nanotoxicology: Characterization, Dosing and Health Effects*, ed. N.A. Monteiro-Riviere and C.L. Tran, 317–37. New York: Informa Healthcare.

Nandedkar, T.D. 2009. Nanovaccines: recent developments in vaccination. *J. Biosci* 34(6):995–1003.

Nano.gov 2012a. Benefits and Applications. www.nano.gov/you/nanotechnology-benefits (accessed on December 19, 2012).

National Nanotechnology Initiative. 2008. Nanotechnology: Big Things from a Tiny World. http:// www.nano.gov/sites/default/files/pub_resource/nanotechnology_bigthingsfromatinyworld-print.pdf (accessed on August 8, 2012).

Pautler, M. and Brenner, S. 2010. Nanomedicine: promises and challenges for the future of public health. *Int J Nanomedicine* 5:803–9.

Project on Emerging Nanotechnologies. 2012a. Inventories. http://www.nanotechproject.org/ inventories/ (accessed December 9, 2012).

Project on Emerging Nanotechnologies. 2012b. Nanoremediation Map. http://www.nanotechproject .org/inventories/remediation_map/ (accessed December 9, 2012).

Raphael, A.P., Prow, T.W., Crichton, M.L. et al. 2010. Targeted, needle-free vaccinations in skin using multilayered, densely-packed dissolving microprojection arrays. *Small* 6(16):1785–93.

Sadrieh, N., Wokovich, A.M., Gopee, N.V. et al. 2010. Lack of significant dermal penetration of titanium dioxide from sunscreen formulations containing nano- and submicron-size TiO_2 particles. *Toxicol Sci* 115(1):156–66.

Scheerlinck, J.P., Gloster, S., Gamvrellis, A. et al. 2006. Systemic immune responses in sheep, induced by a novel nano-bead adjuvant. *Vaccine* 24:1124–31.

Selecta Biosciences 2011. Recent News: Selecta Biosciences Initiates Phase 1 Clinical Study of SEL-068, A First-In-Class Synthetic Nicotine Vaccine for Smoking Cessation and Relapse Prevention. Press release. http://www.selectabio.com/news/recent-news/Selecta-Biosciences-Initiates-Phase-1-Clinical-Study-of-SEL-068.cfm (accessed December 9, 2012).

Shaw, G.K. 2011. Small bites, bigger questions. *New Haven Independent.* http://www.newhaven independent.org/index.php/archives/entry/small_bites_bigger_questions/ (accessed December 11, 2012).

Shi, R., Hong, L., Wu, D. et al. 2005. Enhanced immune response to gastric cancer specific antigen Peptide by coencapsulation with CpG oligodeoxynucleotides in nanoemulsion. *Cancer Biol Ther* 4:218–24.

Sinyakov, M.S., Dror, M., Lublin-Tennenbaum, T. et al. 2006. Nano and microparticles as adjuvants in vaccine design: Success and failure is related to host material antibodies. *Vaccine* 24:6534–41.

Slutter, B., Hagenaars, N. and Jiskoot, W. 2008. Rational design of nasal vaccines. *J Drug Target* 16:1–17.

Taha, M.A., Singh, S.R. and Dennis, V.A. 2012. Biodegradable PLGA85/15 nanoparticles as a delivery vehicle for *Chlamydia trachomatis* recombinant MOMP-187 peptide. *Nanotechnology* 23:325101. doi: 10.1088/0957-4484/23/32/325101.

TATA Swach: Nanotech Water Purifier. 2012. TATA Swach Technology. http://www.tataswach.com/TsrfTechnology.aspx (accessed August 8, 2012).

Tibbals, H.F. 2011. *Medical Nanotechnology and Nanomedicine.* Boca Raton, FL: CRC Press.

United States Department of Agriculture. 2012. Nanotechnology Overview. http://www.csrees.usda.gov/ProgViewOverview.cfm?prnum=16500 (accessed on December 9, 2012).

Weber D.O. 2011. Itty-Bitty Medicine. *Hospitals and Health Networks Daily.* http://www.hhnmag.com/hhnmag/HHNDaily/HHNDailyDisplay.dhtml?id=7510005342 (accessed August 11, 2012).

World Health Organization (WHO). 2007. *Food Safety and Foodborne Illness.* https://apps.who.int/inf-fs/en/fact237.html (accessed December 9, 2012).

World Health Organization (WHO). 2008. WHO Report on the Global Tobacco Epidemic, 2008: The MPOWER Package. http://www.who.int/tobacco/mpower/mpower_report_full_2008.pdf (accessed December 20, 2012).

World Health Organization (WHO). 2010. Media Centre: Hepatitis B. http://www.who.int/mediacentre/factsheets/fs204/en/ (accessed June 18, 2010).

World Health Organization. 2011. The Top 10 Causes of Death: Fact Sheet. http://www.who.int/mediacentre/factsheets/fs310/en/index.html (accessed December 20, 2012).

World Health Organization (WHO). 2012a. Initiative for Vaccine Research: Sexually Transmitted Diseases—*Chlamydia trachomatis.* http://www.who.int/vaccine_research/diseases/soa_std/en/index1.html (accessed December 20, 2012).

World Health Organization (WHO). 2012b. Skin Cancers: Frequently Asked Questions—How Common Is Skin Cancer? http://www.who.int/uv/faq/skincancer/en/index1.html (accessed August 8, 2012).

6

Nanotechnology Applications in Vascular Medicine and Surgery

Manish Mehta, MD; Philip S.K. Paty, MD; W. John Byrne, MD; Yaron Sternbach, MD; John B. Taggert, MD; and Kathleen J. Ozsvath, MD

CONTENTS

6.1 History

The evolution of vascular surgery dates back to somewhere between 800 and 600 BC when Sushruta, a surgeon in ancient India, first described the ligation of a blood vessel. For centuries to come, the art of vascular surgery was limited to ligation of bleeding blood vessels, and amputation or death was the only alternative. It was only in the late eighteenth century that Nikolai Eck, a student in Pavlov's laboratory in Leningrad, was the first to describe the creating of an anastomosis between two blood vessels. This eventually lead to Alexis Carrel's work of pioneering vascular suturing techniques that won him the Nobel Prize in Physiology and Medicine in 1912. Over the first half of the twentieth century, many surgeons experimented through the trials and tribulations of trying to treat vascular patients with a variety of illnesses ranging from aneurysms to critical limb ischemia (CLI) to traumatic injuries during World War I and II. Finally, in 1950s, the specialty of vascular surgery was born, and it is somewhat incomprehensible how evolution in medicine and technology has lead us to where we are in just a little over half a century.

In as little as 5 years from today, the U.S. Census expects 87 million people to be 55 years of age or older. This is the age group that is primarily affected by vascular disease. Current socioeconomic conditions have started to move the U.S. healthcare pendulum toward the patient's ability to have personal control and accountability. Unfortunately, most Americans are exposed to risk factors such as diabetes, obesity, hypertension, hypercholesterolemia, and smoking without awareness of the implications to their vascular health. This lack of awareness leads to years of neglect through early adulthood, resulting in the high prevalence of vascular disease in the aging population.

The baby boomers' thirst for information and their ability to acquire it via understanding information technology, such as the Internet and text messaging, however, are redefining culture and allowing personal control and accountability regarding their health.

Over the past decade, advances in vascular healthcare, driven by innovation and technology, have had a significant impact on providers' ability to diagnose and treat complex vascular and cardiac problems. From the days of significant morbidity and mortality related to surgical procedures for aneurysm repair, to treatment of peripheral arterial occlusive disease, to heart bypass, healthcare providers and facilities have developed infrastructures that are technologically advanced and minimally invasive. Over the next decade, these systems will have significant impact on reshaping vascular and cardiac health-care delivery and improving patient outcomes.

To keep up with this rapidly increasing demand from the aging population, technology, innovation, and vascular awareness will pave the road that will lead to the "new tipping point" for improving vascular health. Healthcare consumers will have the tools to better educate themselves and the younger generations in healthier dietary and lifestyle habits. Patients will be empowered to make choices that focus on risk factor modification earlier in life, and they will have more control over diabetes, obesity, hypertension, and hypercholesterolemia. Innovation will focus not just on better drugs and devices for treating vascular disease but also on information technology (e.g., the Apple iPhone) that allows portability, remote monitoring, and self-management of health information.

The question remains, how will nanotechnology evolve and impact the care of a vascular patient; from tissue engineering that develops prosthetic implants to utilization of genomics in creating "autogenous implants." This introductory chapter in vascular medicine and surgery will highlight several critical vascular diseases, including aortic aneurysms, atherosclerosis and peripheral arterial disease (PAD), renal occlusive disease, venous thromboembolism (VTE), and chronic venous insufficiency (CVI).

6.2 Abdominal Aortic Aneurysm

6.2.1 Description of Disease

An aneurysm is a permanent focal dilatation of an artery. An artery is considered to be aneurysmal when its diameter increases by 50%. The normal infrarenal aortic mean diameter is 2.1 cm in men and 1.9 cm in women. Arterial ectasia is defined as arterial enlargement of less than 50%, whereas arteriomegaly refers to diffuse arterial enlargement of greater than 50% that involves multiple arterial segments and is associated with multiple aneurysms. The majority of the abdominal aortic aneurysm (AAA) are fusiform that result in a generalized enlargement of the aorta. Saccular AAA results from focal aortic wall weakness that leads to an out pouching or bulge in a segment of the aortic wall that historically have been considered more prone to rupture. The overwhelming majority of AAA is degenerative. Inflammatory AAA is characterized by an intense sterile retroperitoneal fibrosis around the aorta and surrounding tissues that can often be associated with ureteric obstruction. Mycotic AAA is caused by bacterial infection, most commonly *Salmonella*.

6.2.2 Rupture Risk

Although true estimation of rupture risk is difficult to define, largely due to the fact that large aneurysms have been repaired, not observed, over the past three decades, the current gold standard was published in 2003.[1] According to these recommendations, AAA diameter is the best predictor of rupture (see Table 6.1). Variability in the rupture risk reported for a given AAA diameter likely reflects other patient-specific variables. Specifically, female gender, chronic obstructive pulmonary disease (COPD), continued smoking, hypertension, severe cardiac disease (CAD), stroke, advanced age, and thrombus content within the AAA have all been identified as predictors of more rapid expansion.

6.2.3 Epidemiology

Approximately 15,000 people die annually in the United States from ruptured AAA (r-AAA), making it the 15th leading cause of death overall and the 10th leading cause of death in men over the age of 55. It is the third leading cause of sudden death in men.

TABLE 6.1

Estimated Annual Rupture Risk

AAA Diameter (cm)	Rupture Risk (%/year)
<4	0
4–5	0.5–5
5–6	3–15
6–7	10–20
7–8	20–40
>8	30–50

An estimated 1.5 million Americans have an AAA, but only 20% are diagnosed per year, and only 3% are treated per year. Of patients with r-AAA, 30%–50% will die before reaching a hospital, and of those who do reach a hospital, 30%–40% will die without undergoing surgery. Current operative mortality for r-AAA is between 40% and 50%. Overall, the mortality for r-AAA is 80%–90%. This high mortality has remained unchanged for the past three decades, although recent evidence suggests that endovascular repair of r-AAA may have a reduced mortality.[2,3] The incidence of AAAs is 36.5–49.3 per 100,000 person-years, with a male-to-female ratio of 4:1. In men aged 65–80 years, 4.3%–7.1% have an AAA. White men are two to three times more likely to have AAA than are black men.

6.2.4 Etiology

The two main pathologic processes are proteolysis and inflammation. In AAA, the adventitia has an intense lymphocytic infiltration, whereas the media is thin with elastin fragments. A subset of enzymes called matrix metalloproteinases (MMPs) are implicated in expansion, specifically MMP-2, MMP-3, and MMP-9.[4] MMPs are normally involved in connective tissue repair. There is also an inherited component to AAA, as the incidence in first-degree male relatives is six times greater than expected.

The major risk factors for AAA include male sex, a history of ever smoking (defined in surveys as 100 cigarettes in a person's lifetime), and age 65 years or older.[5] Other lesser risk factors include family history, coronary heart disease, claudication, hypercholesterolemia, hypertension, cerebrovascular disease (CBVD), and increased height. Factors associated with decreased risk include female sex, diabetes mellitus, and black race.

6.2.5 Physical Examination

Most clinically significant AAA is palpable on physical examination. However, the sensitivity of physical examination depends on the following factors: AAA size, obesity of the patient, focus of the examination, and skill of the examiner. Thorough examination requires the patient be supine while the examiner palpates the abdomen with one hand on each side moving from lateral to medial. Clinical examination in a thin patient with a large aneurysm is frequently diagnostic, while it can often be negative in overweight patients, even in the presence of sizable aneurysms. Based on physical examination alone, the diagnosis can be made in 29% of AAAs 3–3.9 cm, 50% of AAAs 4–4.9 cm, and 75% of AAAs 5 cm or larger. If a focused examination in performed, up to 50% of AAAs greater than 3.5 cm can be detected. However, a nonfocused examination is likely to miss the diagnosis. It is important to note that AAA may be falsely suspected in thin patients with prominent aortic pulsations, hypertension, or tortuous aortas. The positive predictive value of physical examination for the identification of AAA is reported to be only 15%. Physical examination is not a reliable method for measuring aneurysm size.

6.2.6 Associated Conditions: Popliteal Aneurysms

AAAs are associated with aneurysmal disease in other locations, and all patients with this diagnosis should be screened for additional pathology. Among patients with AAA, approximately 35% will have an aneurysm of the iliac, femoral, or popliteal arteries. Most of these are common iliac aneurysms, which will be repaired in conjunction with the AAA. Ten percent of patients with AAA will have some combination of femoral or popliteal aneurysms. Femoral and popliteal aneurysms can usually be readily felt on physical examination. They can be diagnosed by ultrasonography, as well. For patients noted to have a popliteal aneurysm, 40%–60% of them will be bilateral and can be synchronous or metachronous. In general, both femoral and popliteal aneurysms over 2.5 cm should be repaired.

6.2.7 Diagnostic Methods

6.2.7.1 Duplex Ultrasound

Due to the unreliability of routine physical examinations, most AAAs are diagnosed by imaging studies. B-mode duplex ultrasonography (d-US) is a cost-effective, noninvasive, frequently used modality for the screening, initial diagnosis, and follow-up of AAA. It is a reliable and an accurate examination in the hands of experienced operators, with an interobserver variability of 5 mm in 84% of studies. It is more accurate when the patient is in the anteroposterior position than in the lateral position. The study can be limited by bowel gas and may be difficult in obese patients. Ultrasonography is not reliable for the detecting AAA rupture or in determining the extent of abdominal visceral vessel involvement. It tends to underestimate the diameter of the aneurysm by 2–4 mm in the anteroposterior position when compared to computed tomography (CT).

6.2.7.2 Computed Tomography Scan

CT scans are considered the reference standard of imaging when it comes to evaluating aortic aneurysms' size and the extent of involvement from the thoracic aorta, abdominal aorta, and iliac arteries. CT is considered the optimal method of preplanning for endovascular or open surgical AAA repair. It provides more accurate measurement of diameter, with 91% of studies reporting less than 5 mm interobserver variability.[6] Standard axial views may overestimate diameter if the aorta is tortuous and the axial cut is not perpendicular to the center line of the aorta. Advancements in radiological software now allow for detailed three-dimensional reconstructed views that further allow for precise measurements and case planning (Figure 6.1). Furthermore, CT is also considered to be the gold standard in patient follow-up post-endovascular AAA repair (EVAR). There are some limitations: CT involves exposure to radiation as well as iodinated contrast material that could be nephrotoxic, particularly for patients with preexisting chronic renal insufficiency.

6.2.8 Screening

The U.S. Preventative Services Task Force (USPSTF) recommends one-time screening for AAA by ultrasonography in men aged 65–75 years who have ever smoked (grade B recommendation). The USPSTF recommends against routine screening for AAA in women (grade D recommendation).

The Society of Vascular Surgery and the Society for Vascular Medicine recommend AAA screening in all men aged 60–85 years, women aged 60–85 years with cardiovascular risk factors, and men and women aged 50 years and older with a family history of AAA. These groups further recommend the following courses of action after screening: no further testing if aortic diameter is less than 3.0 cm, yearly ultrasonographic screening if aortic diameter is between 3.0 and 4.0 cm, ultrasonography every 6 months if aortic diameter is between 4.0 and 4.5 cm, and referral to a vascular specialist if aortic diameter is greater than 4.5 cm.[7]

FIGURE 6.1
Abdominal aortic aneurysm. Red indicates blood; yellow indicates thrombus.

6.2.9 Indications for Repair of AAA

6.2.9.1 Recommendations for Repair

The indications for repair of AAA are based on several factors: the risk of AAA rupture, the risk of EVAR or open repair, the patient's life expectancy, and the patient's personal preference. The risk of AAA rupture based on size has been defined earlier in this chapter. The U.K. small aneurysm trial[8] and the Aneurysm Dissection and Management Trial (ADAM) trial[9] both describe the progression of aneurysm size and the likelihood of ultimate repair in those patients who were initially assigned to observation. In the U.K. small aneurysm trial, more than 60% of patients initially randomized to observation underwent repair within a mean of 2.9 years. Patient follow-up was rigorous and included an ultrasound every 6 months for AAA less than 5 cm, and every 3 months for AAA 5–5.5 cm. Even with this careful surveillance, the risk of rupture was 1% per year. The operative mortality in the U.K. small aneurysm study was twice (5.8%) than that used in the power calculations when the trial was designed, which suggests that these results cannot be generalized. Furthermore, the risk for AAA rupture was 4.5 times higher in women and this led to the recommendation of earlier AAA repair in women (at AAA 5 cm). In the ADAM study, more than 60% of patients in the surveillance arm underwent repair; 27% of patients with AAA 4–4.5 cm, 53% of patients with AAA 4.5–4.9 cm, and 81% of those with AAA 5–5.4 cm underwent repair during follow-up. Based on these trials, the recommendations for treating asymptomatic AAA based on size alone is gender dependent: men 5.5 cm or greater, and women 5 cm or greater. Also, this decision analysis has to take into account that patients will need careful follow-up via duplex ultrasound, and a significant percentage of patients (27%–81%, based on the AAA size at presentation as stated above) will undergo repair within a mean of 2.9 years. These, coupled with local and regional expertise, determine the risks of EVAR and open surgical repair (OSR). Practice guidelines for the care of patients with AAA were recently published by the Society for Vascular Surgery.[10]

6.2.9.2 Who Is at High Risk for AAA Repair?

The definition of high-risk patients is evolving. Patients traditionally felt to be at high risk for open operative repair were those with advanced coronary artery disease, unstable angina or angina at rest, congestive heart failure, ejection fraction less than 30%, chronic renal insufficiency with a serum creatinine greater than 3 mg/dL, and significant pulmonary disease with a room air PO_2 of less than 50 mm Hg, elevated PCO_2, or both. However, even these high-risk patients have been reported to undergo successful elective OSR with a postoperative mortality of less than 6%.[11]

The advent of endovascular repair of AAAs has reduced the postoperative mortality from 5% to less than 2%.[2,12] EVAR can be performed under spinal anesthesia, obviating the risks associated with general anesthesia. While this can be perceived as a panacea for those high risk-patients, studies have shown otherwise. The EVAR 2 trial[13] showed that when EVAR was performed on patients deemed unsuitable (i.e., excessive risk) for open repair, the 30-day postoperative mortality was 9% (see Table 6.1). When the intervention patients were compared with matched controls, there was no significant difference in all-cause or aneurysm-associated mortality.[2] Both open and endovascular repair can be performed in high-risk patients with acceptable mortality, but it does not seem that aneurysm repair confers any survival benefit. Thus, ultimate decisions as to a patient's fitness for surgery must be an individualized analysis of his or her overall health and the perceived benefit of aneurysm repair. The recommended guidelines for assessment of patient fitness for open repair and suitability for EVAR Trial 1 or 2 are described in Table 6.2. An endoleak is characterised by persistent blood flow within the aneurysm sac following EVAR, and reduction of endoleak incidence is desired to improve surgical outcomes. The classification scheme for endoleaks is described in Table 6.3.

6.2.9.3 Patient Selection for Open versus Endovascular AAA Repair

In determining patient selection for open versus EVAR, there are several considerations including comorbid risk factors, whether the aortoiliac anatomy is suitable for EVAR, and patient preference. Most trials have indicated a lower morbidity and mortality of EVAR over OSR in the short term. There is some controversy as to the long-term benefit of EVAR when compared to OSR. Results from EVAR 1, the largest prospective randomized controlled trial to evaluate outcomes in patients who were medically and anatomically suitable for open and EVAR, suggested that EVAR offers at least a midterm survival advantage when compared to OSR.[14] With recent advancements in endovascular technology, the surgeons'/interventionists' ability to perform EVAR and ability to better detect complications of these procedures and manage them postoperatively, EVAR has become the preferred treatment.[15] In patients who have suboptimal aortoiliac morphology for EVAR, the decision for OSR versus EVAR is based on the balance of patient risk factors for surgical versus endovascular repair and the need to be individualized based on local expertise and patient preferences.

Finally, the patient who agrees to an endovascular approach must be willing to comply with a rigorous follow-up protocol involving serial imaging via either CT or ultrasound, for the rest of his or her life. They also must understand that there is up to a 20% chance that they will require a second intervention for maintenance of their repair within 5 years of the initial repair. Patients who are not willing or able to do this probably should not undergo an endograft repair of their AAA.

TABLE 6.2

Recommended Guidelines for Assessment of Patient Fitness for Open Repair and Suitability for EVAR Trial 1 or 2

Patient fitness for open repair is decided at the local level; however, these guidelines may provide some assistance

Cardiac status

Normally, patients presenting with the following cardiac symptoms would not be recommended for any surgical intervention:

- MI within last 3 months
- Onset of angina within the last 3 months
- Unstable angina at night or at rest

Normally, patients presenting with the following symptoms would be unsuitable for open repair (EVAR Trial 1) but may be suitable for EVAR Trial 2:

- Severe valve disease
- Significant arrhythmia
- Uncontrolled congestive cardiac failure

Respiratory status (no constraints for EVAR Trial 2)

Open repair (EVAR Trial 1) would not be recommended for patients presenting with the following respiratory symptoms:

- Unable to walk up a flight of stairs without shortness of breath (even if there is some angina on effort).
- $FEV_1 < 1.0$ L
- $PO_2 < 8.0$ kPa
- $PCO_2 > 6.5$ kPa

Renal status (no constraints for EVAR Trial 2)

Open repair might not be recommended for patients presenting with serum creatinine levels greater than 200 µmol/L. These patients may be suitable for EVAR Trial 2.

TABLE 6.3

Classification of Endoleaks

Type of Endoleak	Etiology
I	Inadequate fixation at the proximal (Type Ia) and distal (Type Ib) stent graft attachment sites
II	Retrograde flow from patent inferior mesenteric and/or lumbar arteries
III	Stent graft component separation or tears within the stent graft fabric
IV	Porosity of the graft material
V	Persistent aneurysm sac enlargement in the absence of an identifiable endoleak

6.2.9.4 Open Surgical AAA Repair: Transperitoneal versus Retroperitoneal Approach

The transperitoneal approach is advantageous when exposure of the distal segments of the visceral vessels, right renal artery, or right internal iliac artery is needed, or when there is associated intraabdominal pathology that needs surgical treatment. The retroperitoneal approach is considered advantageous in patients with prior intraperitoneal operations, infections, horseshoe/ectopic kidneys, or inflammatory aneurysms. Depending on local expertise and practice patterns, vascular surgeons generally prefer one technique or the other and use it routinely in their practice.

The transperitoneal approach is via a midline laparotomy from the xyphoid to the symphysis pubis, the small bowel is mobilized at the ligament of Treitz and retracted laterally to expose to AAA in the retroperitoneum. This approach provides easy access to both iliac arteries and intraperitoneal organs if needed.

We prefer the retroperitoneal approach because of several advantages including lower incidence of postoperative ileus, reduced pulmonary complications, and decreased hospital length of stay.[16,17] For the retroperitoneal approach, the patient is positioned in the left lateral decubitus position, and the commonly utilized incision is from the tenth to the twelfth intercostal space extending toward the umbilicus. Because this approach allows for exposure of the aorta up to the diaphragmatic hiatus, mesenteric and juxtarenal reconstruction can be performed without necessitating a thoracotomy, and this approach can be utilized for all infrarenal and juxtarenal AAAs. One technical consideration of the retroperitoneal approach is that control of the right iliac artery may be difficult. This can be controlled with balloon occlusion.

6.2.9.5 Technical Considerations

The proximal clamp to control aortic inflow should be placed just distal to the renal arteries. This position allows for uninterrupted blood flow to the kidneys during the aortic repair, while still giving the surgeon sufficient length of aortic cuff to fashion the proximal anastomosis. There are situations, such as in juxtarenal aneurysms, where it is not possible to clamp below the renals and have adequate length of aorta for an anastomosis. In this case, the proximal clamp may be placed above the renal arteries but below the superior mesenteric artery (SMA), or sometimes above the celiac axis. The more proximal the clamp placement, the higher the morbidity and mortality risk due to potential showering of emboli to the viscera and kidneys.

In the case of AAAs, the proximal anastomosis is always performed in an end-to-end manner. This allows for the complete exclusion of the aneurysmal portion of the aorta. A running prolene suture is used to anastomose the prosthetic graft to the cut portion of the proximal aorta. In the case of a straightforward infrarenal aneurysm with adequate neck, this simply involves sewing one cylinder to another. In juxtarenal aneurysms, a beveled anastomosis can be created that includes one renal, and the second renal can be separately reimplanted, or both renals may have to be separately reimplanted.

The distal anastomosis can be a tube graft to the aortic bifurcation, a bifurcated graft to the common iliac arteries or a bifurcated graft to the common femoral arteries. The tube graft is the simplest, quickest, and therefore most desirable, especially in cases of rupture. However, if the bifurcation is too degenerated or if there is known or suspected significant iliac occlusive disease, then a bifurcated graft is required.

6.2.9.6 Postoperative Care

Routine postoperative care of patients who have undergone open aortic aneurysm repair usually involves at least the first overnight in the intensive care unit. Patients with more comorbidities may require longer stays in intensive care. Patients are transitioned to step down units. After resolution of the expected postoperative ileus, the diet is advanced. Physical therapy is frequently used to assist with early mobilization and to identify those patients who would benefit from a stay in a rehabilitation facility. Once the patients are ambulatory and tolerating a diet, they are discharged.

There are several common postoperative complications that occur after open repair of AAAs. These are cardiac events, bleeding, pulmonary insufficiency, and venous insufficiency. Most patients with vascular disease also have significant coronary disease. Any signs of cardiac ischemia should be triaged early and managed appropriately. Those patients who take β-blockers at home should be continued on them during the perioperative period. The discontinuation of β-blockers in the perioperative period is associated with a significant risk of myocardial infarction (MI) and cardiovascular mortality.[18]

Patients are at increased risk for hemorrhagic complications in the postoperative period. Close attention should be paid to serial hematocrits, as well as to hemodynamic stability. Certainly, there are large fluid shifts after this operation and some abnormalities in hematocrit and electrolytes are expected. The need for multiple transfusions should raise concern, and surgeons should have a low threshold for re-exploration. The common sources of postoperative hemorrhage are unrecognized back bleeding from lumbar and intercostal arteries, and from retractor injuries to the spleen.

6.2.9.7 Long-Term Surveillance and Secondary Intervention

Despite the increased morbidity involved compared to endovascular repair, an open approach is durable and rarely requires reintervention. Long-term surveillance is based on the patient's risks for developing concomitant aneurysms in other arterial beds; patients with AAA have lifetime 15%–20% risk of developing thoracic aortic and iliac artery aneurysms, and most vascular practices recommend CT scan surveillance every 5 years following open surgical AAA repair. After OSR, secondary interventions may be required for postoperative bleeding, surgical incisional wound complications, bowel obstruction, incisional hernia, and claudication.[19]

6.2.10 Endovascular Repair of AAA

6.2.10.1 Assessment of Anatomy

Preoperative CT scans are used to assess anatomy and determine whether a patient is a candidate for endovascular repair. There are several areas of anatomic consideration including (1) the length, diameter, shape, angulation, and thrombus at the proximal stent graft landing zone, defined as the area between the takeoff of the renal arteries and the top of the aneurysm sac; (2) the distal stent graft landing zone at the common or external iliac artery length, diameter, or thrombus; and (3) the access for stent graft delivery system via the femoral and external iliac arteries (Side bar 2). When aortoiliac aneurysms involve the common iliac arteries, the internal iliac artery is interrupted by embolization procedures, and the stent graft distal landing zone is extended into the external iliac artery. Unilateral internal iliac artery interruption is generally well tolerated. Although there are several large single-center reports of the relative safety of bilateral internal iliac artery interruption during endovascular aneurysm repair, bilateral internal iliac artery interruption can lead to pelvic ischemia and is generally avoided.[20,21]

6.2.10.2 Graft-Related Technical Issues

6.2.10.2.1 Fixation and Seal

Stent grafts are designed to accommodate the aortoiliac segments and provide adequate fixation and seal above and below the aneurysms. Currently available stent graft sealing zones are below the lower most renal artery and are generally based on stent graft to aortic wall opposition that prevents Type I endoleaks. Stent graft fixation may be suprarenal or infrarenal, and either active or passive. Active fixation is based on metal barbs that are placed into the aortic wall, and passive fixation is based on the stent graft radial forces without barbs or hooks (Figures 6.2 and 6.3). Active fixation improves the ability of the stent graft to anchor to the healthy aortic neck and limits migration. The stent grafts have to withstand the continuous aortic pulsation and maintain the fixation seal and its integrity for successful long-term aneurysm exclusion.

6.2.10.2.2 Challenges with Difficult Neck Anatomy

Challenging aortic neck morphology remains the "Achilles' heel" to successful endovascular aneurysm repair. Generally speaking, there are several ways of looking at the aortic neck anatomy including its shape (cylindrical, conical, flared), length, and diameter

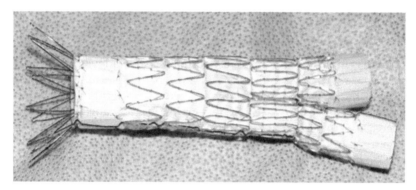

FIGURE 6.2
Stent graft with bare metal stent for suprarenal fixation. (Zenith, Cook Inc.)

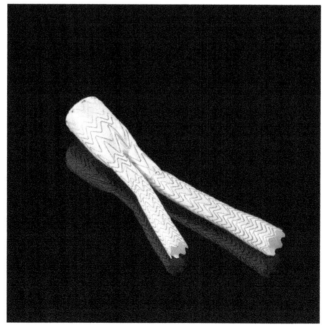

FIGURE 6.3
Stent graft with infrarenal fixation and seal. (Excluder, WL Gore & Ass.)

suitable for particular devices (currently available devices' indications for use include 10–15 mm neck length and 18–32 mm neck diameter); neck angulation (defined as angle between the suprarenal aorta and the aortic neck as well as another angle between the aortic neck and the AAA); and neck thrombus (poorly defined, but generally >3 mm of thrombus that is circumferential at the aortic neck is considered hostile). Aortic neck anatomy is challenging when there are multiple factors that come into play; for instance, an aortic neck that is flared, has >60% angulation, and has significant thrombus would be considered hostile. In some cases, these patients can be treated using fenestrated and branched stent grafts, which are currently available outside the United States.

6.2.10.2.3 Access

Stent graft systems need to be delivered through the femoral arteries, iliac arteries, and the aorta. There are several factors to consider, including vessel size, calcification, and tortuosity. Currently available device delivery systems require the iliac arteries to be approximately 6 mm in minimum diameter, and although size in itself is not a significant limitation, the presence of significant calcification and tortuosity adds to the complexity of access during these procedures. Rarely, adjunctive surgical procedures such as femoral artery endarterectomy and iliac conduits to facilitate stent graft delivery are needed.

6.2.10.3 Long-Term Surveillance

Endovascular repair of AAAs is associated with unique complications including endoleaks, stent graft migration, limb thrombosis, sac enlargement, infection, and aneurysm rupture. The postoperative surveillance protocols following EVAR have generally mirrored what was used in the U.S. pivotal trials that lead to Food and Drug Administration (FDA) approval of various stent grafts. Patients need life-long follow-up after EVAR. Most clinicians see patients in office at 1, 6, and 12 months and obtain CT scans and/or duplex ultrasound at those time intervals; subsequently most clinicians evaluate patients at every 6–12 months via duplex ultrasound, and obtain CT scans yearly thereafter. Although these are generally accepted guidelines, patient's surveillance should be individualized and modified depending on the presence of endoleaks, stent graft migration, and sac enlargement.

6.2.11 Management of Endoleaks

6.2.11.1 Classification

Endoleaks are defined by radiographic evidence of flow outside the stent graft and within the aneurysm sac, and these are categorized according to its source. Type I endoleaks occur due to the inability of stent grafts to provide a seal at the proximal or distal fixation sites in the infrarenal aortic neck or iliac arteries, respectively. Blood flow escapes into the aneurysm and this occurs in <5% of EVARs and, with few exceptions, requires prompt treatment. Type II endoleaks occur due to retrograde flow into the aortic aneurysm sac from collateral branches such as the lumbar arteries and/or the inferior mesenteric artery (arteries that arise from the aneurysm, which upon aneurysm exclusion via the stent graft provide a channel for retrograde flow back into the aneurysm). These Type II endoleaks are noted in approximately 30% of patients following EVAR, and the majority of these endoleaks resolve spontaneously within 6 months following the procedure; one-third persist and might require future intervention. Type III endoleaks occur due to stent graft junctional separation and/or fabric tear, are rare, and require prompt treatment. Type IV endoleaks are generally noted during the initial EVAR and occur due to initial stent graft porosity. These Type IV endoleaks almost always resolve spontaneously within 1 month of the procedure.

6.2.11.2 When and How to Treat Endoleaks

Type I endoleaks transmit systemic pressure to the aneurysm sac and increase the risk of AAA rupture. Type I endoleaks are classified as intraoperative or postoperative. Intraoperative endoleak is treated by placing an aortic cuff or a Palmaz stent to maximize stent graft seal with the aortic wall. For example, when there is inaccurate placement of the proximal end of the graft with a gap of uncovered aorta below the renal arteries, a proximal aortic cuff can be placed. A Palmaz stent positioned in the proximal graft can help seal Type I endoleaks when the neck is short, angulated, or wide. If there is insufficient proximal neck, placement of chimney stents into one or both renal arteries will allow for more proximal aortic coverage. Occasionally, there will be a small, persistent Type I endoleak that cannot be managed by these techniques. These can sometimes be observed, particularly in non-r-AAA. In this scenario, vigilant follow-up is required, but many will seal spontaneously. It is thought that they may be due to small endoleak channels that thrombose over time in the absence of anticoagulation.

In the postoperative period, a Type I endoleak is almost always due to graft migration, neck dilatation, or formation of iliac artery aneurysms and requires prompt treatment. The treatment options are the placement of an aortic cuff with or without renal chimney stents or a limb extension. When these maneuvers fail, the remaining options include embolization of the sac or conversion to an open repair.

The treatment options for Type II endoleaks are evolving. Early research indicated that most Type II endoleaks were associated with systemic pressurization of the aneurysm sac and recommended transfemoral and/or translumbar embolization.[22] However, larger multicenter pivotal trials have suggested that most Type II endoleaks are benign and not associated with AAA sac expansion.[23] Recent reports suggested that when there is a persistent Type II endoleak, even if the sac size remains stable, there is an increased risk of rupture and other secondary complications.[24] Thus, the treatment algorithm for Type II endoleak is moving toward more aggressive embolization in the setting of persistent leak in patients with expanding or nonshrinking AAA.

Type III endoleaks are junctional separation or fabric tears that disrupt the stent graft integrity and are generally treated with stent graft relining. Type IV endoleaks seal spontaneously within a month and do not require any further treatment.

Results of open and endovascular trials are summarized in Table 6.4. The Lifeline registry[25] was established to evaluate long-term outcomes of patients who have undergone endovascular repair of infrarenal AAAs. The registry followed 2664 patients who received EVAR and compared them with 323 open surgical controls. Both methods were successful at preventing rupture, and endovascular repair showed no increased risk of late rupture. At 4 years of follow-up, there was no difference between the endovascular and open groups in terms of overall survival. Women in the endovascular group had a higher risk of rupture and open surgical conversion than did men. Freedom from secondary interventions at 5 years was 78%. At 10 years, significant predictors of mortality after open repair were age, renal failure, and gender. Women had a 2.6-fold higher 1-year all-cause mortality (13.2%) than did men (5.4%).

TABLE 6.4

Summary of Registries/Trials

Registry/Trial	Year Published	Number of Patients	EVAR Mortality	Open Mortality
Lifeline Registry	2008	2,987	1.7% at 30 days	6.7% at 10 years
EVAR 1	2004	1,252	1.7% at 30 days	4.7% at 30 days
EVAR 2	2005	338	9%	n/a
DREAM 1	2004	345	1.2% at 30 days	4.6% at 30 days
DREAM 2	2005	351	2.1% at 2 years	5.7% at 2 years
OVER	2009	881	0.5% at 30 days	3% at 30 days
U.S. Medicare data	2008	22,830	1.4% at 30 days	4.8% at 30 days

6.2.11.3 EVAR 1 and 2, DREAM, OVER, and the U.S. Medicare Patient EVAR Data

The EVAR 1 trial was designed to compare long-term results of EVAR versus open repair with regards to mortality, durability, quality of life, and cost. A total of 1252 patients with at least 5.5 cm AAAs who were fit for either approach were randomized to endovascular or OSR. The mean age of patients in both groups was 74 years and the mean aneurysm size was 6.4 cm. EVAR patients had a significant advantage in 30-day aneurysm-related mortality. At 6 years of follow-up, the early mortality advantage of EVAR was lost. There was no difference between the two groups in terms of all-cause mortality at 4 years or health-related quality of life at 1 year. However, EVAR had a significant disadvantage to OSR in terms of complications (41% EVAR vs. 9% OSR) and secondary interventions (20% EVAR vs. 6% OSR).[13]

The results of EVAR 1 cannot be generalized for several reasons. There were significant delays in patients receiving their assigned treatments. The average time from randomization to treatment was 44 days for the EVAR group and 35 days for the open group, despite the fact that the average aneurysm size was 6.4 cm. A total of 23 patients died prior to EVAR treatment, 10 from aneurysm rupture, and these were assigned to the EVAR treatment arm.

EVAR 2[13] was undertaken to evaluate whether EVAR offered a survival advantage for patients who were medically unfit for open repair. Patients with at least a 5.5 cm aneurysm who were rejected from EVAR 1 due to high risk or unfitness for open repair were randomized to either EVAR or no intervention. The perioperative mortality for the EVAR group was 9%. The mortality rate for the observation group at 4 years was 64%. An intention-to-treat analysis reached several conclusions: there was no difference in all-cause mortality or in AAA-related death after 4 years of follow-up; the rate of secondary interventions was higher in the EVAR group (46% vs. 26%); and EVAR was associated with significantly higher cost. The trial concluded that there was no advantage for EVAR over observation in high-risk patients.

There are several important criticisms of EVAR 2. Like EVAR 1, there were excessive delays in time to treatment for the EVAR group. There were also an excessive number of protocol violations with 47 (27%) of patients assigned to observation crossing over to EVAR (35) or open repair (12). That means in the intention-to-treat analysis, the results of these 47 patients were credited to the no-treatment arm: 2% operative mortality, a 4-year mortality of 23% (half that of the patients who were truly observed), and a nearly 60% reduction in AAA-related death with EVAR.

The DREAM trial[12] was designed to evaluate whether there was a mortality benefit to EVAR in the first 30 days. Three hundred forty-five patients with at least a 5 cm AAA and medically fit for both endovascular and open repair were randomized. The perioperative mortality was lower in the EVAR group. When the combined rate of operative mortality and serious complications was tallied, it was 9.8% in the open group and 4.7% in the endovascular group. The authors concluded that EVAR was preferable to open repair in patients with at least a 5 cm AAA in the short term.

DREAM 2[26] analyzed the 2-year survival of 351 patients with at least a 5 cm infrarenal AAA who were suitable for either EVAR or open repair. It reported cumulative survival rates of 89.6% for open repair and 89.7% for endovascular repair. The advantage of endovascular repair was explained entirely by the lower rate of perioperative events. Beyond the 30-day perioperative period, there were no differences between open and endovascular groups in terms of aneurysm-related mortality.

The OVER trial compared 2-year outcomes of EVAR and open repair of AAA in a randomized multicenter trial of 881 veterans with at least 5 cm AAAs and who were candidates for either approach. This trial's reported 30-day mortality of only 0.5% for EVAR and 3% for open repair was better than that reported by both EVAR 1 and DREAM. This improvement can be credited to improvements in endovascular techniques and technology. No differences were found between the two groups after 2 years in major morbidity, procedure failure, or secondary therapeutic interventions. Overall results of this trial suggest that there is a lower mortality for EVAR than for open repair.

The U.S. Medicare patient EVAR data published in 2008[3] reported on 22,830 patients with non-r-AAA who underwent either open repair or EVAR. This report found that EVAR was utilized more in older and sicker patients. EVAR had a lower morbidity and mortality compared with open repair. EVAR offered a survival advantage in older and sicker patients. It found that the late reinterventions after EVAR were mostly minor. Overall, it concluded that mortality rates for elective and r-AAA are decreasing.

In summary, when interpreting these trials and registries, it is important to bear in mind the time frame in which the trial occurred. The EVAR 1, EVAR 2, and DREAM trials were initiated a decade ago. Stent graft techniques and technologies have evolved significantly over that time. Patient selection and personal experience also have evolved, and our ability to manage complications has significantly improved. Thus, it is reasonable to conclude that current EVAR outcomes are better than what has been reported by these older randomized, controlled trials. Furthermore, many EVARs performed today are done in high-risk patients who fall outside the standard indications for use, due either to significant comorbidities or hostile aortic anatomy. Ultimately, the decision to proceed with EVAR depends on the well-informed patient and a skilled surgeon/interventionalist.

6.2.12 Ruptured Aneurysm

6.2.12.1 Endovascular versus Open Repair

Endovascular management of r-AAA is also gaining acceptance as contemporary data show survival benefits for EVAR over conventional open surgery.[27] The implications of improvements in our technical ability to offer EVAR to patients presenting emergently with AAA rupture are significant: to date, no other therapy has offered such a survival advantage to these high-risk patients. When considering these endovascular techniques for treating r-AAA, one has to prepare for the challenges of streamlining patient care from the emergency room to the operating room (OR) and the subsequent endovascular procedure that often requires a multidisciplinary approach.[28,177]

6.2.12.2 *Technique of Endovascular Management of Ruptured AAA*

All OR and hybrid endovascular/OR suites should be set up to facilitate both endovascular and OSR. As long as the patients maintain a measurable blood pressure, the technique of hypotensive hemostasis (limiting resuscitation to maintain a detectable blood pressure) can help minimize ongoing hemorrhage. Through a percutaneous or femoral artery cutdown, ipsilateral access is obtained using a needle, a floppy guidewire, and a guiding catheter. The floppy guidewire is exchanged for a super-stiff wire that can be used to place a 12-Fr sheath in the ipsilateral femoral artery, and the sheath is advanced to the juxtarenal abdominal aorta to support the aortic occlusion balloon, if needed. In hemodynamically unstable patients, the occlusion balloon is advanced through the ipsilateral sheath over the super-stiff wire into the supraceliac abdominal aorta under fluoroscopic guidance, and the balloon is inflated as needed. Access is subsequently obtained from the contralateral femoral artery (percutaneous or cutdown) in a similar manner, and a marker flush-catheter is advanced to the juxtarenal aorta for an arteriogram. The placement of the stent graft main body is planned based on the aortoiliac morphology that is best suited for EVAR. In hemodynamically stable patient, after the initial arteriogram, the aortic occlusion balloon is removed from the initial ipsilateral side, and the stent graft main body is advanced under fluoroscopic guidance, which limits the number of catheter exchanges. In hemodynamically unstable patients who require inflation of the aortic occlusion balloon, the marker flush-catheter is exchanged for the stent graft main body, which is delivered up to the renal arteries. An arteriogram is done through the sheath that is used to support the aortic occlusion balloon. The tip of the stent graft main body is aligned with the lowermost renal artery, the occlusion balloon is subsequently deflated and withdrawn back with the delivery sheath into the AAA, and the stent graft main body is deployed. The rest of the EVAR procedure is performed the same as in elective circumstances: (1) the tip of the stent graft main body is aligned with the lowermost renal artery, (2) the contralateral gate is aligned to facilitate expeditious gate cannulation, and (3) the ipsilateral and contralateral iliac extensions are planned and deployed as needed. Due to the emergent nature of ruptured EVAR, preoperative planning is limited and this can lead to additional procedures and complications.

Although beyond the scope of discussion here, abdominal compartment syndrome (ACS) merits discussion. ACS after EVAR for r-AAA results from (1) the space-occupying retroperitoneal hematoma, (2) ongoing bleeding from lumbar and inferior mesenteric arteries into the disrupted aneurysm sac in the setting of severe coagulopathy, and (3) the shock state associated with r-AAA, which is associated with alterations in microvascular permeability that can lead to visceral and soft tissue edema. Up to 20% of patients undergoing EVAR for r-AAA develop ACS. Associated factors include the need for an aortic occlusion balloon, the need for massive blood transfusions (mean 8 units of packed red blood cells), and the presence of coagulopathy with elevated activated partial thromboplastin time (PTT) at completion of the case. In one series, patients with ACS had a significantly increased mortality (67%) compared with those without ACS (10%).[29]

Regardless of all the improvements in endovascular techniques, on-table conversion to OSR is sometimes needed, and this approach should be in the armamentarium of all surgeons/interventionists involved in treating r-AAA patients. Endovascular repair of r-AAA is evolving and offers the potential for improved patient survival. Unlike elective EVAR, the time for preoperative planning during emergent EVAR is limited, and the preoperative imaging is often less than ideal. Flexibility and creativity of approach are important to address challenging issues that might arise during these emergent circumstances. The techniques for treating rAAA are similar to primary open repair, with the key difference being that r-AAA requires immediate aortic clamping and control at the suprarenal or supraceliac level.

6.2.13 Key Points

1. Aneurysm repair—either open or EVAR—should be performed on patients with symptomatic aneurysms, patients with AAA > 5.5 cm and few comorbidities, all saccular AAA, and patients whose aneurysm grows at an increased rate.

2. The major trials of EVAR have concluded it to be a safe procedure with a lower perioperative mortality rate than open repair, but with equivalent survival rates between the two approaches in the medium and long term.

3. Patients who undergo EVAR require rigorous follow-up to evaluate for stent graft complications such as endoleak, migration, limb thrombosis, sac enlargement, infection, and aneurysm rupture.

4. Endovascular repair of r-AAA offers the potential for improved patient survival but requires a streamlined effort to coordinate the various specialties involved in getting the patient to the hybrid operating suite. In addition, these patients need to be monitored for ACS.

6.3 Peripheral Arterial Disease

6.3.1 Introduction

PAD, as it is described in this chapter, refers to a spectrum of diseases that causes occlusions or severe stenosis of arteries throughout the body. In general, this is a systemic disease that predominantly affects the lower extremities but can also be seen in the upper extremities. Although cardiologists and interventional radiologists may occasionally treat patients as affected, vascular surgeons are often the first-line providers of definitive diagnosis and management of these patients. Therefore, the discipline of vascular surgery requires a comprehensive knowledge of the pathophysiology, which may cause this disease state.

The management of PAD requires a thorough understanding of the natural history of vascular disease, nonoperative modalities, interventional percutaneous treatment, surgical procedures and combinations, or so-called hybrid solutions. This necessitates at least 5–7 years of specialized training and often a lifetime of experience to appreciate the subtleties and complexity of this patient population and disease. For the management of PAD, a thorough knowledge of medical management of hypertension, diabetes, lipid disorders, renal dysfunction, and risk factor modification is needed. This often engenders cross-medical discipline cooperation between primary care physicians, cardiologists, nephrologists, pulmonologists, intensivists, infectious disease specialists, physiatrists, podiatrists, and others.

The population afflicted by PAD may affect any age group, but as it is a degenerative disease, it generally affects older patients—for example, those aged 60 years or older. As atherosclerosis is a systemic disease, patients with PAD often have arterial disease in other areas of the body such as the heart, brain, kidneys, and mesenteric arterial beds. Disease states and risk factors such as diabetes mellitus, hypertension, cigarette smoking, COPD, renal dysfunction, and lipid disorders are highly prevalent.

6.3.2 Epidemiology

Top 10 causes of mortality and morbidity in patients with PAD are as follows:

1. CAD
   Coronary artery disease
   MI
   Congestive heart failure
2. Stroke
   Embolic stroke
   Hemorrhagic stroke
3. Renal failure
4. Sepsis
5. Limb loss
6. Multisystem organ failure
7. Bleeding
8. Wound infection
9. Failure to thrive
10. Functional impairment/disability

The vast majority of complications related to PAD are a consequence of the systemic nature of atherosclerosis. PAD is more likely associated with CAD than is CAD with PVD. CAD is detectable in over 60% of patients using stress tests and over 90% if coronary angiography is used. This relationship similarly holds true for PAD and CBVD. Significant carotid stenosis (>70%) was present in over 25% of patients with PAD, whereas only 11% of patients with a diagnosis of PAD had significant PAD. Thus, the major causes of death in patients with a diagnosis of claudication are MI, congestive heart failure, and stroke.

The remaining complications are consequences of the operations/interventions. They all relate to magnitude of the procedures, comorbidities of the population, and the consequences of failed procedures. The varied causes of morbidity after procedures used to treat PAD point to the need to have a cross-disciplinary approach in managing these patients.

Renal failure may occur as a consequence of nephrotoxic dye given for a diagnostic angiogram or therapeutic intervention. PAD may affect flow to kidneys as a consequence of renal artery occlusive disease. Alternatively, there may be intrinsic renal disease. The final common pathway is a reduced glomerular filtration rate that makes the kidneys susceptible to dye or medication-induced renal dysfunction. Adequate hydration and nephrology consultation is necessary in this circumstance and if utilized preprocedure may limit the complication rate.

Infection or sepsis may occur as a preexisting condition due to tissue loss and poor perfusion. These complications may also occur in the postoperative period with wound incision issues or infection of a prosthetic conduit. Infection during the perioperative/interventional period can extend the length of hospital stay.

Functional impairment and disability often occur after lower extremity revascularization as a result of several issues. These patients' preoperative functional status is often impaired especially in patients with CLI. The addition of a minor foot amputation, which may be necessary to control infection, can further impair functional ability and make postoperative physical therapy and rehabilitation services a necessity.

6.3.2.1 Major Diagnoses and Procedures

6.3.2.1.1 Diagnoses

Acute occlusion
    Acute embolic occlusion
        Macroembolic
        Microembolic
    Acute thrombotic occlusion
Chronic occlusion
    Claudication
    Rest pain
    Tissue loss
        Nonhealing ulcer/wounds
        Gangrene

In general, arterial ischemia secondary to PAD can be separated into those clinical situations due to acute or chronic occlusions. Causes of acute macroembolic occlusion are usually cardiac in origin specifically from atrial, ventricular, or valvular sources. Microembolic sources are usually arteroarterial in origin and may come anywhere from thoracic aorta to popliteal/tibial sources. The most common source in patients with unilateral disease is the superficial femoral artery at the level of the adductor hiatus. Acute thrombosis usually occurs in the setting of chronic disease that may or may not have been symptomatic. Alternatively, acute thrombosis may occur without preexisting disease when a hypercoaguable state is present.

Chronic occlusive disease when symptomatic is either lifestyle limiting or potentially limb threatening. Claudication is lifestyle limiting and associated with low rates of long-term major limb amputation (3% at 10 years). The symptoms are reproducible and relate to exercise-induced discomfort in major muscle groups of the lower extremities that are relieved with rest. The major diagnostic determination relates to identifying and differentiating vascular from neurogenic causes such as spinal stenosis. Other vascular causes of claudication aside from occlusive disease involve etiologies such as popliteal entrapment or adventitial cystic disease.

Limb threatening or CLI manifests as either rest pain or tissue loss (nonhealing wounds or gangrene). The chance of limb loss in these situations approaches 40%–60% and has prompted an aggressive stance toward revascularization.

6.3.2.1.2 Procedures

Acute
    Lytic therapy/thrombus retrieval
    Angioplasty/stent
    Embolectomy
    Thrombectomy
    Endarterectomy
    Bypass
    Combination/hybrid therapy

Chronic
    Inflow procedures
        Iliac angioplasty/stenting
        Aortobifemoral bypass
        Aortoiliac endarterectomy
        Iliofemoral bypass
        Iliofemoral endarterectomy
        Femoral to femoral bypass
        Axillo femoral bypass
        Endovascular aortoiliac replacement
    Outflow procedures
        Angioplasty/stent
        Bypass
            Autogenous vein
            Prosthetic
            Prosthetic (±heparin-bonded grafts)
            Prosthetic with distal vein cuff
            Prosthetic with distal arteriovenous fistula
            Sequential
            Combination inflow/outflow procedures

The procedures used to treat acute ischemia are targeted toward the etiology and pathogenesis of type of ischemia. In most cases, an aggressive interventional approach is taken first when time allows. Initially, arteriography with an intraarterial catheter-directed approach is performed. The location and possible embolic versus thrombotic cause may be identified. Thrombus retrieval systems and/or lytic therapy of the acute embolus or thrombus with tissue plasminogen activator (TPA) may be possible especially if a guidewire can be directed intraarterially beyond the occlusive lesion. After an initial bolus of TPA, a continuous infusion with rearteriography after 12–24 hours is performed to check the progress of revascularization. If a chronic underlying arterial stenosis is revealed, this may be treated with catheter-directed therapies such as angioplasty and/or stent placement. Close clinical monitoring of the degree of ischemia as regards failure of improvement or worsening of signs and symptoms is critical. This may require abandonment of catheter-directed therapies for an open surgical approach.

Open surgical approaches, if hybrid facilities are not available in the OR, involve an initial arteriography to identify the anatomy followed by direct revascularization. This may involve significant delay in definitive treatment. The presence of hybrid suites in the OR allows minimal delay between the diagnostic imaging and revascularization. If an embolus is suspected, balloon catheter retrieval of embolic debris is performed. Completion arteriography can then confirm the adequacy and completeness of treatment. If underlying chronic stenotic or occluded arteries are found, these may be treated with catheter-directed therapies (angioplasty and/or stent), direct thromboendarterectomy (removal of plaque and thrombus from discrete arterial regions usually in the ileofemoral artery segment), or arterial bypass with prosthetic grafts or autogenous vein (see below).

Following revascularization, lower leg fasciotomy may be necessary based on the degree and duration of ischemia to prevent compartment syndrome. Long-term anticoagulation therapy is also initiated to prevent subsequent future arterial embolic events. In patients with arterial thrombosis secondary to a hypercoaguable disorder, long-term anticoagulation and hematology consultation will be necessary.

In the setting of inflow artery issues, arterial occlusion in the aortic or iliac arteries, when lytic, balloon catheter or angioplasty/stent therapies are not able to reestablish adequate perfusion of the legs, arterial bypass may be necessary. The choices are aortic, iliac, femoral, or axillary inflow sources with a prosthetic bypass graft. The abdominal aortic-based inflow procedures have the best long-term patency but also have the greatest associated mortality and morbidity rates. Femoral and axillary arteries are often chosen as inflow in the acute setting to offset the greater associated risk of more definitive inflow procedures. Accordingly, when perfusion of the foot is limited, despite adequate inflow, and catheter-directed therapies are not possible, it may be necessary to perform distal arterial bypass. In these settings, more for considerations of the time necessary to perform definitive reconstructions, prosthetic bypass may be performed to obtain initial salvage with more definitive autogenous reconstruction reserved for a later date. Due to the magnitude of combined inflow and outflow procedures, these are often staged to reduce the overall physiologic stress and subsequent mortality and morbidity.

In the setting of chronic arterial ischemia, the approach to reconstruction is based on considerations of patient risk/expectations, anatomic options/patterns of disease, and conduit durability/availability. The major cause of mortality and morbidity after arterial reconstruction for chronic disease not surprisingly relates to cardiac complications; therefore, preoperative risk assessment is paramount. Recognition of these facts has prompted an endovascular first approach especially for patients with limited disease who are more likely to present with quality of life indications such as claudication. When the requirement for arterial perfusion is maximal as in tissue loss, the concerns of patient risk, procedure cost, and durability have more equal weight. Our approach is an interventional/endovascular approach first when possible, followed by open surgical revascularization when disease is more extensive and patient risk more acceptable.

In terms of lower extremity arterial reconstruction, percutaneous/endovascular treatment has worked very well for discrete ileofemoral and limited femoral popliteal disease (TASC II A, B, C, and selected D lesions) and in the setting of claudication. When tissue loss is extensive (large area ulceration, gangrene, and invasive infection) and/or long-segment arterial occlusive disease occurs, especially in the popliteal/tibial regions, arterial bypass with autogenous vein is the gold standard in terms of durability and limb salvage. We have found that vein prepared as an in situ conduit has functioned best and is our preferred bypass reconstruction when vein is present and of adequate quality.[178]

When autogenous vein is not available, alternatives, sometimes second best alternatives, for revascularization are considered. Although percutaneous options do not have the same durability and limb salvage as vein bypass, interventions at the below-knee popliteal and tibial level are considered. If the disease is more extensive, bypasses with prosthetic grafts may be the only option. The patency of these reconstructions is so dismal that alternative strategies with heparin bonding of polytetrafluoroethylene (PTFE) grafts and adjunctive procedures with prosthetic grafts are utilized. Some of these adjunctive procedures use attempts to remodel the flow at the distal end of the bypass with vein cuffs or attempts to increase flow through the graft with creation of a distal arteriovenous fistula. Alternatively, a sequential bypass using a combination of prosthetic and residual vein can be constructed.

6.3.3 Cost and Economic Burden

In a recent review of data from the Surveillance, Epidemiology, and End Results (SEER) database, Hirsch et al. found that a total of $4.37 billion was spent on PAD-related treatment. Inpatient care accounted for 88% of expenditures. Medicare program outlays totaled $3.87 billion, while enrollees (or their supplemental insurance) spent the remaining $500 million. Treatment increased with age at rates of 4.5%, 7.5%, and 11.8% for individuals aged 65–74, 75–84, and >85 years, respectively. They concluded that earlier use of preventive care measures such as atherosclerosis risk reduction, outpatient claudication treatments (e.g., supervised exercise), and durable revascularization treatments should be evaluated as strategies to both improve health and limit national healthcare costs.

In a recent paper by Sachs et al.,[179] the costs and limb salvage rates of percutaneous revascularization were compared with those of surgical revascularization. Although the mortality was less, the major amputation rates and costs were greater. This is in opposed to the finding of De Vries et al.,[172] who found that the benefits of open surgical revascularization for claudication are outweighed by the costs, prompting the recommendation of nonoperative or percutaneous means when treatment is considered. Data from previous studies have shown that revascularization in general with attempts at salvage is more effective in terms of overall cost, patient mortality, and functional outcome than is major limb amputation. The costs of angioplasty and surgical revascularization as reported in the TASC II Consensus Statement are approximately $10,000 and $20,000, respectively, per patient. These costs double if the procedure fails and the patient requires repeat treatment. The cost for major lower extremity amputation is $40,000 per patient and does not include the cost of physical therapy and rehabilitation.

6.3.4 Current Standard of Care

Management of patients with acute ischemia is dictated by the severity of presenting ischemia. If the limb is not immediately threatened (severe sensory or motor dysfunction), an interventional approach with catheter-directed thrombolysis is always offered first. The results of the STILES (a prospective randomized trial evaluating surgery versus thrombolysis for ischemia of the lower extremity) trial showed limb salvage rates of greater than 82% at 1 year. In this trial, the patient mortality and morbidity of catheter-directed treatment was also less than the corresponding rates of open surgery. The authors in this study additionally concluded that an interventional first approach also allows a more direct, limited, and potential lower risk treatment strategy. These findings have largely been accepted and incorporated into the practice of vascular surgery.

The most important part of care of the patient with chronic PAD involves the stratification of the patient in terms of the presenting clinical problem and the patient's medical risk. A thorough patient history and physical examination are critical to determine the rapidity to which revascularization must be achieved and the ancillary testing necessary to stratify the medical risk. In the setting of CLI with infection, consultation with infectious disease specialists with use of intravenous antibiotics may be necessary during the pre- and perioperative/interventional period. In addition, cardiac, renal, and pulmonary risk may require additional consultation and testing. Most patients will need cardiac risk stratification with some sort of noninvasive testing such as stress testing or echocardiography.

After the initial clinical evaluation, noninvasive tests such as pulse volume recording are performed to get some physiologic parameters of lower extremity blood flow and perfusion. This will help determine whether the flow issues are related to inflow or outflow problems or some combination of both. The testing is often performed simultaneously with patient medical risk evaluation so as to minimize the length of pre-revascularization delay. Arterial imaging with CT angiography of the aorta and bilateral lower extremities is performed to further evaluate the level of disease. Alternatively, intraarterial angiography is performed. The advantage is that intervention can be performed immediately following the diagnostic part of the study. Our preference is, once diagnostic arteriography is performed, to attempt an interventional treatment of hemodynamically inflow occlusive disease with angioplasty/stenting. The standard of care in the past was aortobifemoral bypass for aortoiliac occlusive disease. Currently, bilateral iliac stents or unilateral stents are placed with planned bypass to the femoral artery in the affected extremity as necessary. When CT angiogram has identified that a hybrid approach may be necessary, the angiogram and surgical reconstruction may be performed in the hybrid operating suite.

Angiography as a diagnostic or therapeutic procedure is usually performed on an outpatient basis. When patients require hybrid procedures or direct inflow revascularization procedures such as endarterectomy, aortobifemoral, iliofemoral, or femoral-femoral bypasses, patients require inpatient hospitalization of 1–5 days depending on the magnitude of the procedure.

Patency rates of iliac interventional procedures are approximately 90% and 66% at 1 and 5 years, respectively. These results depend on the stent design and indication for the procedure. In general, the results for claudication (71% at 5 years) are better than that for CLI and most likely relate to the extent and severity of the occlusive disease. The results of open procedures range from 90% to 95%, 85%, 75%, and 70% 5-year patency rates for aortobifemoral, ileofemoral, femoral-femoral, and axillobifemoral bypasses, respectively.

The most important tool in the management of outflow lower extremity occlusive disease after initial clinical assessment is the diagnostic intraarterial angiogram. An interventional approach is chosen first if feasible. Although initial success at 1 year is greater than 90%, pooled results at 3 years are about 60%–65%. Again, the patency results are better with claudication than with CLI.[173]

The results of open bypass are better than percutaneous methods for treatment of disease extending below the knee. When at all possible, autogenous vein should be used preferentially for bypass. Information about the presence and quality of the vein conduit prior to surgery is extremely important. This can be assessed with duplex scan and confirmed at the time of surgery. Atraumatic preparation of the vein in our hands has allowed for patency rates of greater than 75%–80% at 5 years for bypasses to the below-knee popliteal and tibial arteries. Prosthetic bypass to the same levels allows patency rates of 35% at best at 5 years. The limb salvage rates for patients with CLI treated with bypass exceed 90% when vein is used and prepared in an atraumatic manner.

The current limitations for achieving the standard of care relate to the overall medical risk of the PAD patient population, deficiencies of the revascularization options, and cost of care in a political environment focused on cost containment and not necessarily provision of quality. The population of patients afflicted with PAD is, in general, an older population with significant medical problems and overall risk for any intervention. Coordination of care between vascular surgery and other consultants and subsequent optimization of the patient's medical status prior to and during treatment is extremely important and often requires a well-coordinated approach.

Issues with revascularization options relate to design issues of the metallic stents and pathologic interactions between inserted devices and the host patient. Other issues involve the availability and quality of an autogenous vein conduit for bypass. Vein is a precious resource in the body and may be not available due to prior harvest for coronary artery or other arterial reconstruction. In addition, due to intrinsic vein disease or prior venous ablative procedures, it may not be available. Prosthetic reconstructions may be required but have poor durability and utility.

Cost issues for PAD refer to the need for reimbursement to be based on fiscally responsive models of healthcare. The focus needs to be continually targeted on quality of care and efficient models of delivery. This will require open communication and ongoing dialogue between vascular surgeons, third-party payers, and governmental regulatory agencies.

6.3.5 Potential Nanomedicine Applications

1. Drugs and drug delivery: Nanoparticles (<100 nm in at least one dimension) can be used as drug delivery vehicles. These particles are taken up by cells more efficiently and could be used as effective transport and delivery systems.[30] Nanoparticles can be used in drug delivery targeted at the site of disease to improve the uptake of poorly soluble drugs. Polylactic/glycolic acid- and polylactic acid-based nanoparticles have been formulated to encapsulate dexamethasone, a glucocorticoid with an intracellular site of action. Dexamethasone is a chemotherapeutic agent that has antiproliferative and antiinflammatory effects. This complex then binds to the cytoplasmic receptors in macrophages, and the subsequent drug–receptor complex is transported to the nucleus resulting in the expression of certain genes that control cell proliferation. These drug-loaded nanoparticles formulations that release higher doses of drug for prolonged periods of time may completely inhibit proliferation of vascular smooth muscle cells (SMCs). This may reduce rates of neointimal hyperplasia and subsequent restenosis and thrombosis rates.[171]

2. The diagnosis of disrupted plaque by detection of small deposits of fibrin in erosions or microfractures could allow characterization of a potential "culprit" lesion before a high-grade stenosis has been formed that is detectable.[31] Nanoparticle-targeted fibrin imaging with either ultrasound or paramagnetic MR contrast agents was first demonstrated in the 1990s. Such tissue factor imaging has been demonstrated *in vivo* for molecular imaging with ultrasound and *in vitro* with magnetic resonance imaging (MRI). This technology is still under development and has yet to be characterized and used in clinical practice.

3. Microfabrication technologies have improved in conjunction with increased demand for minimally invasive surgical treatments. Microfabrication technology utilizing laser has been used for placement of drug reservoirs such as those utilizing Paclitaxel in medicated coronary artery and most recently, femoropopliteal artery stents (Zilver stent). Clinical studies have shown reduced rates of restenosis in patients with the Zilver stent in the superficial femoral artery as compared to arteries treated with balloon angioplasty alone. Such nanotextured coatings fit into a category of design concepts that enhance endothelialization of stent struts and may reduce late thrombosis.

6.3.6 Current Medical Needs

There has been no consistent genetic or molecular marker that has preidentified and/or predicted those patients at risk for the development of and failure of treatments for PAD. Identification of genomic sites and development of so-called medical magic bullets similar to those developed in the field of oncology will revolutionize the approach to vascular disease.

The limitations in terms of revascularization options deal with failures of percutaneous and open treatments. Failures of percutaneous treatment often are a consequence of neointimal hyperplasia and stent fracture and relate to stent design. Bioscaffolding developments, as mentioned above, may allow development of more effective medicated stents. In addition, development of dissolvable stents has already been implemented in treatment of femoropopliteal occlusive disease. Further development of these technologies and extension to tibial interventions will allow extension of interventions more reliable to infrageniculate sites.

Failures of open treatments are often a consequence of conduit limitation and relate to mechanisms of vein and prosthetic graft failure. Neointimal hyperplasia and failure of prosthetic grafts to develop and a functional endothelial monolayer are two such issues. Development of bioscaffolding technology to allow extracellular matrix deposition within prosthetic grafts may potentially allow endothelial seeding and a sustainable and functional endothelial monolayer.

6.3.7 Conclusions

The future of vascular surgery for management of PAD will be strongly dependent on development of quality models that are fiscally sound and patient centric. In addition, new technologies utilizing nanotechnology will be needed to meet the needs of conduit and revascularization limitations.

6.4 Renal Occlusive Disease

6.4.1 Introduction

Atherosclerosis is a systemic disease, so vascular surgeons need a working knowledge of many areas of medicine. However, there are limits. While nanomedicine is an exciting and emerging field, most vascular surgeons' impression (and knowledge) of it is confined to the 1966 movie *Fantastic Voyage*, where a microscopic submarine travels through a human body to perform intracranial thrombolysis. Coincidentally, this is also the level of most elementary schoolers' knowledge. The same premise is used in the *Magic School Bus* series of children's books about the human body.

The renal artery and renal physiology rarely feature in popular culture, for good reason. Of all vascular territories, the renal artery may be the most difficult to decipher. Renal artery stenoses (and even chronic occlusion of a single artery) usually produce few symptoms. There are no renal versions of angina or intermittent claudication, and there are no external signs such as ulcers or gangrene. Renal artery stenosis (RAS) defies simple categorization. It can cause dialysis dependence in some, but most patients have normal renal function. RAS causes 2% of all cases of hypertension,[32] but most patients with hypertension have normal renal arteries (Figure 6.4).

Management is hotly debated. Until recently, in the absence of randomized control trials and national guidelines, lesions were treated frequently (but often, to little effect). The development of renal artery stenting[33] in 1991 simplified management (compared to complex open surgical options) but turned into something of a double-edged sword. Stenting can be performed in any "cath lab" and this led to its fairly arbitrary and "liberal" use,

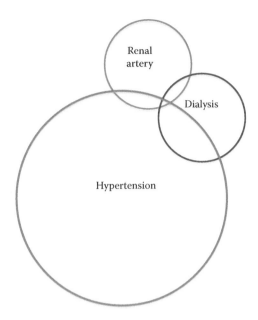

FIGURE 6.4
Schematic to illustrate the place of RAS in the etiology of hypertension and end-stage renal disease (dialysis dependence).

theoretically to preserve renal function. The last 20 years saw an exponential rise in renal artery stenting. From 1996 to 2000, claims for Medicare beneficiaries for renal artery stenting increased from 7,660–18,520 rising to 35,000 in 2005.[34] It was hoped that two large randomized control trials would confirm the role of stenting in preserving renal function and correcting hypertension.[35,36] The first of these, the Angioplasty and Stenting Trial for Renal Artery Lesions, was reported in 2009.[37] Despite concerns about its methodology,[38,39] it showed that patients with severe RAS did not benefit from stenting and did just as well from medical therapy (Figure 6.5). As might be expected, this was highly contentious.

6.4.2 Epidemiology

RAS is usually due to atherosclerosis (>90% of lesions) and occurs in 0.1% of the general population. This might seem an underestimate to vascular specialists, but our patients are not typical. Overall, 2% of all hypertensive patients have RAS.[32] Its incidence rises to 16% in patients with hypertension and coronary artery disease.[40] Among patients undergoing cardiac catheterization, 15% will have RAS greater than 50%.[41] In PAD patients, the prevalence is 27%–50%.[42,43] Predictably, RAS is a marker for life expectancy (Figure 6.6).[44] In patients with a 75% RAS at the time of coronary angiography, 57% were still alive at 4 years, compared to 89% of those without RAS. Patients with bilateral RAS fared even worse with only 47% alive at 4 years.[45]

6.4.3 Normal Renal Physiology and Renal Artery Hemodynamics

At rest, healthy kidneys have blood flow of 1200 mL/min, far in excess of that needed for normal renal cell function. Less than 10% of the blood flow is used for metabolism. The majority is filtered. Each kidney is usually supplied by a single renal artery (Figure 6.7).

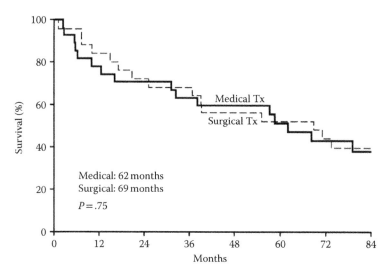

FIGURE 6.5

The Angioplasty and Stenting Trial for Renal Artery Lesions published in 2009 showed no differences between renal artery stenting and medical therapy for RAS. (From ASTRAL Investigators et al., *N. Engl. J. Med.*, 361, 1953–62, 2009.)

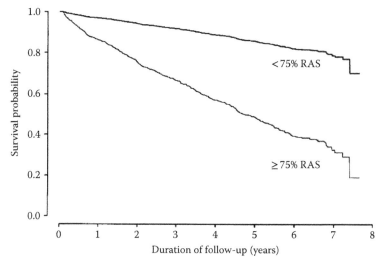

FIGURE 6.6

Kaplan–Meier survival curve showing unadjusted survival according to the presence or absence of significant renal artery stenosis (RAS; $p < .001$). Symbols are as follows: thick line, <75% RAS; thin line, 75% RAS. (From Conlon, P.J. et al., *J. Am. Soc. Nephrol.*, 9, 252–6, 1998.)

Renal nerve supply comprises sympathetic fibers, which control vasomotor function, and parasympathetic fibers, whose function is unknown.

The functional unit of the kidney is the nephron (Figure 6.8) and renal corpuscle. In a healthy individual, each kidney contains a million nephrons. As with neurons, the number decreases with age, but there is considerable redundancy. Renal impairment becomes apparent only when more than 50% of nephrons have been lost. If arterial perfusion

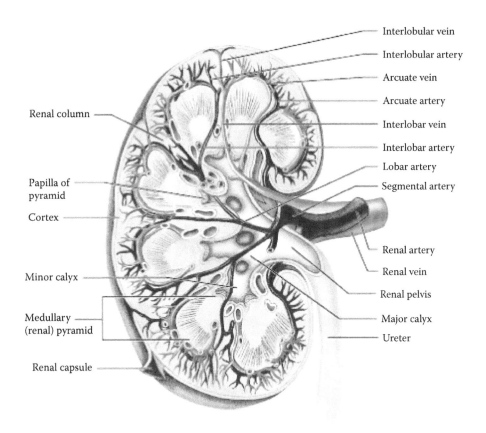

FIGURE 6.7
Renal artery anatomy. (Copyright 2001 Benjamin Cummings, an imprint of Addison Wesley Longman Inc.)

abruptly ceases, nephrons remain viable only for 30 minutes at body temperature ("warm ischemia time"). At the cellular level, an afferent arteriole delivers oxygenated blood to the renal corpuscle. It is a tuft of capillaries surrounded by Bowman's capsule. A smaller efferent arteriole allows the still-oxygenated blood to leave the corpuscle. As it is of smaller diameter, the efferent artery maintains high hydrostatic pressure within the glomerulus for more effective filtering. One-fifth of the perfusing plasma, containing water and solutes, is filtered through the glomerulus.

From Bowman's capsule, the filtered fluid drains into the nephron. In the proximal convoluted tubule, two-thirds of it is reabsorbed. The remaining filtrate then passes down the loop of Henle. The water-permeable descending limb allows water to leave the filtrate. The ascending limb is waterproof; it actively pumps sodium into the interstitial space. The filtrate then passes through the distal convoluted tubule, whose cells are rich in mitochondria. Finally, the urine drains into the collecting duct system. Urine then leaves via the medullary collecting ducts.

Finally, there is the juxtaglomerular apparatus, which produces renin. This stimulates production of angiotensin II (the renin–angiotensin–aldosterone mechanism) (Figure 6.9).

In parallel to the RAA mechanism, there is an autonomic component to renal autoregulation. In the 1940s and 1950s, extensive surgical sympathectomy was employed in the belief that systemic vasodilatation could reduce blood pressure.[46] This involved removal of most of the thoracic sympathetic ganglia, half the celiac ganglion, and a variable number of lumbar ganglia. All patients required thoracotomies, and operative mortality was 4%. Almost

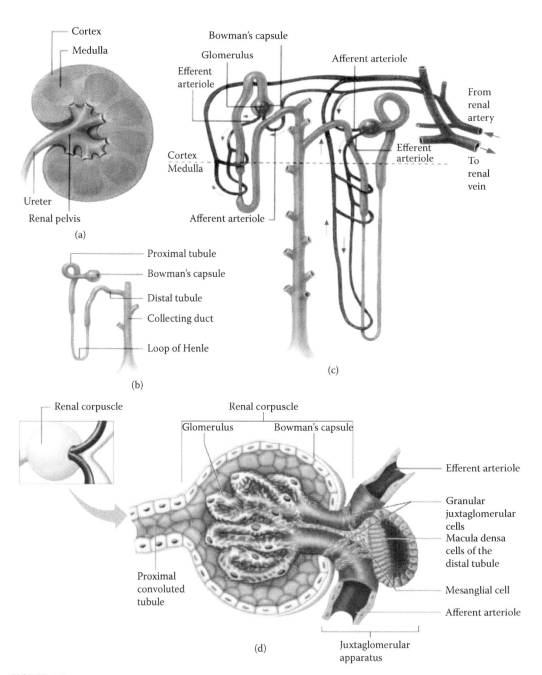

FIGURE 6.8
The functional unit of the kidney comprises the renal corpuscle, juxtaglomerular apparatus, and the nephron. (a) Kidney, (b) urinary tubule and collecting system, (c) arterial and venous supply to the urinary tubule and collecting system, and (d) renal corpuscle. (Copyright 2001 Benjamin Cummings, an imprint of Addison Wesley Longman Inc.)

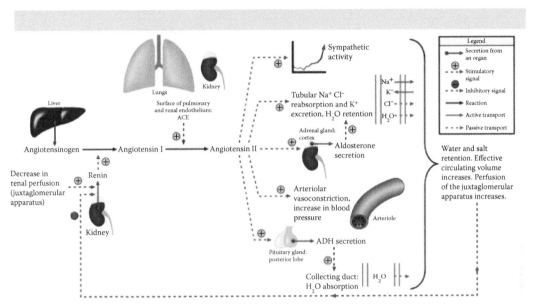

FIGURE 6.9
The renin–angiotensin–aldosterone system. (Courtesy of Wikipedia.)

all males undergoing first lumbar ganglion resection were rendered impotent. Although many[47,48] reported profound improvements in blood pressure control, it never really caught on and by the 1970s had been largely abandoned. Sympathectomy may have been effective due to renal sympathetic neurectomy. This comes from observations in nephrectomy and transplant patients[49] and is supported by animal experiments where renal sympathetic stimulation causes hypertension.[50] Anatomical pathways also support the renal sympathetic nerves and the central nervous system as regulators of blood pressure (Figure 6.10).

6.4.4 Pathophysiology

Complete bilateral renal artery occlusion results in loss of all renal function and dialysis dependence. On the other hand, RAS can cause (1) chronic hypertension ("renovascular hypertension"), (2) ischemic nephropathy, and (3) fluid overload ("flash" pulmonary edema). Or, it may produce no pathology at all.

6.4.4.1 Renovascular (Goldblatt) Hypertension

In 1827, Richard Bright[51] described an association between "hardness of the pulse," proteinuria, "dropsy," and "hardening of the kidney." Over 70 years later, in 1898, Tigerstedt and Bergmann,[52] working in the Karolinska Institute in Stockholm, discovered renin. In 1934, Goldblatt[53] published his seminal paper on RAS-induced hypertension (still referred by some as "Goldblatt hypertension"). When renal perfusion diminishes, the juxtaglomerular apparatus is activated, producing renin and ultimately angiotensin II (Figure 6.11). Angiotensin II is a potent vasoconstrictor that promotes the release of aldosterone. When RAS affects only one renal artery, hypertension is "renin dependent" due to systemic vasoconstriction and increased peripheral resistance. Renin also causes increased aldosterone release, which causes increased salt and water retention. The healthy kidney excretes all excess salt and water retention, but renin levels remain high due to reduced renal

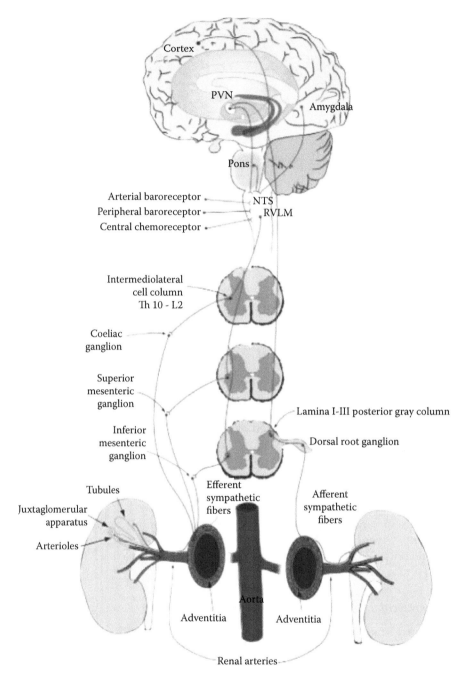

FIGURE 6.10
The renal sympathetic nervous system. The kidney is supplied by sympathetic fibers closely applied to the surface of each renal artery. Also located in the renal artery adventitia are afferent sympathetic nerves. *Abbreviations:* NTS, solitary tract nucleus; PVN, paraventricular nucleus; RVLM, rostral ventrolateral medulla.

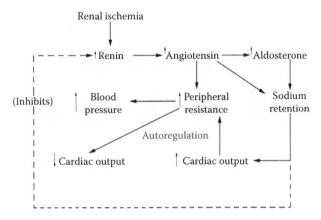

FIGURE 6.11
Role of renin–angiotensin–aldosterone system RAS in the pathogenesis of hypertension. (E-medicine figure on renovascular hypertension. Courtesy of Wikipedia.)

perfusion pressure. Correction of the renal artery lesion should normalize blood pressure, especially if renin levels are high in the renal vein draining the affected kidney. These can be sampled using a catheter inserted through the femoral vein.

But when renal artery stenoses are bilateral, or there is only a single kidney, this cannot be compensated for. As well as vasoconstriction, there is now a hypervolemic component to the patient's hypertension. There is a negative feedback on the production of renin, so that renin levels are low. Again, correcting the renal artery lesions should correct blood pressure.

However, there is a so-called *third phase* of renovascular hypertension, when blood pressure is unremitting, even after removal of the provocative lesions. In this setting, recalcitrant hypertension likely represents ischemic nephropathy in one or both kidneys.

6.4.4.2 Ischemic Nephropathy

Ischemic nephropathy is "impairment of renal function beyond occlusive disease in the main arteries." The histopathological finding is interstitial fibrosis, where collagen is laid down around the renal tubules with tubular atrophy. When atherogenesis and hypercholesterolemia are combined with RAS, an inflammatory cascade is produced (Figure 6.12), which results in fibrogenesis. Initial repeated insults result in recovery in animal models, but prolonged insults result in a chronically fibrotic kidney. If ischemic nephropathy can be treated early in its course, therefore, it may be possible to reverse it. And although there are reports of recovery of renal function in chronically ischemia kidneys following open or endovascular revascularization,[54,55] this is not usual and would not be expected. The best that could be hoped for in patients with established interstitial fibrosis is stabilization and maintenance of existing function.

6.4.4.3 Flash Pulmonary Edema

Flash pulmonary edema is acute congestive heart failure that develops rapidly, often over a matter of minutes. It is a rare manifestation of RAS found in patients with significant bilateral RAS or severe stenosis in a solitary kidney. There is an increased hemodynamic burden or hypervolemia, increased vascular permeability caused by elevated angiotensin levels and defective natriuresis.[56] There is no mechanism, however, to compensate for the excess sodium and water. Even though most of these patients have normal left ventricular function, they endure repeated hospitalizations with "flash" pulmonary edema. Such a

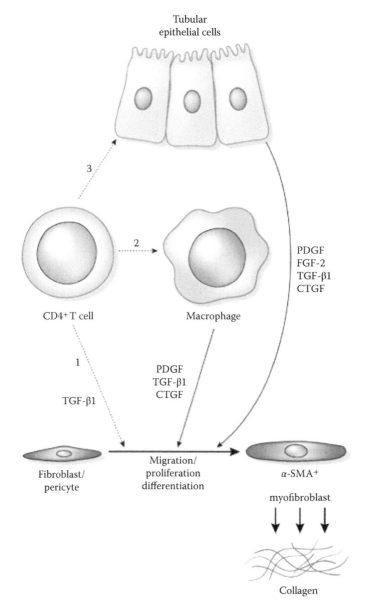

FIGURE 6.12
Schematic representation of pathogenesis of ischemic nephropathy.

clinical picture should prompt a search for renal artery lesions. The finding of bilateral RAS or a stenosis in a solitary functioning kidney mandates revascularization. In one report, this reduced the need for further hospitalizations in 77% of patients.[57]

6.4.5 Natural History of RAS in Normotensive Patients with Normal Renal Function

We often diagnose RAS serendipitously in asymptomatic patients who are normotensive with normal renal function. But, what is its significance? In 1998, a cohort study[58] of 170 patients with RAS affecting at least one renal artery monitored with serial duplex scanning for 33 months reported disease progression in 35%. Only 3% of arteries progressed to occlusion. At 2 years, renal atrophy and loss of parenchyma occurred in 20% of those patients with $\geq$60% stenosis. In 2000, Iglesias et al.[59] showed that RAS in asymptomatic RAS patients had the same risk of decline in renal function as matched controls with normal renal arteries. A Mayo Clinic study[60] looked at the outcome for 68 patients with high-grade (>70%) lesions who were treated medically over a mean of 39 months. Only 4 of the 68 (5.8%) patients required revascularization. By study end of the study, 28% of patients had died of nonrenal causes and 7.4% were on dialysis for non-RAS-related reasons. Most patients did not require dialysis and remained normotensive. Most at risk were those with bilateral RAS or a single functioning kidney. Finally, in 2006, a study of 834 "free-living elderly Americans" in North Carolina[61] showed that asymptomatic RAS is relatively harmless in patients older than 65 years with none progressing to occlusion. The advice from this study was that incidental asymptomatic RAS in older patients should not be treated due to its benign prognosis.

6.4.6 Current Management of Renal Artery Stenosis

6.4.6.1 Balloon Angioplasty and Stenting

Balloon angioplasty of the renal arteries was first performed by Grüntzig and coworkers[62] in 1978. It is effective for nonatheromatous renal artery lesions such as fibromuscular dysplasia. Renal artery ostial lesions are really an extension of aortic plaque into the renal artery origins and do not respond well to balloon angioplasty alone. Stenting is more difficult than simple balloon angioplasty but more effective for ostial and proximal renal artery stenoses (Figure 6.13). Unfortunately, manipulation of the renal artery origin in an atheromatous aorta has its problems. Despite careful technique, renal function may worsen after intervention as angioplasty and stenting of renal arteries may cause significant atheroemboli.

6.4.6.1.1 Morbidity, Mortality, and Patency Rates of Endovascular Therapy (Table 6.5)

In 1993, a Swedish study[63] compared balloon angioplasty with renal endarterectomy in 58 patients.[64] This was prior to renal stenting. The technical success rate was higher in the surgical groups (97% vs. 83%), but secondary patency rates were similar. The authors suggested balloon angioplasty be first-line treatment, provided adequate surveillance was performed. In 1999, a randomized controlled trial compared renal stents with balloon angioplasty.[65] Stents had a superior technical success rate (88% vs. 55%), and 6-month primary patency was significantly better (75% vs. 29%). The trial was stopped early following an interim analysis of the data.

How effective such procedures are at reversing renovascular hypertension and ischemic nephropathy? A 2008 study from Wake Forest[66] reported that hypertension was cured in 1.1%, improved in 20.5%, and unchanged in 78.4%. In ischemic nephropathy, the goal is to prevent disease progression. Although initial reports suggested that after renal artery stenting renal function improved or stabilized in the majority (69%) of patients,[67] a more recent meta-analysis of the available renal artery stenosis studies comparing stenting to medical therapy found no significant improvement in renal function following intervention.[68]

6.4.6.2 Surgical Therapy

Bypass for RAS can be divided into *anatomic* and *extra-anatomic* (Table 6.6). *Anatomic bypasses* arise from the aorta or aortic bypass grafts but include thromboendarterectomy and renal artery reimplantation (Figure 6.13). These reconstructions have both high flow and

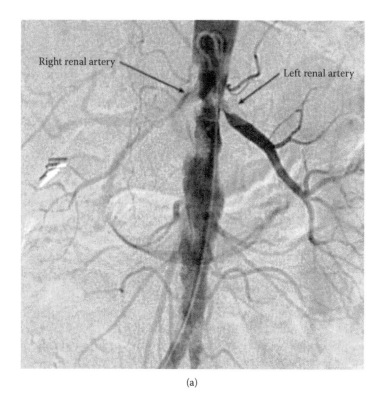

(a)

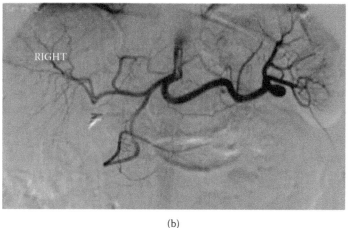

(b)

FIGURE 6.13

Angiogram demonstrating severe atherosclerosis of the aorta with severe stenosis of both renal arteries (a). The right kidney was atrophic, and the patient had significant comorbidities that made her a good candidate for splenorenal bypass. It is also important to assess inflow by selective celiac angiography (b).

excellent patency. *Extra-anatomic bypasses* are hepatorenal, splenorenal, and rarely iliorenal or mesorenal revascularizations. These are secondary procedures reserved for high-risk individuals, challenging anatomy, and patients with significant comorbidities. Ultimately, there is nephrectomy, which still has a limited role even today.

TABLE 6.5

Endovascular Renal Artery Revascularization with Angioplasty and Stenting: Morbidity, Mortality, and Patency Rates of Surgical Bypass

First Author	Year	Patients	Bilateral Treatment (%)	Preoperative Renal Dysfunction (%)	Renal Function Response (%)			Hypertension Response (%)			Perioperative Outcome (%)	
					Improved	Unchanged	Worsened	Cured	Improved	Failed	Death	Morbidity
Burket	2000	127	NR	29	43	57		NR			2	4
Lederman	2001	300	41	37	9	78	14	70		30	<1	2
Bush	2001	73	16	68	23	51	26	NR			1.4	9
Rocha-Singh	2002	51	55	100	77	18	5	91		9	0	14
Kennedy	2003	261	NR	36	61		39	NR			NR	NR
Gill	2003	100	26	75	31	38	31	4	79	17	2	18
Zeller	2003	215	23	52	52	48		76		24	0	5
Henry	2003	56	14	32	14	66	0	18	59	23	1.8	NR
Zeller	2004	456	NR	52	34	39	27	46		54	<1	NR
Nolan	2005	82	NR	59	23	53	24	NR	81	NR	0	7
Kayshap	2007	125	36	100	42	23	25	NR			1.6	6
Holden	2006	63	32	100	97		3	0	55	45	NR	NR
Corriere	2008	99	11	75	28	65	7	1	21	78	0	5.5
Mean%[b]			30	55	31	38	31	18	54	28	1	6.2

Source: Edwards, M.S., & Corriere, M.A., *J. Vasc. Surg.*, 2009.

NR, not reported.

[a] Series selected based on publication in 2000 or later, use of angioplasty and stenting, and inclusion of ≥50 patients, and categorical reporting of renal function and/or hypertension responses.

[b] Weighted mean based on number of patients with reported data categorized according to column headings; references where data were not reported or categorical response categories were combined were not included in calculation.

TABLE 6.6

Options for "Open" Surgical Treatment of Renal
Artery Stenosis

Anatomic	Extraanatomic
Transaortic renal endarterectomy	Hepatorenal bypass
Aortorenal bypass	Splenorenal bypass
Synchronous aortic surgery/renal artery bypass	Iliorenal bypass
Renal reimplantation	SMA-renal bypass
Other	
Ex *vivo* renal artery reconstruction	
Nephrectomy	

6.4.6.2.1 Morbidity, Mortality, and Patency Rates of Surgical Bypass (Table 6.7)

To date, there have been no randomized trials directly comparing surgery with stenting, but there are large series on renal artery reconstruction. In 2002, Cherr et al.[69] reported the results of operative management of 500 patients (776 kidneys) undergoing renal artery surgery for atherosclerosis (384 aortorenal bypass, 267 endarterectomy, 56 nephrectomy, 56 reimplantation, and 13 splanchnorenal bypass). Perioperative mortality was 4.6%. Of the deaths, 23 of the 24 occurred following bilateral renal artery reconstruction combined with aortic or mesenteric repair. Mortality for isolated renal artery repair was 0.8%. They reported an 85% improvement in blood pressure control and 58% improvement in ischemic nephropathy.

6.4.7 Other Renal Artery Treatment Strategies

6.4.7.1 Renal Artery Denervation

The autonomic and central nervous systems have a substantial role in blood pressure regulation. The sympathetic nerves to the kidney closely adhere to the renal artery adventitia, and a radiofrequency probe that transmits energy through the renal artery wall can ablate them. This is introduced through the femoral artery and guided to each renal artery. The Symplicity-1 trial[70] was a multicenter, proof-of-principle cohort study. It recruited 45 patients and showed promising reductions in both systolic and diastolic blood pressure. The follow-up study was the Symplicity HTN-2 trial.[71] In this, 106 patients with severe hypertension were randomized to either medical treatment only or medical treatment plus renal denervation. The end point was systolic blood pressure at 6 months. The renal denervation group did significantly better than the medical group in terms of systolic and diastolic blood pressure reduction. The concern is damage to the healthy renal artery. However, medium-term follow-up of these patients has not shown this to be the case. But, not everybody responds. In the Symplicity HTN-2 trial, 13% of patients showed no improvement at all in blood pressure.

6.4.8 Current Nanomedicine Applications

A PubMed search in December 2012 for the terms "nanomedicine" and "renal artery stenosis" produced no results. Nanomedicine is a young discipline. The first cited paper on "nanomedicine" appeared in 1999[72] (the unambiguous title was "Nanomedicine") and the journal *Nanomedicine* published its first issue only in 2005. Vascular surgery is also a relatively young specialty. In 1949, an *Index Medicus* search for successful surgical treatment

TABLE 6.7

Selected Series of Surgical Renal Artery Revascularization: Morbidity, Mortality, and Patency Rates of Surgical Bypass[a]

Reference	Year	Patients	Bilateral Repair (%)	Preoperative Renal Dysfunction (%)	Renal Function Response (%)			Hypertension Response (%)			Perioperative Outcome (%)	
					Improved	Unchanged	Worsened	Cured	Improved	Failed	Death	Morbidity
Fergany[b]	1995	175	2.3	92.3	35	47	18	46	54	0	2.9	NR
Cambria	1996	139	13	77	73		27	8	71	21	8	7.2
Darling	1999	568	18	NR	26	68	6	NR			5.5	15.9
Hansen	2000	232	64	100	58	35	7	11	76	13	7.3	30
Paty	2001	414	NR	4	97		3	NR			5.5	11.4
Cherr	2002	500	59	48.8	43	47	10	73	12	15	4.6	16
Marone	2004	96	27	100	42	41	16	NR			4.1	NR
Mean%[b]			31	38	38	47	14	12	73	15	5.4	21

Source: Edwards, M.S., & Corriere, M.A., *J. Vasc. Surg.*, 2009.

NR, not reported.

[a] Renal function and hypertension responses expressed percentage of patients surviving operation. Series selected based on publication in 1995 or later, inclusion of ≥50 patients, and categorical reporting of renal function and/or hypertension responses.

[b] Weighted mean based on number of patients with reported data categorized according to column headings; references where data were not reported or categorical response categories were combined were not included in calculation.

of PAD, aortic occlusive disease, or carotid artery stenosis would have also produced no returns. In 1973, searches for balloon angioplasty of coronary or peripheral arteries would have also produced no results. As recently as 1985, searches for peripheral artery stenting would have proven fruitless. Nanomedicine has been used in three main areas of cardiovascular medicine: imaging, diagnosis, and monitoring disease progression and treatment.

In renal disease, ultrasound is a particularly challenging examination. The renal arteries are located deep inside the abdomen and it is not unusual to have duplicate or multiple accessory renal arteries supplying each kidney. Even when a single renal artery supplies each kidney, patient "habitus" can make imaging particularly difficult. There are contrast agents ("microbubbles") to enhance diagnosis, but they have not been widely embraced. Intraarterial injection of gelatin-encapsulated nitrogen bubbles to enhance ultrasound diagnosis was first described[73] in 1980. Newer microbubbles comprise a high-molecular-weight gas such as perflourocarbon contained in a thin coating of albumin or a lipid.[74,75] They are highly echogenic. Although this is interesting, what makes them especially useful is that when they are phagocytosed by neutrophils and macrophages, they remain acoustically active.[76] This can be used to detect atherosclerosis.

The pathogenesis of atherosclerosis is quite complex, but that at its most basic involves an inflammatory process. In healthy individual, the endothelial cells that form the intima of all arteries are resistant to leukocyte adhesion. However, under conditions of arterial stress (e.g., smoking, diabetes, hypertension, or dyslipidemia), leukocytes adhere to the intima. Oxidized lipids and proinflammatory cytokines such as interleukin-1β and tumor necrosis factor-α stimulate vascular cell adhesion molecule-1 (VCAM-1) production. The VCAM-1 receptor binds monocytes and T-lymphocytes and is abundant in young atheromatous plaques. VCAM-1 recruits further cells and aids plaque evolution. VCAM-1 is only present in endothelial cells during plaque formation, so it is a good marker for pathology. Kaufman et al.[77] described the use of biotinylated, lipid-shelled decafluorobutane microbubbles containing antibodies to VCAM-1 to localize plaques rich in VCAM-1. They suggested that this could identify particularly dangerous plaques and has been confirmed by others.[78,79]

CT has also been used to determine plaque morphology. When gold-containing nanoparticles and iodinated contrast agents are infused, the nanoparticles are selectively taken up by macrophages within the plaque and the iodine remains intravascular.[80] This has been proposed as a means of determining particularly bioactive plaques.[81]

Superparamagnetic iron oxide particle accumulates in macrophages in atherosclerotic plaques. As with ultrasound, nanoparticles containing antibodies to low-density lipoproteins have been used to target plaques using MRI. An obstacle to the advance of cardiac MRI is the constant motion of the heart. In 2012, Alam et al.[82] from Edinburgh described a potential mechanism to overcome this. They used superparamagnetic nanoparticles to detect infracted cardiac muscle following acute MI.

Monitoring renal disease progression is already a reality. In 2012, Marom et al.[83] from Haifa applied nanomedicine techniques (gold nanoparticle sensors) to detect chronic kidney disease and disease progression in patients. This technique relies on the detection of volatile organic compounds in the exhaled breath of patients.

Targeted therapy has been the holy grail of nanotechnology. Combining diagnostic tests with targeted therapy is called *theranostics*. Most research has focused on targeted tumor therapy, but there has also been work on plaque stabilization. Plaques may be rendered less virulent by targeted destruction of their macrophages and also by targeting antiinflammatory drugs (glucocorticoids) at atherosclerosis. In 2010, McCarthy et al.[84] described a "theranostic nanoagent" based on magnetofluorescent nanoparticles. In a twenty-first-century version of the Trojan horse, these nanoparticles are modified with fluorophores (fluorescent chemical compounds that can reemit light upon excitation or absorb light at one wavelength and reemit it at a longer one) with light-activated chemicals. The nanoparticles with their "payload" are absorbed by macrophages in the plaque. Irradiation of the atheroma with 650 nm light activates the therapeutic component and kills the inflammatory macrophages. Lobatto et al.[85] also reported in 2010 on the use of targeted liposomal formulation of glucocorticoid on atherosclerotic plaques in a rabbit model. Significant antiinflammatory effects were seen in the plaques as soon as 2 days following administration. No changes were observed in animals treated with systemic glucocorticoids only.

Despite initial success, the bugbear of all stents is neointimal hyperplasia. Neointimal hyperplasia is a natural response to endothelial injury. When the intima is injured—for example, by a stent or graft—endothelial cells release inflammatory mediators. They, in turn, trigger platelet aggregation, fibrin deposition, and leukocyte recruitment. This cascade causes growth factor expression that results in SMC migration from the arterial wall media to the intima. They proliferate and lay down an extracellular matrix. The process is similar to scar tissue formation, and the end result is the formation of a neo-intima over the area. This is a normal response. However, when this becomes disordered, it results in neointimal hyperplasia and arterial stenosis. Coating stents and balloons with paclitaxel has been shown to be effective in reducing this.[86,87] These SMCs have also been targeted by paclitaxel using nanoparticles of perflourocarbon emulsions containing gadolinium.[88] Others have targeted extracellular chondroitin sulfate proteoglycans with prednisolone.[89]

6.4.9 Current Medical Needs

Despite a profusion of research, few applications have yet made it to "primetime." Some of the problem is that major manufacturers currently observed little advantage in making bespoke drugs targeted at specific sites when the unadorned molecule does a pretty adequate job. In the area of RAS and ischemic nephropathy, there are no FDA-approved nanomedicines yet. In patients with RAS, there is a role for plaque stabilization to prevent progression, but as most renal artery plaques already seem to have relatively benign outlook, "job one" is to identify malign plaques and to stabilize them. In patients with ischemic nephropathy, it is clear that early identification of disease is critical as once interstitial fibrosis is established and nephron mass is reduced, no amount of revascularization will reverse the process. There is a role for targeted anti-inflammatory therapy at the pathophysiological processes involved in ischemic nephropathy.

Ultimately, of course, for those patients who are already dialysis dependent, the foregoing treatments may be too late. However, peritoneal dialysis and hemodialysis remain cumbersome and restrict the patient. In 2009, Nissenson[90] from UCLA described the development of a human nephron filter (HNF) comprising two membranes in series within a cartridge, the first mimicking the glomerulus and the second the renal tubules. Operating 12 hours a day, 7 days a week, the HNF provides the equivalent of 30 mL/min glomerular filtration rate (half that of conventional hemodialysis) but uses no diasylate. However, as of yet, artificial implantable dialysis units are still a long way off.

6.4.10 Conclusions

Renal physiology is complex and treatment continues to evolve since the first successful hemodialysis treatment in 1945 by Kolff.[91] While initially seen as a temporary fix in extreme situations, hemodialysis can now be offered to most patients with end stage renal disease (ESRD). It comes at a huge economic and human cost, however. In 2004, according to the U.S. Renal Data System, 335,963 patients received dialysis treatment at a cost of $18.1 billion.[92] However, the annual mortality rate for patients on dialysis in the United States is high (23%) with few surviving 10 years or more. By the end of 2009, the number had raised to 398,861 with 172,553 surviving renal transplant patients at a cost of $42.5 billion (Figure 6.14).[93] Added to the cost of renal revascularization for renovascular hypertension and ischemic nephropathy, the burden on the healthcare system is enormous. While RAS and ischemic nephropathy account for the minority of patients developing ESRD and requiring hemodialysis, it is clear that many renal pathologies are linked, and that many of the pathologies have the same end result of an inflammatory pathway that ends up in interstitial fibrosis and nephron death. For these conditions, vascular surgery, while it has made some impact, remains a crude tool. The time has probably come for a newer generation of vascular specialists with much more delicate and infinitely smaller tools, even if they do not get to travel in microscopic submarines (Figure 6.15).

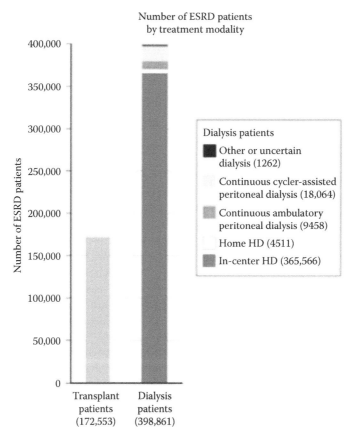

FIGURE 6.14

U.S. patients with end-stage renal disease and their treatment modalities in 2009. (From Kidney Disease Statistics for the United States National Kidney and Urologic Disease Information Clearinghouse. http://www.kidney.niddk.nih.gov/kudiseases/pubs/kustats/.)

FIGURE 6.15
Future vascular specialists probably would not get to ride in microscopic submarines.

6.5 Venous Thromboembolism

6.5.1 Introduction

VTE encompasses an array of disorders that includes deep vein thrombosis (DVT) and pulmonary embolism (PE). Bound by a common process that results in abnormal formation of thrombus, DVT most often affects the lower extremities involving the veins from below the knee to the inguinal ligament as well as the iliac veins, the inferior and superior vena cava, and occasionally, veins of the upper extremities, mesenteric circulation, kidneys, and other pelvic organs. PE results from the propagation of dislodged clot to the pulmonary circulation, where manifestations may range from mild symptoms to significant cardiopulmonary instability and even stroke. Recognized for its substantial public health implications, VTE was the subject of the Surgeon-General's call to action in 2008 in an effort to raise awareness both about the disease process and the importance of prevention. While VTE may occur in the community, hospitalization is acknowledged as a substantial risk factor both for DVT and PE. Consequently, the broad implementation of prophylactic measures has been a major focus for the Agency for Healthcare Research and Quality, while the Institute of Medicine has designated suboptimal preventive and screening measures for hospitalized patients as a medical error. The number of cases in the United States has been estimated at up to 600,000 or beyond with associated expenditures that run into the billions annually,[94] while average single episodes of postoperative thromboembolic complications may exceed $18,000 per event.[95] Given an exponential increase in incidence with age in an aging population, such a burden is likely to become exacerbated in the coming years. From the perspective of healthcare institutions, compliance with accepted screening methods and preventive measures is essential to ensuring optimal clinical outcomes and fiscal responsibility.

Beyond the acute episodes, long-term sequela of VTE have come to the forefront of medical care as physicians recognize both the high cost and the compromised quality of life associated with long-term anticoagulation, CVI, the postthrombotic syndrome, and even pulmonary hypertension.

6.5.2 Epidemiology

In the United States, the overall incidence of VTE slightly exceeds 100 episodes per 100,000 people annually,[96] eclipsing the parallel rates of breast, colon, or lung cancer.[97] Of all events, two thirds are DVT, with the balance PE. Although these diagnoses may coexist in approximately one third of PEs, no DVT is identified. The frequency of such disorders increases with age. An initial rise is noted in the third decade of life with a steeper rate of rise beyond middle age and the highest rates toward the eighth and ninth decades when the incidence may reach up to 700 per 100,000.[98] Men may have a slightly higher incidence compared with women,[99] although differences have not been consistently observed. With increasing age, the relative frequency of PEs appears to increase in both genders. Moreover, the incidence of VTE has increased over time with a 33% jump between 2002 and 2006.[100] Differences in incidence may also be dependent on race. Caucasians and African-Americans appear to have similar rates, although the higher frequency of episodes noted in blacks by one study[98] was associated with secondary events rather than primary ones and may reflect differences in screening or prophylaxis rather than biology. Rates in the Asian, Latino and Pacific Islander, and Native American populations appeared lower.[101,102] Little information is available about the incidence of VTE in Africa or east Asia, which would confirm inherent racial differences in genetics or biology of thrombosis.

The high incidence of VTE recurrence highlights the chronic nature of the disease. After an initial event, recurrent episodes may be seen in approximately 6%–10% of patients at 6 months and up to 30% over the ensuing 5–10 years with cancer patients showing some early excess risk.[103–106] The overwhelming majority of recurrences after DVT were in fact DVT, whereas two thirds of episodes after PE were PE.[107]

The identification of risk factors associated with VTE (Table 6.8) is vital to effective screening and prophylaxis. Although in some patients it may be difficult to isolate specific factors, hospitalization confers a 100-fold risk of VTE with risk being equally distributed between those with medical illness and those undergoing surgery.[108] The majority of overall episodes are noted to occur in patients who are either hospitalized or in nursing homes.[100] PE may be related to 10% of all in-hospital deaths.[109]

Cancer appears to confer a hypercoagulable state on those with active disease and may account for up to 25% of all cases in the community.[103] Among malignancies, those affecting the gastrointestinal tract, pancreas and liver, as well as lymphoma and malignant brain tumors, appear to pose the highest risk. Incidence may be heightened further in patients actively undergoing immunosuppressive therapy or treatment with cytotoxic agents as well as those with metastatic disease.[110]

TABLE 6.8

Risk Factors for Venous Thromboembolism

Congenital	Acquired
Anticoagulant deficiencies (protein C, protein S, antithrombin III)	Hospitalization
	Advanced age
	Obesity
Excess clotting factors (I, II, V, VII, VIII, IX, XI)	Cancer
	Hormone replacement
Altered genetics (factor V Leiden, prothrombin 20210A)	Pregnancy
	Immobility
Elevated D-dimer	Trauma/surgery
	Acute medical illness
	Immune diseases
	Inflammatory disorders
	Elevated homocysteine

Surgery and trauma present a significant risk for VTE occurrence, particularly since multiple risk factors may coalesce in a single clinical setting. Both advanced age and extended procedure duration are associated with excess risk. High event rates are recognized in those undergoing major orthopedic procedures such as total joint replacements and repair of hip fractures. Similarly, increased risk in neurosurgery patients is noted as well as those undergoing major abdominal and pelvic procedures, oncologic surgery, and renal transplants. Morbidly obese patients, as well as those with evidence of varicose veins or CVI, may be at additional risk for events. Perioperative risks are further accentuated by an extended need for critical care and immobility.

The list of medical conditions associated with VTE is extensive as well. Beyond those recognized in surgical patients, the following diagnoses have been associated as well: stroke with paralysis, myeloproliferative diseases, inflammatory bowel disease, renal syndromes (nephritic and nephrotic), systemic lupus, disseminated intravascular coagulation, heparin-induced thrombocytopenia (HIT), thrombotic thrombocytopenic purpura homocysteinuria and homocyteinemia, paroxysmal nocturnal hemoglobinuria, Buerger's disease (thrombangitis obliterans), and Behcet's disease.[111,112]

Although the impact of gender in VTE remains unclear, oral contraceptives (OCs) as well women's unique role in childbearing exposes them to additional risks. Both OCs and hormone replacement therapy (HRT) are associated with a higher VTE incidence. Combined estrogen–progestin HRT poses greater DVT risk that unopposed estrogen alone,[113] while third-generation OCs similarly appear to elevate risk.[114] Pharmacologic agents that modulate estrogen activity would seem to present similar challenges. Raloxifene, indicated for the attenuation or prevention of osteoporosis, has been implicated in the development of excess DVT or PE. Pregnancy elevates the risk of events fourfold, however, ostensibly due to mechanical venous compression, although this may be potentiated by a rise in circulating clotting factors and impaired endogenous fibrinolysis. An even sharper rise is seen in the postpartum period as compared with pregnancy.[115] Clues as to the risk of antiphospholipid syndrome may be gleaned from an obstetrical history that includes unexplained miscarriages, stillborn or growth-impaired infants, or even toxemia-associated premature births. There is an immune response against a circulating anticoagulant or phospholipid-binding proteins, which may persist beyond the postpartum period.

Travel has often been implicated in the genesis of VTE. The immobility in prolonged flights spawned the "economy class syndrome," although travel by air, automobile, train, or bus for periods greater than 4 hours appears to double the risk of events, most notably in the first week after travel. Demonstrating the additive impact of multiple risk factors, rates were higher in travelers with Factor V Leiden, obesity with a body mass index (BMI) > 30, and those on OCs.[116]

Underlying hematologic abnormalities contribute to the incidence of VTE. Crowther and Kelton[117] identified thrombophilic states associated with abnormal venous thrombosis and classified them accordingly. Deficiencies in endogenous anticoagulants such as protein C, protein S, and antithrombin III occur infrequently but may present a significant risk for thrombosis. Supranormal amounts of certain clotting factors (I, II, V, VII, VIII, IX, and XI) appear to be more frequent but seem to exert a weaker influence on thrombus formation. Factor V Leiden and Prothrombin 20210A represent genetic alterations of factors that contribute to hypercoagulability. Although multiple other single nucleotide polymorphisms (SNPs) are associated with thrombosis, their relative importance is still unclear. Elevated levels of D-dimer, a degradation product of fibrin, may elevate the risk of thrombotic events compared with median and low values, although the underlying mechanism remains unclear. Finally, although the role of the amino acid homocysteine has yet to be clearly defined, elevated levels appear to be associated with increased risk of VTE. Evidence for risk reduction as a result of lowering homocysteine is, however, lacking.

An aggregate risk assessment of VTE is critical in ensuring appropriate prophylaxis. A number of models have been proposed, which incorporate many of the clinical risk factors mentioned. While a detailed analysis of each is beyond the scope of this chapter, an updated system for risk stratification of individual patients with recommendations for preventive measures has been provided by the American College of Chest Physicians (ACCP).[118] On screening for multiple thrombosis-associated SNPs, de Han et al. proposed a genetic risk score based on the number of alleles identified. They proposed that even greater accuracy may be realized on combining genetic and clinical risk models.[119]

6.5.3 Methods of Prophylaxis

The importance of VTE has been recognized across specialties, both for people in the community and those hospitalized or in nursing homes. Although Rudolf Virchow had identified the eponymous triad of venous stasis, hypercoagulability, and endothelial damage in the nineteenth century, one may consider that a biological shift, which, during transition from the community to inpatient setting, promotes thrombosis by shifting the equilibrium maintained between the endogenous procoagulant and thrombolytic activity. It is evident that the average person tolerates extended periods of decreased mobility during long stretches of work, rest, and sleep, despite the potential association with venous dilatation, stasis, and perhaps endothelial compromise that would promote thrombosis. The impact of similar risks in hospital is substantial and necessitates a well-conceived system of VTE prophylaxis. Currently accepted methods (Table 6.9) may be employed alone or in conjunction with others.

General preventive measures for all people at risk involve emphasis on preserving activity level. Ambulation is essential and should be encouraged regardless of setting. Postoperatively, the resumption of activity should be expedited within the bounds of safety as determined by the treating surgeon. For those limited by extended travel, repeated and regular flexion and extension exercises of the ankles, knees, elbows and wrists, and shoulders enhance both arterial and venous flow. Similar maneuvers can be performed actively by hospital inpatients whose activity is curtailed for other reasons. Those in need of assistance benefit from passive exercises directed by physical therapists or other ancillary

TABLE 6.9

VTE Methods of Prophylaxis

I General
Activity
Ambulation
Range of motion exercises
Hydration
Leg elevation
II Mechanical
Graduated compression stockings
Intermittent pneumatic compression devices
Foot compression devices
IVC filters (PE prophylaxis)
II Pharmacologic
Unfractionated heparin
Low-molecular-weight heparin
Vitamin K antagonists
Antiplatelet agents
Factor Xa inhibitors
Direct
Indirect
Thrombin inhibitors

healthcare providers as well as family members. Leg elevation employs gravity to enhance venous return, reduces venous pressure, and facilitates the reversal of lower extremity edema and perhaps any associated endothelial changes. Adequate hydration is vital in any setting in order to limit an elevation in blood viscosity, which may contribute to thrombus formation in a patient with static venous flow.

The principles of mechanical VTE prophylaxis are based on increasing venous return and the intermittent acceleration of venous blood flow. Whether or not there is an associated accentuation of endogenous fibrinolytic activity remains a matter of some debate.[120,121] Typically, such methods are used alone in settings where the risk of bleeding with anticoagulants may be excessive such as the perioperative period or combined with pharmacologic methods. Initiation of VTE preventive measures should be expedited upon hospitalization.

Elastic stockings are generally manufactured to provide and maintain a pressure gradient from the foot to the stocking's proximal extent: knee or thigh. For those employed in prophylaxis, this generally extends from 18–20 mmHg distally to 8–10 mmHg proximally. Compliance with knee-high stockings tends to be greater with no compromise in VTE incidence.[122] Their efficacy in preventing DVT in moderate-risk surgical patients has been documented,[123] but this effect cannot necessarily be extrapolated to the highest risk patients or to PE prevention. In the hospitalized patient, compliance with stockings may be impaired in morbidly obese patients and those with significant leg edema, extremity dressings, or wounds.

Intermittent sequential compression devices (ISCDs) further extend the concept of mechanical. Aimed at ensuring compliance in hospitalized patients, this is a garment that encases a series of air bladders that can be wrapped around each patient's extremities. The garment is then attached to an electrically powered or battery-operated pneumatic pump. Inflation of each bladder in a distal to proximal sequence accelerates lower extremity venous blood flow cephalad. Both the pressure gradient and the cycle speed may be controlled by healthcare providers. Evidence suggests effective DVT risk reduction overall compared with no prophylaxis[124] and efficacy in some of the highest risk groups.[125,126] Beyond any obvious condition affecting the lower extremities that would preclude device application, the most notable contraindication to ISCDs is an acute DVT with concerns dislodging the thrombus and inducing a pulmonary embolus. Screening of patients prior to application of ISCDs is generally not indicated unless clinical suspicion for DVT exists.

Pneumatic foot pumps (PFPs) enact the compression concept distally. With an air chamber on the plantar surface, more rapid inflation to higher pressures results in augmented venous outflow. Passengers after long flights did not appear to have benefited from PFPs compared with independent exercises when assessed for venous hemodynamics;[127] however, VTE prophylaxis appears to be equivalent to low-molecular-weight heparin (LMWH) in orthopedic patients undergoing knee or hip arthroplasty.[128] Beyond elimination of the hemorrhage risk, early range-of-motion activities and ambulation may be facilitated, although such devices are not uniformly well tolerated.

In the setting of innate or induced hypercoagulability, anticoagulants have formed a vital part of both prevention and treatment of thromboembolic disorders. Their use raises the specter of bleeding complications, whose risk must be balanced against their demonstrated efficacy.

Unfractionated heparin (UH) had long been the mainstay of prophylactic anticoagulation. The reduction in the incidences of both DVT and PE has long been recognized in patients undergoing a variety of procedures. Administered subcutaneously in divided doses, it was noted to reduce the risk of DVT and PE both in surgical and medical patients and those with malignancy.[129] There does not appear to be clinical difference between dosing regimens requiring administration two or three times daily.[130] The short half-life of heparin makes it favorable for use, where a high potential for reversal is anticipated, although the same pharmacokinetics generate significant expense due to dosing frequency. Inherent in the use of heparin is the risk of bleeding and HIT. The latter diagnosis occurs in up to 5% of patients and necessitates intermittent monitoring of the platelet count.

LMWHs have substantially altered the landscape of parenteral anticoagulation. These are a structurally variable group of drugs with activity against Factor Xa. They exhibit a longer half-life and more reliable bioavailability, dosing frequency has been reduced, and the need for monitoring of bleeding parameters has been eliminated, facilitating outpatient management and reducing the cost of patient care. Multiple trials and meta-analyses have reaffirmed the efficacy of appropriately dosed LMWH with UH and warfarin in VTE prevention,[131,132] while bleeding complications appear less frequent. It should be noted that LMWHs are associated with a significantly reduced incidence of HIT[133] but do not eliminate it completely. Clinical suspicion of HIT remains an absolute contraindication to prescription of LMWH. Overall, LMWHs have assumed the role of the parenteral anticoagulant of choice and are highlighted in the current ACCP guidelines.

Vitamin K antagonists (VKA) have been the mainstay of oral anticoagulation over the last six decades. Coumarin analogues such as warfarin are hampered by a long half-life that necessitates a period of days to initiate or stop anticoagulation. Monitoring of the prothrombin time is essential, although activity may be expressed as a normalized ratio (INR) with the goal being 2–3 times the control value with an ongoing need for monitoring. Bridging both prior to and after invasive procedures with fractionated heparin may be undertaken in the outpatient setting. VKA are effective in preventing DVT compared with no prophylaxis but lag behind LMWH while increasing the risk of associated hemorrhage.

When necessary, reversal is accomplished with fresh frozen plasma concentrates or concentrated vitamin K.

There is conflicting evidence regarding the benefits of antiplatelet agents in the prevention of venous thrombosis. Acetylsalicylic acid (ASA/aspirin) is the prototypical agent. In blocking thromboxane A2 production for the life of the platelet, typically 7–10 days, aggregation is impaired and vasoconstriction is attenuated. Currently, there is no recommendation promoting the use of ASA alone in VTE prophylaxis, although it appears to be widely used as an adjunct to accepted methods in clinical practice. Beyond formal anticoagulation after an initial unprovoked VTE event, data are emerging regarding the benefits of extended secondary prevention with ASA.[134,135] There is limited information on clopidogrel for this indication and no formal recommendation for its use has been issued.

A number of novel anticoagulants are being used with increasing frequency for VTE prevention and treatment. These are summarized in Table 6.10. Direct thrombin inhibitor (DTI) acts on both free and bound thrombin to attenuate the clotting cascade and associated prothrombotic effects. They are indicated for anticoagulation in the setting of HIT as no interaction with platelet factor IV occurs. An oral DTI agent, Dabigatran, appears promising as well, based on the results of phase 3 trial for treatment of acute VTE.[136] Factor Xa inhibitors (FXaI) such as the polysaccharide fondaparinux appear as effective as LMWH in VTE prophylaxis after both orthopedic and general surgery procedures[137,138] and have garnered appropriate indications from the FDA with broader use. Rivaroxiban is an oral FxAI that appears to be equally effective in DVT prophylaxis preoperatively.

The use of indwelling vena caval filters for VTE prophylaxis remains somewhat controversial. Typically, such devices are reserved for those at high risk for acute VTE with contraindications to anticoagulation such as excess bleeding risk or previous failure of medical therapy. Temporary filters obviate the long-term risks associated with permanent devices; however, scant data exist regarding their true efficacy. The new generation of anticoagulants appears to be shifting the risk of hemorrhage and may necessitate a reassessment of filter use.

A combination of mechanical and pharmacologic methods of VTE prevention has been evaluated in a number of settings. There appears to be a fairly consistent agreement regarding an apparently synergistic effect when multiple modalities are employed.[139] Even with the use of IVC filters, many clinicians may recommend concomitant use of anticoagulants, unless prohibitive bleeding risks are noted. Individual recommendations are determined according to clinician assessment of each patient's risk profile (Table 6.11). Ultimately, a program of successful prevention is limited by physician adherence to evidence-based guidelines and patient compliance with the prescribed plan of care.[140]

TABLE 6.10

New Anticoagulants

Agent	Mechanism of Action	Administration	Half-Life	Notes
Hirudin/ Lepirudin	DTI	Parenteral	1–2 hours	Indication for HIT
Dabigatran	DTI	Oral	8–10 hours	Currently indicated for atrial fibrillation
Argatroban	DTI	parenteral	39–51 minutes	Indicated for HIT
Fondaparinux	FXaI-I	Parenteral	17 hours	VTE prophylaxis
Rivaroxiban	FXaI-D	Oral	6–8 hours	No indication yet

Abbreviations: DTI, direct thrombin inhibitor; FXaI, factor Xa inhibitor; D/I, direct/indirect.

TABLE 6.11

Basic Guidelines for Perioperative VTE Prophylaxis

	Early Mobilization	Mechanical	Pharmacologic	Combined Methods
Low risk	Yes	Stockings = IPC	No	No
Intermediate risk	Yes	IPC > Stockings	Low dose	No
Elevated risk	Yes	IPC > Stockings	Higher dose	Yes
Highest risk	Yes	IPC > Stockings	Higher dose, extended period	Yes

6.5.4 Clinical Diagnosis

The importance of diagnosing VTE cannot be overstated. DVT remains an important etiologic factor in the postthrombotic syndrome and ulceration, with cumulative incidences of up to 40% and 4%, respectively, over 20 years.[103,141] Recurrence rates approach 30% at 10 years, although the vast majority occur early. The significant short-term mortality risk associated with DVT diagnosis is higher in PE when compared with DVT. Up to 25% of PE patients may present with sudden death.[142]

The clinical diagnosis of VTE may be difficult and hinges on an appropriate index of suspicion, given the clinical setting and individual patient risk factors. Patients may be completely asymptomatic or present with vague complaints attributable to multiple causes. The frequently cited signs and symptoms include leg pain or tenderness, edema, erythema or cyanosis, and bulging superficial veins. These may become apparent on examination but lack both sensitivity and specificity. Homan's sign (calf pain with dorsiflexion) is notoriously unreliable in facilitating diagnosis. All complaints should be viewed through the prism of an appropriate history and risk assessment. Lack of asymmetry should never exclude the diagnosis, particularly since bilateral DVTs may occur. In its most severe form, severe edema may be associated with purple discoloration due to severe venous engorgement (phlegmasia cerulea dolens) or blanching associated with flow compromise (phlegmasia alba dolens), which constitute limb-threatening emergencies. This should be considered if the diagnosis of DVT is suspected.

d-US combines B-mode imaging with color-flow Doppler to evaluate the venous system. Widely available with standard protocols, this study is relatively quick, noninvasive, and inexpensive, assessing the iliac to tibial veins for patency, compressibility, and flow augmentation with distal compression. Although operator dependent and limited proximally in obese patients, this modality is both sensitive and specific in the diagnosis of DVT. In selected cases, repeated studies may be employed to assess for extension or resolution at predetermined intervals.

The ascending venogram remains the gold standard for VTE diagnosis. Although rarely used as an initial screening tool, it provides a reliable "map" of the veins, evaluating both flow and the lumen for abnormal contours. Its expense and invasiveness have largely relegated it to a role in secondary confirmation or to cases where other modalities may be deemed suboptimal.

Additional imaging modalities may be of use as well but are less frequently employed. Magnetic resonance venography (MRV) can be both sensitive and specific, although it is hampered by expense, length of examination, and limited availability. It may be optimal in evaluating other locations of thrombosis including pelvic, mesenteric, portal, and renal veins. Impedance plethysmography is less accurate than the above methods and is rarely used, while studies employing radioisotopes have yet to demonstrate clear advantage over existing modalities.

D-Dimer is reflective of circulating fibrin degradation products in the blood. Measured by enzyme immunoassay, it is a highly sensitive marker for hypercoagulability and VTE outside of some high-risk conditions including malignancy, pregnancy, and recent surgery. Combined with inconsistent specificity, its use in VTE diagnosis is primarily a part of diagnostic algorithms for outpatients and in conjunction with other imaging modalities.[143,144]

The risk factors for PE parallel those of DVT, and patients may range from asymptomatic to moribund. A majority of patients with DVT have silent PEs. While clinical suspicion is critical to diagnosis, important signs and symptoms include unexplained tachycardia, tachypnea, chest pain, dyspnea, hypoxia, and hemodynamic compromise. Computed tomographic angiography is the diagnostic test of choice with high sensitivity and specificity. When iodinated contrast is contraindicated (renal impairment or hypersensitivity), radioisotope ventilation/perfusion scanning of the lungs (V/Q scan) may be performed and interpreted within the framework of existing clinical information. Pulmonary angiograms are rarely performed due to invasiveness and expense and are generally relegated to those patients who undergo intervention. As with DVT, the D-dimer remains a useful adjunct as well.

For both these entities, multiple models have been constructed to optimize diagnostic accuracy in an expedient manner. In incorporating clinical criteria, imaging studies, and D-dimer assays, a variety of algorithms have been generated to risk stratify patients and assist in determining appropriate treatment and follow-up.

6.5.5 Treatment

The goal of any treatment for acute VTE remains threefold: to prevent extension, embolization, and recurrence. The nature of treatment therefore overlaps with long-term (secondary) prophylaxis. While anticoagulation remains fundamental to both PE and DVT therapy, adjunctive therapies are emerging for particular patient subgroups.

6.5.5.1 Anticoagulation

Initiation of anticoagulation is indicated in all patients diagnosed with VTE, unless a prohibitive risk exists. Even if there is strong but unconfirmed clinical suspicion, treatment should be considered until definitive diagnosis is made. Initial therapy is undertaken with full-dose UH or LMWH and continued until therapeutic anticoagulation is achieved (INR 2.0–3.0) with oral agents, typically VKA that are started concurrently. Tight control of the INR is vital to preventing recurrent events and avoiding bleeding complications that rise substantially beyond an INR of 3.[145,146] When UH is used, a bolus is administered (80 U/kg) with subsequent continuous infusion thereafter (5–20 U/kg/hr). The PTT is verified initially 6 hours after the bolus and subsequently adjusted accordingly to maintain a PTT ratio of approximately twice the control value. High-dose subcutaneous administration of UH is effective and has been obviated by the availability of fractionated heparins.

The emergence of LMWH has simplified anticoagulation with once (or twice) daily dosing on a weight-based regimen (150–200 U/kg), which requires no monitoring other than a platelet count. LMWH are as effective as UH for treatment of both DVT and PE,[147] with associated reduction in bleeding complications. They are now the primary parenteral anticoagulant recommended for VTE treatment, according to the most recent ACCP guidelines. Economic benefits may be derived from treatment in the outpatient setting, shortening hospitalization, or in some cases obviating the need completely. For patients with intolerance to VKA or in whom the INR is difficult to control, LMWH use on a chronic basis is a viable alternative.

In the setting of possible HIT, Fondaparinux is a reasonable alternative. It has been evaluated in the treatment of both acute DVT and PE, with outcomes comparable to controls treated with LMWH.

6.5.5.2 Mobility and Compression

Conventional wisdom has previously restricted patients' activity until therapeutically anticoagulated for a number of days. While such measures may benefit some symptomatic DVT patients, compression and early ambulation may relieve symptoms without excess risk of PE. Activity may be allowed, once therapeutic anticoagulation has been initiated.

6.5.5.3 Intervention

When symptoms related to VTE are apparent, intervention may be indicated for stabilization, relief, and ultimately for prevention of long-term sequela.

Thrombolysis with agents such as recombinant TPA may be effective in clearing thrombus and preventing the postthrombotic syndrome. Initially evaluated with systemic intravenous infusion, limited success in clearing thrombus was hampered by significant risk of hemorrhage. A subsequent shift toward catheter-directed lysis has shown substantially improved results while limiting complication rates. Most notably, major bleeding complications were relatively infrequent and PE was rare.[148] Mounting evidence regarding the benefits of early removal of thrombus with restoration of flow[149] has led to standard recommendations for intervention in patients with symptomatic iliofemoral DVT. Adjuncts to thrombolysis include balloon angioplasty and selective use of stents when underlying stenoses are identified. An elevated risk of bleeding and recent surgery are contraindications to thrombolysis. Patients are typically monitored in a critical care setting with careful blood pressure control.

Numerous devices aimed at facilitating thrombectomy are currently on the market as well (Table 6.12). They are geared toward mechanical thrombus disruption and extraction, although one employs ultrasound to enhance ongoing thrombolytic infusion as well. Improved results have been described when modalities are combined, resulting in increased clearance of thrombus burden, fewer follow-up studies, shorter hospital stays, and improved vein patency. For patients with symptomatic limbs that are not imminently threatened, our preference is for the initiation of catheter-directed thrombolysis, with subsequent repeat angiography within 12–24 hours and intervention for residual thrombus or stenosis. Anticoagulation is continued after discontinuation of the lytic agent.

TABLE 6.12

Devices for Venous Thrombectomy

Mechanical Thrombectomy	Ultrasound-Assisted Thrombolysis	Segmental Pharmacomechanical Thrombolysis
Oasis (Boston Scientific) Angiojet (Possis Medical) Tretorola (Arrow International)	Ekosonic Endowave (Ekos Corp)	Trellis (Bacchus Vascular)

In the acute setting, surgical intervention may be indicated for severe symptomatic patients in whom thrombolysis is contraindicated or in the setting of phlegmasia, where a limb is threatened. The procedure, optimally performed under fluoroscopy, involves distal to proximal extrinsic compression of the extremity with balloon-assisted thrombus extraction proximally. An important adjunct is the creation of a small arteriovenous fistula to assist in maintaining patency. Postoperative anticoagulation remains an essential component.

The management of acute massive PE is tailored to the patient's clinical status. Stabilization in a critical care setting is essential. Hypoxemia and hypotension are ominous signs, particularly when superimposed on significant cardiopulmonary illness or malignancy. As with the threatened extremity, pharmacologic and mechanical thrombolysis may be reasonable options with surgery reserved for rare patients and associated with significant risks of morbidity and mortality.

Secondary prophylaxis implies prevention of VTE recurrence. Guidelines are provided according to risk factors assessment. Decisions regarding anticoagulation generally account for precipitating events, extent of thrombosis, clinical course over time, and perceived risk of hemorrhagic complications. In general, an initial event associated with transient hypercoagulability merits 3–6 months of formal anticoagulation. Recurrences in such instances are treated for 6–12 months. An initial idiopathic VTE event is similarly anticoagulated for 6 months, although a recurrence mandates indefinite anticoagulation. In patients at high risk for recurrence, those with demonstrated thrombophilias or incompletely treated malignancy, long-term anticoagulation should be considered.

6.5.6 Future Directions and Possible Nanotechnology Applications

The diagnosis and management of venous thromboembolic disease remains a significant public health challenge. As a disease of the elderly, VTE growth may be expected to parallel our aging society, while its potential impact will be further amplified with the likelihood of a corresponding increase in health-care expenditures. Such stressors will continue to provide openings for innovation in the way we approach this disease.

Although there appears to be an ongoing evolution in the realm of anticoagulants, little has changed in thrombolysis. The agents that are currently available have a relatively narrow therapeutic window. They may be delivered locally but can generate a systemic response with substantial bleeding risk. Reports regarding successful molecular targeting of plasminogen activators to thrombus are exciting, as they appear to exhibit improved therapeutic efficacy and may offer protection from adverse events.[150,151] Similarly, other agents may be able to be directed toward the various components of thrombus, potentially destabilizing the structure and rendering it more vulnerable to thrombolysis.[152] Whether such treatment can ultimately be used as a standalone modality or as part of a protocol to accelerate therapy in conjunction with catheter-based intervention remains to be seen.

The other glaring opening in the realm of thrombosis pertains to detection. For patients at higher risk, it may be useful to develop modalities for recognizing early thrombus formation. In theory, thrombus likely begins to form before the onset of symptoms. With nanoparticle-enhanced imaging, early detection may provide an opportunity to initiate or resume anticoagulation in patients with preclinical disease or evolving recurrence. While the cost-effectiveness of such an approach may be questioned today, shifting the use of resources to prevention may be beneficial in the long run.

It is likely that as the molecular basis of thrombosis becomes understood more completely, diagnostic and therapeutic modalities such as these will succeed in altering our approach to a complex and heterogeneous group of patients.

6.6 Chronic Venous Insufficiency

6.6.1 Epidemiology

The clinical manifestations of CVI are among the most commonly encountered medical conditions worldwide. Problems due to CVI are not new, and initial reports of varicose veins appeared as early as 1550 BC.[153] Treatment with both ligation and cauterization appeared as early as 270 BC. The treatment principles of ligation and cauterization are still employed. Although CVI is not a limb or life-threatening disorder, it negatively impacts individual quality of life measures and has a large societal cost.

CVI is primarily a disease process of the lower extremity and can lead to problems such as varicose veins, skin changes, and venous ulcers. CVI may also cause chronic leg swelling, symptoms of leg heaviness, and leg fatigue. Erect posture is thought to play a role in the development of CVI, as it occurs exclusively in man and spares all other species.

The significant risk factors for developing CVI include female gender, prior phlebitis, DVT, and family history. DVT carries the highest relative risk of developing CVI, increasing the odds ratio by 25 times.[154] Other risk factors include obesity, older age, pregnancy, and occupations requiring prolonged standing.[155] Currently, CVI is reported to be more prevalent in Western countries. It is likely that venous disease is also highly prevalent in Eastern societies but is less studied and documented. An estimated 27% of American adults have some form of CVI. The annualized risk of developing CVI is 2.6% in American women and 1.9% in men.[156]

Venous ulceration is the most serious clinical manifestation of CVI. Patients with venous ulceration report reduced quality of life, lost time from work, depression, isolation, and reduced earning ability.[157,158] The societal costs of venous ulceration are also high. An estimated 1–5 billion U.S. healthcare dollars are spent treating venous ulceration annually and 2 million days are lost from work. The prevalence of venous ulceration in American population is 1%.[159]

Modern treatment of complex CVI cases often requires coordinated efforts from multiple medical disciplines. Centers of excellence in vein care combine disciplines with specialization in more discreet aspects of treatment where any one discipline by itself may not be able to provide the full complement of care needed. Disciplines involved in complex cases include vascular surgery, wound care, hematology, plastic surgery, interventional radiology, and dermatology.

6.6.2 Venous Anatomy and Physiology of the Lower Extremity

The lower extremity venous system in the normal physiologic state allows one-way flow of blood from the foot through the leg and into the central circulation. The increased venous pressure associated with erect posture and muscular contraction is controlled with a series of bicuspid check valves within the vein lumen, which retard the reversal of blood flow. In normal physiology, approximately 95% of blood returns to the central circulation through the deep leg veins. The deep leg veins are located within muscular fascial compartments and course are adjacent to the major arteries. These veins are referred to as the deep system. Muscular contraction increases compartment pressure, especially at the calf, thus increases pressure within the deep venous system. Repetitive muscular contraction of the calf with activities such as vigorous walking creates a pump system and is even referred to as the calf muscle pump. The calf muscle pump assists blood flow out of the leg against the forces of gravity.[176]

The deep veins have names similar to the arteries each accompanies. There are paired anterior tibial, posterior tibial, and peroneal veins in the calf, which form a confluence at the proximal calf called the popliteal vein. The popliteal vein becomes the femoral vein once passing through the adductor canal. The femoral vein was previously called the "superficial femoral vein," but this caused confusion among healthcare providers, since it was part of the deep system and the word "superficial" has been dropped. The deep femoral and femoral veins join to form the common femoral vein just below the inguinal ligament.

A complex of veins, referred to as the superficial system, resides in the soft tissues outside of the lower extremity muscular compartments and fascia. The superficial and deep systems are connected at multiple locations throughout the leg and calf. There is an anatomically constant connection of the superficial to the deep system in the upper thigh called the saphenofemoral junction. There are a series of other more variable connections between the superficial and deep systems lower in the thigh and calf, which are referred to as perforator veins. Check valves within the saphenofemoral junction and perforators maintain unidirectional blood flow from the superficial into deep system. The check valves protect the superficial system from the higher pressure and larger volume flow of the deep system. The check valves prevent reflux of blood into the superficial system during calf muscle contraction and gravitation forces when in an erect posture. There is great capacity and redundancy within the superficial venous system of the lower extremity allowing these veins to be removed or ablated as a form of treatment, so long as the deep system is functional.

The principal and relatively constant superficial veins of the leg are named the long and short saphenous veins (aka greater and lesser saphenous vein, respectively). The long saphenous is the longest vein in the body. It courses from the dorsum of the foot anterior to the medial malleolus following a medial course and is closely associated with the saphenous nerve at this level. It continues medially in the thigh until terminating at the fossa ovalis and joining the common femoral vein of the deep system 5–10 cm below the inguinal ligament. The lesser saphenous vein begins posterior to the lateral malleolus and courses along the middle of the posterior calf before emptying into the deep system popliteal vein behind the knee. The short saphenous vein is closely associated with the sural nerve and medial sural cutaneous nerve.

The iliac veins drain blood from the leg and into the inferior vena cava. The right and left iliac veins converge to the right side of the vertebral column at the lumbar level forming the inferior vena cava. The aorta lies to the left side of the vertebral column. The right iliac artery crosses over the ventral surface of the left iliac vein, placing it at risk of external compression. Central vein compression, stenosis, occlusion, or heart failure may increase venous pressures and manifest as CVI of the lower extremity.

6.6.3 Pathophysiology of CVI

CVI is caused by venous hypertension. In most cases, this is due to failure of the check valves (or one-way valves). Failed or incompetent check valves allow a prolonged reversal in the direction of blood flow, a medical condition called reflux. In most cases, no etiologic cause of valvular dysfunction can be identified and is therefore termed primary valvular incompetence. When there is an identifiable cause of valvular incompetence, the term secondary valvular incompetence is used. Most cases of secondary valvular incompetence are due to previous episodes of thrombosis, especially DVT.

Obstructive lesions also play a role in the pathogenesis of CVI, particularly when located at the iliac or vena cava level. Highly symptomatic subjects often have obstructive lesions as well as valvular reflux contributing to CVI. Previous episodes of DVT are the most frequent identifiable cause of obstructive lesions at this level. Intrinsic chronic vein abnormalities are common after acute DVT. Chronic total occlusion may persist after DVT, or luminal webs and membranes may remain in venous segments that recanalize. External compression is another cause of obstruction. The classic location at risk for external compression is the left iliac vein, where the right common iliac artery crosses anteriorly. When this situation presents with CVI or DVT, the condition is referred to as May–Thurner syndrome. More recently, other locations of iliac vein compression are recognized including crossing of the left hypogastric artery over the left iliac vein, the right common iliac artery crossing the right common iliac vein and the right external iliac artery crossing the adjacent external iliac vein.[160]

As stated previously, acute DVT is the single strongest factor for developing CVI. Symptomatic CVI following DVT is called postthrombotic syndrome and occurs in 29%–50% of subjects after acute DVT.[103,161] A second episode of acute DVT in the index limb almost guarantees long-term CVI.[162] CVI secondary to acute DVT results from valvular destruction, venous stenosis, and persistent chronic total occlusion or obstruction. Unrecognized and untreated iliac vein obstruction will predispose to a second episode of DVT. The diagnosis of postthrombotic syndrome is typically withheld for at least 3 months of clinical follow-up as leg swelling from acute DVT may improve over time.

A complex series of cellular and molecular changes are triggered by CVI. When chronic, a cascade of events results in impaired oxygen delivery, disrupted lymphatic flow, capillary microthrombosis, and tissue hypoxia. Fibrinogen has been demonstrated at high concentrations in the extracellular matrix and may attract fibroblasts and mast cells.[163] Thickening around tissue capillaries is observed and is referred to as a fibrin cuff. The fibrin cuff is believed to impair nutrient delivery. MMPs are released in the interstitium and alter Type I and III collagen levels. Scar tissue develops with collagen bundles and degraded elastic fibers.[164] These morphologic changes are seen clinically and referred to as lipodermatosclerosis, a precursor to skin ulceration.

6.6.4 Classification of Chronic Venous Disease: CEAP Classification

The CEAP classification system defines the clinical class (C), etiology (E), anatomic distribution (A), and pathologic mechanism (P) of abnormal physical findings associated with CVI (see Table 6.13).[165] The CEAP classification is useful for stratifying patients to various treatment options. The CEAP classification system also allows for comparison group and subgroup analysis when outcomes data and treatments are evaluated. From a terminology standpoint, chronic venous disease (CVD) includes any subject C_1 or above, while the term CVI is reserved for subjects with more advanced disease, C_3, or above.[166]

Duplex ultrasound is the most utilized diagnostic test for assessing the anatomy and pathophysiology of subjects with CVD. Both anatomic and functional abnormalities can be demonstrated with this imaging modality. It is of minimal risk, noninvasive, inexpensive, and can be performed in an office setting. B-mode imaging reveals valve leaflet abnormalities such as thickening, frozen position, and shortening. Vein diameters, vein wall thickening, thrombus, major branch points, and anatomic variations are identified. Doppler is used to assess direction of flow and to measure reflux times. In the normal state, valve closure occurs after a brief episode of flow reversal. Flow reversal beyond 0.5 seconds is considered pathologic and is termed venous reflux (Figure 6.16).

Obstructive lesions in the iliac veins and vena cava are most reliably identified with intravascular ultrasound (IVUS). Ascending venography as a standalone modality fails

TABLE 6.13

CEAP Classification System for Chronic Venous Disease of the Lower Extremity

Classification (C)		Etiology (E)	
C_0	No signs venous disease	E_c	Congenital
C_1	Telangiectasias and spider veins	E_p	Primary
C_2	Varicose veins	E_s	Secondary
C_3	Edema		
C^4	Skin changes and lipodermatosclerosis		
C_5	Healed ulcers		
C_6	Active venous ulcer		
Anatomic Segment (A)		**Pathologic Mechanism (P)**	
A_S	Superficial vein	P_R	Reflux
A_D	Deep vein	P_O	Obstruction
A_P	Perforator vein	$P_{R,O}$	Reflux and obstruction

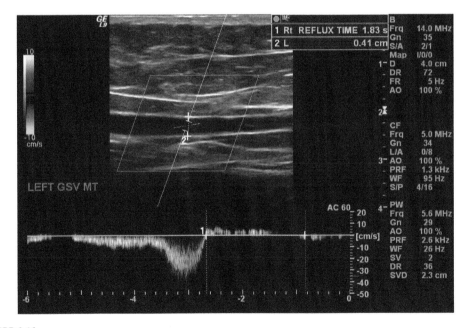

FIGURE 6.16
Venous reflux ultrasound study of left long saphenous vein at the mid-thigh. Flow reversal and reflux is depicted along the horizontal axis at the bottom of the image. A vertical dashed line between –4 and –2 demonstrates the onset of flow reversal. Flow reversal ends at the second vertical dashed line to the right. Reflux time measured at 1.83 seconds. The vein is seen in long axis in the gray scale image. The diameter is 4.1 mm.

to identify up to 50% of iliac vein lesions due to flooding of contrast in webbed lesions coupled with lesions that frequently lie parallel to the plane of imaging. Evaluation and intervention can be performed entirely with IVUS and fluoroscopy without the use of iodinated contrast, an option for subjects with anaphylactic dye allergies. Emerging technologies in CT and MR venography show promise but, at this time, are less sensitive and specific than IVUS. Additionally, these two modalities are exclusively diagnostic requiring a second medical visit for intervention.

Complete identification of the etiology, anatomy, and pathologic mechanism of CVD is essential for the appropriate selection of intervention and treatment. While duplex scanning is the most frequently employed imaging method for initial venous evaluation, other imaging modalities are needed for certain clinical situations including contrast venography, IVUS, MRV, and CT venography. Functional testing with plethysmography and venous pressure measurement are also used in certain circumstances but currently are of limited clinical value.

6.6.5 Treatment of Superficial and Perforator Venous Insufficiency

During the past 15 years, treatment of superficial vein reflux has evolved from surgical ligation and stripping procedures to percutaneous endovenous thermal ablation. The short-term advantages of endovenous thermal ablation over surgical stripping include reduced postprocedural pain, earlier return to work, and better early quality of life scores.[167] Additionally, endovenous thermal ablation can be performed in an office setting under local anesthesia. In contrast, ligation and stripping are typically performed in an OR under regional or general anesthesia. Longer term follow-up comparing these two approaches demonstrates both to have equivalent freedom from limb reflux and similar rates of varicose vein recurrence. Endovenous thermal ablation patients, however, continue to report higher quality of life scores on longer term follow-up than matched stripping patients.[168]

Endovenous thermal ablation of the saphenous vein can be performed with one of the two currently available methods: radiofrequency ablation and endovenous laser treatment (Table 6.14). Both methods create a thermal injury to the vein wall with destruction of the inner endothelial lining, or intima, and cause collagen denaturation in the medial vein wall layer. The end goal of thermal ablation is fibrotic occlusion of the vein over time. Since the saphenous vein is left in situ in thermal ablation procedures in contrast to excision

TABLE 6.14

FDA-Approved Devices and Injectables for Venous Disorder Treatment

Endovenous Thermal Ablation	ClosureFast	Covidien
	Vari-Lase Bright Tip and Platinum Bright Tip	Vascular Solutions
	VenaCure	Angiodynamics
	ELVeS PL Laser System	Biolitec
	D FlexiPulse	Dornier MedTech
Sclerotherapy	Asclera (polidocanol)	Chemische Fabrik Kreussler & Company
	Sotradecol (sodium tetradecyl sulfate)	Angiodynamics
	Sodium morrhuate[a]	
	Ethanolamine oleate[a]	

Abbreviation: FDA, Food and Drug Administration.

[a] Exempted from approval by FDA; no longer commonly used.

during stripping, failures occur where the injury was insufficient to permanently close the vein allowing for recanalization. Even with occasional failures, endovenous patients demonstrate similar or better endpoints on long-term follow-up when compared with patients who have undergone stripping.

The endovenous thermal ablation procedure is initiated using standard angiography vascular access devices such 21-gauge needles, access wires, and vascular sheaths. Once vascular access is obtained, the specific ablation catheter is placed. Ultrasound is used for procedural imaging including initial vein cannulation, navigation of wires and catheters through the superficial vein, positioning of the thermal device, and the administration of tumescent solution along the course of the vein to be treated.

The procedure begins by placing the patient in a supine position for treatment of the long saphenous vein or in a prone position for treatment of the small saphenous vein. The index leg is prepped and draped in sterile manner, and the table is tilted in a reverse Trendelenbergh position (leg down). This position augments venous filling and engorgement, making cannulation of the saphenous vein easier. Venous access is obtained at a distal point in the leg, typically at the level of the knee or calf. The vein is located with ultrasound and the overlying tissues and skin are anesthetized with local anesthetic. An access needle is placed into the vein with ultrasound guidance and a wire is placed into the vein through the needle in standard Seldinger technique. The needle is then removed over the wire and replaced with a vascular access sheath. Placement and positioning of the thermal ablation catheter is then performed with ultrasound visualization. Most devices recommend positioning the catheter several centimeters back from the junction of the superficial vein with the deep venous system. The bed is then tilted into a Trendelenbergh position to empty the vein of blood. The tissues surrounding the vein are then infused with a dilute anesthetic agent, or tumescent, which serves several purposed aside from pain control. The tumescent fluid compresses the vein around the catheter, so heat is better delivered to the vein wall increasing severity of the heat-induced injury. The tumescent also acts as a heat sink reducing exposure of the surrounding tissues to thermal injury. The thermal catheter is activated and either drawn back continuously or incrementally depending on the specific device being used. The current version of the available radio frequency ablation (RFA) device is drawn back incrementally and all laser-based devices are drawn back continuously.

The currently available RFA catheter heats the vein wall to 120°C for 20 seconds over a length of 3 or 7 cm depending on the catheter selected. The catheter is drawn back incrementally with 0.5 cm of overlap of each treated segment. The first treated increment is treated with two complete cycles before the catheter is repositioned. Each increment is thereafter treated with one cycle. The catheter is suitable veins with a diameter ranging from 2 to 15 mm.

A number of laser-based thermal ablation catheters are available (Table 6.14). Available lasers have either a hemoglobin-specific wavelength (810, 940, and 980 nm) or a water-specific wavelength (1470, 1320, and 1319 nm). Ideally, laser ablation catheters create injury to the vein wall without causing full-thickness perforation. Full-thickness perforation is associated with increased hematoma formation, bruising, and postprocedural discomfort.

When comparing RFA to laser-based saphenous ablation, both techniques are highly effective. In general, there may be somewhat less discomfort associated with RFA during the first 10 days of recovery as well as reduced bruising. On long-term follow-up, there may be slightly lower rates of failure and vein recanalization when laser is used.

A number of major but infrequent complications may occur when treating saphenous vein reflux. Endovascular thermal ablation can cause skin burns, DVT, and PE. Less-significant complications include superficial thrombophlebitis, transient paresthesias, and skin hyperpigmentation.

Perforator vein incompetence is usually treated after saphenous incompetence is addressed if symptoms persist. It is aggressively pursued in advanced disease states such as active ulceration. Current percutaneous treatment options include RFA and sclerotherapy. Laser use is not FDA approved for perforator vein treatment. Several surgical options are also available including subendofascial perforator ligation. This procedure is not performed commonly due to the development of percutaneous techniques.

In cases of combined superficial and deep system reflux, the superficial system is first addressed followed by perforators. In some patients, the deep system reflux resolves once superficial reflux is eradicated. In other subjects, there may be significant symptom improvement, despite continued deep system reflux once superficial and perforator veins are treated. If significant symptoms persist after eradication of superficial and perforator reflux, the iliac veins and vena cava are evaluated for an obstructive lesion.

Surgical ligation and stripping are typically reserved for very large vein diameters, tortuous, or obstructed veins that will not permit thermal catheter passage, significant local anesthetic allergy, and vein recanalization after failed endovenous treatment (although repeat endovenous treatment can be performed in some cases). The procedure is typically combined with phlebectomy of varicose veins. The traditional surgical approach is performed infrequently, given the minimally invasive options available.

6.6.6 Treatment of Varicose Veins

The most frequently used techniques for the treatment of varicose veins of the leg are ambulatory phlebectomy and sclerotherapy. Most ambulatory phlebectomy procedures are performed with local anesthetic in an office setting. Some centers offer additional sedation with oral or IV agents, but these are frequently unnecessary. Sclerotherapy is a versatile procedure used to treat a variety of venous disorders. It is typically performed in an office setting without the need for local anesthetic or sedation. Sclerotherapy is typically used to treat small varicose veins, incompetent perforator veins, spider veins, and reticular veins.

Ambulatory phlebectomy is the typical treatment modality for veins larger than 3–4 mm in diameter. The procedure is performed by first marking skin over the varicose veins with the patient in an upright position. The patient is then placed in either a supine or prone position with the index leg elevated to reduce venous pressure and filling. The planned areas of treatment are prepared and draped in sterile manner and the marked areas are anesthetized. Small incisions are made and segments of varicose veins are pulled out through the incision with either small hooks or fine clamps. The exteriorized vein segments "snap" or tear free from the underlying venous connections. The remaining free ends are not ligated, as this would necessitate a larger skin incision and negatively impact cosmesis. Brief manual compression at the site of treatment followed by a compressive dressing provides hemostasis. Skin incisions are often closed with a simple adhesive strip. Ambulatory phlebectomy allows for expeditious and cost-effective treatment with excellent symptom relief and cosmetic outcome.

Sclerotherapy agents are classified by the mechanism through which vein injury and closure occurs. There are three classes of sclerotherapy agents: detergents, hypertonic solutions, and chemical irritants. The current FDA-approved sclerotherapy agents are all detergents and include polidocanol, sodium tetradecyl sulfate, sodium morrhuate, and ethanolamine oleate (Table 6.14). Detergents cause injury to the vein wall by disrupting surface lipids of the endothelial cells lining the vein lumen.

Sclerotherapy is commonly used to treat spider and reticular veins. Spider veins by definition are <1 mm in diameter and reticular veins between 1 and 3 mm in diameter. These veins are too small for phlebectomy. Injections are done through a 30-gauge needle with the patient in Trendelenbergh position. Sclerotherapy is more successful when sources of reflux are treated first.

Complications of sclerotherapy include allergic reactions, anaphylaxis, skin necrosis, embolization, skin pigmentation, and telangiectatic matting. Of the FDA-approved sclerosants, polidocanol has the lowest incidence of allergy and anaphylaxis. Skin necrosis is associated with larger volumes of sclerosant injection at a single site and use of higher concentration sclerosant mixtures. Embolization has been reported with sclerotherapy, more so with foam sclerotherapy, where an agent such as sodium tetradecyl sulfate is aerated and injected. Although rare, devastating embolic complications may occur when a patent foramen ovale is present in the heart and the embolus passes into the arterial circulation. Stroke and vision loss have been reported in these rare situations. Skin pigmentation occurs when red blood cells extravasate from the vessel lumen and deposit hemosiderin in the dermis or soft tissues. Reducing injection pressure and sclerosant concentration reduces this risk. Telangiectatic matting describes the appearance of small and fine red vessels at the sites of prior treatment. Risk is reduced by using a lower concentration sclerosant injected at lower pressure and lower volume.

6.6.7 Treatment of Iliac and Vena Cava Obstruction

Mounting evidence from a limited number of centers demonstrates successful and durable treatment options for chronic obstructive lesions of the iliac veins and vena cava.[169] These problems are associated with worse CEAP scores and frequently with ulceration. Treatment of these lesions improves symptoms and reduces ulceration.[175,180]

Pioneering work by Drs. Raju and Neglen has led to the following recommendations when treating obstructive iliac and vena cava lesions: (1) use IVUS routinely to improve sensitivity over venography alone, (2) use high pressure angioplasty prior to stent placement, (3) treat all areas of obstruction, (4) stent into the inferior vena cava as needed for stenosis at the iliac vein confluence, (5) extended stents below the inguinal ligament into the common femoral vein as needed, (6) use large-size stents with 2 mm over size beyond the vein diameter, and (7) consider performing the procedures under general anesthesia to control the pain experienced during high-pressure angioplasty.[170]

6.6.8 Emerging Technology

Replacement therapies for thermal ablation are currently under investigation. Cyanoacrylate adhesive may offer the advantage of reduced intraprocedural and postprocedural pain when used for saphenous vein closure. A proprietary blend of cyanoacrylate is being tested by Sapheon Inc. The adhesive would obviate the use of tumescent infusion along the entire course of the vein to be treated. Tumescent infusion is a source of discomfort during the procedure and the first several days of recovery. With an adhesive, only local anesthetic infiltration at the vein access site is needed. Thermal injury of adjacent tissue, vein perforation, and hematoma formation would be eliminated. Saphenous and sural nerve injury would also be reduced or avoided. Concerns of durability, embolization, DVT, allergic reaction, and infection need to be evaluated.

Another emerging technology utilizes mechanicochemical ablation for treatment of saphenous vein reflux. One experimental device, the ClariVein catheter, has an angulated rotating tip that elicits vein spasm concentrating the infused polidocanol sclerosant into a smaller volume vessel lumen. The single catheter delivers the injury and sclerosant. This technique does not require tumescent infusion and would have advantages similar to adhesives when compared to standard thermal ablation.

Currently, the only FDA-approved stents for use in veins is restricted to arteriovenous hemodialysis access and portal vein to hepatic vein shunt creation (transjugular intrahepatic portosystemic shunt [TIPS]). Both bare metal and covered stents have approvals for these indications. The early and limited data regarding bare metal stent use for treating iliac vein obstructing lesions are promising but are off-label use. Data for bare metal stents used in femoral and popliteal veins are very limited. Evaluation of drug deliver with coated stents or balloons has yet to be done in the venous system. Data regarding covered stent use in the venous system for indications other than hemodialysis access and TIPS are lacking.

Current therapies for reversing lipodermatosclerosis associated with long-standing CVI are minimally effective and poorly studied. Horse chestnut extract taken orally may reduce pain associated with more severe forms of CVI. Phlebotonics are a diverse class to drugs often used to treat CVI. When taken orally, there may be some improvement in leg edema. Topical capsaicin may improve early or acute changes of lipodermatosclerosis. Stanozolol, a synthetic testosterone derivative, when taken orally, may also be effective in this setting. Intralesion injection with triamcinolone also has some reported efficacy. None of these treatments are generally accepted or considered standard.

Finally, DVT prevention through prophylaxis at times of high subject risk is an essential component of CVI care. DVT is the strongest risk factor for the development of CVI. Furthermore, prior DVT is a marker for more advanced forms of CVI. Many subjects would never develop CVI if DVT was avoided. Medications used for DVT prophylaxis until recently have been injectable only. Oral medications such as rivaroxaban are becoming available with indications for DVT prophylaxis. Oral dosing may improve patient compliance and comfort.

Overall, presently accepted therapies for CVI treat at the macrovascular level. No cellular or molecular-targeted treatments are available. This is an area wide open to research and development.

References

1. Brewster DC, Cronenwett JL, Hallett JW Jr, Johnston KW, Krupski WC, and Matsumura JS. Joint Council of the American Association for Vascular Surgery and Society for Vascular Surgery. Guidelines for the treatment of abdominal aortic aneurysms. Report of a subcommittee of the Joint Council of the American Association for Vascular Surgery and Society for Vascular Surgery. *J Vasc Surg*. 2003 May;37(5):1106–17.
2. The United Kingdom EVAR trial investigators, Greenhalgh RM, Brown LC, Powell JT, Thompson SG, Epstein D, and Sculpher MJ. Endovascular versus open repair of abdominal aortic aneurysm. *N Engl J Med*. 2010;362:1863–71.
3. Schermerhorn ML, O'Malley AJ, Jhaveri A, Cotterill P, Pomposelli F, and Landon BE. Endovascular vs. open repair of abdominal aortic aneurysm in Medicare population. *N Engl J Med*. 2008 Jan 31;358(5):464–74.
4. Rizas KD, Ippaqunta N, and Tilson MD. Immune cells and molecular mediators in the pathogenesis of the abdominal aortic aneurysm. *Cardiol Rev*. 2009;17:201–10.
5. Wilmink AB and Quick CR. Epidemiology and potential for prevention of abdominal aortic aneurysm. *Br J Surg*. 1998;85:155–62.
6. Jaakkola P, Hippelainen M, Farin P, Rytkönen P, Kainulainen S, and Partanen K. Interobserver variability in measuring the dimensions of the abdominal aorta: comparison of ultrasound and computed tomography. *Eur J Vasc Endovasc Surg*. 1996;12:230–7.
7. Kent KC, Zwolak RM, Jaff MR, Hollenbeck ST, Thompson RW, Schermerhorn ML et al. Screening for abdominal aortic aneurysm: a consensus statement. *J Vasc Surg*. 2004;39:267–9.

8. The United Kingdom Small Aneurysm Trial Participants. Long-term outcomes of immediate repair compared with surveillance of small abdominal aortic aneurysms. *N Engl J Med.* 2002;346:1445–52.

9. Lederle FA, Wilson SE, Johnson GR, Reinke DB, Littooy FN, Acher CW et al. Immediate repair compared with surveillance of small abdominal aortic aneurysms. *N Engl J Med.* 2002;346:1437–44.

10. Chaikof EL, Brewster DC, Dalman RL, Makaroun MS, Illig KA, Sicard GA, Timaran CH, Upchurch GR, and Veith FJ. SVS practice guidelines for the care of patients with an abdominal aortic aneurysm: executive summary. *J Vasc Surg.* 2009;50:880–96.

11. Hollier LH, Reigel MM, Kozmier FJ, Pairolero PC, Cherry KJ, and Hallett JW Jr. Conventional repair of abdominal aortic aneurysm in the high risk patient: a plea for abandonment of nonresective treatment. *J Vasc Surg.* 1986;3:712.

12. Prinssen M, Verhoeven ELG, Buth J, Philippe WM, Cuypers PWM, van Sambeek MRHM et al. A randomized trial comparing conventional and endovascular repair of abdominal aortic aneurysms. *N Engl J Med.* 2004;351:1607–18.

13. EVAR Trial Participants. Endovascular aneurysm repair and outcome in patients unfit for open repair of abdominal aortic aneurysm (EVAR trial 2): randomized controlled trial. *Lancet.* 2005;365:2187–92.

14. Greenhalgh RM, Brown LC, Kwong GP, Powell JT, Thompson SG, and EVAR participants. Comparison of endovascular aneurysm repair with open repair in patients with abdominal aortic aneurysm (EVAR trial 1), 30-day operative mortality results: randomized controlled trial. *Lancet.* 2004;364:843–8.

15. Wyss TR, Dick F, Brown LC, and Greenhalgh RM. The influence of thrombus, calcification, angulation, and tortuosity of attachment sites on the time to the first graft-related complication after endovascular aneurysm repair. *J Vasc Surg.* 2011;54:965–71.

16. Leather RP, Shah DM, Kaufman JL, Fitzgerald KM, and Chang BB. Comparative analysis of retroperitoneal and transperitoneal aortic replacement for aneurysm. *Surg Gynecol Obstet.* 1989;168:387–93.

17. Sicard GA, Reilly JM, Rubin BG, Thompson RW, Allen BT, Flye MW, Schechtman KB, Young-Beyer P, Weiss C, and Anderson CB. Transabdominal versus retroperitoneal incision for abdominal aortic surgery: report of a prospective randomized trial. *J Vasc Surg.* 1995;21:174–83.

18. Feringa HH, Bax JJ, Boersma E, Kertai MD, Meij SH, Galal W et al. High dose beta blockers and tight heart rate control reduce myocardial ischemia and troponin T release in vascular surgery patients. *Circulation.* 2006;114:I344–9.

19. Lederle FA, Freischlag JA, Kyriakides TC, Padberg FT, Matsumura JS, Kohler TR et al. Outcomes following endovascular vs open repair of abdominal aortic aneurysm: A randomized trial. *JAMA.* 2009;302:1535–42.

20. Mehta M, Veith FJ, Darling RC, Roddy SP, Ohki T, Lipsitz EC et al. Effects of bilateral hypogastric artery interruption during endovascular and open aortoiliac aneurysm repair. *J Vasc Surg.* 2004 Oct;40(4):698–702.

21. Criado FJ, Wilson EP, Velazquez OC, Carpenter JP, Barker C, Wellons E, Abul-Khoudoud O, and Fairman RM. Safety of coil embolization of the internal iliac artery in endovascular grafting of abdominal aortic aneurysms. *J Vasc Surg.* 2000 Oct;32:684–8.

22. Baum RA, Carpenter JP, Cope C, Golden MA, Velazquez OC, Neschis DG, Mitchell ME, Barker CF, and Fairman RM. Aneurysm sac pressure measurements after endovascular repair of abdominal aortic aneurysms. *J Vasc Surg.* 2001;33:32–41.

23. Veith FJ, Baum RA, Ohki T, Amor M, Adisishiah M, Blankensteijn JD et al. Nature and significance of endoleaks and endotension: summary of opinions expressed at an international conference. *J Vasc Surg.* 2002;35:1029–35.

24. Jones JE, Atkins MD, Brewster DC, Chung TK, Kwolek CJ, LaMuraglia GM, Hodgman TM, and Cambria RP. Persistent type 2 endoleak after endovascular repair of abdominal aortic aneurysm is associated with adverse late outcomes. *J Vasc Surg.* 2007 Jul;46:1–8.

25. Zwolak RM, Sidawy AN, Greenberg RK, Schermerhorn ML, Shackelton RJ, Siami FS, and Society for Vascular Surgery Outcomes Committee. Lifeline registry of endovascular aneurysm repair: open repair surgical controls in clinical trials. *J Vasc Surg*. 2008;48:511–8.
26. Blakensteijn JD, de Jong SECA, Prinssen M, van der Ham AC, Buth J, van Sterkenburg SMM, Verhagen HJM, Buskens E, and Grobbee DE. Two-year outcomes after conventional or endovascular repair of abdominal aortic aneurysms. *N Engl J Med*. 2005;352:2398–405.
27. Rayt HS, Sutton AJ, London NJ, Sayers RD, and Bown MJ. A systematic review and meta-analysis of endovascular repair (EVAR) for ruptured abdominal aortic aneurysm. *Eur J Vasc Endovasc Surg*. 2008;36(5):536–44.
28. Mehta M. Endovascular aneurysm repair for ruptured abdominal aortic aneurysms: the Albany Vascular Group approach. *J Vasc Surg*. 2010;52:1706–12.
29. Mehta M, Darling RC III, Roddy SP, Fecteau S, Ozsvath KJ, Kreienberg PB, Paty PS, Chang BB, and Shah DM. Factors associated with abdominal compartment syndrome complicating endovascular repair of ruptured abdominal aortic aneurysms. *J Vasc Surg*. 2005;42(6):1047–51.
30. Panyam J and Labhasetwar V. Sustained cytoplasmic delivery of drugs with intracellular receptors using biodegradable nanoparticles. *Mol Pharm*. 2004;1(1):77–84.
31. Wickline SA, Neubauer AM, Winter P, Caruthers S, and Lanza G. Applications of nanotechnology to atherosclerosis, thrombosis, and vascular biology. Arterioscler Thromb Vasc Bio. 2006;26:435–41.
32. Piecha G, Wiecek A, and Januszewicz A. Epidemiology and optimal management in patients with renal artery stenosis. *J Nephrol*. 2012 Nov–Dec;25(6):872–8.
33. Rees CR, Palmaz JC, Becker GJ, Ehrman KO, Richter GM, Noeldge G, Katzen BT, Dake MD, and Schwarten DE. Palmaz stent in atherosclerotic stenoses involving the ostia of the renal arteries: preliminary report of a multicenter study. *Radiology*. 1991 Nov;181(2):507–14.
34. Textor SC. Atherosclerotic renal artery stenosis: overtreated but underrated? *J Am Soc Nephrol*. 2008 Apr;19(4):656–9.
35. Sarac TP. Influence and critique of the ASTRAL and CORAL Trials. *Semin Vasc Surg*. 2011 Sep;24(3):162–6.
36. Cooper CJ, Murphy TP, Matsumoto A, Steffes M, Cohen DJ, Jaff M et al. Stent revascularization for the prevention of cardiovascular and renal events among patients with renal artery stenosis and systolic hypertension: rationale and design of the CORAL trial. *Am Heart J*. 2006 Jul;152(1):59–66.
37. ASTRAL Investigators, Wheatley K, Ives N, Gray R, Kalra PA, Moss JG et al. Revascularization versus medical therapy for renal-artery stenosis. *N. Engl. J. Med*. 2009;361(20):1953–62.
38. White CJ. Kiss my astral: one seriously flawed study of renal stenting after another. *Catheter Cardiovasc Interv*. 2010 Feb 1;75(2):305–7.
39. Weinberg MD and Olin JW. Stenting for atherosclerotic renal artery stenosis: one poorly designed trial after another. *Cleve Clin J Med*. 2010 Mar;77(3):164–71.
40. Yorgun H, Kabakçi G, Canpolat U, Aytemir K, Fatihoglu G, Karakulak UN, Kaya EB, Sahiner L, Tokgözoglu L, and Oto A. Frequency and predictors of renal artery stenosis in hypertensive patients undergoing coronary angiography. *Angiology*. 2013 Jul; 64(5): 385–90.
41. Harding and MB, Smith LR, Himmelstein SI, Harrison K, Phillips HR, Schwab SJ, Hermiller JB, Davidson CJ, and Bashore TM. Renal artery stenosis: prevalence and associated risk factors in patients undergoing routine cardiac catheterization. *J Am Soc Nephrol*. 1992 May;2(11):1608–16.
42. Missouris CG, Buckenham T, Cappuccio FP, and MacGregor GA. Renal artery stenosis: a common and important problem in patients with peripheral vascular disease. *Am J Med*. 1994 Jan;96(1):10–4.
43. Swartbol P, Thorvinger BO, Pärsson H, and Norgren L. Renal artery stenosis in patients with peripheral vascular disease and its correlation to hypertension. A retrospective study. *Int Angiol*. 1992 Jul–Sep;11(3):195–9.
44. Conlon PJ, Athirakul K, Kovalik E, Schwab SJ, Crowley J, Stack R, McCants CB Jr, Mark DB, Bashore TM, and Albers F. Survival in renal vascular disease. *J Am Soc Nephrol*. 1998 Feb;9(2):252–6.

45. Conlon PJ, Little MA, Pieper K, and Mark DB. Severity of renal vascular disease predicts mortality in patients undergoing coronary angiography. *Kidney Int*. 2001 Oct;60(4):1490–7.

46. Newcombe CP, Shucksmith HS, and Suffern WS. Sympathectomy for Hypertension: follow-up of 212 patients. *Br Med J*. 1959 Jan 17;1(5115):142–4.

47. Longland CJ and Gibb WE. Sympathectomy in the treatment of benign and malignant hypertension: a review of 76 patients. *Br J Surg*. 1954 Jan;41(168):382–92.

48. White PD. Severe hypertension: study of one hundred patients with cardiovascular complications: follow-up results in fifty controls and fifty patients subjected to Smithwick's lumbodorsal sympathectomy, 1941 to 1946. *J Am Med Assoc*. 1956 Mar 24;160(12):1027–8.

49. Bertog SC, Sobotka PA, and Sievert H. Renal denervation for hypertension. *JACC Cardiovasc Interv*. 2012 Mar;5(3):249–58.

50. Oparil S. The sympathetic nervous system in clinical and experimental hypertension. *Kidney Int*. 1986 Sep;30(3):437–52.

51. Bright R. *Reports of medical cases with a view of illustrating symptoms and the cure of diseases by reference to morbid anatomy*. London: Longman, 1827.

52. Tigerstedt R and Bergmann PG. Niere und Kreislauf. Scand. *Arch Physiol*. 1898;4:223–71.

53. Goldblatt H. Studies on experimental hypertension. I. The production of persistent elevation of systolic blood pressure by means of renal ischemia. *J. Exp Med*. 1934;59:347–80.

54. Modrall JG, Timaran CH, Rosero EB, Chung J, Arko FA 3rd, Valentine RJ, Clagett GP, and Trimmer C. Predictors of outcome for renal artery stenting performed for salvage of renal function. *J Vasc Surg*. 2011 Nov;54(5):1414–21.

55. Simeoni S, Girelli D, Lino M, Olivieri O, and Corrocher R. Recovery of renal function after 3 months of dialysis in a patient with atherosclerotic renovascular disease following aortoiliac bypass and left renal artery reimplantation. *Eur J Vasc Endovasc Surg*. 2004 Nov;28(5):562–4.

56. Gandhi SK, Powers JC, Nomeir AM, Fowle K, Kitzman DW, Rankin KM, and Little WC. The pathogenesis of acute pulmonary edema associated with hypertension. *N Engl J Med*. 2001 Jan 4;344(1):17–22.

57. Gray BH, Olin JW, Childs MB, Sullivan TM, and Bacharach JM. Clinical benefit of renal artery angioplasty with stenting for the control of recurrent and refractory congestive heart failure. *Vasc Med*. 2002;7(4):275–9.

58. Caps MT, Perissinotto C, Zierler RE, Polissar NL, Bergelin RO, Tullis MJ, Cantwell-Gab K, Davidson RC, and Strandness DE Jr. Prospective study of atherosclerotic disease progression in the renal artery. *Circulation*. 1998 Dec 22–29;98(25):2866–72.

59. Iglesias JI, Hamburger RJ, Feldman L, and Kaufman JS. The natural history of incidental renal artery stenosis in patients with aortoiliac vascular disease. *Am J Med*. 2000 Dec 1;109(8):642–7.

60. Chábová V, Schirger A, Stanson AW, McKusick MA, and Textor SC. Outcomes of atherosclerotic renal artery stenosis managed without revascularization. *Mayo Clin Proc*. 2000 May;75(5):437–44.

61. Pearce JD, Craven BL, Craven TE, Piercy KT, Stafford JM, Edwards MS, and Hansen KJ. Progression of atherosclerotic renovascular disease: a prospective population-based study. *J Vasc Surg*. 2006 Nov;44(5):955–62; discussion 962–3.

62. Kuhlmann U, Grüntzig A, Vetter W, Furrer J, Lütolf U, and Siegenthaler W. [Renovascular hypertension: therapy by means of percutaneous transluminal dilatation of renal artery stenoses]. *Schweiz Med Wochenschr*. 1978 Nov 25;108(47):1847–50.

63. Weibull H, Bergqvist D, Bergentz SE, Jonsson K, Hulthén L, and Manhem P. Percutaneous transluminal renal angioplasty versus surgical reconstruction of atherosclerotic renal artery stenosis: a prospective randomized study. *J Vasc Surg*. 1993 Nov;18(5):841–50; discussion 850–2.

64. Edwards MS and Corriere MA. Contemporary management of atherosclerotic renovascular disease. *J Vasc Surg*. 2009 Nov; 50(5): 1197–210.

65. van de Ven PJ, Kaatee R, Beutler JJ, Beek FJ, Woittiez AJ, Buskens E, Koomans HA, and Mali WP. Arterial stenting and balloon angioplasty in ostial atherosclerotic renovascular disease: a randomised trial. *Lancet*. 1999 Jan 23;353(9149):282–6.

66. Corriere MA, Pearce JD, Edwards MS, Stafford JM, and Hansen KJ. Endovascular management of atherosclerotic renovascular disease: early results following primary intervention. *J Vasc Surg.* 2008 Sep;48(3):580–7; discussion 587–8.

67. Harden PN, MacLeod MJ, Rodger RS, Baxter GM, Connell JM, Dominiczak AF, Junor BJ, Briggs JD, and Moss JG. Effect of renal-artery stenting on progression of renovascular renal failure. *Lancet.* 1997 Apr 19;349(9059):1133–6.

68. Shetty R, Biondi-Zoccai GG, Abbate A, Amin MS, and Jovin IS. Percutaneous renal artery intervention versus medical therapy in patients with renal artery stenosis: a meta-analysis. *EuroIntervention.* 2011 Nov;7(7):844–51.

69. Cherr GS, Hansen KJ, Craven TE, Edwards MS, Ligush J Jr, Levy PJ, Freedman BI, and Dean RH. Surgical management of atherosclerotic renovascular disease. *J Vasc Surg.* 2002 Feb;35(2):236–45.

70. Symplicity HTN-1 Investigators. Catheter-based renal sympathetic denervation for resistant hypertension: durability of blood pressure reduction out to 24 months. *Hypertension.* 2011 May;57(5):911–7.

71. Symplicity HTN-2 Investigators, Esler MD, Krum H, Sobotka PA, Schlaich MP, Schmieder RE, and Böhm M. Renal sympathetic denervation in patients with treatment-resistant hypertension (The Symplicity HTN-2 Trial): a randomised controlled trial. *Lancet.* 2010 Dec 4;376(9756):1903–9.

72. Weber DO. Nanomedicine. *Health Forum J.* 1999 Jul–Aug;42(4):32, 36–7.

73. Carroll BA, Turner RJ, Tickner EG, Boyle DB, and Young SW. Gelatin encapsulated nitrogen microbubbles as ultrasonic contrast agents. *Invest Radiol.* 1980 May–Jun;15(3):260–6.

74. Porter TR, Xie F, Kricsfeld A, and Kilzer K. Noninvasive identification of acute myocardial ischemia and reperfusion with contrast ultrasound using intravenous perfluoropropane-exposed sonicated dextrose albumin. *J Am Coll Cardiol.* 1995 Jul;26(1):33–40.

75. Fritz TA, Unger EC, Sutherland G, and Sahn D. Phase I clinical trials of MRX-115. A new ultrasound contrast agent. *Invest Radiol.* 1997 Dec;32(12):735–40.

76. Lindner JR, Dayton PA, Coggins MP, Ley K, Song J, Ferrara K, and Kaul S. Noninvasive imaging of inflammation by ultrasound detection of phagocytosed microbubbles. *Circulation.* 2000 Aug 1;102(5):531–8.

77. Kaufmann BA, Sanders JM, Davis C, Xie A, Aldred P, Sarembock IJ, and Lindner JR. Molecular imaging of inflammation in atherosclerosis with targeted ultrasound detection of vascular cell adhesion molecule-1. *Circulation.* 2007 Jul 17;116(3):276–84.

78. Nahrendorf M, Jaffer FA, Kelly KA, Sosnovik DE, Aikawa E, Libby P, and Weissleder R. Noninvasive vascular cell adhesion molecule-1 imaging identifies inflammatory activation of cells in atherosclerosis. *Circulation.* 2006 Oct 3;114(14):1504–11.

79. Wu J, Leong-Poi H, Bin J, Yang L, Liao Y, Liu Y, Cai J, Xie J, and Liu Y. Efficacy of contrast-enhanced US and magnetic microbubbles targeted to vascular cell adhesion molecule-1 for molecular imaging of atherosclerosis. *Radiology.* 2011 Aug;260(2):463–71.

80. Cormode DP, Roessl E, Thran A, Skajaa T, Gordon RE, Schlomka JP et al. Atherosclerotic plaque composition: analysis with multicolor CT and targeted gold nanoparticles. *Radiology.* 2010 Sep;256(3):774–82.

81. Baturin P, Alivov Y, and Molloi S. Spectral CT imaging of vulnerable plaque with two independent biomarkers. *Phys Med Biol.* 2012 Jul 7;57(13):4117–38.

82. Alam SR, Shah AS, Richards J, Lang NN, Barnes G, Joshi N et al. Ultrasmall superparamagnetic particles of iron oxide in patients with acute myocardial infarction: early clinical experience. *Circ Cardiovasc Imaging.* 2012 Sep 1;5(5):559–65.

83. Marom O, Nakhoul F, Tisch U, Shiban A, Abassi Z, and Haick H. Gold nanoparticle sensors for detecting chronic kidney disease and disease progression. *Nanomedicine (Lond).* 2012 May;7(5):639–50.

84. McCarthy JR, Korngold E, Weissleder R, and Jaffer FA. A light-activated theranostic nanoagent for targeted macrophage ablation in inflammatory atherosclerosis. *Small.* 2010 Sep 20;6(18):2041–9.

85. Lobatto ME, Fayad ZA, Silvera S, Vucic E, Calcagno C, Mani V et al. Multimodal clinical imaging to longitudinally assess a nanomedical anti-inflammatory treatment in experimental atherosclerosis. *Mol Pharm.* 2010 Dec 6;7(6):2020–9.

86. Tepe G, Zeller T, Albrecht T, Heller S, Schwarzwälder U, Beregi JP, Claussen CD, Oldenburg A, Scheller B, and Speck U. Local delivery of paclitaxel to inhibit restenosis during angioplasty of the leg. *N Engl J Med.* 2008 Feb 14;358(7):689–99.

87. Dake MD, Ansel GM, Jaff MR, Ohki T, Saxon RR, Smouse HB et al. Paclitaxel-eluting stents show superiority to balloon angioplasty and bare metal stents in femoropopliteal disease: twelve-month Zilver PTX randomized study results. *Circ Cardiovasc Interv.* 2011 Oct 1;4(5):495–504.

88. Lanza GM, Yu X, Winter PM, Abendschein DR, Karukstis KK, Scott MJ, Chinen LK, Fuhrhop RW, Scherrer DE, and Wickline SA. Targeted antiproliferative drug delivery to vascular smooth muscle cells with a magnetic resonance imaging nanoparticle contrast agent: implications for rational therapy of restenosis. *Circulation.* 2002 Nov 26;106(22):2842–7.

89. Joner M, Morimoto K, Kasukawa H, Steigerwald K, Merl S, Nakazawa G et al. Site-specific targeting of nanoparticle prednisolone reduces in-stent restenosis in a rabbit model of established atheroma. *Arterioscler Thromb Vasc Biol.* 2008 Nov;28(11):1960–6.

90. Nissenson AR. Bottom-up nanotechnology: the human nephron filter. *Semin Dial.* 2009 Nov–Dec;22(6):661–4.

91. Kolff WJ. First clinical experience with the artificial kidney. *Ann Int Med.* 1965;62:608–19. 2006.

92. United States Renal Data System. http://www.usrds.org/2006/pdf/11_econ_06.pd. 2006.

93. Kidney Disease Statistics for the United States National Kidney and Urologic Disease Information Clearinghouse. http://www.kidney.niddk.nih.gov/kudiseases/pubs/kustats/. 2006.

94. Hirsh J and Hoak J. Management of deep vein thrombosis and pulmonary embolism. A statement for healthcare professionals from the Council on (in consultation with the Council on Cardiovascular Radiology), American Heart Association. *Circulation.* 1996;93:2212–45.

95. Dimick JB, Chen SL, Taheri PA, Henderson WG, Khuri SF, and Campbell DA Jr. Hospital costs associated with surgical complications: a report from the private-sector National Surgical Quality Improvement Program. *J Am Coll Surg.* 2004;199:531–7.

96. White RH. The epidemiology of Venous Thromboembolism. *Circulation.* 2003;107:I4–8.

97. Siegle R, Naishadham D, and Jemal A. Cancer statistics for Hispanics/Latinos, 2012. *CA Cancer J Clin.* 2012 Sep-Oct; 62(5): 283–98.

98. Tsai AW, Cushman M, Rosamond WD, Heckbert SR, Polak JF, and Folsom AR. Cardiovascular risk factors and venous thromboembolism incidence: the longitudinal investigation of thromboembolism etiology. *Arch Intern Med.* 2002 May 27;162(10):1182–9.

99. Silverstein MD, Heit JA, Mohr DN, Petterson TM, O'Fallon WM, and Melton LJ 3rd. Trends in the incidence of deep vein thrombosis and pulmonary embolism: a 25-year population-based study. *Arch Intern Med.* 1998;158:585–93.

100. Heit J, O'Fallon W, Petterson T, Lohse C, Silverstein M, Mohr D, and Melton L. Relative impact of risk factors for deep vein thrombosis and pulmonary embolism: a population-based study. *Arch Intern Med.* 2002;162:1245–8.

101. White RH, Zhou H, and Romano PS. Incidence of idiopathic deep venous thrombosis and secondary thromboembolism among ethnic groups in California. *Ann Intern Med.* 1998 May 1;128(9):737–40.

102. Hooper WC, Holman RC, Heit JA, and Cobb N. Venous thromboembolism hospitalizations among American Indians and Alaska Natives. *Thromb Res.* 2002 Dec 15;108(5–6):273–8.

103. Cushman M, Tsai A, Heckbert SR, White R, Rosamund W, Enright P et al. Incidence rates, case fatality, and recurrence rates of deep vein thrombosis and pulmonary embolus: the Longitudinal Investigation of Thromboembolism Etiology (LITE). *Thromb Haemost.* 2001;86 (suppl 1).

104. Prandoni P, Lensing AW, Cogo A, Cuppini S, Villalta S, Carta M, Cattelan AM, Polistena P, Bernardi E, and Prins MH. The long-term clinical course of acute deep venous thrombosis. *Ann Intern Med.* 1996 Jul 1;125(1):1–7.

105. Hansson PO, Sörbo J, and Eriksson H. Recurrent venous thromboembolism after deep vein thrombosis: incidence and risk factors. *Arch Intern Med.* 2000;160:769–74.

106. Heit JA, Mohr DN, Silverstein MD, Petterson TM, O'Fallon WM, and Melton LJ 3rd. Predictors of recurrence after deep vein thrombosis and pulmonary embolism: a population-based cohort study. *Arch Intern Med*. 2000;160:761–8.

107. Murin S, Romano PS, and White RH. Comparison of outcomes after hospitalization for deep venous thrombosis or pulmonary embolism. *Thromb Haemost*. 2002;88:407–14.

108. Heit J, Melton L, Lohse C, Petterson T, Silverstein M, Mohr D, and O'Fallon W. Incidence of venous thromboembolism in hospitalized patients versus community residents. *Mayo Clinic Proc*. 2001;76:1102–10.

109. Baglin TP, White K, and Charles A. Fatal pulmonary embolism in hospitalised medical patients. *J Clin Pathol*. 1997 Jul;50(7):609–10.

110. Heit J, Silverstein M, Mohr D, Petterson T, O'Fallon W, and Melton L. Risk factors for deep vein thrombosis and pulmonary embolism: a population-based case-control study. *Arch Intern Med*. 2000;160:809–15.

111. Key N and McGlennen R. Hyperhomocyst(e)inemia and thrombophilia. *Arch Path Lab Med*. 2002;126:1367–75.

112. Heit J, Farmer S, Petterson T, Bailey K, and Melton L. Novel risk factors for venous thromboembolism: a population-based, case-control study. *Blood*. 2005;106:463A.

113. Douketis JD, Julian JA, Kearon C, Anderson DR, Crowther MA, Bates SM et al. Does the type of hormone replacement therapy influence the risk of deep vein thrombosis? A prospective case-control study. *J Thromb Haemost*. 2005 May;3(5):943–8.

114. Gomez M and Deitcher S. Risk of venous thromboembolic disease associated with hormonal contraceptives and hormone replacement therapy: a clinical review. *Arch Intern Med*. 2004;164:1965–76.

115. Heit J, Kobbervig C, James A, Petterson T, Bailey K, and Melton LI. Trends in the incidence of deep vein thrombosis and pulmonary embolism during pregnancy or the puerperium: a 30-year population-based study. *Ann Intern Med*. 2005;143:697–706.

116. Cannegieter SC, Doggen CJ, van Houwelingen HC, and Rosendaal FR. Travel-related venous thrombosis: results from a large population-based case control study (MEGA study). *PLoS Med*. 2006 Aug;3(8):e307.

117. Crowther MA and Kelton JG. Congenital thrombophilic states associated with venous thrombosis: a qualitative overview and proposed classification system. *Ann Intern Med*. 2003 Jan 21;138(2):128–34.

118. Guyatt GH, Norris SL, Schulman S, Hirsh J, Eckman MH, Akl EA et al. Methodology for the development of antithrombotic therapy and prevention of thrombosis guidelines: Antithrombotic Therapy and Prevention of Thrombosis, 9th ed: American College of Chest Physicians Evidence-Based Clinical Practice Guidelines. *Chest*. 2012 Feb;141(suppl 2):53S–70S. doi: 10.1378/chest.11–2288.

119. de Haan HG, Bezemer ID, Doggen CJ, Le Cessie S, Reitsma PH, Arellano AR et al. Multiple SNP testing improves risk prediction of first venous thrombosis. *Blood*. 2012 Jul 19;120(3):656–63.

120. Killewich LA, Cahan MA, Hanna DJ, Murakami M, Uchida T, Wiley LA, and Hunter GC. The effect of external pneumatic compression on regional fibrinolysis in a prospective randomized trial. *J Vasc Surg*. 2002 Nov;36(5):953–8.

121. Comerota AJ, Chouhan V, Harada RN, Sun L, Hosking J, Veermansunemi R, Comerota AJ Jr, Schlappy D, and Rao AK. The fibrinolytic effects of intermittent pneumatic compression: mechanism of enhanced fibrinolysis. *Ann Surg*. 1997 Sep;226(3):306–13; discussion 313–4.

122. Sajid MS, Desai M, Morris RW, and Hamilton G. Knee length versus thigh length graduated compression stockings for prevention of deep vein thrombosis in postoperative surgical patients. *Cochrane Database Syst Rev*. 2012 May 16;5:CD007162.

123. Amaragiri SV and Lees TA. Elastic compression stockings for prevention of deep vein thrombosis. *Cochrane Database Syst Rev*. 2000;(3):CD001484.

124. Falck-Ytter Y, Francis CW, Johanson NA, Curley C, Dahl OE, Schulman S, Ortel TL, Pauker SG, Colwell CW Jr, and American College of Chest Physicians. Prevention of VTE in orthopedic surgery patients: Antithrombotic Therapy and Prevention of Thrombosis, 9th ed: American College of Chest Physicians Evidence-Based Clinical Practice Guidelines. *Chest*. 2012 Feb;141(suppl):e278S–325S. doi: 10.1378/chest.11–2404.

125. Epstein NE. Intermittent pneumatic compression stocking prophylaxis against deep venous thrombosis in anterior cervical spinal surgery: a prospective efficacy study in 200 patients and literature review. *Spine (Phila PA 1976).* 2005 Nov 15;30(22):2538–43.

126. Sobieraj-Teague M, Hirsh J, Yip G, Gastaldo F, Stokes T, Sloane D, O'Donnell MJ, and Eikelboom JW. Randomized controlled trial of a new portable calf compression device (Venowave) for prevention of venous thrombosis in high-risk neurosurgical patients. *J Thromb Haemost.* 2012 Feb;10(2):229–35.

127. Pitto RP, Hamer H, Heiss-Dunlop W, and Kuehle J. Mechanical prophylaxis of deep-vein thrombosis after total hip replacement a randomised clinical trial. *J Bone Joint Surg Br.* 2004 Jul;86(5):639–42.

128. Pour AE, Keshavarzi NR, Purtill JJ, Sharkey PF, and Parvizi J. Is venous foot pump effective in prevention of thromboembolic disease after joint arthroplasty: a meta-analysis. *J Arthroplasty.* 2013 Mar;28(3):410–17.

129. Alikhan R and Cohen AT. Heparin for the prevention of venous thromboembolism in general medical patients (excluding stroke and myocardial infarction). *Cochrane Database Syst Rev.* 2009 Jul 8;(3):CD003747.

130. Phung OJ, Kahn SR, Cook DJ, and Murad MH. Dosing frequency of unfractionated heparin thromboprophylaxis: a meta-analysis. *Chest.* 2011 Aug;140(2):374–81. doi: 10.1378/chest.10-3084.

131. Geerts WH, Jay RM, Code KI, Chen E, Szalai JP, Saibil EA, and Hamilton PA. A comparison of low-dose heparin with low-molecular-weight heparin as prophylaxis against venous thromboembolism after major trauma. *N Engl J Med.* 1996 Sep 5;335(10):701–7.

132. van Den Belt AG, Prins MH, Lensing AW, Castro AA, Clark OA, Atallah AN, and Burihan E. Fixed dose subcutaneous low molecular weight heparins versus adjusted dose unfractionated heparin for venous thromboembolism. *Cochrane Database Syst Rev.* 2000;(2):CD001100.

133. Unqueira DR, Perini E, Penholati RR, and Carvalho MG. Unfractionated heparin versus low molecular weight heparin for avoiding heparin-induced thrombocytopenia in postoperative patients. *Cochrane Database Syst Rev.* 2012 Sep 12;9:CD007557.

134. Becattini C, Agnelli G, Schenone A, Eichinger S, Bucherini E, Silingardi M et al. Aspirin for preventing the recurrence of venous thromboembolism. *N Engl J Med.* 2012 May 24;366(21):1959–67.

135. Brighton TA, Eikelboom JW, Mann K, Mister R, Gallus A, Ockelford P et al. Low-dose aspirin for preventing recurrent venous thromboembolism. *N Engl J Med.* 2012 Nov 22;367(21):1979–87.

136. Schulman S, Kearon C, Kakkar AK, Mismetti P, Schellong S, Eriksson H, Baanstra D, Schnee J, Goldhaber SZ, and RE-COVER Study Group. Dabigatran versus warfarin in the treatment of acute venous thromboembolism. *N Engl J Med.* 2009 Dec 10;361(24):2342–52.

137. Agnelli G, Bergqvist D, Cohen AT, Gallus AS, Gent M, and PEGASUS investigators. Randomized clinical trial of postoperative fondaparinux versus perioperative dalteparin for prevention of venous thromboembolism in high-risk abdominal surgery. *Br J Surg.* 2005 Oct;92(10):1212–20.

138. Tran AH and Lee G. Fondaparinux for prevention of venous thromboembolism in major orthopedic surgery. *Ann Pharmacother.* 2003 Nov;37(11):1632–43.

139. Turpie AG, Bauer KA, Caprini JA, Comp PC, Gent M, Muntz JE, and Apollo Investigators. Fondaparinux combined with intermittent pneumatic compression vs. intermittent pneumatic compression alone for prevention of venous thromboembolism after abdominal surgery: a randomized, double-blind comparison. *J Thromb Haemost.* 2007 Sep;5(9):1854–61.

140. Cohen AT, Tapson VF, Bergmann JF, Goldhaber SZ, Kakkar AK, Deslandes B et al. Venous thromboembolism risk and prophylaxis in the acute hospital care setting (ENDORSE study): a multinational cross-sectional study. *Lancet.* 2008 Feb 2;371(9610):387–94.

141. Schulman S, Lindmarker P, Holmström M, Lärfars G, Carlsson A, Nicol P et al. Post-thrombotic syndrome, recurrence, and death 10 years after the first episode of venous thromboembolism treated with warfarin for 6 weeks or 6 months. *J Thromb Haemost.* 2006 Apr;4(4):734–42.

142. Heit JA. The epidemiology of venous thromboembolism in the community: implications for prevention and management. *J Thromb Thrombolysis.* 2006 Feb;21(1):23–9.

143. Gosselin RC, Wu JR, Kottke-Marchant K, Peetz D, Christie DJ, Muth H, and Panacek E. Evaluation of the Stratus CS Acute Care D-dimer assay (DDMR) using the Stratus CS STAT Fluorometric Analyzer: a prospective multisite study for exclusion of pulmonary embolism and deep vein thrombosis. *Thromb Res*. 2012 Nov;130(5):e274–8.

144. Kearon C, Ginsberg JS, Douketis J, Crowther MA, Turpie AG, Bates SM, Lee A, Brill-Edwards P, Finch T, and Gent M. A randomized trial of diagnostic strategies after normal proximal vein ultrasonography for suspected deep venous thrombosis: D-dimer testing compared with repeated ultrasonography. *Ann Intern Med*. 2005 Apr 5;142(7):490–6.

145. Saour JN, Sieck JO, Mamo LA, and Gallus AS. Trial of different intensities of anticoagulation in patients with prosthetic heart valves. *N Engl J Med*. 1990;322:428–32.

146. Palareti G, Leali N, Coccheri S, Poggi M, Manotti C, D'Angelo A et al. Bleeding complications of oral anticoagulant treatment: an inception-cohort, prospective collaborative study (ISCOAT). Italian Study on Complications of Oral Anticoagulant Therapy. *Lancet*. 1996;348:423–8.

147. Andras A, Sala Tenna A, and Crawford F. Vitamin K antagonists or low-molecular-weight heparin for the long term treatment of symptomatic venous thromboembolism. *Cochrane Database Syst Rev*. 2012 Oct 17;10:CD002001.

148. Enden T, Kløw NE, Sandvik L, Slagsvold CE, Ghanima W, Hafsahl G et al. Catheter-directed thrombolysis vs. anticoagulant therapy alone in deep vein thrombosis: results of an open randomized, controlled trial reporting on short-term patency. *J Thromb Haemost*. 2009 Aug;7(8):1268–75.

149. Mewissen MW, Seabrook GR, Meissner MH, Cynamon J, Labropoulos N, and Haughton SH. Catheter-directed thrombolysis for lower extremity deep venous thrombosis: report of a national multicenter registry. *Radiology*. 1999 Apr;211(1):39–49.

150. Bi F, Zhang J, Su Y, Tang YC, and Liu JN. Chemical conjugation of urokinase to magnetic nanoparticles for targeted thrombolysis. *Biomaterials*. 2009;30:5125–30.

151. Ma YH, Wu SY, Wu T, Chang YJ, Hua MY, and Chen JP. Magnetically targeted thrombolysis with recombinant tissue plasminogen activator bound to polyacrylic acid-coated nanoparticles. *Biomaterials*. 2009;30:3343–51.

152. McCarthy JR, Sazonova IY, Erdem SS, Hara T, Thompson BD, Patel P et al. Multifunctional nanoagent for thrombus-targeted fibrinolytic therapy. *Nanomedicine (Lond)*. 2012 Jul;7(7):1017.

153. Magno, G. *The healing hand: man and wound in the ancient world*. Harvard University Press, Cambridge, MA, 1975.

154. Scott TE, LaMorte WW, Gorin DR, and Menzoian JO. Risk factors for chronic venous insufficiency: a dual casecontrol study. *J Vasc Surg*. 1995;22(5):622–8.

155. Michael H. Criqui MH, Denenberg JO, Bergan J, Langer, RD, and Fronek A. Risk factors for chronic venous disease: the San Diego population study. *J Vasc Surg*. 2007;46:331–7.

156. Brand FN, Dannenberg AL, Abbott RD, and Kannel WB. The epidemiology of varicose veins: the Framingham Study. *Am J Prev Med*. 1988;4:96–101.

157. Hareendran A, Bradbury A, Budd J, Geroulakos G, Hobbs R, Kenkre J, and Symonds T. Measuring the impact of venous leg ulcers on quality of life. *J Wound Care*. 2005 Feb;14(2):53–7.

158. Rabe E and Pannier F. Societal costs of chronic venous disease in CEAP C4, C5, C6 disease. *Phlebology*. 2010;25 (suppl 1):64–7.

159. Cullum N, Nelson E, Fletcher A, and Sheldon T. Compression for venous leg ulcers (review). *Cochrane Database Syst Rev*. 2001;(2):CD000265.

160. Seshadri Raju S and Neglen P. High prevalence of nonthrombotic iliac vein lesions in chronic venous disease: a permissive role in pathogenicity. *J Vasc Surg*. 2006;44:136–44.

161. Kahn SR, Shbaklo H, Lamping DL, Holcroft CA, Shrier I, Miron MJ et al. Determinants of health-related quality of life during the 2 years following deep vein thrombosis. *J Thromb Haemost*. 2008;6(7):1105.

162. Kahn SR, Shrier I, Julian JA, Ducruet T, Arsenault L, and Miron MJ. Determinants and time course of the postthrombotic syndrome after acute deep venous thrombosis. *Ann Intern Med*. 2008;149:698–707.

163. Pappas PJ, DeFouw DO, Venezio LM, Gorti R, Padberg FT Jr, Silva MB, Goldberg MC, Durfin WN, and Hobson RW. Morphometric assessment of the dermal microcirculation in patients with chronic venous insufficiency. *J Vasc Surg*. 1997;26:784–95.

164. Phillips II LJ and Rajabrata Sarkar R. Molecular characterization of post-thrombotic syndrome. *J Vasc Surg.* 2007;45:116A–22A.

165. Porter JM, Moneta GL, and International Consensus Committee on Chronic Venous Disease. Reporting standards in venous disease: an update. *J Vasc Surg.* 1995;21:635–45.

166. Eklof B, Perrin M, Delis KT, Rutherford RB, and Gloviczki P. Updated terminology of chronic venous disorders: The VEIN-TERM transatlantic interdisciplinary consensus document. *J Vasc Surg.* 2009;49:498–501.

167. Rautio T, Ohinmaa A. Perala J, Ohtonen P, Heikkinen T, Wiik H, Karjalainen P, Haukipuro K, and Juvonen T. Endovenous obliteration versus conventional stripping operation in the treatment of primary varicose veins: a randomized controlled trial with comparison of the costs. *J Vasc Surg.* 2002;35:958.

168. Lurie F, Creton D, Eklof B, Kabnick LS, Kistner RL, Pichot O, Sessa C, and Schuller-Petrovic S. Prospective randomised study of endovenous radiofrequency ablation (closure) versus ligation and vein stripping (EVOLVeS): 2 year follow-up. *Eur J Vasc Endovasc Surg.* 2005;29:67.

169. Raju S, Darcey R, and Neglen P. Unexpected major role for venous stenting in deep reflux disease. *J Vasc Surg.* 2010;51:401–9.

170. Ascher E. *Haimovici's vascular surgery 6th Edition.* Blackwell Publishing, John Wiley & Sons, West Sussex, UK, 2012.

171. Caves JM and Chaikof EL. The evolving impact of microfabrication and nanotechnology on stent design. *J Vasc Surg.* 2006:44(6):1363–8.

172. De Vries SO, Visser K, de Vries JA, Wong JB, Donaldson MC, and Hunink MG. Intermittent claudication: cost-effectiveness of revascularization versus exercise therapy. *Radiology.* 2002;222:25–36.

173. Golomb BA, Dang TT, and Criqui MH. Peripheral arterial disease: morbidity and mortality implications. *Circulation.* 2006; 114:688–99.

174. Laiho MK, Oinonen A, Sugano N, Harjola VP, Lehtola AL, Roth WD, Keto PE, and Lepäntalo M. Preservation of venous valve function after catheter-directed and systemic thrombolysis for deep venous thrombosis. *Eur J Vasc Endovasc Surg.* 2004 Oct;28(4):391–6.

175. Lurie F, Kistner RL, Eklof B, and Tsukamoto JK. Prevention of air travel-related deep venous thrombosis with mechanical devices: active foot movements produce similar hemodynamic effects. *J Vasc Surg.* 2006 Oct;44(4):889–91.

176. Mehta M, Byrne J, Darling RC III, Paty PSK, roddy SP, Kreienberg PB, Taggert JB, and Feustel P. Endovascular repair of ruptured infrarenal abdominal aortic aneurysms is associated with lower 30-day mortality and better 5-year survival rates than open surgical repair. *J Vasc Surg.* 2013 Feb;57(2):368–75.

177. Norgren L, Hiatt WR, Dormandy JA, Nehler MR, Harris KA, and Fowkes FG. Trans-Atlantic Inter-Society Consensus for the management of peripheral arterial disease (TASC II). *J Vasc Surg.* 2007;45 (suppl S):S5–67.

178. Sachs T, Pomposelli F, Hamdan A, Wyers M, and Schermerhorn M. *J Vasc Surg.* 2011;54(4):1021–31.

179. Weaver FA, Comerota AJ, Youngblood M, Froehlich J, Hosking JD, and Papanicolaou G. Surgical revascularization versus thrombolysis for nonembolic lower extremity native artery occlusions: Results of a prospective randomized trial. *J Vasc Surg.* 1996;24(4):513–23.

7

Therapeutic Applications and Targeted Delivery of Nanomedicines and Nanopharmaceutical Products

Heidi M. Mansour, PhD and Chun-Woong Park, PhD

CONTENTS

7.1 Introduction: Nanomedicines and Their Delivery Systems

In past decades, there have been remarkable achievements with nanotechnology in the research and development fields of chemical engineering,[1] biotechnology,[2] and medicinal sciences.[3] A size ranging from 1 to 1000 nm has frequently been used in both the nanotechnology science and pharmaceutical science because of its important meaning in biopharmaceutical aspects.[4–7]

The European Science Foundation (ESF) defined nanomedicines as "the science and technology of diagnosing, treating, and preventing disease and traumatic injury, of relieving pain, and of preserving and improving human health, using molecular tools and molecular knowledge of the human body."[8] With this comprehensive definition, five main categories of nanomedicines had been established by ESF as "analytical tools"; "nanoimaging"; "nanomaterials and nanodevices"; "novel therapeutics and drug delivery systems"; and "clinical, regulatory, and toxicological issues."[9] In nanomedicine, as "an offshoot of nanotechnology, refers to highly specific medical interventions at the molecular scale for curing disease or repairing damaged tissues, such as bone, muscle, or nerve." In summary, nanomedicines have recently intensive investigations in the fields of treatment, diagnosis, monitoring, and control of biological systems.[10,11]

Nanomedicine delivery systems can have their essentials as targeted drug delivery at the site of disease, the improved uptake of poorly soluble drugs, enhancement of drug bioavailability, reduced drug toxicity, and controlled release property of the drug.[12–17] Considering the nanocarrier design, nanomedicines can be defined as two categories of nanocarriers (i.e., "hard type" and "soft type").[18] Hard type of nanomedicines, comprising polymeric nanoparticles[19–22] and lipid nanoparticles (LNs),[23–25] have specific physicochemical properties with low flexibility and elasticity. In contrast, soft type of nanomedicines can have more free deformation and reformation by external or internal stress, compared with "hard type."[26] There are a few cases of "soft type" nanomedicines (i.e., liposomes,[27–31] nanoemulsions,[27,32,33] submicron lipid emulsions,[23,24,34–37] nanogels,[25,28,29] and polymeric micelles[30,31,38]).

In summary, the ideal features of nanomedicines are to achieve medical needs using nanobiotechnology (i.e., to increase drug accumulation in target site, to protect biological unstable drugs against potential enzymatic or hydrolytic degradation in the body, and to confirm the low toxicity with biocompatible and biodegradable nanomaterials, to optimize therapeutical effects with high drug loading capacity, extended circulation time, controlled drug release profiles, and the ability to efficiently carry poorly soluble pharmaceuticals).[39] The chemical structures and therapeutic areas of selected example drugs used in approved active agents for the approved marketed nanomedicine delivery systems are shown in Figure 7.1. Table 7.1 shows examples of therapeutic drug categories, diseases states, and marketed nanomedicine products used.

7.2 Administration Routes of Nanomedicines

7.2.1 Injectible Administration

The parenteral administration route of nanomedicines enables direct access to the systemic circulation. Injectible delivery systems include intramuscular (IM), intravenous (IV), and subcutaneous (SC) routes. Tables 7.2 and 7.3 list the therapeutic areas, drug/nanotechnology, and approved marketed nanomedicine-injectible products.

7.2.2 Oral Administration

The oral route is a noninvasive route with convenience and simplicity. Table 7.4 lists the therapeutic areas, drug/nanotechnology, and approved marketed nanomedicine products for oral administration.

Moreover, for the oral administration, there are promising effects of nanomedicines, such as the enhanced water solubility and increasing bioavailability of hydrophobic drugs.[26,40–43]

NanoCrystal® technology (ELAN Cooperation, Ireland) is the most successful technology for the approved marketed nanomedicines for oral application with the remarkable advantages as a commercially confirmed technology for new chemical entities with poor water solubility.[44] Furthermore, this technology comprises selected generally regarded as safe (GRAS) stabilizers.[45]

Therapeutic application	Drug	Chemical structure
Anticancer/ Chemotherapeutic/ Oncology	Cytarabine	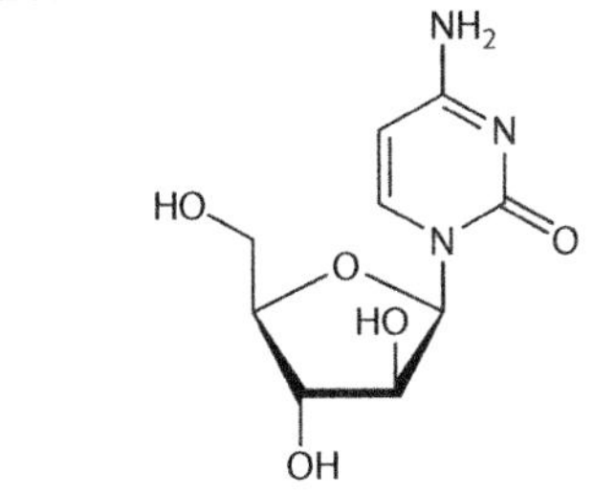
	Daunorubicin	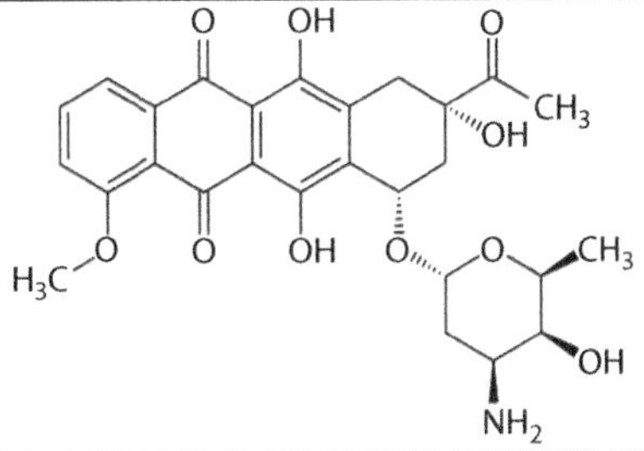
Immunosuppression/ Transplantation/ Auto-immune diseases	Cyclosporin A	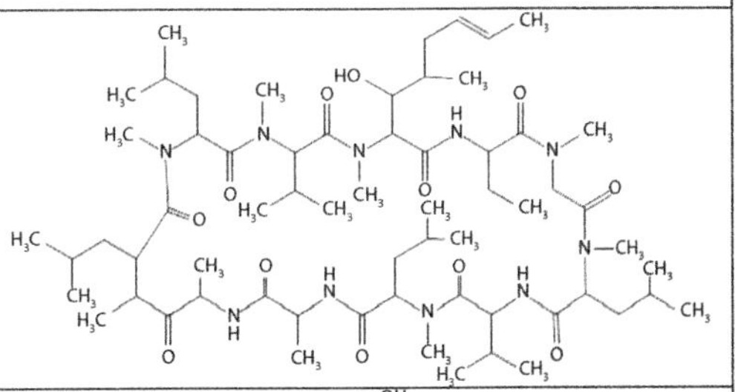
Antifungal/ Anti-infective/ Infectious diseases	Amphotericin B	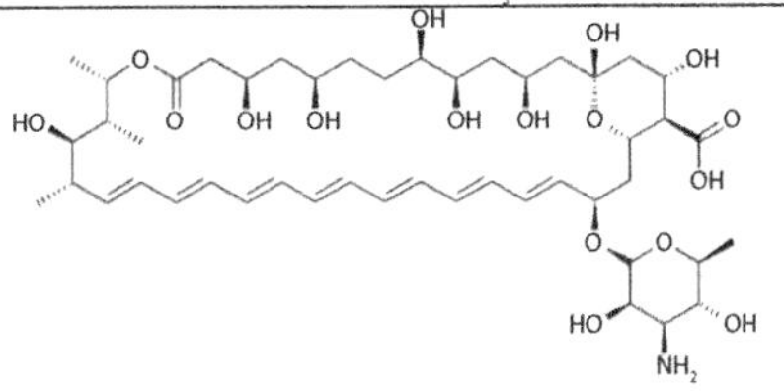

FIGURE 7.1

Chemical structures and therapeutic applications of selected example drugs used in approved marketed nanomedicines.

7.2.3 Pulmonary Administration

The lung is an alternative route for nanomedicine delivery systems due to avoidance of first-pass metabolism, fast onset of therapeutic action, and the availability of huge surface area. Furthermore pulmonary delivery can be noninvasive.[46] Nanomedicines have many advantages for pulmonary drug delivery.[47] Table 7.5 lists the therapeutic areas, drug/ nanotechnology, and approved marketed nanomedicine products for pulmonary administration by intratracheal instillation.

7.2.4 Transdermal and Dermal Administration

Phospholipids and cholesterol are innate to skin composition and can form bilayers, as seen in skin.

TABLE 7.1

Examples of Disease States Treated by FDA-Approved Nanomedicine Marketed Products

Therapeutic Drug Categories	Disease State	FDA-Approved Nanomedicine/ Nanotechnology
Cancer chemotherapeutic agent	Kaposi's sarcoma	DaunoXome®/encapsulated liposomes
		Doxil® (United States); Caelyx(others)/PEGylated liposomes
	Lymphomatous meningitis	DepoCyt®/sustained-release liposomes
	Metastatic breast cancer	Myocet®/liposome-encapsulated complex
		Abraxane®/Albumin bound nanoparticles
	Metastatic ovarian cancer	Doxil® (United States); Caelyx(others)/PEGylated liposomes
	Multiple myeloma	Doxil (United States); Caelyx(others)/PEGylated liposomes
Anti-fungal agent	*Aspergillus* species, *Candida* species and/or *Cryptococcus* species infections	AmBisome®/liposomes Abelcet®/phospholipid complex
	Invasive aspergillosis	Amphotec®/colloidal suspension of lipid-based formulation
	Cryptococcal meningitis in HIV-infection	AmBisome/liposomes Abelcet/phospholipid complex
Immunosuppressant	Inflammation due to chronic dry eye	Restasis®/lipid emulsion
	Organ rejection after a kidney, liver, or heart transplant	Neoral®/Self-emulsifying drug delivery systems RAPAMUNE®/nanocrystal
	Severe psoriasis or severe rheumatoid arthritis	Neoral/self-emulsifying drug delivery systems

LNs (solid lipid nanoparticles [SLNs] and nanostructure lipid carriers [NLCs]) have been used as a common nanocarrier in transdermal/dermal delivery.[48,49]

Submicron emulsions are one of the focused topical/transdermal drug delivery system for nanomedicines.[50] Table 7.6 lists the therapeutic areas, drug/nanotechnology, and approved marketed nanomedicine products for transdermal and dermal administration.

7.2.5 Ocular Administration

Nanocarrier systems represent promising drug carriers for ocular drug delivery. Main clinical and medical needs in ocular drug delivery are to increase drug bioavailability and to prolong the drug residence time on the cornea, conjunctiva, and corneal epithelia.[26] The size of nanomedicines can be suitable for ocular administration. Moreover, nanomedicines as a nanosized colloidal formulation can be administered in liquid form (i.e., nanosuspension and nanoemulsion), which can be applied as eye-drop solutions. Nanomedicines have controlled-release properties and can facilitate bioavailability of drugs by increasing drug residence time in the eye or by protecting drugs against enzymatic inactivation.[51–54]

TABLE 7.2

Examples of Approved Marketed Injectable Nanomedicines by Intravenous (IV) Route of Administration

Administration Route	Therapeutic Application	Drug/Nanotechnology	Brand Name/Company
IV	Oncology/advanced HIV-related Kaposi's sarcoma	Daunorubicin citrate/encapsulated liposomes	DaunoXome®/Gilead Sciences
	Oncology/metastatic ovarian cancer and AIDS-related Kaposi's sarcoma	Doxorubicin/PEGylated liposomes	Doxil® (United States), Caelyx (others)/OrthoBiotech Schering-Plough
	Oncology/late-stage metastatic breast cancer	Doxorubicin/liposome-encapsulated complex	Myocet®/Zeneus Pharma
	Oncology/cancers	Paclitaxel/albumin-bound nanoparticles	Abraxane®/Abraxis BioScience AstraZeneca
	Oncology/cancers	Paclitaxel/polymeric micelles	Genesol-PM®/Samyang
	Enzyme replacement therapy	Adenosine deaminase/PEGylation	Adagen®/Enzon
	Infectious disease/fungal infections	Amphotericin B/liposomes	AmBisome®/NeXstar Pharmaceuticals Inc.
	Infectious disease/invasive fungal infections	Amphotericin B/phospholipid complex	Abelcet®/Enzon
	Imaging/organ-specific MRI contrast agent	Iron oxide/ SPIO nanoparticles coated with carboxydextran	Resovist®/Bayer-Schering Pharma AG
	Imaging/MRI contrast agent (liver lesions)	Iron oxide/SPIO nanoparticles	Feridex IV®/AMAG Pharmaceuticals, Inc.
	Anesthetic	Propofol/lipid emulsion	Diprivan®/AstraZeneca

Abbreviations: PEG, polyethylene glycol; SPIO, superparamagnetic iron oxide; MRI, magnetic resonance imaging.

TABLE 7.3

Examples of Approved Marketed Injectable Nanomedicines for Subcutaneous (SC or SQ), Intramuscular (IM), and Intrathecal (IT) Routes of Administration

Administration Route	Therapeutic Applications	Drug/Nanotechnology	Brand Name/Company
SC or SQ	Acromegaly	hGH/PEGylation	Somavert®/Nektar Pfizer
	Chronic hepatitis C virus infection	Interferon alfa-2a/PEGylation	Pegasys®/Nektar Hoffmann-La Roche
	Chronic hepatitis C virus infection	Interferon alfa-2b/PEGylation	PEGIntron®/Enzon Schering-Plough
	Febrile neutropenia	Recombinant methionyl human G-CSF/PEGylation	Neulasta®/Amgen
	Invasive aspergillosis	Amphotericin B/colloidal suspension of lipid-based formulation	Amphotec®/Sequus
	Leukemia	Asparginase/PEGylation	Oncaspar®/Enzon
	Relapsing-remitting multiple sclerosis	Glatiramer acetate/copolymer of L-glutamic acid, L-alanine, L-tyrosine, and L-lysine	Copaxone®/TEVA
IT	Leukemias, lymphomas, meningeal neoplasms	Cytarabine/sustained release liposomes	DepoCyt®/SkyePharma Enzon
IM	Active immunization (Hepatitis A)	Hepatitis A vaccine/virosomes	Epaxal®/Berna Biotech

Abbreviations: PEG, polyethylene glycol; hGH, human growth hormone; G-CSF, granulocyte colony-stimulating factor.

TABLE 7.4

Examples of Approved Marketed Nanomedicines for Oral Route of Administration (PO)

Administration Route	Therapeutic Application	Drug/Nanotechnology	Brand Name/Company
PO	Antianorexic	Megestrol acetate/nanocrystal	Megace® ES/Par Pharma, Elan
	Attention deficit hyperactivity	Dexmethyl-phenidate HCl/Nanocrystal	Focalin® XR/Novartis, Elan
	Attention deficit hyperactivity	Methylphenidate HCl/Nanocrystal	Ritalin® LA/Novartis, Elan
	Antiretroviral	Ritonavir/SMEDDS	Norvir®/Abbott Labs
	Antiviral	Saquinavir/SMEDDS	Forovase®/Roche
	Immunosuppressant	Cyclosporine A/SMEDDS	Neoral®/Norvatis
	Immunosuppressant for kidney transplants	Sirolimus/nanocrystal	Rapamune®/Wyeth, Elan
	Lipid disorders	Fenofibrate/nanocrystal	Triglide®/SkyePharma First Horizon
	Muscle relaxant	Tizanidine HCl/nanocrystal	Zanaflex® Capsules/Acorda Inc., Elan
	Nausea in chemotherapy patients	Aprepitant/nanocrystal	Emend®/Merck, Elan
	Primary hypercholesteremia	Fenofibrate/nanocrystal	TriCor®/Abbott
	Psychostimulantants	Morphine sulfate/nanocrystal	Avinza®/King Pharma, Elan

Abbreviation: SMEDDS, self-microemulsifying drug delivery systems.

TABLE 7.5

Examples of Approved Marketed Nanomedicines for the Pulmonary Route of Administration by Intratracheal (Endotracheal) Instillation

Administration Route	Therapeutic Application	Drug/Nanotechnology	Brand Name/Company
Pulmonary	Respiratory distress syndrome (RDS)	Lung surfactant-bovine extract/phospholipid	Alveofact®/Boehringer Ingelheim
		Lung surfactant-porcine extract/phospholipid	Curosurf®/Dey
		Lung surfactant-synthetic (lipid only)/phospholipid	Exosurf®/ GlaxoSmithKline
		Lung surfactant-synthetic (recombinant human polypeptide)/phospholipid	Surfaxin®/Drug Discovery Labs
		Lung surfactant-bovine extract/phospholipid	Survanta®/Abbott Labs

TABLE 7.6

Examples of Approved Marketed Nanomedicines by Transdermal/Dermal Route of Administration

Administration Route	Therapeutic Application	Drug/Nanotechnology	Brand Name/Company
Transdermal/Dermal	Local anesthesia	Lidocaine/liposomal topical delivery	LMX®-4/Ferndale Labs
	Endocrinology (female)	Estradiol/micellar nanoparticles	Estrasorb®/Novavax
	Endocrinology/ menopause (female)	Estradiol/calcium phosphate nanoparticles	Elestrin®/BioSanté

TABLE 7.7

Examples of Approved Marketed Nanomedicines for Ocular Drug Delivery by Ophthalmic Route of Administration

Administration Route	Therapeutic Application	Drug/Nanotechnology	Brand Name/Company
Ophthalmic (non-invasive/topical)	Anti-inflammatory/ corticosteroid	Difluprednate/lipid emulsion	Durezol®/Siron Therapeutics
	Immunosuppressant	Cyclosporin A/lipid emulsion	Restasis® Allergan
Ophthalmic (invasive Intravitreal)	Neovascular age-related macular degeneration	Anti-VEGF aptamers/ PEGylated aptamers	Macugen®/OSI Pharmaceuticals, Pfizer

Abbreviations: PEG, polyethylene glycol; VEGF, vascular endothelial growth factor.

Liposomal formulations can increase the potential of ocular drug absorption.[54] Recently, there have been several studies in ocular nanomedicine delivery using commonly used polymers (i.e., poly[lactic-*co*-glycolic acid] [PLGA],[55] poly [ε-caprolactone] [PCL],[56] and poly[alkylcyanoacrylate]),[57] and natural polymers (i.e., chitosan, sodium alginate,[58] and albumin[59]). Table 7.7 lists the therapeutic areas, drug/nanotechnology, and approved marketed nanomedicine products for ocular administration.

7.3 Nanomedicine Delivery Systems

7.3.1 Biomaterials for the Nanomedicine Delivery Systems

7.3.1.1 Polymers

For the pharmaceutical applications, polymers should have both biodegradability and biocompatibility into biological condition *in vivo*. Biodegradable property undergoes two main steps[60]: (1) water penetration into polymeric matrix, for the breakdown of the chemical bonds by hydrolysis, following as decrease of chain length and molecular weight of the polymer and (2) surface erosion of the polymer when increasing the rate of conversion of the polymer into water soluble-structures. Biocompatibility can be defined as a specific property of nontoxicity or noninjurious effects on biological systems.

Polyester-based polymers (Figure 7.2a) are commonly used in nanomedicine delivery systems, with biodegradability and biocompatibility and can provide controlled drug release from weeks to months. PLGA is present in several commercially available pharmaceutical products.[61] PLGA heteropolymer degrades relatively faster than PLA which releases drugs over months.[62] Degradation and release rate of PLGA-based systems can be affected by molecular weight, size and size distribution of systems, and the lactide/glycolide ratio of PLGA.[63]

PEG is well known and intensively researched for its "stealth" properties, which prolongs circulation times. Polyvinyl alcohol (PVA) has been used in controlled-release nanomedicine delivery systems, such as gene delivery,[64] pulmonary delivery,[65] parenteral delivery,[66] and transdermal and dermal delivery.[67]

Chitosan (Figure 7.2b) is a type of polyaminosaccharide, derived from chitin.[68] There are strong intramolecular hydrogen bonding between hydroxyl and amino groups in the

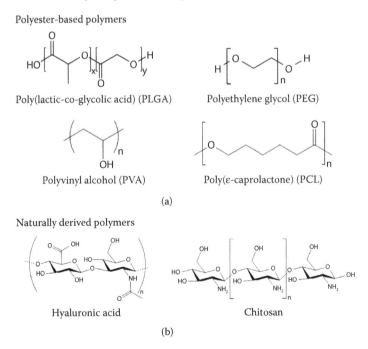

FIGURE 7.2
Chemical structures of selected polymers for the biomaterials of nanomedicines. (a) Polyester-based polymers and (b) naturally derived polymers.

molecule of chitosan, which make it stable at elevated temperatures. Alginate is an anionic block copolymer and can form a stimuli-responsive hydrogel for nanomedicine delivery systems.[69] Hyaluronic acid is a major carbohydrate component of the extracellular matrix found in synovial fluids and on cartilage surfaces,[70] and is used in gene delivery because of its non-immunological properties.[70]

7.3.1.2 Lipids

Lipids as biomaterials in nanotechnology are attractive carriers for nanomedicines, and possess the potential to develop as the new generation of drug delivery systems for the organ, tissue, and cellular level of human body in the future. Liquid lipids usually show a higher solubility for drugs than solid lipids, due to a higher drug-loading capacity, lower drug expulsion during storage, and higher level of modulation of the drug release profiles.[71]

The amounts and kinds of phospholipids for liposomal nanomedicines is related to the physical stability of the liposome formulation both *in vitro* and *in vivo* and the drug to be loaded into the liposomes.[72] The chemical structures of lipids and phospholipids are shown in Figure 7.3.

7.3.2 Multifunction of Nanomedicine Delivery Systems

7.3.2.1 Long-Circulating Nanomedicine Delivery Systems

Long-circulating nanomedicine delivery systems have been studied by coating the nanomedicines with the inert, biocompatible polymers, such as PEG, which form a protective layer over the surface and slow down their recognition by opsonins.[73–75] Such PEG-coated nanomedicines (i.e., PEG-coated liposomes) are also called "stealth" or "sterically stabilized" liposomes.[76] One of the important aspects of protective polymers is their flexibility to create an impermeable layer over the surface with a relatively small number of surface-grafted polymer molecules.[77,78] DOXIL® has been a successful FDA-approved marketed nanomedicine of doxorubicin incorporated into long-circulating PEGylated liposomes (Stealth®). PEG surface-coated liposomes are capable of long circulation enough to accumulate in tumors, as tumors have compromised or leaky vasculature.

PEG is an amphiphilic polymer composed of repeating ethylene oxide subunits and can dissolve in organic solvents as well as in water. The physicochemical and biopharmaceutical properties of PEG particularly relevant to pharmaceutical applications are (1) long circulation time due to lack of renal or cellular clearance mechanisms, (2) reduced antigenicity and escape from phagocytosis and proteolysis, (3) increased solubility and stability, and (4) reduced dosage frequency and low toxicity.[79] PEG degradation is related to the molecular weight and concentration of PEG. The degradation mechanism based on the strong hydrophilicity of PEG along with the hydrogen-bonding interaction between PEG and water molecules.[79]

7.3.2.2 Targeted Nanomedicine Delivery Systems

Targeted delivery systems have spatial control in delivery to specific organs, tissues, and/or cell types. Characteristic advantages include enhanced therapeutic efficacy and reduced side-effects. Targeted delivery systems include pulmonary delivery by inhalation aerosols to specific lung regions based on aerodynamic deposition, transdermal delivery, ocular delivery, vaginal delivery, rectal delivery, and so on. Targeted nanomedicine delivery

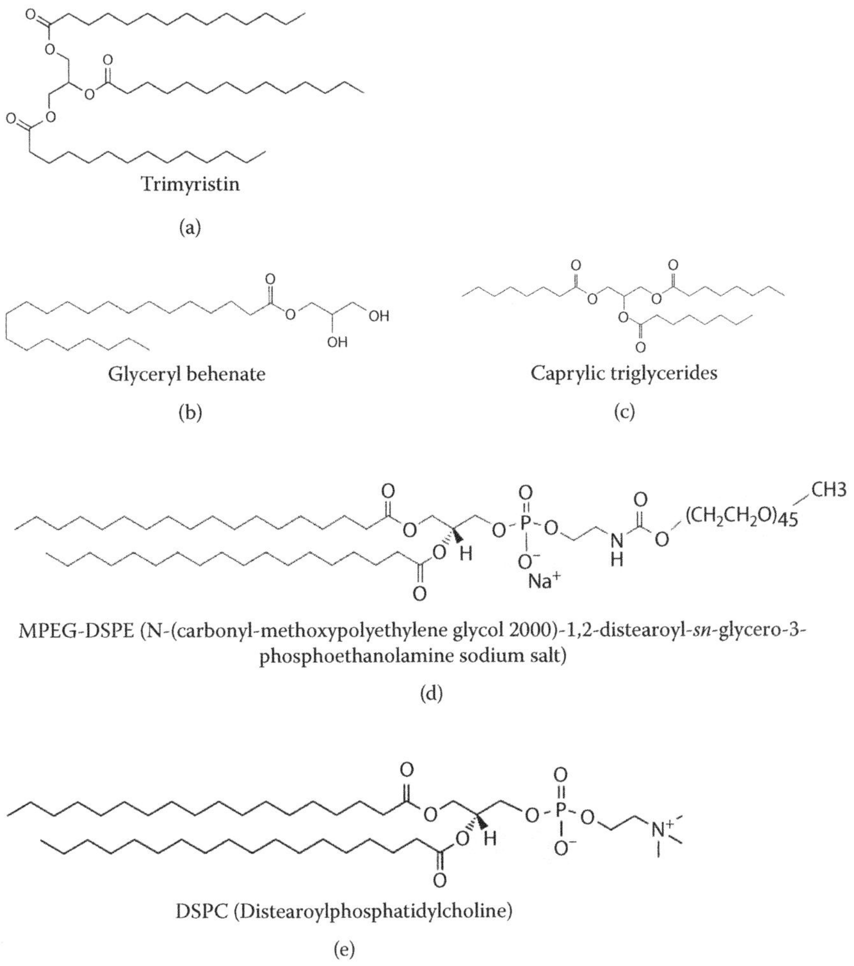

Trimyristin

(a)

Glyceryl behenate

(b)

Caprylic triglycerides

(c)

MPEG-DSPE (N-(carbonyl-methoxypolyethylene glycol 2000)-1,2-distearoyl-*sn*-glycero-3-phosphoethanolamine sodium salt)

(d)

DSPC (Distearoylphosphatidylcholine)

(e)

FIGURE 7.3
Chemical structures of selected lipids and phospholipids for the biomaterials of nanomedicines. Lipids (a-c), pegylated phospholipid (d), and saturated phospholipid (e).

systems can have site-specific moieties (i.e., antibodies, peptides, sugar derivates, folates, and other ligands to the target organs and tissues, especially cancer tissues).[80]

7.3.2.3 Stimuli-Responsive Nanomedicine Delivery Systems

Although passive and active targeted drug delivery has a main potential to enhance the bioavailability of drugs at the disease site, and especially on cellular internalization, additional properties can be included in nanomedicine delivery systems, incorporated with stimuli-responsive property. With the advancement in material science, development of nanocarrier systems can be designed to respond to biological stimuli such as temperature, pH, and hypoxia as triggers at the targeted disease site. For the successful design of stimuli-responsive nanomedicine delivery systems, it should be based on the understanding of the difference between normal and pathological tissues and cells and the biotechnological application to material science.[81] Stimuli-responsive drug delivery systems can be prepared by including labile linkages to the particle components.[82]

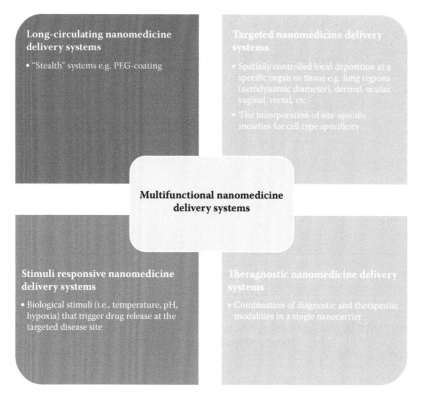

FIGURE 7.4
Illustrative summary of multifunctional nanomedicine delivery systems.

7.3.2.4 Theragnostic Nanomedicine Delivery Systems

For the comprehensive understanding of theragnostic nanomedicines, the basic principle of nanocarriers with attached contrast agents need to be understood. Contrast agents can absorb energy more readily than surrounding environment such as normal biological tissues. SPIO (superparamagnetic iron oxide) nanoparticle has been intensively investigated for MRI (magnetic resonance imaging).[83–86] Table 7.2 lists some of the marketed nanomedicine products using SPIO nanoparticles for IV administration. There have been efforts to prepare nanomedicine delivery systems that combine diagnostic and therapeutic modalities in a single nanocarrier.[87]

As shown by Figure 7.4, "smart" nanomedicine delivery systems are being designed with multifunctionality. These systems are included into a single nanocarrier as long-circulating, active targeted, stimuli responsive, and theragnostic nanomedicine delivery systems.

7.4 Product Development and Clinical Trials

Nanomedicine delivery systems have been used with remarkable advantages for reduced toxicity, enhanced bioavailability, physicochemical and biological stabilization, and enhancement of cellular uptake into tissues of interest. In addition to these advantages, there are FDA-approved marketed nanomedicines in some but not all disease-state categories (Table 7.1) because of the high level of hurdles in translational science (i.e., regulation and scale-up). Tables 7.2 through 7.7 list the marketed nanomedicine products for different administration routes along with the disease state and the nanotechnology used. Examples of nanomedicines in clinical trials are listed in Table 7.8.

TABLE 7.8

Examples of Nanomedicines in Clinical Trials

Therapeutic Area	Drug/Nanotechnology	Company	Status
Age-related macular degeneration	2′-ribo purine and 2′-fluoro pyrimidine/PEGylated aptamers[a]	Ophthotech	Phase I
Age-related macular degeneration	DNA/PEGylated aptamers[a]	Ophthotech	Phase I
Anticoagulation	DNA/PEGylated aptamers[a]	ARCA biopharma	Phase II
Anticancer	Thymectacin/nanocrystal[b]	Elan	Phase I/II
Anticancer	Vincristine/liposomes[a]	Enzon	Phase II
Anticancer	Lurtotecan/liposomes[a]	OSI	Phase II
Anticancer	Platinum compounds/liposomes[a]	Aronex Pharma	Phase I/II
Anticancer	siRNA/cyclodextrin-siRNA nanoparticles[a]	Calando Pharma	Phase I
Acute myeloid leukemia	DNA/G-rich DNA aptamers[a]	Antisoma	Phase II
Lung cancer	INGN-401/nanoparticles[a]	Introgen	Phase I
Multiple myeloma	l-RNA/ PEGylated aptamers[a]	NOXXON Pharma	Phase I
Ovarian cancer, prostate cancer	2-methoxy estradiol/nanocrystal[b]	Elan	Phase II
Recurrant glioblatoma	Undisclosed/nanoedge[a]	Baxter	Phase II
Solid tumor	Gold compounds/colloidal gold-TNF conjugates[a]	Cytimmune Science	Phase I
Solid tumor	p53 gene/liposomes[a]	SynerGene Therapeutics	Phase I
Crohn's disease	Cytokine inhibitor/nanocrystal[b]	Elan	Phase II
Cystic fibrosis	Amikacin/sustained-release liposomes[a]	Transave Inc.	Phase I/II
Detection of blood pooling using MRA	Iron oxide/carboxy dextran-coated ultrasmall SPIO[a]	Bayer Schering	Phase III
Detection of blood pooling using MRA	Iron oxide/citrate-coated SPIO[a]	Charite -Universitatsmedizin Berlin	Phase I
Differentiation of cancerous from noncancerous lymph nodes	Iron oxide/dextran-coated ultrasmall SPIO[a]	AMAG Pharma	Phase III
Nervous system disease, brain neoplasms, peripheral artery	Iron oxide/polyglucose sorbitol carboxymethyl ether-coated SPIO[a]	AMAG Pharma	Phase II
Tumor imaging	Iron oxide/SPIO nanoparticles[a]	Advanced Magnetics	Phase III
Local antifungal	Nystatin/liposomes[c]	Aronex Pharma	Phase III
Percutaneous coronary intervention	2′-Ribo purine and 2′-fluoro pyrimidine/PEGylated aptamers[a]	Regado Biosciences	Phase II
Thrombotic microangiopathies	DNA/PEGylated aptamers[a]	Archemix	Phase II
Type 2 diabetes	l-RNA/PEGylated aptamers[a]	NOXXON Pharma	Phase I
Vaginal microbicide	SPL7013/dendrimer gel[c]	Star Pharma	Phase II

[a] Parenteral administration.
[b] Oral administration.
[c] Topical administration.

7.5 Conclusions and Future Directions

Currently, there is an increasing number of successful nanomedicine products marketed for human use. Routes of administration for these available nanomedicine products have expanded in recent years to include the injectible route (i.e. IV, IM, SQ, and IT), oral route, ocular route, pulmonary route, and transdermal route. Disease states treated include various cancers, infectious diseases, transplants/immunosuppression, respiratory distress syndrome, ocular diseases, hormone replacement, and psychiatric disorders. In addition, nanotechnology is currently being used in marketed products for vaccination/immunization, pain, lipid disorders, inflammation, and imaging. There is an increasing number of human clinical trials involving nanomedicine delivery systems for various diseases. This area of growth is expected to continue for the next several years. This future growth will not only expand the nanomedicine product market but also provide improved therapies available for patients in the treatment of a wide array of disease states.

References

1. Farokhzad, O. C. and Langer, R. Impact of nanotechnology on drug delivery, *ACS Nano* 3 (1), 16–20, 2009.
2. Roco, M. and Bainbridge, W. S. Converging technologies for improving human performance: nanotechnology, biotechnology, information technology and cognitive science, *Computing Reviews* 45 (8), 480–481, 2004.
3. Kubik, T., Bogunia-Kubik, K., and Sugisaka, M. Nanotechnology on duty in medical applications, *Current Pharmaceutical Biotechnology* 6 (1), 17–33, 2005.
4. Brigger, I., Dubernet, C., and Couvreur, P. Nanoparticles in cancer therapy and diagnosis, *Advanced Drug Delivery Reviews* 54 (5), 631–651, 2002.
5. Tiwari, S. B. and Amiji, M. M. A review of nanocarrier-based CNS delivery systems, *Current Drug Delivery* 3 (2), 219–232, 2006.
6. Kaur, I. P., Bhandari, R., Bhandari, S., and Kakkar, V. Potential of solid lipid nanoparticles in brain targeting, *Journal of Controlled Release* 127 (2), 97–109, 2008.
7. Davis, M. E. Nanoparticle therapeutics: an emerging treatment modality for cancer, *Nature Reviews Drug Discovery* 7 (9), 771–782, 2008.
8. ESF, E. S. F. Nanomedicine: An ESF–European Medical Research Councils (EMRC) Forward Look Report. Strasbourg cedex, France, 2004.
9. Webster, T. J. Nanomedicine: what's in a definition? *International Journal of Nanomedicine* 1 (2), 115, 2006.
10. Park, J. H., Lee, S., Kim, J. H., Park, K., Kim, K., and Kwon, I. C. Polymeric nanomedicine for cancer therapy, *Progress in Polymer Science* 33 (1), 113–137, 2008.
11. Resnik, D. B. and Tinkle, S. S. Ethical issues in clinical trials involving nanomedicine, *Contemporary Clinical Trials* 28 (4), 433–441, 2007.
12. Koo, O. M., Rubinstein, I., and Onyuksel, H. Role of nanotechnology in targeted drug delivery and imaging: a concise review, *Nanomedicine: Nanotechnology, Biology and Medicine* 1 (3), 193–212, 2005.
13. Jain, K. Nanotechnology-based drug delivery for cancer, *Technology in Cancer Research & Treatment* 4 (4), 407, 2005.
14. Park, K. Nanotechnology: what it can do for drug delivery, *Journal of Controlled Release* 120 (1–2), 1, 2007.

15. Ould-Ouali, L., Noppe, M., Langlois, X., Willems, B., Te Riele, P., Timmerman, P., Brewster, M. E., Arien, A., and Preat, V. Self-assembling PEG-p (CL-co-TMC) copolymers for oral delivery of poorly water-soluble drugs: a case study with risperidone, *Journal of Controlled Release* 102 (3), 657–668, 2005.

16. Kipp, J. The role of solid nanoparticle technology in the parenteral delivery of poorly water-soluble drugs, *International Journal of Pharmaceutics* 284 (1–2), 109–122, 2004.

17. Suri, S. S., Fenniri, H., and Singh, B. Nanotechnology-based drug delivery systems, *Journal of Occupational Medicine and Toxicology* 2 (1), 16, 2007.

18. Florence, A. T. Pharmaceutical aspects of nanotechnology, *Modern Pharmaceutics* 2, 453–492, 2009.

19. Von Werne, T. and Patten, T. E. Preparation of structurally well-defined polymer-nanoparticle hybrids with controlled/living radical polymerizations, *Journal of the American Chemical Society* 121 (32), 7409–7410, 1999.

20. Pridgen, E. M., Langer, R., and Farokhzad, O. C. Biodegradable, polymeric nanoparticle delivery systems for cancer therapy, *Nanomedicine* 2 (5), 669–680, 2007.

21. Hussain, F., Hojjati, M., Okamoto, M., and Gorga, R. E. Review article: polymer-matrix nanocomposites, processing, manufacturing, and application: an overview, *Journal of Composite Materials* 40 (17), 1511, 2006.

22. van Vlerken, L. E. and Amiji, M. M. Multi-functional polymeric nanoparticles for tumour-targeted drug delivery, *Expert Opinion on Drug Delivery* 3 (2), 205–16, 2006.

23. Jamaty, C., Bailey, B., Larocque, A., Notebaert, E., Sanogo, K., and Chauny, J. M. Lipid emulsions in the treatment of acute poisoning: a systematic review of human and animal studies, *Clinical Toxicology* 48 (1), 1–27, 2010.

24. Pouton, C. W. Lipid formulations for oral administration of drugs: non-emulsifying, self-emulsifying and 'self-microemulsifying' drug delivery systems, *European Journal of Pharmaceutical Sciences* 11, S93–S98, 2000.

25. Vinogradov, S. V. Colloidal microgels in drug delivery applications, *Current Pharmaceutical Design* 12 (36), 4703, 2006.

26. Rhee, Y. S. and Mansour, H. M. Nanopharmaceuticals I: nanocarrier systems in drug delivery, *International Journal of Nanotechnology* 8 (1), 84–114, 2011.

27. Wu, W., Wang, Y., and Que, L. Enhanced bioavailability of silymarin by self-microemulsifying drug delivery system, *European Journal of Pharmaceutics and Biopharmaceutics* 63 (3), 288–294, 2006.

28. Vinogradov, S. V., Zeman, A. D., Batrakova, E. V., and Kabanov, A. V. Polyplex Nanogel formulations for drug delivery of cytotoxic nucleoside analogs, *Journal of Controlled Release* 107 (1), 143–157, 2005.

29. Raemdonck, K., Demeester, J., and De Smedt, S. Advanced nanogel engineering for drug delivery, *Soft Matter* 5 (4), 707–715, 2008.

30. Kataoka, K., Harada, A., and Nagasaki, Y. Block copolymer micelles for drug delivery: design, characterization and biological significance, *Advanced Drug Delivery Reviews* 47 (1), 113–131, 2001.

31. Nasongkla, N., Bey, E., Ren, J., Ai, H., Khemtong, C., Guthi, J. S., Chin, S. F., Sherry, A. D., Boothman, D. A., and Gao, J. Multifunctional polymeric micelles as cancer-targeted, MRI-ultrasensitive drug delivery systems, *Nano Letters* 6 (11), 2427–2430, 2006.

32. Lawrence, M. J. and Rees, G. D. Microemulsion-based media as novel drug delivery systems, *Advanced Drug Delivery Reviews* 45 (1), 89–121, 2000.

33. Jadhav, K., Shaikh, I., Ambade, K., and Kadam, V. Applications of microemulsion based drug delivery system, *Current Drug Delivery* 3 (3), 267–273, 2006.

34. Hashida, M., Kawakami, S., and Yamashita, F. Lipid carrier systems for targeted drug and gene delivery, *Chemical and Pharmaceutical Bulletin* 53 (8), 871–880, 2005.

35. Davis, S. S., Washington, C., West, P., Illum, L., Liversidge, G., Sternson, L., and Kirsh, R. Lipid emulsions as drug delivery systems, *Annals of the New York Academy of Sciences* 507 (1), 75–88, 1987.

36. Collins-Gold, L., Lyons, R., and Bartholow, L. Parenteral emulsions for drug delivery, *Advanced Drug Delivery Reviews* 5 (3), 189–208, 1990.
37. Charman, W. N. Lipids, lipophilic drugs, and oral drug delivery—some emerging concepts, *Journal of Pharmaceutical Sciences* 89 (8), 967–978, 2000.
38. Maeda, H., Bharate, G., and Daruwalla, J. Polymeric drugs for efficient tumor-targeted drug delivery based on EPR-effect, *European Journal of Pharmaceutics and Biopharmaceutics* 71 (3), 409–419, 2009.
39. Torchilin, V. P. Lipid-core micelles for targeted drug delivery, *Current Drug Delivery* 2 (4), 319–327, 2005.
40. Yoncheva, K., Guembe, L., Campanero, M., and Irache, J. Evaluation of bioadhesive potential and intestinal transport of pegylated poly (anhydride) nanoparticles, *International Journal of Pharmaceutics* 334 (1–2), 156–165, 2007.
41. Desai, M. P., Labhasetwar, V., Amidon, G. L., and Levy, R. J. Gastrointestinal uptake of biodegradable microparticles: effect of particle size, *Pharmaceutical Research* 13 (12), 1838–1845, 1996.
42. Foger, F., Hoyer, H., Kafedjiiski, K., Thaurer, M., and Bernkop-Schnurch, A. In vivo comparison of various polymeric and low molecular mass inhibitors of intestinal P-glycoprotein, *Biomaterials* 27 (34), 5855–5860, 2006.
43. Francis, M. F., Cristea, M., and Winnik, F. M. Exploiting the vitamin B_{12} pathway to enhance oral drug delivery via polymeric micelles, *Biomacromolecules* 6 (5), 2462–2467, 2005.
44. Bawa, R. Nanopharmaceuticals, *European Journal of Nanomedicine* 3 (1), 34–39, 2010.
45. Junghanns, J. U. A. H. and Muller, R. H. Nanocrystal technology, drug delivery and clinical applications, *International Journal of Nanomedicine* 3 (3), 295, 2008.
46. Hickey, A. J. and Mansour, H. M. Delivery of drugs by the pulmonary route, in *Modern Pharmaceutics*, 5 ed. Informa Healthcare, New York, 2009, pp. 191–219.
47. Bailey, M. M. and Berkland, C. J. Nanoparticle formulations in pulmonary drug delivery, *Medicinal Research Reviews* 29 (1), 196–212, 2009.
48. Schafer-Korting, M., Mehnert, W., and Korting, H. C. Lipid nanoparticles for improved topical application of drugs for skin diseases, *Advanced Drug Delivery Reviews* 59 (6), 427–443, 2007.
49. Pardeike, J., Hommoss, A., and Muller, R. H. Lipid nanoparticles (SLN, NLC) in cosmetic and pharmaceutical dermal products, *International Journal of Pharmaceutics* 366 (1–2), 170–184, 2009.
50. Santos, P., Watkinson, A., Hadgraft, J., and Lane, M. Application of microemulsions in dermal and transdermal drug delivery, *Skin Pharmacology and Physiology* 21 (5), 246–259, 2008.
51. Gaudana, R., Jwala, J., Boddu, S. H. S., and Mitra, A. K. Recent perspectives in ocular drug delivery, *Pharmaceutical Research* 26 (5), 1197–1216, 2009.
52. Sahoo, S. K., Dilnawaz, F., and Krishnakumar, S. Nanotechnology in ocular drug delivery, *Drug Discovery Today* 13 (3–4), 144–151, 2008.
53. Nagarwal, R. C., Kant, S., Singh, P., Maiti, P., and Pandit, J. Polymeric nanoparticulate system: a potential approach for ocular drug delivery, *Journal of Controlled Release* 136 (1), 2–13, 2009.
54. Sultana, Y., Jain, R., Aqil, M., and Ali, A. Review of ocular drug delivery, *Current Drug Delivery* 3 (2), 207–217, 2006.
55. Bejjani, R. A., BenEzra, D., Cohen, H., Rieger, J., Andrieu, C., Jeanny, J. C., Gollomb, G., and Behar-Cohen, F. F. Nanoparticles for gene delivery to retinal pigment epithelial cells, *Molecular Vision* 11 (2), 124–32, 2005.
56. Calvo, P., Vila Jato, J. L., and Alonso, M. J. Comparative in vitro evaluation of several colloidal systems, nanoparticles, nanocapsules, and nanoemulsions, as ocular drug carriers, *Journal of Pharmaceutical Sciences* 85 (5), 530–536, 1996.
57. Harmia, T., Speiser, P., and Kreuter, J. A solid colloidal drug delivery system for the eye: encapsulation of pilocarpin in nanoparticles, *Journal of Microencapsulation* 3 (1), 3–12, 1986.
58. Motwani, S. K., Chopra, S., Talegaonkar, S., Kohli, K., Ahmad, F. J., and Khar, R. K. Chitosan-sodium alginate nanoparticles as submicroscopic reservoirs for ocular delivery: formulation, optimisation and in vitro characterisation, *European Journal of Pharmaceutics and Biopharmaceutics* 68 (3), 513–525, 2008.

59. Irache, J., Merodio, M., Arnedo, A., Camapanero, M., Mirshahi, M., and Espuelas, S. Albumin nanoparticles for the intravitreal delivery of anticytomegaloviral drugs, *Mini Reviews in Medicinal Chemistry* 5 (3), 293–305, 2005.

60. Mansour, H. M., Sohn, M. J., Al-Ghananeem, A., and DeLuca, P. P. Materials for pharmaceutical dosage forms: molecular pharmaceutics and controlled release drug delivery aspects, *International Journal of Molecular Sciences* 11 (9), 3298–3322, 2010.

61. Rhee, Y. S., Park, C. W., DeLuca, P. P., and Mansour, H. M. Sustained-release injectable drug delivery, *Pharmaceutical Technology Special Issue Supplement-Drug Delivery* S6–S13, 2010.

62. Wischke, C. and Schwendeman, S. P. Principles of encapsulating hydrophobic drugs in PLA/ PLGA microparticles, *International Journal of Pharmaceutics* 364 (2), 298–327, 2008.

63. Houchin, M. and Topp, E. Chemical degradation of peptides and proteins in PLGA: a review of reactions and mechanisms, *Journal of Pharmaceutical Sciences* 97 (7), 2395–2404, 2008.

64. Prabha, S., Zhou, W. Z., Panyam, J., and Labhasetwar, V. Size-dependency of nanoparticle-mediated gene transfection: studies with fractionated nanoparticles, *International Journal of Pharmaceutics* 244 (1–2), 105–115, 2002.

65. Salama, R., Hoe, S., Chan, H. K., Traini, D., and Young, P. M. Preparation and characterisation of controlled release co-spray dried drug-polymer microparticles for inhalation 1: Influence of polymer concentration on physical and in vitro characteristics, *European Journal of Pharmaceutics and Biopharmaceutics* 69 (2), 486–495, 2008.

66. Keegan, M. E., Falcone, J. L., Leung, T. C., and Saltzman, W. M. Biodegradable microspheres with enhanced capacity for covalently bound surface ligands, *Macromolecules* 37 (26), 9779–9784, 2004.

67. Davaran, S., Rashidi, M. R., Khandaghi, R., and Hashemi, M. Development of a novel prolonged-release nicotine transdermal patch, *Pharmacological Research* 51 (3), 233–237, 2005.

68. Kumar, R. and Majeti, N. A review of chitin and chitosan applications, *Reactive and Functional Polymers* 46 (1), 1–27, 2000.

69. Rowley, J. A., Madlambayan, G., and Mooney, D. J. Alginate hydrogels as synthetic extracellular matrix materials, *Biomaterials* 20 (1), 45–53, 1999.

70. Liao, Y. H., Jones, S. A., Forbes, B., Martin, G. P., and Brown, M. B. Hyaluronan: pharmaceutical characterization and drug delivery, *Drug Delivery* 12 (6), 327–342, 2005.

71. Muller, R., Radtke, M., and Wissing, S. Solid lipid nanoparticles (SLN) and nanostructured lipid carriers (NLC) in cosmetic and dermatological preparations, *Advanced Drug Delivery Reviews* 54, S131–S155, 2002.

72. Brandl, M. Liposomes as drug carriers: a technological approach, *Biotechnology Annual Review* 7, 59–85, 2001.

73. Moghimi, S. M., Hunter, A. C., and Murray, J. C. Nanomedicine: current status and future prospects, *The FASEB Journal* 19 (3), 311, 2005.

74. Romberg, B., Hennink, W. E., and Storm, G. Sheddable coatings for long-circulating nanoparticles, *Pharmaceutical Research* 25 (1), 55–71, 2008.

75. Gref, R., Minamitake, Y., Peracchia, M. T., Trubetskoy, V., Torchilin, V., and Langer, R. Biodegradable long-circulating polymeric nanospheres, *Science* 263 (5153), 1600, 1994.

76. Lasic, D. D. and Martin, F. *Stealth Liposomes* CRC Press, Boca Raton, FL, 1995.

77. Torchilin, V. P., Omelyanenko, V. G., Papisov, M. I., Bogdanov Jr, A. A., Trubetskoy, V. S., Herron, J. N., and Gentry, C. A. Poly (ethylene glycol) on the liposome surface: on the mechanism of polymer-coated liposome longevity, *Biochimica et Biophysica Acta (BBA)-Biomembranes* 1195 (1), 11–20, 1994.

78. Torchilin, V. P. and Trubetskoy, V. S. Which polymers can make nanoparticulate drug carriers long-circulating? *Advanced Drug Delivery Reviews* 16 (2–3), 141–155, 1995.

79. Esposito, P., Barbero, L., Caccia, P., Caliceti, P., D'antonio, M., Piquet, G., and Veronese, F. PEGylation of growth hormone-releasing hormone (GRF) analogues, *Advanced Drug Delivery Reviews* 55 (10), 1279–1291, 2003.

80. Elbayoumi, T. A. and Torchilin, V. P. Enhanced cytotoxicity of monoclonal anticancer antibody 2C5-modified doxorubicin-loaded PEGylated liposomes against various tumor cell lines, *European Journal of Pharmaceutical Sciences* 32 (3), 159–168, 2007.

81. Ganta, S., Devalapally, H., Shahiwala, A., and Amiji, M. A review of stimuli-responsive nano-carriers for drug and gene delivery, *Journal of Controlled Release* 126 (3), 187–204, 2008.
82. Kale, A. A. and Torchilin, V. P. Design, synthesis, and characterization of pH-sensitive PEG-PE conjugates for stimuli-sensitive pharmaceutical nanocarriers: the effect of substitutes at the hydrazone linkage on the ph stability of PEG-PE conjugates, *Bioconjugate Chemistry* 18 (2), 363–370, 2007.
83. Lu, Y., Yin, Y., Mayers, B. T., and Xia, Y. Modifying the surface properties of superparamagnetic iron oxide nanoparticles through a sol-gel approach, *Nano Letters* 2 (3), 183–186, 2002.
84. Thorek, D. L. J., Chen, A. K., Czupryna, J., and Tsourkas, A. Superparamagnetic iron oxide nanoparticle probes for molecular imaging, *Annals of Biomedical Engineering* 34 (1), 23–38, 2006.
85. Simberg, D., Zhang, W. M., Merkulov, S., McCrae, K., Park, J. H., Sailor, M. J., and Ruoslahti, E. Contact activation of kallikrein-kinin system by superparamagnetic iron oxide nanoparticles in vitro and in vivo, *Journal of Controlled Release* 140 (3), 301–305, 2009.
86. Weinstein, J. S., Varallyay, C. G., Dosa, E., Gahramanov, S., Hamilton, B., Rooney, W. D., Muldoon, L. L., and Neuwelt, E. A. Superparamagnetic iron oxide nanoparticles: diagnostic magnetic resonance imaging and potential therapeutic applications in neurooncology and central nervous system inflammatory pathologies, a review, *Journal of Cerebral Blood Flow & Metabolism* 30 (1), 15–35, 2009.
87. Jain, T. K., Richey, J., Strand, M., Leslie-Pelecky, D. L., Flask, C. A., and Labhasetwar, V. Magnetic nanoparticles with dual functional properties: drug delivery and magnetic resonance imaging, *Biomaterials* 29 (29), 4012–4021, 2008.

Index

Intramuscular (IM) route of administration,
    injectable nanomedicines, 322, 325
Intraocular encapsulated cells, 204
Intraurethral prostaglandin-E1, 227
Intravascular ultrasound (IVUS), 304
Intravenous contrast media, 199
Intravenous (IV) route of administration,
    injectable nanomedicines, 322, 325
iPledge program, 91
ISCDs, *see* Intermittent sequential compression
    devices
Ischemic nephropathy, 281, 282
Isoflavones, 129
Isotretinoin, 91, 99

J

Junctional nevus, 140, 141

K

Keloids, 142
Keratinocytes, 132
Keratocytes, 196
Keratolytics, 153, 154
Keratotic crust, 118
Kerion, 155

L

Laser-activated gold nanoparticles, 207
Laser surgery, 113
Lateral flow immunochromatography (LFIC),
    25, 34
Lectin-functionalized CdSe QDs, 21
Lens fibers, 197
Lens refilling, 207
Lesion-targeted therapy, 112
Leukocytes extravasation, 110
LFIC, *see* Lateral flow immunochromatography
Lichenified plaques, 103
Lifesaver™, 245
Light scattering, 36
    detection, 31–32
Limbus, 196
Lipids, 330, 331
Lipophilic medications, 203
Liposomal amphotericin B, 60
Liposomal formulations, 328
Liposomes, 57–60, 93, 203, 217
    antiandrogenic therapy, 97–98
    benzoyl peroxide, 94–95
    combination therapy through, 98

nanobiosensor, 48–49
    retinoids, 95–97
Liquid lipids, 330
Liquid nitrogen cryosurgery, 113
Liver disease, due to HBV and HCV, 20
LNCaP prostate cancer cells, 219, 221
Long-circulating nanomedicine delivery
    systems, 330
Low-income countries, causes of death in, 237
Luminescence, emitted by QDs, 16
Lymph nodes, 214
Lymphotropic superparamagnetic
    nanoparticles in lymph nodes, 214

M

Macromolecular antitumor drugs, 128
Macrophage phagocyte system (MPS), 57
Magnetic Gram stain, 44
Magnetic nanoparticles (MNPs), 222–223
    diameter of, 3
    and IDs biosensing, 40–45
Magnetic relaxation nanoswitch (MRnS),
    42–44
Magnetic remanence, 41–42
Magnetic resonance imaging (MRI), 44,
    214, 273
Magnetospirillum gryphiswaldense, 45
Malaria, 29
Matrix metalloproteinases (MMPs), 253
MC1R gene, 121
Melanoblasts, 139
Melanocytic nevi, 139, 140, 152
Melanoma
    clinical features, 121–122
    diagnosis, 124–127
    epidemiology, 120–121
    treatment, 122–124, 127
Melanoma-associated mRNAs, 128–129
Methicillin-resistant *Staphylococcus aureus*
    (MRSA), 33
Methotrexate (MTX), 94
    benefits of, 146, 147
    chemical addition of, 130
Miconazole nitrate, 159–160
Microarrays, 20, 113, 127
Microbial-based microbial biosensing, 49–50
Microcantilever-based infectious diseases
    biosensing, 45–47
Microcantilevers, linear array of, 45
Microcomedones, 89
Microemulsions, 201–202
Microencapsulation, 101

Printed and bound by CPI Group (UK) Ltd, Croydon, CR0 4YY

18/10/2024

01776254-0007